Electrolytes for Energy Storage Applications

The book offers detailed progress and challenges in energy storage technologies with respect to various electrolyte chemistries including energy storage devices such as batteries and supercapacitors. It introduces energy storage systems and explains the selection of electrolytes for energy storage systems, aqueous- and non-aqueous-based electrolytes, metal-air batteries, and multivalent chemistries.

Key features:

- Provides a better understanding of electrolytes for electrochemical energy storage devices
- Focuses on electrolytes and electrolytes valence
- Presents challenges and opportunities in energy storage systems
- Explains supercapacitors with respect to electrolyte systems
- Summarizes multivalent-based energy storage systems

This book is written for graduate students and researchers in electrochemistry, physical chemistry, and renewable energy.

Electrolytes for Energy Storage Applications
Fundamentals and Advances

Edited by
Ramesh K. Guduru,
Saravanakumar Balasubramaniam,
Venkateswarlu Manne,
Ananthakumar Ramadoss, and
Rambabu Bobba

CRC Press
Taylor & Francis Group
Boca Raton London New York

CRC Press is an imprint of the
Taylor & Francis Group, an **informa** business

Designed Cover Image: Ricurve Patel

First edition published 2025
by CRC Press
2385 NW Executive Center Drive, Suite 320, Boca Raton, FL 33431

and by CRC Press
4 Park Square, Milton Park, Abingdon, Oxon, OX14 4RN

CRC Press is an imprint of Taylor & Francis Group, LLC

ISBN: 978-1-032-45263-0 (hbk)
ISBN: 978-1-032-57826-2 (pbk)
ISBN: 978-1-003-44118-2 (ebk)

DOI: 10.1201/9781003441182

Typeset in Times
by SPi Technologies India Pvt Ltd (Straive)

Contents

About the Editors

Dr. Ramesh K. Guduru (Ph.D. in Nanomaterials from NC State University) is a member of faculty at Pandit Deendayal Energy University, India, and founded a start-up, GreenXcel Technologies, that specializes in circular economy with a focus on green energy solutions. Prior to that, he was a full-time member of faculty at Texas State University for five years and also worked as an adjunct faculty member at the University of Michigan and as a Chief Scientist at CSquared Innovations LLC. Dr. Guduru has more than 12 years of experience in rechargeable batteries and supercapacitors, has published more than 110 articles, seven book chapters, and holds four international patents and has two Indian patents pending, and has served on review panels for NSF-USA, QNRF-QATAR, Skolkovo Foundation-RUSSIA, and SNF-Switzerland.

Dr. Saravanakumar Balasubramaniam is a Junior Scientist at SARP-LARPM, Central Institute of Petrochemicals Engineering and Technology (CIPET), Bhubaneswar, Odisha, India. He received a doctoral degree from Jeju National University, Jeju, Republic of Korea. His research focuses on developing new materials and fabrication techniques for energy harvesting applications, as well as new types of self-powered devices. Similarly, he is working in the field of energy storage devices such as supercapacitors and batteries in the area of aqueous-based electrolytes. He has published more than 70 articles in peer-reviewed journals with approximately 4100 citations and an H-index of 37.

Dr. Venkateswarlu Manne, MSc, M.Phil, Ph.D., served as a Director at Customized Energy Solutions India, leading cutting-edge storage technology development for E-Mobility and ESS, and is presently working as Chief Technology Officer (CTO), Kushmanda Power Ltd at Chandigarh. Prior to that, he worked with Amara Raja Batteries Ltd, Electrotherm, HBL, and conducted post-doctoral research on next-gen lithium-ion batteries at NTUST, Taiwan. His areas of expertise include solid electrolyte research, advanced batteries, technology licensing, project management, and pilot-scale production. Collaborating closely with research institutions, he optimizes resource utilization to deliver groundbreaking solutions in the energy domain. As a co-author of numerous articles, he significantly contributes to electrochemical and power sources research. He has received prestigious awards, including Best Research Scientist, Industrial Researcher of the Year, and Leadership in Clean Mobility awards for his unwavering dedication to advancing Li-ion battery tech for e-mobility and ESS applications.

Dr. Ananthakumar Ramadoss is a Scientist at the Advanced Research School for Technology & Product Simulation (ARSTPS), Central Institute of Petrochemical Engineering and Technology (CIPET), Chennai, India. He received his PhD degree in Mechanical Engineering from Jeju National University, South Korea (2015), and PhD degree in Physics from Bharathidasan University, India (2024). He was a Scientist (Jr) at CIPET: SARP-LARPM, Bhubaneswar, India (2016–2022); Alexander Von Humboldt postdoctoral fellow at the University of Konstanz, Germany (2016);

a postdoctoral fellow at the KAIST, Korea (2015–2016); and a research fellow at the CSIR-CECRI, India (2009–2011). His research interests mainly focus on developing nanostructured materials and their applications in supercapacitors, nanogenerators, self-powered systems, 3D printing, thin films, corrosion protection, and bio-implant applications. He has an accomplished track record of publications in high-impact journals (more than 100), more than 20 book chapters, and two patents. Also, he has received more than 20 awards and prestigious fellowships. In addition, he has guided three PhD theses, 12 Masters theses, four Bachelors theses and trained eight research personnel. In addition, he has completed two ongoing sponsored national and four sponsored international projects in the respective areas. Further, three mentor projects and eight national and international internships/workshops/PAC meetings have been accomplished.

Dr. Rambabu Bobba is a professor of physics at Southern University and A&M College, Baton Rouge (SUBR). He earned his Ph.D. in Physics in 1982 from the Indian Institute of Technology-Madras/Chennai. After completing his post-doctoral assignments at the University of Minnesota and the University of New Orleans, he became a faculty member in the Department of Physics at SUBR in 1986. During his tenure at SUBR, he published 125 papers and several extended abstracts and successfully completed visiting professor assignments in AT&T Bell Laboratories, Lawrence Livermore National Laboratory, Dow Chemical Company, University of West Indies, US-DOD, Office of Naval Research, and finally as USIEF-Fulbright Fellow in India. His research encompasses the design and development of low-cost, more efficient nanoionics materials that can facilitate widespread commercialization of electrochemical energy storage and conversion technologies such as lithium batteries, supercapacitors, fuel cells (HT-PEMFCs and LT-SOFCs), and solar hydrogen to address the world's energy and environmental challenges.

Contributors

S. Aafrin Hazaana
Materials Research Centre
Coimbatore, India
and
Research Centre of Physics
Fatima College (affiliated with Madurai Kamaraj University)
Madurai, India

Ahmed I. Abdel-Salam
Nanotechnology Research Centre (NTRC)
The British University in Egypt (BUE)
Cairo, Egypt

Aya Mohamed Abuelftooh
Chemistry Department, Faculty of Women for Arts, Science and Education
Ain Shams University
Cairo, Egypt

S. Arumugam
Centre for High-Pressure Research, School of Physics
Bharathidasan University
Tiruchirappalli, India

Y. Bhaskara Rao
Department of Physics
SRM University AP
Amaravati, India
and
SRM-Amara Raja Center for Energy Storage Devices
SRM University AP
Amaravati, India

Prakash Chandra
Department of Chemistry, School of Technology
Pandit Deendayal Petroleum University
Gujarat, India

Murthy Chavli
Office of the Dean (Research)
Dr. Vishwanath Karad MIT World Peace University (MIT-WPU)
Kothrud, Pune, Maharashtra, India

Mahmoud A. M. Elsaid
Metallurgy Department, Nuclear Research Center (NRC)
Egyptian Atomic Energy Authority (EAEA)
Cairo, Egypt

Mohammad Etesami
Department of Chemical Engineering, Faculty of Engineering
Chulalongkorn University
Bangkok, Thailand

Guozhao Fang
School of Materials Science and Engineering
Central South University
Hunan, PR China

B. C. Gireeshkumar
Centre for Incubation, Innovation, Research and Consultancy (CIIRC)
Jyothy Institute of Technology
Bangalore, India
and
Department of P.G. Studies in Chemistry, Jnana Sahyadri, Shanakaraghatta
Kuvempu University
Shimoga, India

Mohan Gopalakrishnan
Department of Chemical Engineering Faculty of Engineering
Chulalongkorn University
Bangkok, Thailand

Ramesh K. Guduru
Department of Mechanical Engineering
School of Technology
Pandit Deendayal Energy University
Gandhinagar, India

Insik In
Department of Polymer Science and Engineering
Department of IT-Energy Convergence, (BK21 FOUR)
Chemical Industry Institute
Korea National University of Transportation
Chungju, Republic of Korea

R. Jagadeesh
Centre for High-Pressure Research
School of Physics
Bharathidasan University
Tiruchirappalli, India

P. W. Jaschin
Department of Materials Science & Engineering
and
Maryland Energy Innovation Institute
University of Maryland
College Park, Maryland

Jitti Kasemchainan
Department of Chemical Technology
Faculty of Science
Chulalongkorn University
Bangkok, Thailand
and
Center of Excellence on Advanced Materials for Energy Storage
Chulalongkorn University
Bangkok, Thailand

Dawid Kasprzak
School of Engineering Faculty of Applied Science
University of British Columbia
Vancouver, British Columbia, Canada
and
Institute of Chemistry and Technical Electrochemistry
Poznan University of Technology
Poznan, Poland

Sumit Khatua
Department of Physics
and
SRM-Amara Raja Center for Energy Storage Devices
SRM University AP
Amaravati, India

Soorathep Kheawhom
Department of Chemical Engineering
Faculty of Engineering
and
Center of Excellence on Advanced Materials for Energy Storage
and
Bio-Circular-Green-Economy Technology & Engineering Center (BCGeTEC)
Faculty of Engineering
Chulalongkorn University
Bangkok, Thailand

Ramin Khezri
Department of Chemical Engineering
Faculty of Engineering
Chulalongkorn University
Bangkok, Thailand

Manoj Kumar
Department of Physics School of Technology
Pandit Deendayal Energy University
Gujarat, India

Anoop Kumar Kushwaha
School of Basic Sciences
Indian Institute of Technology Bhubaneswar
Odisha, India

Seung Jun Lee
Department of Polymer Science and Engineering
Department of IT-Energy Convergence, (BK21 FOUR)
Chemical Industry Institute
Korea National University of Transportation
Chungju, Republic of Korea

Shuquan Liang
School of Materials Science and Engineering
Central South University
Hunan, PR China

Jian Liu
School of Engineering
Faculty of Applied Science
University of British Columbia
Vancouver, British Columbia, Canada

Zhexuan Liu
School of Materials Science and Engineering
Central South University
Hunan, PR China

G. Mahesh
Department of Mechanical Engineering
JCT College of Engineering & Technology
Coimbatore, India

Venkateswarlu Manne
R & D Customized Energy Solutions India Private Limited
Pune, India

M. Manoj Cristopher
Multifunctional Materials & Devices Lab
School of Electrical & Electronics Engineering
SASTRA Deemed University
Thanjavur, India

R. Meera Naachiyar
Materials Research Centre
Coimbatore, India
and
Research Centre of Physics
Fatima College (affiliated with Madurai Kamaraj University)
Madurai, India

Ahmad Azmin Mohamad
School of Materials and Mineral Resources Engineering
Universiti Sains Malaysia
Pinang, Malaysia

Saad G. Mohamed
Mining and Metallurgy Engineering Department
Tabbin Institute for Metallurgical Studies (TIMS)
Cairo, Egypt

Shiva Rezaei Motlagh
Department of Chemical Engineering
Faculty of Engineering
Chulalongkorn University
Bangkok, Thailand

N. Muniraj Vignesh
Materials Research Centre
Coimbatore, India
and
Research Centre of Physics
Mannar Thirumalai Naicker College (affiliated with Madurai Kamaraj University)
Madurai, India

G. Murali
Department of Polymer Science and Engineering
Department of IT-Energy Convergence (BK21 FOUR)
Chemical Industry Institute
Korea National University of Transportation
Chungju, Republic of Korea

H. B. Muralidhara
Centre for Incubation, Innovation, Research and Consultancy (CIIRC)
Jyothy Institute of Technology
Bangalore, India

J. Niresh
Department of Automobile Engineering
PSG College of Technology
Coimbatore, India

Mostafa M. Omran
Chemistry Department
Faculty of Science
Cairo University
Cairo, Egypt

Young Ho Park
Department of Polymer Science and Engineering
Department of IT-Energy Convergence, (BK21 FOUR)
Chemical Industry Institute
Korea National University of Transportation
Chungju, Republic of Korea

L. N. Patro
Department of Physics
SRM University AP
Amaravati, India
and
SRM-Amara Raja Center for Energy Storage Devices
SRM University AP
Amaravati, India

P. Periasamy
Nehru Institute of Engineering and Technology
T.M. Palayam
Coimbatore, India

Ananthakumar Ramadoss
School for Advanced Research in Petrochemicals (SARP) – Advanced Research School for Technology and Product Simulation (ARSTPS)
Central Institute of Petrochemicals Engineering & Technology (CIPET)
Chennai, India

K. Ramakrushna Achary
Department of Physics, SRM University AP
Amaravati, India
and
SRM-Amara Raja Center for Energy Storage Devices
SRM University AP
Amaravati, India

Jeevan Kumar Reddy Modigunta
Department of Polymer Science and Engineering
Department of IT-Energy Convergence, (BK21 FOUR)
Chemical Industry Institute
Korea National University of Transportation
Chungju, Republic of Korea

Malaya K. Sahoo
School of Advanced Materials Science and Engineering
Sungkyunkwan University (SKKU)
Suwon, Republic of Korea

Saravanakumar Balasubramaniam
Materials Development Lab
School for Advanced Research in Petrochemicals (SARP) – Laboratory for Advanced Research in Polymeric Materials (LARPM)
Central Institute of Petrochemicals Engineering & Technology
Chennai, India

Prasanna Sekar
Department of Physics
Ravenshaw University
Odisha, India
and
Materials Development Lab
School for Advanced Research in Petrochemicals (SARP) – Laboratory for Advanced Research in Polymeric Materials (LARPM)
Central Institute of Petrochemicals Engineering & Technology (CIPET)
Bhubaneswar, India

S. Selvasekarapandian
Materials Research Centre
Coimbatore, India
and

Department of Physics
Bharathiar University
Coimbatore, India

S. Settu
Department of Mechanical Engineering
JCT College of Engineering & Technology Coimbatore
Pichanur, India

A. M. Shanmugharaj
Centre for Energy and Alternative Fuels
Vels Institute of Science Technology and Advanced Studies
Chennai, India

B. Sharath Kumar
Centre for Incubation, Innovation, Research and Consultancy
Jyothy Institute of Technology
Bangalore, India

Robin Singh
Department of Chemistry
School of Energy Technology
Pandit Deendayal Energy University
Gandhinagar, India

T. Sri Devi Kumari
Arunachala College of Engineering for Women
Nagercoil, Tamil Nadu, India

P. Suresh Kumar
Department of Mechanical Engineering
JCT College of Engineering & Technology
Coimbatore, India

Saroj Kumar Nayak
School of Basic Sciences
Indian Institute of Technology Bhubaneswar
Odisha, India

G. Tamilselvan
Department of Automobile Engineering
PSG College of Technology
Coimbatore, India

Neha Tiwari
Department of Physics
School of Energy Technology
Pandit Deendayal Energy University
Gandhinagar, India

Brijesh Tripathi
Department of Physics
School of Energy Technology
Pandit Deendayal Petroleum University
Gujarat, India

Hariharan Udhayakumar
Department of Polymer Science and Engineering
Department of IT-Energy Convergence, (BK21 FOUR)
Chemical Industry Institute
Korea National University of Transportation
Chungju, Republic of Korea

Krishna Venkatesh
Centre for Incubation, Innovation, Research and Consultancy (CIIRC)
Jyothy Institute of Technology
Bangalore, India

Preface

As we move towards a sustainable future, it becomes increasingly important to address the global energy demand and transition to renewable energy sources. While renewable energy technologies have made significant progress over the past few decades, they face a critical challenge of intermittency, and this is where energy storage technologies can play a crucial role in addressing several challenges. As we continue to rely more and more on energy storage technologies in our daily lives, it becomes increasingly important to understand the critical role of different components in energy storage devices, especially electrolytes, in enhancing the performance and lifetime of batteries and supercapacitors.

This book provides a comprehensive resource for researchers, engineers, and students working in the field of energy storage, highlighting the importance of electrolytes and their impact on the performance and efficiency of energy storage devices. It covers the latest advances in designing and developing both aqueous and non-aqueous electrolytes, polymer electrolytes, solid electrolytes, and electrolyte additives that help reduce degradation and enable wide potential windows, enhance the lifetime and performance of energy storage devices, increase their energy and power densities, and more.

The book delves into the criticality of different types of energy storage devices, including batteries, various chemistries of batteries, such as Li-ion and Na-ion batteries, Zn-air, Al-ion, redox flow batteries, supercapacitors, solid-state batteries, 2D MXenes, and more. It also explores the role of electrolytes in the above chemistries for energy storage, and the theoretical and modeling approaches used to select appropriate electrolyte chemistries for specific chemistry or application, taking into account different factors such as electrochemical stability, energy density, potential windows, and electrode material compatibility, etc.

In addition, the book addresses the challenges facing electrolyte chemistry in energy storage applications, including issues with electrode and electrolyte degradation, as well as minimizing solid electrolyte interface and cathode interface issues. It also explores the future outlook of electrolyte chemistry in energy storage applications, including emerging technologies and new materials, and the development of new theoretical and modeling approaches to accelerate the discovery and optimization of novel electrolyte chemistries.

The contents of this book are presented in three sections following the introduction about energy storage technologies. In the first section, the focus is on Li-ion batteries and related electrolyte chemistries, along with theoretical approaches in designing electrolytes for Li-ion batteries. The second section discusses chemistries beyond Li-ion chemistries, such as Na-ion, Zn-ion, Al-ion, Zn-air, Redox Flow Batteries, and so on. Finally, the third section emphasizes supercapacitors with different types of electrolytes ranging from aqueous to non-aqueous, ionic liquids and hybrid electrolytes, etc.

Overall, this book provides a comprehensive understanding of the critical role electrolytes play in energy storage devices and their impact on the performance and efficiency of these devices. By exploring the latest advances in electrolyte chemistry and identifying future research directions, this book will provide invaluable insights into the design and optimization of next-generation energy storage devices.

Dr. Ramesh K. Guduru
Pandit Deendayal Energy University, India

Dr. Saravanakumar Balasubramaniam
CIPET: SARP-LARPM, India

Dr. Venkateswarlu Manne
Amara Raja Batteries Limited

Dr. Ananthakumar Ramadoss
CIPET: SARP-ARSTPS, India

Dr. Rambabu Bobba
Southern University and A&M College, USA

1 Introduction to Energy Storage Systems

P. Suresh Kumar, J. Niresh, G. Mahesh, S. Settu, G. Tamilselvan, Robin Singh, Neha Tiwari, and Ramesh K. Guduru

1.1 INTRODUCTION

The scarcity of non-renewable resources is a growing concern as the world heavily relies on fossil fuels [1]. This has led to an energy crisis and a pressing need for energy security. One solution to this problem is to shift towards renewable energy sources that include the wind, hydro, solar power, etc. However, these intermittent sources require efficient energy storage systems to provide a reliable energy supply. Energy storage systems are devices that store energy for later use and play a critical role in modern power systems. These systems help balancing the overall supply as well as complete the demand of electrical energy and enable the integration of intermittent renewable energy sources.

Energy storage systems come in many forms, including batteries, flywheels, pumped hydro storage, compressed air energy storage, and thermal energy storage. Each storage system has its own unique characteristics, such as energy density, power density, cycle life, and efficiency, making it suitable for certain applications. Batteries are the most common energy storage system and are used in every application—from smart phones to electric cars to grid-scale energy storage systems. They store energy electrochemically, using chemical reactions to store and release the energy. In contrast, other systems like flywheels store energy mechanically by spinning a rotor at high speed and then using that stored energy to generate electricity when needed. Hydro storage systems utilize excess electrical energy to pump water at a certain height, which, when released, produces on-demand electrical energy using a turbine [2]. With compressed air energy storage, air is compressed and stored in underground caves as well as caverns—and in the later stages, the compressed air rotates the turbine, resulting in electricity generation.

In some cases excess heat is stored in different materials in various forms including salt or phase-change materials, and this stored energy is used later for different purposes including electricity production or cooling-heating processes. Among these, electrochemical energy storage approaches, such as batteries as well as supercapacitors, are a promising solution owing to their high efficiency and energy density [3]. By optimizing the composition and properties of battery components, their performance and safety can be improved. The primary components of a battery system include electrodes, electrolyte, separator, and the current collectors. Electrolytes play a critical role in the overall performance and safety of these batteries. In addition, the use of additives in the electrolyte can also improve the battery's performance and extend its lifespan.

DOI: 10.1201/9781003441182-1

This chapter discusses the basic principles of electrochemical energy storage systems and explores various types of electrolytes, as well as the additives for a variety of battery and supercapacitor systems, including Lithium-ion batteries, non-Lithium-ion batteries, and metal air batteries [4]. This chapter also covers the challenges and opportunities in the development of electrochemical energy storage systems, including cost, safety, and sustainability concerns. Overall, it provides a basic overview of the current state of the electrochemical energy storage systems, their potentials for addressing the energy crisis, and the challenges and opportunities in their development and deployment. This chapter concludes by highlighting the importance of energy storage systems in achieving a sustainable and secure energy future and the need for continued research and development in energy storage applications.

1.2 HISTORICAL DEVELOPMENTS: AN INSIGHT INTO HOW ENERGY STORAGE BEGAN

The history of batteries dates back more than 2,000 years to ancient civilizations that used simple electrochemical cells made of copper, zinc, and lemon juice [5, 6]. The first known battery-like device was the Baghdad Battery, a clay jar with an iron rod in the center and a copper cylinder surrounding it, which was discovered in Iraq in the 1930s and believed to date back to the Parthian period, around 250 BC to 224 AD [7, 8]. In the early 1800s Italian physicist Alessandro Volta invented the first battery, known as the "Voltaic Pile." Over the years, batteries have become increasingly important as a source of portable power for a wide range of applications, from early telegraph machines to modern electric vehicles and smartphones [9, 10]. Figure 1.1 shows the overall developments in the domain of electrochemical energy storage systems [11].

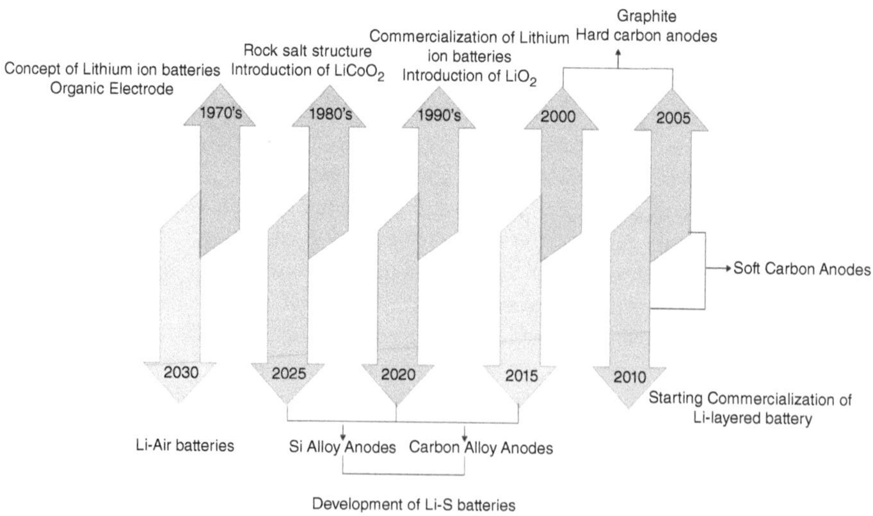

FIGURE 1.1 Historical developments in Li-ion batteries over the years.

Source: Reproduced with permission from (Copyright – Open Access Available under Creative Common CC by 4.0 License) Ref. [11].

In terms of battery materials, a wide variety of materials have been used in batteries over the years, including lead, nickel-cadmium, nickel-metal hydride, Li-ion, and many others. Each material has its own strengths and weaknesses in terms of energy density, cost, safety, and environmental impact, and the choice of materials for a particular application depends on multiple factors. New battery materials—such as solid-state electrolytes, silicon anodes, and lithium-sulfur cathodes, which could potentially lead to even higher energy density and longer cycle life than current battery technologies—are being investigated. There has also been a lot of research and development into alternative battery materials, such as solid-state batteries, sodium-ion (Sodium-ion) batteries, and zinc-air batteries [12].

Here are some significant developmental milestones of batteries over the years:

- ***1800*:** The first true battery, Voltaic Pile, consisting of alternative layers of zinc and copper separated by a saltwater-coated cardboard was invented by Alessandro Volta [13].
- ***1836*:** John F. Daniell invented the Daniell Cell, which consisted of a copper sulfate electrolyte and a zinc anode. This design greatly improved the stability and reliability of batteries [13].
- ***1866*:** Georges Leclanché invents the Leclanché Cell, which uses a manganese dioxide cathode and a zinc anode. This design is still used in many modern batteries, including dry-cell batteries [14, 15].
- ***1887*:** Carl Gassner invented the dry-cell battery, which used a paste electrolyte instead of a liquid electrolyte. This design made batteries much more portable and reliable [16].
- ***1949*:** The alkaline battery was invented by Lewis Urry. It used a potassium hydroxide electrolyte and a zinc-manganese dioxide cathode, which greatly improved the energy density of batteries [17].
- ***1970s–1990s*:** Rechargeable batteries became increasingly popular, with the development of nickel-cadmium (NiCd), nickel-metal hydride (NiMH), and Li-ion batteries [18].
- ***2010s–2030s*:** New battery materials and technologies are in progression to improve battery performance while reducing cost. These include solid-state batteries, lithium-sulfur batteries, and Na-ion batteries, among others [19].

Over the past two decades, efforts toward developing batteries with more sustainable and abundant materials have been put forth. For example, research is being done on Na-ion batteries, which use sodium as the electrode material rather than lithium. Sodium is much more abundant than lithium, which could help reduce the cost of batteries and make them more sustainable in the long term. Other materials being investigated for battery use include aluminum, magnesium, and zinc. Conventional lead-acid batteries are still widely used in cars and other applications, while Li-ion batteries are used in smartphones, laptops, and electric vehicles.

Other materials used in batteries include nickel, cadmium, cobalt, manganese, and iron [3]. The availability and cost of these materials significantly impact the price of the batteries. The choice of battery chemistry depends on the specific application and requirements, such as energy density, power output, and cycle life. As battery technology continues to evolve, we can expect to see new battery materials and designs emerge that will further improve the performance and sustainability of various battery systems. Overall, the history of batteries is a story of constant innovation and development, as scientists and engineers seek to create better, more efficient, and more sustainable energy storage solutions.

1.3 THE NEED FOR ENERGY STORAGE SYSTEMS

Energy is generated by different methods; however, due to different limitations including intermittent weather conditions, energy storage systems are very much essential. Regardless of the type of energy generation—solar, hydro, wind, hydrothermal, or chemical—the need for energy storage systems is of the utmost importance. Also, when it comes to the automobile sector, storage is paramount, and new technological advancements make it possible for energy storage to grow in a very favorable way. As costs decrease and the availability of renewable energy technologies increases, energy storage will gain a huge market share. Figure 1.2 shows the potential benefits of various energy storage systems [20].

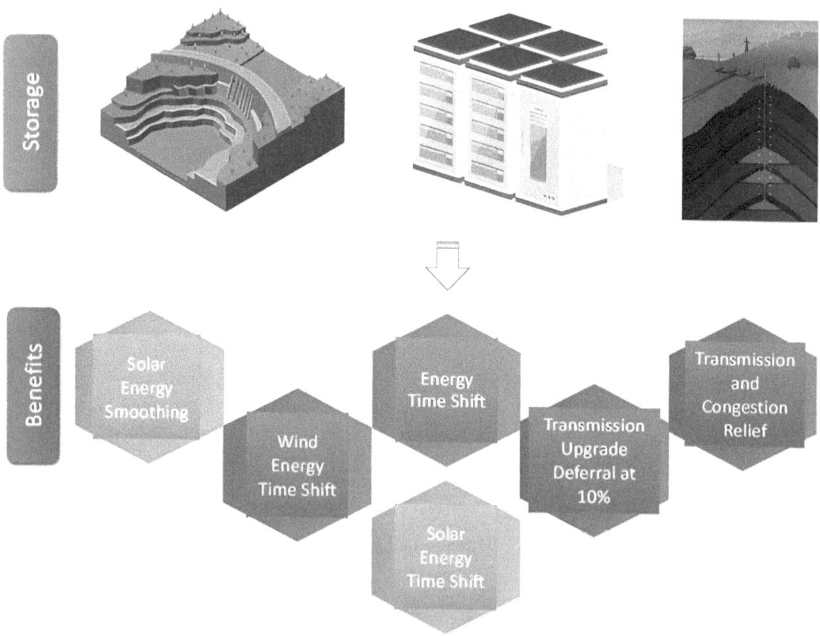

FIGURE 1.2 The potential benefits and applications of energy storage systems.

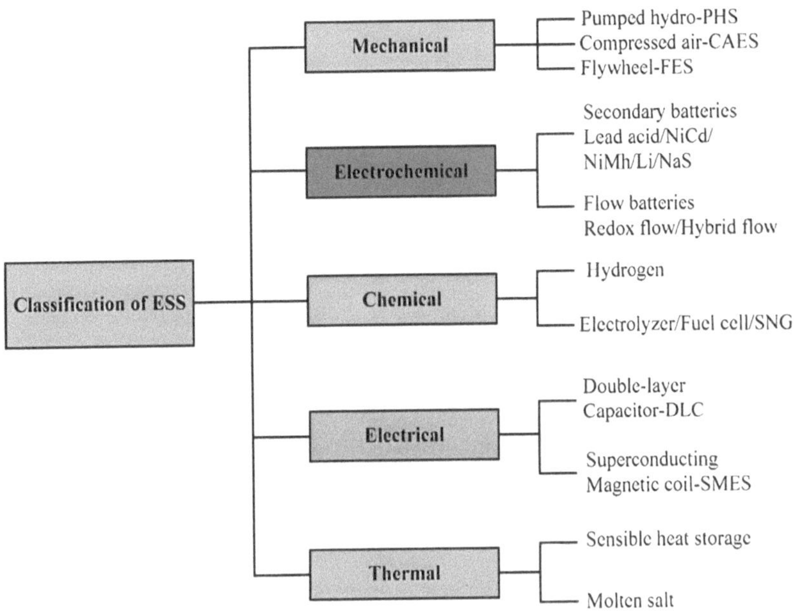

FIGURE 1.3 Different energy storage systems.

Source: Copyright – Open Access Available under Creative Common CC by 4.0 License, Ref. [21].

1.4 DIFFERENT TYPES OF ENERGY STORAGE SYSTEMS

Chemical, mechanical, thermal, and kinetic energies are all represented by a variety of energy storage systems, as shown in Figure 1.3 [21]. Chemical energy may be produced by batteries through chemical processes, including present battery technologies. Pumped storage and compressed air are two sources of mechanical energy. Thermal energy is used in power plants as well as for heating and cooling by drastically varying their temperatures to store extra energy. Kinetic energy is demonstrated through flywheels [13, 14].

1.5 ELECTROCHEMICAL ENERGY STORAGE SYSTEMS

Batteries, which are electrochemical-based energy storage devices, can be classified as primary and secondary batteries, and Figure 1.4 shows the classification of primary and secondary batteries. Primary batteries may be defined as single-use batteries and cannot be recharged. Examples of primary batteries are dry cells and alkaline batteries. Secondary batteries are rechargeable and can be used again and again with certain cycle life. Examples for rechargeable secondary batteries are nickel-cadmium (Ni-Cd), lead acid, and Li-ion batteries [22].

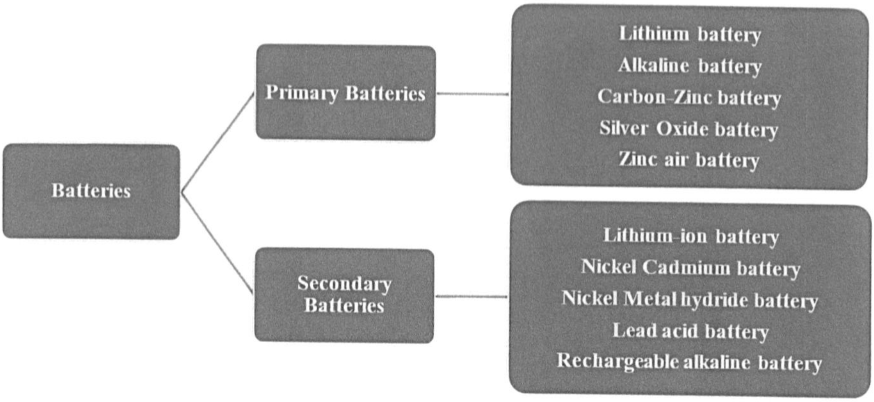

FIGURE 1.4 Classification of different types of batteries.

Various types of battery systems are available in the market, each with its own advantages and disadvantages. The basics of each type of battery system are mentioned in this chapter in brief and then discussed further in subsequent chapters with major focus on chemistries and the impact of electrolytes on the respective battery systems. The basics for some of the most common types of battery systems are described below.

1.5.1 Lithium and Lithium-Ion Batteries

Lithium batteries and Li-ion batteries are two distinct types of energy storage devices that utilize lithium in different ways. Lithium batteries are rechargeable and feature metallic lithium as the anode and paired with various cathode materials [23]. They were popular for applications requiring long-lasting power, such as cameras, calculators, and medical equipment. However, the dendrite formation of lithium during the process of charging-discharging cycles and short circuiting of electrodes resulted in many failures and accidents, which led to the development of Li-ion batteries.

Li-ion batteries revolutionized the rechargeable battery market by using graphite anodes instead of metallic lithium, allowing for repeated charging and discharging cycles [24, 25]. Although the chemistry of Li-ion batteries varies, they always move lithium ions between the cathode and anode, with graphite serving as the most typical anode and the cathode typically being a metal oxide. The schematic for a Li-ion battery is shown in Figure 1.5 [26]. These batteries offer numerous advantages, including cost-effectiveness, environmental friendliness, and higher energy density. However, the choice between these batteries depends on the specific needs of the application. As technology advances, further improvements and innovations in battery technology will enable new possibilities for energy storage and powering our increasingly connected world [25].

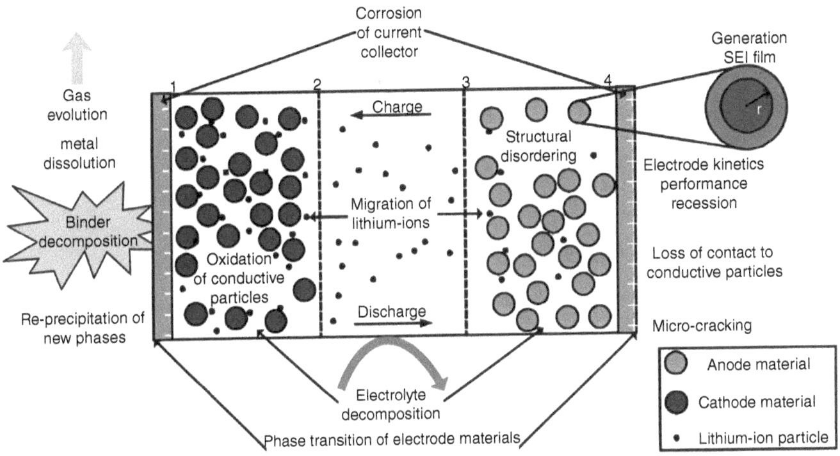

FIGURE 1.5 Working and functionality of Li-ion battery.

Source: Copyright – Open Access Available under Creative Common CC by 4.0 License, Ref. [26].

There are two categories of electrolytes for Li-ion batteries—liquid and solid—with advantages and disadvantages to each depending on chemical variances, and some are more effective than others [11, 26]. Li-ion batteries have a high energy density, are less costly, have extended lifespan cycles, are rechargeable, require less maintenance, and have the potential to discharge at rapid rates. Electric cars, mobile phones, and computers have all successfully used Li-ion batteries. There is an increase in large-scale project research, design, and production for grid scale storage applications [11]. Table 1.1 consists of different materials used for different purposes for anode, cathode, and separator in Li-ion batteries.

All the Li-ion battery systems follow the same basic principle from the operation point of view; however, the chemistry may vary depending upon the type of battery being considered for various applications. Figure 1.6 shows the properties of different materials used as anodes as well as cathodes in Li-ion battery systems.

1.5.2 Lead-Acid Batteries

Lead-acid batteries involve a chemical reaction between lead and lead oxide in a sulfuric acid electrolyte. These batteries are constructed using a plastic or rubber container with lead plates immersed in an electrolyte solution [35]. During charging, an external electric current causes electrochemical reduction, restoring the battery's energy storage capacity. Discharging reverses the process, releasing electrical energy, as shown in Figure 1.7 [36]. Lead-acid batteries are widely used in various applications due to their cost-effectiveness, robustness, and widespread support. However, they have limitations, such as being heavy and having lower energy density

TABLE 1.1
Different Materials Used for Cathode, Anode, and Separator Along with Their Respective Properties

Component	Material	Properties
Cathode	Lithium cobalt oxide (LCO) [27]	High energy density, high specific capacity, expensive [27]
	Lithium manganese oxide (LMO) [28]	Safe, long cycle life, lower energy density than LCO, low cost [28]
	Lithium iron phosphate (LFP) [27]	Safe, long cycle life, lower energy density than LCO, low cost [27]
	Lithium nickel cobalt aluminum oxide (NCA) [27, 29]	High energy density, long cycle life, expensive [27, 29]
Anode	Graphite [30]	High conductivity, low cost, good cycle life [30]
	Silicon [31]	High specific capacity, low conductivity, poor cycle life [31]
	Lithium titanium oxide (LTO) [29]	Safe, stable, good cycle life, low specific capacity [29]
Separator	Polyethylene (PE) [32]	Good chemical and thermal stability, low cost [32]
	Polypropylene (PP) [33]	High mechanical strength, good thermal stability [33]
	Ceramic [33]	High thermal stability, good chemical stability [33]

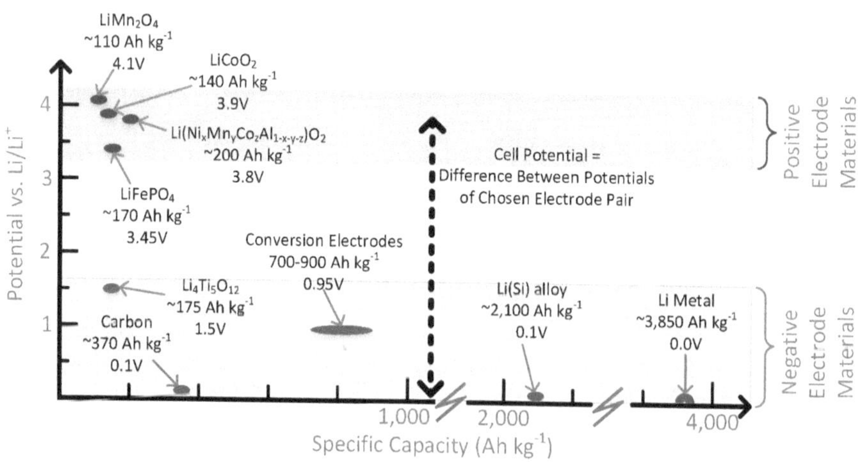

FIGURE 1.6 Different available materials for cathode and anode in different battery systems.

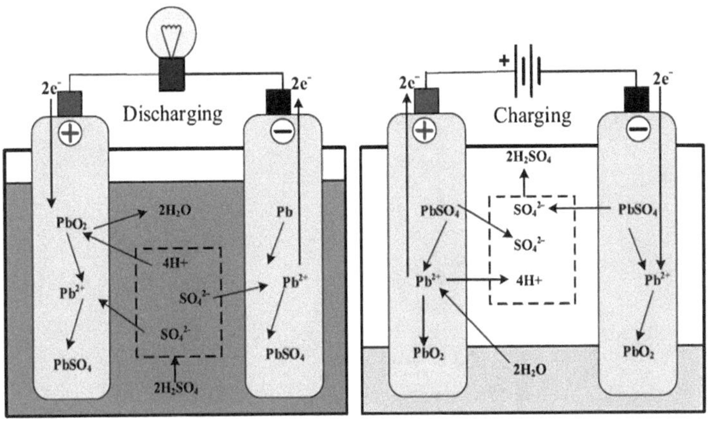

FIGURE 1.7 Schematic for discharging and charging process of a lead-acid-based battery.

Source: Reproduced with permission Copyright MDPI and Elsevier, Ref. [36].

compared to newer technologies. Regular maintenance is essential, and their limited lifespan makes replacements more frequent. Lead-acid batteries contain toxic materials, which pose environmental risks if not managed properly [37]. Proper handling, use, and disposal of lead-acid batteries are crucial, as improper disposal can lead to soil and water contamination.

1.5.3 Nickel-Cadmium Batteries

Ni-Cd batteries are popular rechargeable batteries owing to their dependability, extended cycle life, and high current capabilities. They consist of a positive electrode made up of nickel hydroxide, a negative electrode made up of cadmium, and an electrolyte solution including potassium hydroxide [38], as shown in Figure 1.8. Ni-Cd batteries have a long cycle life, making them suitable for applications requiring frequent recharging. They also deliver high currents with excellent discharge performance and low internal resistance, making them popular in high-energy bursts such as power tools, electric vehicles, and hybrid electric vehicles. However, Ni-Cd batteries have some drawbacks, including the presence of toxic cadmium, which poses environmental concerns. Proper recycling and disposal methods are necessary to prevent cadmium contamination and health risks. The development of more environmentally friendly battery technologies has led to a decline in the utilization of these batteries [39]. In addition, a phenomenon called a "memory effect" occurs when a battery is frequently recharged without being completely depleted, resulting in a decrease in total capacity. Ni-MH and Li-ion batteries are now more eco-friendly, with increased energy density and enhanced performance. However, Ni-Cd batteries continue to find applications in industries that require their unique characteristics, such as high-reliability backup systems and niche industrial applications [40].

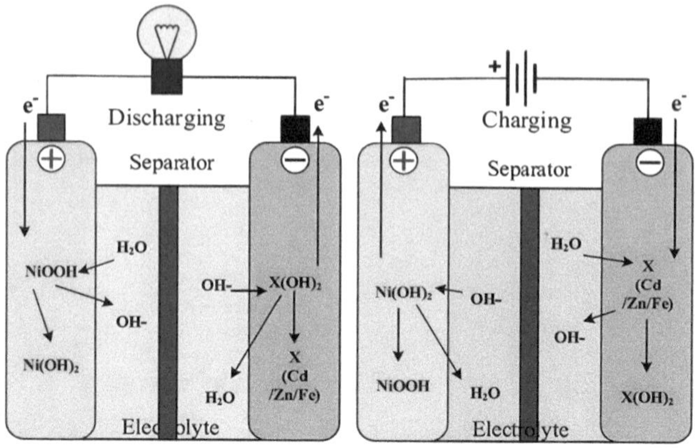

FIGURE 1.8 Schematic showing charging and discharging of a nickel-based battery system.

Source: Reproduced with permission Copyright MDPI and Elsevier, Ref. [36].

1.5.4 Nickel-Metal Hydride (Ni-MH) Batteries

These are a popular and versatile rechargeable battery technology with numerous advantages over their predecessors, such as nickel-cadmium (Ni-Cd) batteries [41]. These batteries offer higher energy density, improved environmental friendliness, and reduced memory effect. Their higher energy density allows for longer runtimes in portable devices, making them popular in consumer electronics, hybrid electric vehicles, and renewable energy systems. Additionally, Ni-MH batteries do not contain toxic cadmium, making them more environmentally conscious. Their reduced memory effect allows for more flexible charging patterns and convenience in applications where frequent charging and discharging occur. They can operate reliably in both high- and low-temperature conditions, making them suitable for diverse environments [42]. However, they have limitations, such as higher self-discharge rates compared to Li-ion batteries. Advancements in Ni-MH battery technology have led to the development of low self-discharge (LSD) Ni-MH batteries, which retain charge for longer durations.

Lastly, when compared to Li-ion batteries, Ni-MH batteries have a lower energy density, making them suitable for compact and lightweight devices. However, they face competition from Li-ion batteries in consumer electronics, such as smartphones and laptops, where size and weight are critical factors [41]. As battery technology advances, Ni-MH batteries will likely maintain their relevance in niche applications that benefit from their unique features.

1.5.5 Na-Ion Batteries

These batteries are a potential alternative to Li-ion batteries and have been developed as well as commercialized with different chemistries for large-scale energy storage applications [43]. They have several advantages over Li-ion batteries, including lower price, improved safety, and similar power delivery characteristics. Both

Na and Li-ion batteries have similar structure; however, the electrode and electrolyte materials utilized differ [44, 45]. The working principle involves sodium ions moving between the anode and cathode during charge and discharge cycles similar to Li-ion batteries. However, challenges such as slower diffusion rates and material instability need to be addressed for widespread adoption of Na-ion batteries. Despite these challenges, Na-ion batteries could play a significant role in addressing growing energy storage needs and accelerating the transition to a more sustainable and renewable energy future [46].

1.5.6 Zinc-Ion Batteries (ZIBs)

ZIBs are an exciting new energy storage technology with potential advantages in cost, safety, and environmental sustainability. These batteries are characterized by their abundant and low-cost zinc resources, high energy density, and inherent safety. The structure of a ZIB is similar to other rechargeable batteries, with an anode, cathode, and an electrolyte [47]. During charge and discharge cycles, Zn ions are inserted and extracted reversibly into and out of the anode and cathode materials, as shown in Figure 1.9 [48].

Zn is a cost-effective and sustainable choice for large-scale battery production, making them a cost-effective and sustainable choice. Additionally, Zn-ion batteries have higher thermal stability, reducing the risk of catastrophic failures and enabling safer battery operation [49]. However, ZIBs also face challenges, such as dendrite formation on the zinc anode during cycling and the need for suitable cathode materials that can reversibly intercalate or alloy with zinc ions. Further research is needed to optimize and explore new materials to unlock the full potential of ZIBs, which show promise in various sectors including stationary energy storage applications,

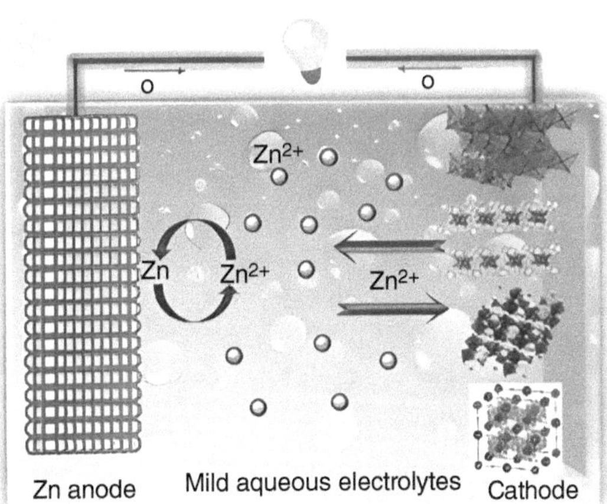

FIGURE 1.9 Working principle of aqueous Zn-ion batteries (AZIBs).

Source: Reproduced with permission Copyright MDPI & Elsevier, Ref. [48].

renewable energy integration, grid-scale storage, portable electronics, backup power systems, and off-grid applications [49, 50].

In conclusion, Zn-ion batteries offer a compelling alternative to traditional Li-ion batteries, leveraging their abundant resources, high energy density, inherent safety, and environmental sustainability. With further advancements and optimization, Zn-ion batteries have the potential to play a significant role in addressing increasing energy storage needs and accelerating the transition to a more sustainable energy future [47, 49–51].

1.5.7 Zinc (Zn) – Air Batteries

With their high energy density, low price, and environmentally favorable nature, Zn-air batteries are a promising solution for efficient and cost-effective energy storage. These batteries are ideal for long-duration energy storage applications, such as electric vehicles and grid-scale storage. The working principle, as shown in Figure 1.10 [52], involves the electrochemical oxidation of zinc at the anode and the reduction of oxygen from the air at the cathode. The electrolyte used in Zn-air batteries is usually an alkaline solution, such as potassium hydroxide (KOH) [51].

The air cathode eliminates the need for expensive cathode materials, making Zn-air batteries a desirable option for large-scale energy storage projects. Additionally, zinc is a non-toxic and recyclable element, aligning with the growing focus on clean and sustainable energy storage solutions [51, 53]. However, challenges such as limited cyclability and shelf life—as well as environmental factors like humidity and carbon dioxide levels—need to be addressed for widespread adoption. Advanced electrolytes and protective measures are being developed to minimize these effects and ensure stable battery operation under different environmental conditions [53].

In conclusion, Zn-air batteries offer a promising solution for efficient and cost-effective energy storage, with the potential to revolutionize the energy storage

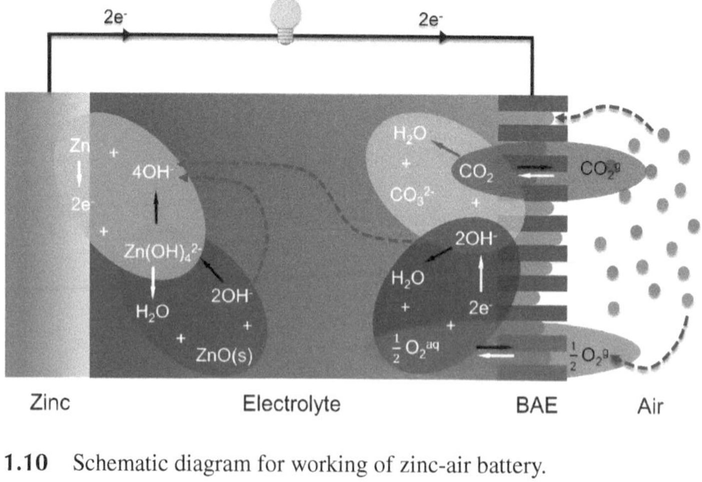

FIGURE 1.10 Schematic diagram for working of zinc-air battery.

Source: Copyright – Open Access Available under Creative Common CC by 4.0 License, Ref. [52].

landscape. As these challenges are overcome, Zn-air batteries will play a significant role in advancing clean energy technologies and enabling a sustainable energy future.

1.5.8 Aluminum (Al) – Ion Batteries

Al-ion batteries are particularly promising for grid-scale energy storage due to the abundance of aluminum [54, 55]. They offer the advantages of low-cost, high-energy density and low flammability. These batteries also function in a similar fashion to other metal-ion batteries, and graphite cathodes in Al-ion batteries can provide sufficient voltage after repeated cycles against Al anode. They are also attractive because of their high theoretical energy density, abundance of aluminum resources, low cost, and environmental benefits. However, widespread adoption faces challenges, such as the development of cathode materials that efficiently intercalate or alloy with ions, resulting in high capacity, long cycle life, and fast kinetics [56, 57].

Research efforts are focused on optimizing cathode materials and stable electrolytes to support reversible electrochemical reactions in aluminum batteries. With further optimizations and technological advancements, aluminum batteries can play a significant role in meeting the increasing demand for efficient, sustainable, and low-cost energy storage solutions [58].

1.5.9 Redox Flow Batteries (RFBs)

RFBs are a promising method for storing energy that utilize the redox reactions of electroactive species in a liquid electrolyte, allowing for high energy density, rapid response times, and extended cycle life [59]. The typical structure consists of two main components: electrochemical cells and a separate storage tank for the electrolyte. Each cell contains two electrodes, an anode and a cathode, made of porous materials with high surface areas to facilitate electrochemical reactions. The electrolyte, which contains electroactive species dissolved in a liquid medium, is stored in external tanks. The working principle of a redox flow battery involves the oxidation and reduction reactions of the electroactive species in the electrolyte, as shown in Figure 1.11 [60].

RFBs offer several advantages, including scalability, long-duration energy storage capabilities, high efficiency, rapid response times, and excellent cycle life and long-term durability. The design allows for quick and efficient charge and discharge processes, enabling fast response to changes in power demand. The ability to replace or replenish the electrolyte contributes to the longevity of redox flow batteries [61]. However, challenges remain, such as the development of cost-effective and efficient electroactive species, which require the selection of suitable species with high redox potentials, good solubility, and long-term stability to achieve high energy density and long cycle life.

RFBs are a rechargeable electrical energy storage device that uses two chemical components dissolved in liquids to store and release energy [59, 62]. They are appropriate for large-scale energy storage because of their technical qualities, flexibility in system design, and cost-effectiveness. Recent progress in redox couples, membranes, and electrode materials has made redox flow batteries a promising alternative to

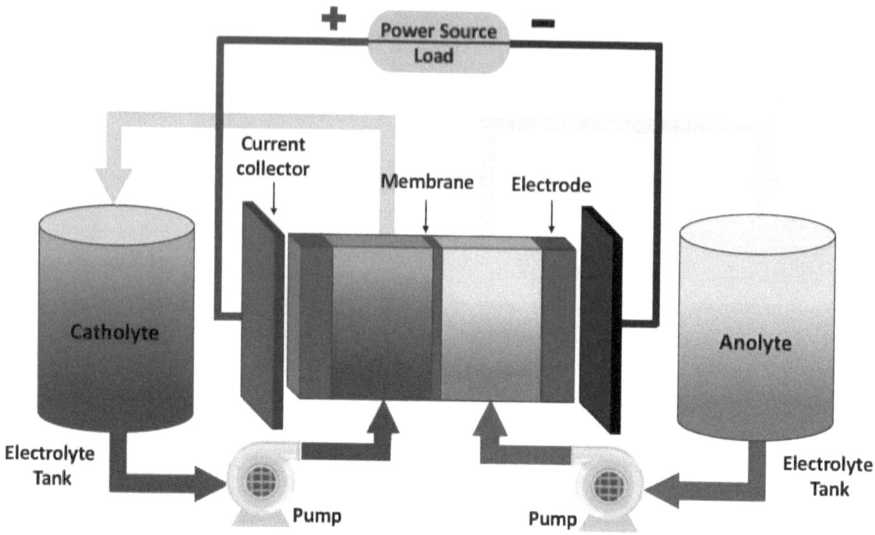

FIGURE 1.11 Working of a redox flow battery system.

Source: Copyright – Open Access Available under Creative Common CC by 4.0 License, Ref. [60].

other types of batteries due to their low vulnerability, flexibility, and cost-effectiveness. Membranes play a crucial role in preventing cross-contamination and selective transport of ions. The design of efficient and stable electrolytes is an ongoing area of research. Redox flow batteries hold great potential in various sectors, such as integrating renewable energy sources into the grid, smoothing out fluctuations in power generation, and providing grid-level storage for enhanced grid stability [59, 61, 62].

1.5.10 Supercapacitor

In addition to batteries, supercapacitors are another system that is continually being investigated and developed for energy storage applications. Electricity is stored in the supercapacitors as static energy rather than chemical energy [63]. While they may not have the energy density of batteries, they do have distinguishing characteristics that secure their place in an energy-efficient future. The evolution of supercapacitors is depicted in Figure 1.12 [64].

A supercapacitor is composed of two porous activated carbon-based electrodes separated by a thin and porous insulator/membrane soaked in electrolyte [65]. When a charge is supplied, the ions in the electrolyte move towards the electrodes and develop electric double layer capacitance (EDLC) and thereby store the energy [64]. Apart from EDLC, pseudo capacitance-based energy storage also exists in the supercapacitors at the electrode surface layers where the ions are adsorbed only into the surface layers of the metal oxide electrodes—unlike batteries where the ions are deeply intercalated into the active electrode materials. In either type (EDLC or

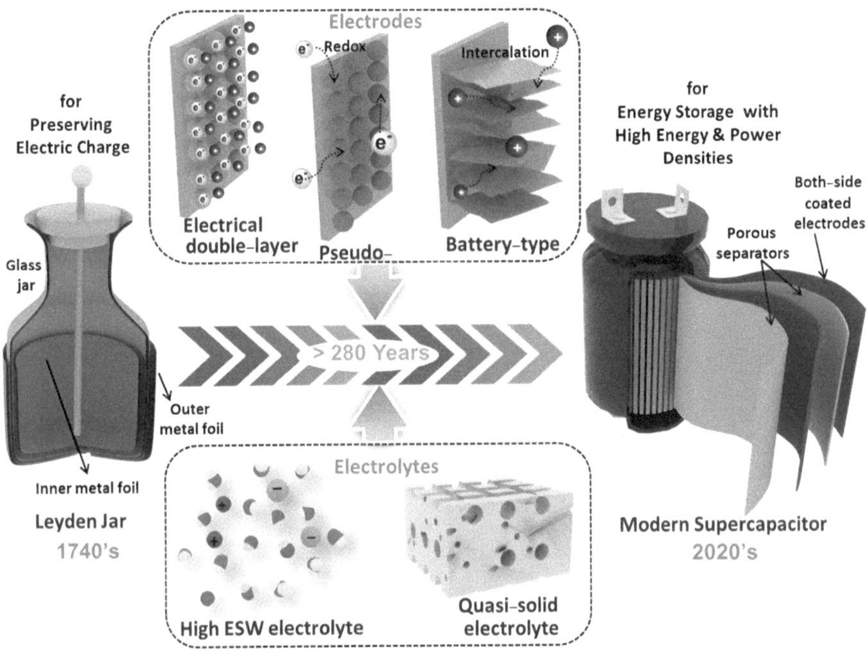

FIGURE 1.12 Schematic showing the developments of supercapacitors.

pseudocapacitive) of supercapacitors the surface area of the electrodes is crucial to obtain large amounts of charge storage.

The structure of a supercapacitor is similar to the battery; however, in EDLC-based supercapacitors both the anode and cathode are usually made of activated porous carbon material, which are very well commercially established in the market. On the other hand, pseudocapacitive-based supercapacitors are still being researched and have yet to find a way toward commercialization. For more details on supercapacitors, refer to the chapters provided in Section III of this book.

1.5.11 Differentiating Between Supercapacitors and Batteries?

Even though batteries and supercapacitors both store energy, they function quite differently. Compared to current Li-ion batteries, which have an energy density of 80–300 Wh/kg [63], supercapacitors have lower energy density of 1–30 Wh/kg [63, 66]. In contrast, supercapacitors exhibit a far higher power density than batteries. As supercapacitors do not rely on intercalation and chemical interactions with the active electrode materials, they may release a considerable amount of energy in a short time. Supercapacitors have a specific power of up to 10,000 W/kg [63, 67], compared to conventional Li-ion batteries (1,000–3,000 W/kg) [63, 67]. The relative comparison of both capacitors and batteries is mentioned in Table 1.2.

TABLE 1.2
Relative Comparison of the Capacitors and a Battery [68, 69]

	Capacitor	Battery
Storing mechanism	Electrical energy	Chemical energy
Component type	Passive	Active
Energy density	Low	High
Charging/discharging rate	Fast	Slow
Voltage type	Unstable voltage	Stable voltage
Operating temperature	Between −3 °C to +125 °C [70]	20 °C to 30 °C during charging and 15 °C to 25 °C during discharging [22]
Cost	Higher	Low
Basic components	Metal sheets	Metals, chemicals

1.6 COMPONENTS OF A COMMERCIAL Li-ION CELL SYSTEM AND THEIR FUNCTIONALITY

A battery system typically consists of several components—each of which has its own functionality—the basic cell schematic is shown in Figure 1.13 [71].

Here are the main components of a battery system and their functionality:

- ***Anode***: The anode is the electrode where oxidation takes place during the discharge of the battery. It releases electrons that flow through the external circuit to the cathode, generating an electric current. During charging, the anode becomes the positive electrode and receives electrons from the external circuit [72].
- ***Cathode***: The cathode is the electrode where reduction takes place during the discharge of the battery. It accepts the electrons from the anode through the external circuit, generating an electric current. During charging, the cathode becomes the negative electrode and releases electrons to the external circuit [72].
- ***Electrolyte***: The electrolyte is the medium through which the ions flow between the anode and cathode. It is typically a liquid or gel that contains

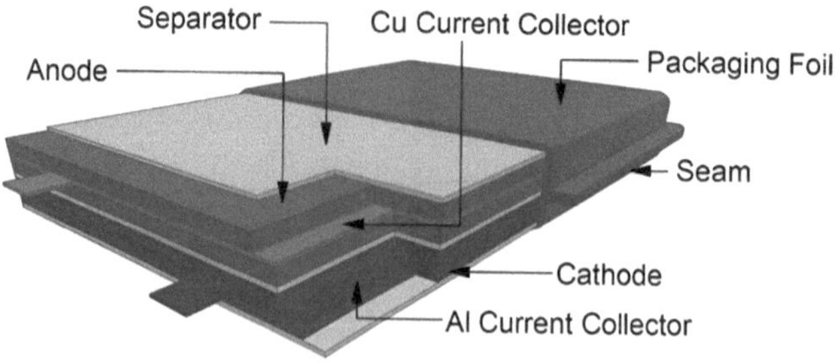

FIGURE 1.13 Basic schematic showing the cell design of a battery system.

Source: Copyright – Open Access Available under Creative Common CC by 4.0 License, Ref. [71].

ions that can move freely between the electrodes. The electrolyte facilitates the electrochemical reaction between the anode and cathode and is a key determinant of the battery's performance [65, 72].

- ***Separator***: The separator is a permeable membrane that separates the anode and cathode and prevents them from coming into direct contact. The separator allows the ions to flow between the electrodes while preventing the electrons from short-circuiting the battery [33, 73, 74].
- ***Anode Current Collector***: The anode current collector is a conductive material (usually copper in Li-ion batteries) that connects the anode to the external circuit. It collects the electrons released by the anode and transfers them to the external circuit during discharge, and collects the electrons from the external circuit during charging and transfers them to the anode [75].
- ***Cathode Current Collector***: The cathode current collector is a conductive material (usually aluminum in Li-ion batteries) that connects the cathode to the external circuit. It collects the electrons from the external circuit during discharge and transfers them to the cathode, and collects the electrons released by the cathode during charging and transfers them to the external circuit [76].
- ***Casing***: The casing is the outer shell of the battery that protects the internal components from damage and provides a safe and secure container for the electrolyte [72].

These components work together to store and release electrical energy in a battery system. Each component plays a crucial role in the electrochemical reactions that take place inside the battery, and the overall performance of the battery system depends on the design and quality of each component [77].

1.7 IMPORTANCE OF ELECTROLYTE IN ENERGY STORAGE SYSTEMS

Electrochemical energy storage systems, such as batteries and supercapacitors, use a variety of electrolytes to enable ion movement and electrochemical reactions that store and release energy. Factors such as the type of energy storage system, desired performance characteristics, safety considerations, and cost-effectiveness all influence the electrolyte used.

Some of the most common electrolytes used in electrochemical energy storage systems are as follows:

- ***Aqueous Electrolytes***: Because of their low cost, strong ionic conductivity, and relatively favorable safety profile, aqueous electrolytes—which are primarily based on water solutions—are widely employed. They frequently use salt compounds dissolved in water, such as lithium salts (e.g., lithium sulphate or lithium hydroxide) or alkaline metal hydroxides [78, 79]. Aqueous electrolytes are widely utilized in applications such as lead-acid batteries, alkaline batteries, and certain flow batteries.

- ***Non-Aqueous Organic Electrolytes*****:** These are commonly employed in Li-ion batteries and other rechargeable battery chemistries. These electrolytes are made out of organic solvents (such as ethylene carbonate and dimethyl carbonate) combined with lithium salts (e.g., lithium hexafluorophosphate or lithium trifluoromethanesulfonate) [80, 81]. Non-aqueous organic electrolytes have excellent ionic conductivity and a wide voltage window, making them suitable for high-energy-density and high-power applications. However, because of their flammability and volatility, they must be handled with attention and safety procedures.
- ***Solid-State Electrolytes*****:** Solid-state electrolytes are a potential class of electrolytes for next-generation energy storage devices. These electrolytes' solid or gel-like form eliminates the need for liquid or gel electrolyte solutions [65, 82]. Ceramics, polymers, and composites are some of the materials used to make solid-state electrolytes. They offer advantages such as enhanced safety, a broader operating temperature range, and possibly higher energy density. Solid-state electrolytes are currently being developed for use in Li-ion batteries, solid-state lithium batteries, and other advanced energy storage systems [65, 83].
- ***Ionic Liquids*****:** Ionic liquids are molten salts that stay liquid even at low temperatures. They have distinct characteristics such as minimal volatility, a broad electrochemical stability window, and good thermal stability. Ionic liquids are being studied as electrolytes for high-temperature energy storage devices, supercapacitors, and new battery technologies. However, their high cost and viscosity may be prohibitive in some applications [84, 85].
- ***Hybrid Electrolytes*****:** Hybrid electrolytes blend multiple electrolyte components to take advantage of their unique properties. A solid-state electrolyte, for example, can be mixed with a liquid electrolyte to improve both safety and ionic conductivity [86, 87]. Hybrid electrolytes provide a balance of the benefits of various electrolyte types and can be customized to suit energy storage system requirements.

1.8 FACTORS AFFECTING THE PERFORMANCE OF ELECTROLYTES

Several factors can significantly impact the performance of electrolytes in energy storage systems. These factors influence the conductivity, stability, safety, and overall efficiency of the electrolyte, ultimately affecting the performance of the entire energy storage system.

Here are some key factors that can affect the performance of electrolytes:

- ***Ionic Conductivity*****:** Ionic conductivity is an important feature because it controls the rate at which ions can travel through the electrolyte. High ionic conductivity is desirable as it allows for faster charge and discharge rates, resulting in increased power output and reaction time of the energy storage system [88]. The composition and selection of electrolyte materials can have a significant impact on ionic conductivity.

- ***Electrochemical Stability***: Electrolytes must be electrochemically stable enough to endure the voltage and potential oscillations that occur during charge and discharge cycles. If the electrolyte is not electrochemically stable, it might cause undesirable side reactions, electrolyte degradation, and even safety issues [79, 89]. To ensure long-term performance and safety in high-voltage applications, electrochemical stability is extremely important.
- ***Viscosity***: The viscosity of the electrolyte influences the ease with which ions can travel through it. Low viscosity allows for faster ion diffusion, lowering internal resistance and increasing the energy efficiency of the energy storage system as a whole. Ion movement can be hampered by high viscosity, resulting in slower charging and discharging rates [90].
- ***Thermal Stability***: Thermal stability is required to ensure the safe operation of energy storage devices. Electrolytes should be stable over a wide temperature range, encompassing both typical working circumstances and potential high-temperature scenarios [91]. Thermal stability avoids thermal runaway reactions, deterioration, or breakdown of the electrolyte, which can lead to safety issues or performance loss.
- ***Safety and Environmental Considerations***: In energy storage systems, safety is of the utmost significance. Electrolytes' flammability, toxicity, and reactivity can pose hazards, particularly in large-scale applications or under extreme conditions. The formulation and materials used in electrolytes should prioritize safety by reducing flammability, toxicity, and hazardous qualities [92, 93]. Environmental issues are also crucial, and electrolytes with little environmental effect or made from sustainable resources are preferred.
- ***Ion Solubility and Mobility***: The solubility and mobility of ions in the electrolyte have a major impact on the performance of the energy storage device. High ion mobility allows for quick ion movement through the electrolyte, lowering internal resistance. Good solubility enables sufficient ion concentration for efficient charge and discharge processes [94, 95].
- ***Cost-Effectiveness***: Electrolyte costs can have an impact on the overall cost-effectiveness of energy storage systems. To ensure the scalability and commercial feasibility of energy storage technology, electrolyte materials should be economically viable, and their availability and production costs should be evaluated.
- ***Electrolyte Compatibility with Electrode Materials***: Electrolytes must be compatible with the electrode materials chosen in order to provide efficient ion transport and electrochemical reactions [96, 97]. Poor performance, lower cycle life, or electrode deterioration might result from mismatched electrode-electrolyte interactions.

Conductivity, stability, safety, and environmental impact are crucial factors in selecting an electrolyte. Safety measures include minimizing thermal runaway, fire, or explosion risks, and ensuring thermal stability, low flammability, and reduced toxicity. Environmental impact is minimized by utilizing environmentally benign electrolytes, such as aqueous electrolytes based on water [95]. Cost-effectiveness impacts the widespread adoption of energy storage systems, with aqueous electrolytes being more cost-effective due to their abundance and low cost of water. High-performance non-aqueous electrolytes, such as those in advanced batteries or emerging technologies, offer enhanced energy density, power capabilities, and long cycle life. Continuous research and innovation in electrolyte development will advance the capabilities and sustainability of these systems, ensuring a cleaner, more sustainable future. For a thorough understanding on the role of electrolytes, their stability, and the relevant side reactions and additives added to electrolytes to minimize the above effects, refer to the following chapters.

1.9 CONCLUSIONS

Electrolytes are critical components of electrochemical energy storage systems, influencing their efficiency, safety, environmental impact, and cost-effectiveness. Conductivity and stability are critical for efficiency and power output, while safety considerations focus on limiting dangers. Environmental compatibility is achieved through investigating environmentally friendly materials and sustainable electrolyte choices. Future prospects for electrolytes in energy storage systems are exciting, with research and development efforts focusing on solid-state electrolytes, additives, and the integration of electrolytes with emerging technologies like nanomaterials and additive manufacturing. These advancements will contribute to high-performance, safe, environmentally friendly, and cost-effective energy storage systems.

Energy storage systems are essential for modern infrastructure, providing reliable and affordable power to communities worldwide. Recent advancements in technology and research have shown promising results, but cost and scalability remain challenges. With continued research, development, and investment, energy storage systems can play a vital role in achieving a sustainable, low-carbon energy system.

This chapter sets the stage for the upcoming chapters by providing an overview of the basics of electrochemical energy storage systems and will serve as a foundation for the detailed discussions that follow. The chapters delve into each electrochemical energy storage system in detail, exploring their working principles, components, and performance characteristics while focusing on the criticality of the electrolyte. By the end of the subsequent chapters, readers will have a comprehensive understanding of the various electrochemical energy storage systems and their battery chemistries along with the factors influencing battery performance, such as energy density, power density, cycle life, safety, and environmental impacts. Readers are advised to refer to the upcoming chapters for detailed battery chemistries and the related developments.

REFERENCES

1. Depletion of Non-renewable Resources: Causes, Effect & Solution – EcoWut at https://ecowut.com/depletion-nonrenewable-resources-causes-effect-solution/
2. An Introduction to Energy Storage Systems – Technical Articles at https://eepower.com/technical-articles/an-introduction-to-energy-storage-systems/

3. R. K. Beniwal, S. Dhundhara, and A. Kalra, Introduction to Energy Storage Systems, in *Energy Storage for Modern Power System Operations*, eds. S. Dhundhara and Y. P. Verma, 2021, 1–31, Wiley.
4. H. Abdi, B. Mohammadi-Ivatloo, S. Javadi, A. R. Khodaei, and E. Dehnavi, Energy Storage Systems, *Distributed Generation Systems: Design, Operation and Grid Integration*, 2017, Butterworth-Heinemann: Oxford, UK, 333–368.
5. A Simple History of Batteries – Panasonic Energy Co., Ltd. at https://www.panasonic.com/global/energy/study/academy/history.html
6. Powering a Generation: Power History #1 at https://americanhistory.si.edu/powering/past/prehist.htm
7. Baghdad Battery|Oldest Battery Known to Man at https://www.globalmarvels.com/baghdad-battery/
8. Was the Baghdad Battery really a battery? BBC Science Focus Magazine at https://www.sciencefocus.com/science/was-the-baghdad-battery-really-a-battery/
9. Alessandro Volta-Inventor of the Voltaic Pile at https://www.thoughtco.com/alessandro-volta-1992584
10. Alessandro Volta|Biography, Facts, Battery, & Invention. Britannica at https://www.britannica.com/biography/Alessandro-Volta
11. S. S. Rangarajan, S. P. Sunddararaj, A. V. V. Sudhakar, C. K. Shiva, U. Subramaniam, E. R. Collins, and T. Senjyu, *Clean Technol.*, 2022, 4, 908–930.
12. An Overview of Energy Storage Systems and Their Applications-Technical Articles at https://eepower.com/technical-articles/an-overview-of-energy-storage-systems-and-their-applications/
13. History and Timeline of the Battery at https://www.thoughtco.com/battery-timeline-1991340
14. Jensen, W. B. The Leclanché Cell. Museum Notes, Oesper Collections, 2014.
15. Leclanché, G. Quelques Observations Sur l'emploi Des Piles Électriques. Pile Constante Au Peroxyde de Manganese à Un Seul Liquide. *Les Mondes*, 1868, 16, 532.
16. National Historic Chemical Landmarks at https://www.acs.org/content/acs/en/education/whatischemistry/landmarks/drycellbattery.html
17. science.ca: Lew Urry at https://www.science.ca/scientists/scientistprofile.php?pID=277
18. A Brief History of Lithium-Ion Battery Development at https://www.hidenanalytical.com/blog/lithium-ion-battery-development/
19. C. P. Grey, and D. S. Hall, *Nat. Commun.*, 2020, 11, 1–4.
20. L. Malka, A. Daci, A. Kuriqi, P. Bartocci, and E. Rrapaj, *Energies (Basel)*, 2022, 15, 4032.
21. A. Riaz, M. R. Sarker, M. H. M. Saad, and R. Mohamed, *Sensors*, 2021, 21, 5041.
22. C. Chukwuka, and K. A. Folly, *IEEE Power and Energy Society Conference and Exposition in Africa: Intelligent Grid Integration of Renewable Energy Resources*, 2012, 1–6.
23. P. M. Attia, A. Bills, F. Brosa Planella, P. Dechent, G. dos Reis, M. Dubarry, P. Gasper, R. Gilchrist, S. Greenbank, D. Howey, O. Liu, E. Khoo Y. Preger, A. Soni, S. Sripad, A. G. Stefanopoulou, and V. Sulzer, *J. Electrochem. Soc.*, 2022, 169, 28.
24. How Lithium-ion Batteries Work|Department of Energy at https://www.energy.gov/energysaver/articles/how-lithium-ion-batteries-work
25. T. Kim, W. Song, D. Y. Son, L. K. Ono, and Y. Qi, *J. Mater. Chem. A*, 2019, 7, 2942–2964.
26. L. Yao, S. Xu, A. Tang, F. Zhou, J. Hou, Y. Xiao, and Z. Fu, *World Electr. Veh. J.*, 2021, 12, 113.
27. Effects of Cathode Material on Lithium-Ion Battery Performance-Battery Skills at https://www.batteryskills.com/effects-of-cathode-material-on-lithium-ion-battery-performance/

28. Cathode Material – An overview|ScienceDirect Topics at https://www.sciencedirect.com/topics/materials-science/cathode-material
29. N. Nitta, F. Wu, J. T. Lee, and G. Yushin, *Mater. Today*, 2015, 18, 252–264.
30. H. Zhang, Y. Yang, D. Ren, L. Wang, and X. He, *Energy Storage Mater.*, 2021, 36, 147–170.
31. C. Zhang, F. Wang, J. Han, S. Bai, J. Tan, J. Liu, and F. Li, *Small Struct.*, 2021, 2, 2100009.
32. K. J. Kim, M. S. Park, H. Kim, and Y. J. Kim, *Jpn. J. Appl. Phys.*, 2012, 51, 09MB03.
33. S. Luiso, and P. Fedkiw, *Curr. Opin. Electrochem.*, 2020, 20, 99–107.
34. Y. Miao, P. Hynan, A. Von Jouanne, and A. Yokochi, *Energies*, 2019, 12, 1074.
35. H. Bode, *Lead-Acid Batteries*, 1977, John Wiley & Sons, New York.
36. M. A. Hannan, M. M. Hoque, A. Mohamed, and A. Ayob, *Renew. Sust. Energ. Rev.*, 2017, 69, 771–789.
37. P. Ruetschi, *J. Power Sources*, 1977, 2, 3–120.
38. S. Petrovic, *Battery Technology Crash Course*, 2021, 73–88.
39. C. J. Rydh, and M. Karlström, Life Cycle Inventory of Recycling Portable Nickel-Cadmium Batteries, *Resour. Conserv. Recycl.*, 2002, 34(4), 289–309. https://doi.org/10.1016/S0921-3449(01)00114-8
40. J. David, *J. Power Sources*, 1995, 57, 71–73.
41. S. R. Ovshinsky, M. A. Fetcenko, and J. Ross, *Science*, 1993, 260, 176–181.
42. M. Assefi, S. Maroufi, Y. Yamauchi, and V. Sahajwalla, *Curr. Opin. Green Sustain. Chem.*, 2020, 24, 26–31.
43. J. Y. Hwang, S. T. Myung, and Y. K. Sun, *Chem. Soc. Rev.*, 2017, 46, 3529–3614.
44. J. F. Peters, A. P. Cruz, and M. Weil, *Batteries*, 2019, 5, 10.
45. J. Peters, D. Buchholz, S. Passerini, and M. Weil, *Energy Environ. Sci.*, 2016, 9, 1744–1751.
46. K. Vignarooban, R. Kushagra, A. Elango, P. Badami, B. E. Mellander, X. Xu, T. G. Tucker, C. Nam, and A. M. Kannan, *Int. J. Hydrog. Energy*, 2016, 41, 2829–2846.
47. J. Ming, J. Guo, C. Xia, W. Wang, and H. N. Alshareef, *Mater. Sci. Eng. R. Rep.*, 2019, 135, 58–84.
48. X. Zeng, J. Hao, Z. Wang, J. Mao, and Z. Guo, *Energy Storage Mater.*, 2019, 20, 410–437.
49. B. Tang, L. Shan, S. Liang, and J. Zhou, *Energy Environ. Sci.*, 2019, 12, 3288–3304.
50. T. Zhang, Y. Tang, S. Guo, X. Cao, A. Pan, G. Fang, J. Zhou, and S. Liang, *Energy Environ. Sci.*, 2020, 13, 4625–4665.
51. V. Caramia, and B. Bozzini, *Mater. Renew. Sustain. Energy*, 2014, 3, 1–12.
52. S. Clark, A. Latz, and B. Horstmann, *Batteries*, 2018, 4, 5.
53. Y. Li, and H. Dai, *Chem. Soc. Rev.*, 2014, 43, 5257–5275.
54. H. Sun, W. Wang, Z. Yu, Y. Yuan, S. Wang, and S. Jiao, *Chem. Commun.*, 2015, 51, 11892–11895.
55. A. G. Olabi, E. T. Sayed, T. Wilberforce, A. Jamal, A. H. Alami, K. Elsaid, S. M. A. Rahman, S. K. Shah, and M. A. Abdelkareem, *Energies*, 2021, 14, 7373.
56. S. K. Das, S. Mahapatra, and H. Lahan, *J. Mater. Chem. A*, 2017, 5, 6347–6367.
57. X. Wang, Y. Li, Y. Gao, Z. Wang, and L. Chen, *Nano Energy*, 2015, 13, 687–692.
58. D. Y. Wang, C. Y. Wei, M. C. Lin, C. J. Pan, H. L. Chou, H. A. Chen, M. Gong, Y. Wu, C. Yuan, M. Angell, Y. J. Hsieh, Y. H. Chen, C. Y. Wen, C. W. Chen, B. J. Hwang, C. C. Chen, and H. Dai, *Nat. Commun.*, 2017, 8, 1–7.
59. M. L. Perry, and A. Z. Weber, *J. Electrochem. Soc.*, 2016, 163, A5064–A5067.
60. A. Clemente, and R. Costa-Castelló, *Energies*, 2020, 13, 4514.
61. W. Wang, Q. Luo, B. Li, X. Wei, L. Li, and Z. Yang, *Adv. Funct. Mater.*, 2013, 23, 970–986.

62. A. Z. Weber, M. M. Mench, J. P. Meyers, P. N. Ross, J. T. Gostick, and Q. Liu, *J. Appl. Electrochem.*, 2011, 41, 1137–1164.
63. Supercapacitors vs. Batteries: What's the Difference? at https://www.howtogeek.com/786195/supercapacitors-vs-batteries-whats-the-difference/
64. K. C. S. Lakshmi, and B. Vedhanarayanan, *Batteries*, 2023, 9, 202.
65. R. K. Guduru, and J. C. Icaza, *Nano*, 2016, 6, 41.
66. C. Liu, Z. Yu, D. Neff, A. Zhamu, and B. Z. Jang, *Nano Lett.*, 2010, 10, 4863–4868.
67. Supercapacitor – Wikipedia at https://en.wikipedia.org/wiki/Supercapacitor
68. R. T. Yadlapalli, R. K. R. Alla, R. Kandipati, and A. Kotapati, *J. Energy Storage*, 2022, 49, 104194.
69. Y. U. de L.-Ul. > D. d'aérospatiale et mécanique > I. des véhicules terrestres Louvigny, J. U. de L.-Ul. > D. d'aérospatiale et mécanique > I. des véhicules terrestres Nzisabira, and P. U. de L.-Ul. > D. d'aérospatiale et mécanique > I. des véhicules terrestres Duysinx, *EET-2008 European Ele-Drive Transportation Conference + IAMF International Advanced Mobility Forum*, 2008 at https://orbi.uliege.be/handle/2268/9423
70. Understanding Ceramic Capacitor Temp-Coefficients – Passives/Capacitors – Electronic Component and Engineering Solution Forum – TechForum | Digi-Key at https://forum.digikey.com/t/understanding-ceramic-capacitor-temp-coefficients/727
71. M. Wentker, M. Greenwood, and J. Leker, *Energies*, 2019, 12, 504.
72. Components of Cells and Batteries at https://depts.washington.edu/matseed/batteries/MSE/components.html
73. P. Arora, and Z. Zhang, *Chem. Rev.*, 2004, 104, 4419–4462.
74. Battery Separators – Types and Importance in the Performance of Battery at https://components101.com/articles/battery-seperators-types-and-importance
75. N. Li, T. Jia, Y. Liu, S. Huang, F. Kang, and Y. Cao, *Front. Chem.*, 2022, 10, 884308.
76. US6403263B1 – Cathode Current Collector for Electrochemical Cells – Google Patents at https://patents.google.com/patent/US6403263B1/en
77. Electric Vehicle News and Analysis | InsideEVs at https://insideevs.com/news/
78. J. Yi, and Y. Xia, *MRS Energy Sustain.*, 2022, 9, 106–128.
79. D. Reber, R. Figi, R. S. Kühnel, and C. Battaglia, *Electrochim. Acta*, 2019, 321, 134644.
80. Y. Xu, M. Zhou, and Y. Lei, *Mater. Today*, 2018, 21, 60–78.
81. Y. Yu, H. Che, X. Yang, Y. Deng, L. Li, and Z. F. Ma, *Electrochem. Commun.*, 2020, 110, 106635.
82. H. M. Woolley, and N. M. Vargas-Barbosa, *J. Mater. Chem. A*, 2023, 11, 1083–1097.
83. Y. Wang, S. Song, C. Xu, N. Hu, J. Molenda, and L. Lu, *Nano Mater. Sci.*, 2019, 1, 91–100.
84. D. Monti, E. Jónsson, M. R. Palacín, and P. Johansson, *J. Power Sources*, 2014, 245, 630–636.
85. D. Kumar, and S. A. Hashmi, *Solid State Ion.*, 2010, 181, 416–423.
86. K. Matsumoto, R. Taniki, T. Nohira, and R. Hagiwara, *J. Electrochem. Soc.*, 2015, 162, A1409–A1414.
87. J. Feng, L. Wang, Y. Chen, P. Wang, H. Zhang, and X. He, *Nano Converg.*, 2021, 8, 1–12.
88. A. Fukunaga, T. Nohira, R. Hagiwara, K. Numata, E. Itani, S. Sakai, and K. Nitta, *J. Appl. Electrochem.*, 2016, 46, 487–496.
89. C. V. Manohar, K. A. Raj, M. Kar, M. Forsyth, D. R. MacFarlane, and S. Mitra, *Sustain. Energy Fuel*, 2018, 2, 566–576.
90. K. Tsunashima, and M. Sugiya, *Electrochem. Commun.*, 2007, 9, 2353–2358.
91. S. Ma, M. Jiang, P. Tao, C. Song, J. Wu, J. Wang, T. Deng, and W. Shang, *Prog. Nat. Sci.: Mater. Int.*, 2018, 28, 653–666.

92. Z. Yang, J. He, W. H. Lai, J. Peng, X. H. Liu, X. X. He, X. F. Guo, L. Li, Y. Qiao, J. M. Ma, M. Wu, and S. L. Chou, *Angew. Chem. Int. Ed.*, 2021, 60, 27086–27094.
93. B. Mosallanejad, S. S. Malek, M. Ershadi, A. A. Daryakenari, Q. Cao, F. Boorboor Ajdari, and S. Ramakrishna, *J. Electroanal. Chem.*, 2021, 895, 115505.
94. D. Kumar, and S. A. Hashmi, *J. Power Sources*, 2010, 195, 5101–5108.
95. S. Gupta, and P. Sappidi, *J. Chem. Eng. Data*, 2023, 68, 162–172.
96. Z. S. Iro, C. Subramani, and S. S. Dash, *Int. J. Electrochem. Sci.*, 2016, 11, 10628–10643.
97. M. Dahbi, N. Yabuuchi, M. Fukunishi, K. Kubota, K. Chihara, K. Tokiwa, X. F. Yu, H. Ushiyama, K. Yamashita, J. Y. Son, Y. T. Cui, H. Oji, and S. Komaba, *Chem. Mater.*, 2016, 28, 1625–1635.

2 Importance of Electrolytes and Their Selection for Energy Storage Systems

Prakash Chandra, Brijesh Tripathi, and Manoj Kumar

2.1 INTRODUCTION

In all forms of electrical energy storage (EES) devices, an electrolyte is a crucial component that is often liquid and aids in conducting electricity by transporting ions rather than electrons. The identification of the electrolyte is essential for creating a safe and effective EES device, as it is positioned between and closely interacts with the anode and the cathode. In EES devices, an electrolyte typically functions in collaboration with a porous membrane or gel and essentially possesses highly ionic conducting and electronic insulation properties, either in a solid or a liquid state [1–6]. An electrolyte can be an aqueous, organic, or liquid-salt (ionic liquid or molten salt) in corroboration with salts, solvents, and functional additions. See Figure 2.1 for different variations in the electrolytes.

The electrolyte must possess a wide electrochemical window to allow for higher operating voltages, which can lead to higher energy density in the electrochemical device. High ionic conductivity is significant for effective charge and mass transport, permitting fast charge/discharge rates and high-power densities. Higher ionic conductivity leads to higher power output and faster charging rates. Ionic conductivity depends on factors such as the electrolyte composition, temperature, and the presence of impurities [7].

Chemical and thermal durability are vital for long-standing device capabilities and toughness, and the chemical inertness of other cell components like the separator, electrode substrates, and cell packaging materials helps to avoid undesirable reactions or degradation of the device. Thermal stability is the ability of the electrolyte to withstand high temperatures without breaking down or decomposing. Electrolytes with high thermal stability are preferred because they can withstand high operating temperatures, which can increase power output and extend the lifetime of the energy storage device [8].

Safety and non-toxicity are the critical property of electrolytes for energy storage like portable electronics and electric vehicles. Electrolytes that are flammable, toxic, or corrosive can pose significant safety risks. Non-flammable and non-toxic electrolytes, such as water-based electrolytes or solid-state electrolytes, are being developed to ameliorate the safety of EES devices.

DOI: 10.1201/9781003441182-2

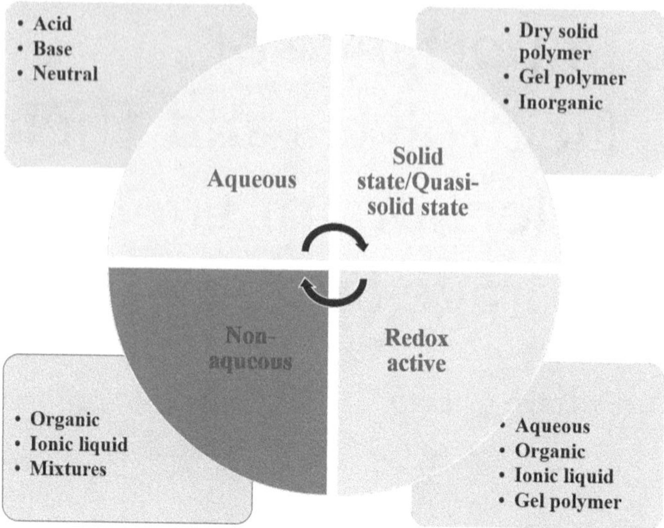

FIGURE 2.1 Miscellaneous electrolyte materials used in different energy storage.

The electrochemical stability window is the voltage range within which the electrolyte remains stable and does not break down or degrade. Electrolytes with a wider electrochemical stability window are preferred because they allow for a larger range of operating voltages, which can increase energy density and power output. Additionally, an electrolyte with low electrochemical stability can cause the battery to overheat or catch fire, presenting a safety hazard [8–10].

Economical production is essential to enable widespread adoption of solid-state electrolyte-based devices. The development of scalable and cost-effective manufacturing methods for these materials is an active area of research and development in the field [11–17]. Viscosity is a measure of how easily an electrolyte flows between the anode and cathode. Low viscosity electrolytes can improve energy density and power output but can also increase the risk of the formation of dendrite that may lead to short-circuit and safety hazards [8].

Electrode and electrolyte interactions and side reactions can significantly impact the performance, safety, and durability of EES devices [8, 10, 18]. The electrolyte can decompose at the electrode surfaces, resulting in the formation of unwanted reaction products and depletion of the electrolyte. This can lead to a loss of battery capacity, reduced cycling stability, and safety issues such as gas generation and thermal runaway. In some battery chemistries, such as lithium-ion batteries, a SEI layer forms at the electrode surface because of electrolyte decomposition. The SEI layer can affect the battery performance by impeding ion transport and increasing the internal resistance of the battery.

Electrolyte additives are often used to improve the performance and stability of batteries; however, these additives can also have side effects, such as increased electrolyte viscosity, reduced ionic conductivity, and increased decomposition reactions. Some electrolytes are sensitive to water, which can cause hydrolysis and lead to the formation of unwanted reaction products. This can reduce the battery performance and stability. Thus, finding an electrolyte that properly meets all these requirements is actually very difficult. To address these challenges, researchers are exploring various strategies such as the development of novel electrolyte systems to ameliorate battery performance. There have been and will continue to be significant and ongoing scientific endeavors.

In this chapter the significance of miscellaneous electrolyte aqueous, organic, ionic liquid and molten electrolyte systems for different EES devices is illustrated in detail. This chapter also provides an insight into some novel electrolytes recently reported for EES, and their interesting findings are discussed.

2.2 AQUEOUS ELECTROLYTES

Aqueous electrolytes are solutions of dissolved salts in water, which can conduct electricity due to the presence of ions for electrochemical energy storage systems. Amongst the diversity of energy storage systems known, aqueous electrolytes for the Li-ion batteries have recently gained attention to replace organic electrolytes because they possess high ionic conductivity, low flammability, low toxicity, and potential for low cost. However, the use of aqueous electrolytes in Li-ion batteries is still a relatively new area of research, and challenges such as limited electrochemical stability and compatibility with high-voltage electrodes need to be addressed.

Several Li-based aqueous electrolytes are well known; for example, LiOH is cheap but unstable; Li_2SO_4 is stable but now suitable to operate at low temperature; $LiNO_3$ is stable and conductive but corrosivity is the major problem with it; LiTSFI is both stable and conductive [19, 20]. The research on aqueous electrolytes for LIBs is ongoing, and there is important interest in emerging new materials and approaches to overcome these restrictions. Aqueous electrolytes have the potential to significantly reduce the cost and environmental impact of LIBs while maintaining or improving their performance, making them an attractive area for further research and development.

2.3 ORGANIC ELECTROLYTES

Organic electrolytes are liquid electrolytes made from organic compounds used in batteries to facilitate the flow of ions between the electrodes. They are typically composed of a lithium salt dissolved in organic solvents like ethylene carbonate (EC) or propylene carbonate (PC). However, aqueous electrolytes are not conventionally used in Li-ion batteries due to their limited voltage window and potential for corrosion.

EC is a frequently utilized solvent in Li-ion batteries because it demonstrates high dielectric constant and capability to dissolve Li-salts. PC is another commonly used solvent in ion batteries, which has a lower melting point than EC and can ameliorate low-temperature performance of batteries. DMC is a solvent with a lower viscosity than EC and PC that can ameliorate battery performance at high temperatures. Ethyl methyl carbonate (EMC) is a solvent commonly used as a co-solvent with EC and DMC to improve the performance of batteries at low temperatures. Diethyl carbonate (DEC) is a solvent commonly used as a co-solvent with EC and DMC to improve the stability and safety of batteries. Organic electrolytes are most extensively investigated electrolytes in LIBs due to their high performance and stability. Ongoing research is focused on developing new materials and strategies to address their limitations and further improve their performance in LIBs.

2.4 CONVENTIONAL ELECTROLYTES

Conventional electrolytes are a type of liquid electrolytes extensively investigated in Li-ion batteries. These conventional electrolyte systems are composed of Li-salts dissolved in a mixture of organic solvents. They are called "conventional" because they have been widely used for many years and considered to be the standard electrolyte for Li-ion batteries. Some important conventional electrolytes used for Li-ion batteries are lithium hexafluorophosphate ($LiPF_6$) in a mixture of ethylene carbonate (EC), and dimethyl carbonate (DMC) is the most widely used electrolyte for Li-ion batteries.

$LiPF_6$ is a Li-salt dissolved in a mixture of EC and DMC solvents. This electrolyte provides good electrochemical performance and is relatively safe to use. Lithium bis(trifluoromethylsulfonyl)imide (LiTFSI) in a mixture of EC and PC LiTFSI is another Li-salt that is dissolved in a mixture of EC and PC solvents. This electrolyte has high ionic conductivity and can operate at high voltages, making it suitable for high-energy-density batteries. Lithium boron tetrafluoride ($LiBF_4$) in a mixture of EC and diethyl carbonate (DEC): $LiBF_4$ is a Li-salt dissolved in a mixture of EC and DEC solvents. This electrolyte has good thermal stability and low viscosity, which can improve the low-temperature performance of batteries. Lithium trifluoromethanesulfonate (LiOTf) is a lithium salt dissolved in a mixture of EC and PC solvents. This electrolyte has good thermal stability and high ionic conductivity, which can improve battery performance. Lithium hexafluoroarsenate ($LiAsF_6$) is dissolved in a mixture of EC and DMC solvents. This electrolyte has high thermal stability and can operate at high voltages, but it is less commonly used due to its potential safety hazards [21–24].

2.5 IONIC LIQUID ELECTROLYTES

Ionic liquid electrolytes behave as liquid salts at room temperature in EES devices and demonstrate negligible vapor pressure. They are often used as an alternative to conventional organic electrolytes because they have some advantages, such as high thermal and electrochemical stability, low flammability, and low volatility. Some common types of ionic liquid electrolytes used in Li-ion batteries are discussed in the section.

Lithium bis(trifluoromethanesulfonyl)imide (LiTFSI) is an ionic electrolyte with high ionic conductivity that can operate at high voltages, making it suitable for high-energy-density batteries. Lithium bis(fluorosulfonyl)imide (LiFSI) is an ionic liquid electrolyte with low viscosity and high ionic conductivity, which can improve battery performance. Lithium tetrafluoroborate ($LiBF_4$) dissolved in an ionic liquid, such as 1-butyl-1-methylpyrrolidinium tetrafluoroborate, serves as an electrolyte and has good thermal stability and low viscosity, which improves the battery's performance at low temperature. Lithium trifalate (LiTf) is a lithium salt that is dissolved in an ionic liquid, such as 1-ethyl-3-methylimidazolium triflate. This electrolyte has high ionic conductivity and good electrochemical stability, making it suitable for high-performance batteries. Lithium bis(oxalate)borate (LiBoB) is dissolved in an ionic liquid, such as N-butyl-N-methylpyrrolidinium bis(oxalate)borate, to act as an electrolyte. This electrolyte has good electrochemical stability and low viscosity, which can enhance the stability of batteries [25].

2.6 ROLE OF ADDITIVES IN ELECTROLYTE PERFORMANCE

Additives improve the performance and stability of EES systems. An additive can suppress unwanted reactions in the electrolyte, such as side reactions that cause electrolyte decomposition, SEI formation, or gas generation. Additives improve electrolyte performance by increasing ionic conductivity and enhancing ion solvation. When ions are dissolved in the electrolyte, they become surrounded by solvent molecules, forming a solvation shell. This solvation process affects the mobility of the ions. Some additives can facilitate the formation of a more stable and efficient solvation shell around the ions, which reduces the energy barrier for ion movement and enhances the overall conductivity.

Additionally, electrolyte additives can reduce the viscosity of the electrolyte. Additives can act as viscosity modifiers, reducing the internal friction of the electrolyte and improving the fluidity of the system. By improving the solvation of ions and reducing electrolyte viscosity, electrolyte additives contribute to faster ion transport within the battery. This, in turn, enhances the battery's performance by allowing for quicker charge and discharge rates, increased power output, and improved overall efficiency. Higher ionic conductivity leads to reduced resistance and improved overall performance. Examples include lithium hexafluorophosphate ($LiPF_6$) or lithium bis(trifluoromethanesulfonyl)imide (LiTFSI), which are commonly used in Li-ion batteries [26–28].

Electrolyte additives contribute to the formation of a shielding layer on the electrode surface, which helps prevent electrode degradation and enhances performance and durability. When the electrolyte additive is present in the electrolyte solution, it can form a protective layer on the surface of the electrode. This layer acts as a barrier between the electrode and the electrolyte, preventing undesirable side reactions and reducing degradation processes such as corrosion, passivation, and formation of solid electrolyte interphase (SEI) layers.

For example, in Li-ion batteries, additives like vinylene carbonate (VC) and fluoroethylene carbonate (FEC) are used to form stable SEI layers on the surface of the anode, preventing further degradation and improving cycling stability. These additives improve the ionic conductivity of the electrolyte, allowing for faster ion transport between the electrode and electrolyte. Some additives are used to stabilize the electrochemical system by reducing unwanted side reactions or preventing the decomposition of active materials. For instance, in redox flow batteries, certain organic additives can help reduce crossover of reactants and improve overall system efficiency [29].

Improvement of the capability of a battery at low temperatures can be achieved either by reducing the freezing point of the electrolyte or by increasing the conductivity of the electrolyte at low temperatures. This is indeed an important consideration for many applications, as cold temperatures can negatively impact their performance. Two common approaches to enhance battery performance at low temperatures are reducing the freezing point of the electrolyte and increasing its conductivity at low temperatures. By reducing the freezing point of the electrolyte or increasing its conductivity at low temperatures, battery performance can be improved even in cold environments, enabling more reliable operation and enhanced energy delivery [30, 31].

The reduction in the electrolyte freezing point helps prevent it from solidifying or forming ice crystals at low temperatures, which can hinder ion transport and decrease battery performance. One way to achieve this is by using electrolyte additives called "cryoprotectants." They help maintain the liquid state of the electrolyte at sub-zero temperatures, enabling better ion mobility and preserving battery performance. Common cryoprotectants include ethylene carbonate (EC), propylene carbonate (PC), and diethyl carbonate (DEC).

Electrolyte additives increase the electrolyte conductivity at low temperatures by decreasing the ionic conductivity of the electrolyte, reducing the ability of ions to move between the electrodes and impeding battery performance. To overcome this, electrolyte formulations can be optimized to enhance conductivity at low temperatures. One approach is to use ionic liquid-based electrolytes, as they typically exhibit better conductivity at low temperatures compared to traditional organic liquid electrolytes. Ionic liquids consist of large organic cations paired with inorganic or organic anions and can offer improved low-temperature performance.

Another approach involves the addition of specific electrolyte additives that can enhance conductivity at low temperatures. For instance, using Li-salts such as lithium bis(trifluoromethanesulfonyl)imide (LiTFSI) or lithium bis(fluorosulfonyl)imide (LiFSI) in the electrolyte can improve conductivity at lower temperatures, thereby enhancing the low-temperature performance of Li-ion batteries. Optimizing the freezing point and conductivity of the electrolyte is a delicate balance, as excessively low freezing points may lead to other issues, such as increased viscosity or reduced safety. Therefore, careful consideration is necessary to achieve an optimal balance between low-temperature performance and other desired electrolyte properties.

Electrolyte additives play an important role in battery technology by assisting in tuning the redox potential of the electrolyte or shifting the redox potential of the electrodes. This can lead to several benefits, including improved energy density, power density, and cycle life of the battery. By modifying the redox potential, electrolyte additives can influence the overall battery performance. These electrolyte additives can optimize the redox potential of the electrolyte, which affects the voltage and energy density of the battery. By adjusting the redox potential, additives can increase the voltage and, consequently, the energy stored in the battery per unit mass or volume.

Electrolyte additives can influence the redox potential of the electrode materials, allowing for faster redox reactions and improved power output. This can be particularly important in applications that require high power bursts, such as electric vehicles or certain portable electronics. Electrolyte additives can help mitigate side reactions, reduce degradation, and improve the stability of the electrolyte and electrode materials. This leads to longer cycle life and increased overall battery durability. Different types of batteries, such as Li-ion, lead-acid, or flow batteries, may require different additives to achieve the desired performance improvements.

Electrolyte additives contribute significantly towards enhancement of the electrolyte stability window, improving electrolyte stability and widening the SEI layer at the cathode and anode surface. Electrolyte additives are utilized because they improve electrolyte stability. At high voltages, the electrolyte can undergo decomposition reactions, leading to the formation of reactive species that attack the electrodes and cause degradation. Additives act as sacrificial species, reacting preferentially with these reactive species and preventing them from damaging the electrode surfaces. This enhances the overall electrochemical stability of the electrolyte and allows the battery to operate at higher voltages without significant degradation.

Electrolyte additives can also suppress the decomposition reactions of the electrolyte at high voltages. They act as scavengers for reactive species, such as peroxides or radicals, that are formed during electrolyte decomposition. By capturing and neutralizing these reactive species, additives reduce their concentration and slow the degradation processes. This helps to maintain the integrity of the electrolyte and extends the voltage window within which the battery can operate safely.

The SEI layer forms on the electrode surfaces during initial cycling of a battery and acts as a passivation layer preventing further reactions between the electrode and the electrolyte. Electrolyte additives can influence the formation and properties of the SEI layer. By introducing appropriate additives, the composition, morphology, and thickness of the SEI layer can be controlled, resulting in improved performance and stability.

Some notable examples for the effect of electrolyte additives in widening the SEI layer are discussed here. VC is a commonly used electrolyte additive that improves the formation of a stable SEI layer on graphite anodes in Li-ion batteries. It promotes the formation of a more compact and uniform SEI layer, reducing side reactions and improving cycling stability. FEC is another additive often used in Li-ion batteries. It helps form a more robust SEI layer on the graphite anode surface, which improves the overall stability and capacity retention of the battery.

FEC can also passivate the surface of transition metal oxide cathodes, reducing their degradation and improving cycling performance. LiBOB is a salt-based additive that has shown promise in enhancing the SEI layer formation. It forms a stable and uniform SEI layer on both graphite anodes and high-voltage cathodes, improving their cycling stability and reducing capacity loss. Certain fluorinated additives, such as lithium difluoro(oxalate)borate (LiDFOB), have been investigated for their ability to improve the SEI layer formation. These additives can enhance the stability and uniformity of the SEI layer, reducing degradation and improving the overall performance of the battery.

The effectiveness of electrolyte additives in enhancing the SEI layer can depend on various factors, including the battery chemistry, electrode materials, electrolyte composition, and cycling conditions. Different additives may have varying effects in different battery systems, and their optimal concentrations may vary as well.

Extensive research and optimization are conducted to identify the most suitable additives for specific battery designs and requirements.

However, the selection of the appropriate additive and its concentration requires careful consideration, as it can affect the performance of the battery and lead to unintended side effects. Many different types of additives can be used in Li-ion battery electrolytes to improve their performance and stability. Additives such as $LiNO_3$, $LiBF_4$, or LiBOB improve electrolyte ionic conductivity and battery performance [32–34]. Fluoroethylene carbonate (FEC) is an electrolyte additive that forms a stable SEI layer on the electrode surface, improving a battery's cycling stability and safety. Vinylene carbonate (VC) is an electrolyte additive that can reduce the decomposition of the electrolyte and prevent the formation of unwanted reaction products; it can also improve the cycling stability of the battery. Electrolytes like FEC and fluoroethyl methyl carbonate (FEMC) improve electrolyte's thermal and electrochemical stability along with reduction of viscosity [35].

Additives such as ferrocene and viologen can tune the redox potential of the electrolyte or the electrode, which can improve the energy density and cycling stability of the battery. Additives such as triphenyl phosphate (TPP) and tris(2,2,2-trifluoroethyl) phosphate (TFEP) can improve the safety of the battery by suppressing thermal runaway and preventing the propagation of fire. These are just a few examples of the many types of additives that can be used in Li-ion battery electrolytes to improve their performance and stability. The choice of additive depends on the specific battery chemistry, performance requirements, and safety considerations.

The safety issues and low Coulombic efficiency associated with Li-containing batteries are due to the uncontrolled growth of dendrites (branching structures that can penetrate the separator and cause short circuits) and mossy lithium, as well as the decomposition of the electrolyte. The application of lithium polysulfide and $LiNO_3$ as additives in an ether-based electrolyte can generate a stable and consistent solid electrolyte interphase layer, which prevents dendrite growth and minimizes electrolyte decomposition. This is noteworthy because the reaction between Li and lithium polysulfide has been considered a critical flaw in Li-S batteries, but in this case, it was leveraged to suppress dendrite growth. [33].

Lithium tetrafluoroborate ($LiBF_4$) has been exploited as an efficient electrolyte additive to expand the cycling performance of a $LiNi_{0.5}Co_{0.2}Mn_{0.3}O_2$/graphite cell at higher operating voltages. The study found that adding 1.0 wt.% of $LiBF_4$ to the electrolyte greatly enhanced the capacity retention of the Li-ion battery after 100 cycles, from 29.2% to 90.1%, in the voltage range of 3.0 V–4.5 V. The research examined the cell performance and electrode interface property to get a better insight into the capacity retention at high voltage. Results indicate $LiBF_4$ involvement at the interface film formation, reducing impedance. This approach could lead to improved cycling performance and stability of Li-ion batteries at higher operating voltages [36].

2.7 IMPEDANCE OF ELECTROLYTE

The impedance of electrolyte in energy storage devices can have a significant impact on their performance, efficiency, and safety. Impedance can arise due to the electrolyte composition, temperature, and impurities [37–39]. In impedance spectroscopy, the effect of AC voltage and the resulting current are investigated, and an example for the same along with an equivalent circuit is shown in Figure 2.2. By varying the frequency of the applied AC voltage, a complex impedance spectrum can be obtained. This spectrum is typically plotted as a Nyquist plot that investigates real vs. imaginary component of impedance. It is characterized by two regions: a semicircle region and a linear Warburg region.

The semicircle region arises due to the capacitances existing in the sample, while the Warburg region is due to the ionic diffusion in the active electrode material. The *x*-intercepts of the semicircle region correspond to resistive and capacitive component of the sample. The resistive component can be correlated to the electrolyte ionic conductance, while the capacitive component is correlated to the dielectric characteristics of the electrode material. The Warburg slope is correlated to the ionic diffusion coefficient in the electrode.

The Warburg region is typically characterized by a straight line with a slope of $1/\sqrt{f}$ (f represents the frequency of applied voltage). The interpretation of impedance spectroscopy data can provide important information about the properties of electrode materials, and particularly electrolytes. For example, impedance spectroscopy can be used to study the transport properties of ion-selective membranes, the behavior of redox-active species in solution, and the kinetics of electrochemical reactions.

In this response, electrolyte impedance on the performance of energy storage devices is critical because electrolyte impedance can reduce the capacity of energy storage devices, such as batteries and supercapacitors, as it hinders the movement of ions between the anode and cathode. This can lead to a decrease in energy density and overall performance. For example, in Li-ion batteries, impedance in the electrolyte can lead to a decrease in the battery's specific capacity and power density [40]. Similarly, impedance in the electrolyte of supercapacitors can result in a decrease in their capacitance and energy density [41, 42].

The division of the Nyquist plot into three major areas as proposed by the authors is a common way to interpret the data obtained from impedance spectroscopy measurements. The Nyquist plot's high frequency region demonstrates the inductive effect generated due to the cell geometry and porosity of the electrode. This region characterized by a small capacitive loop represents the formation of an electric double layer capacitor. The intercept with the real axis, or the *x*-intercept, corresponds to the total value of the ohmic resistance, which is related to the ionic conductivity of the electrolyte.

At low frequency ranges, the Nyquist plot typically shows a semicircle that represents the diffusion of Li^+ cations in the solid-state phase. This semicircle represents the charge transporter relaxation at the SEI generated during the battery's cycling process. The Nyquist's second semicircle plot demonstrates the electrode potential along with capacitance and resistance. The reactions occurring at the junction of the electrode and electrolyte are demonstrated by the semicircle region; see Figure 2.2(a) & (b) [43].

Impedance in the electrolyte can increase the internal resistance of energy storage devices, which reduces their power output. This results in slower charging and discharging rates and decreased efficiency. For example, in Li-ion batteries, high

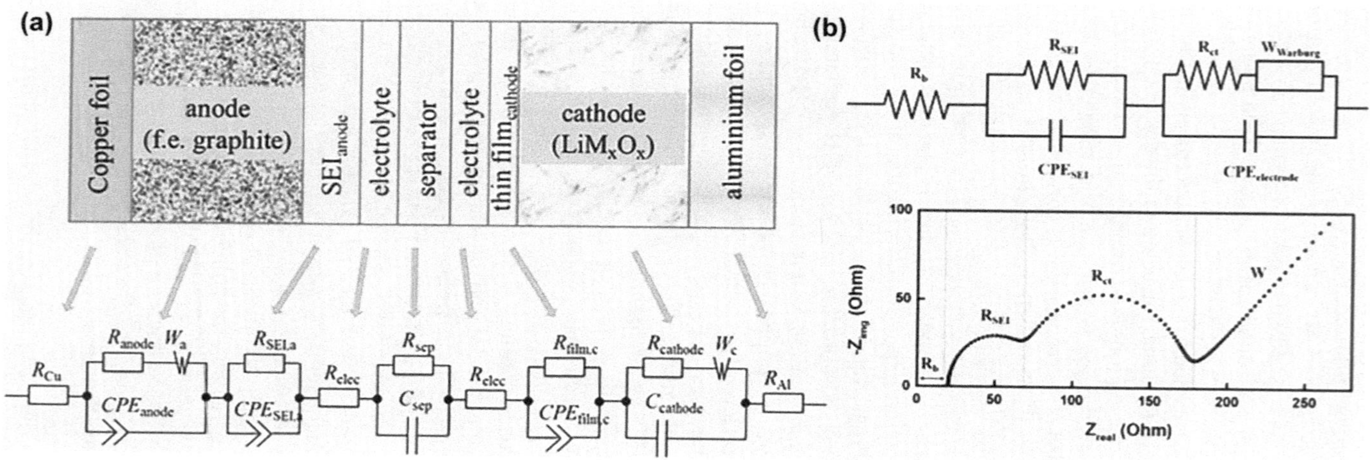

FIGURE 2.2 (a) Equivalent circuit model for a typical Li-ion battery; (b) represents the equivalent circuit model for the Li-ion battery.

Source: Reproduced with permission [43]. Copyright 2021 Wiley.

impedance in the electrolyte can lead to increased voltage hysteresis, which reduces the battery's power output and efficiency. Similarly, in redox flow batteries, impedance in the electrolyte can increase the cell resistance, which decreases the battery's power density and efficiency [37].

Impedance in the electrolyte can cause energy storage devices to heat up, which can accelerate performance degradation and reduce their lifetime. Additionally, high operating temperatures can lead to safety hazards and decrease the overall reliability of the energy storage device. For example, in Li-ion batteries, high impedance in the electrolyte can lead to increased heat generation during charging and discharging, which can accelerate aging and decrease the battery's lifetime. Similarly, in redox flow batteries, high impedance in the electrolyte can lead to increased heat generation, which can cause thermal runaway and safety hazards [44, 45]. Electrolyte impedance can increase the risk of dendrite formation, which occurs when metal ions deposit on the surface of the anode and can cause short-circuits in energy storage devices.

Dendrite formation can lead to a decrease in performance and even safety hazards. For example, in Li-ion batteries, impedance in the electrolyte can promote dendrite formation, which can cause short-circuits and decrease the battery's performance and safety. Similarly, in solid-state batteries, impedance in the solid electrolyte promotes dendrite formation, which can cause short-circuits and decrease the battery's performance and safety [46]. To mitigate the effects of impedance on energy storage device performance, various techniques can be employed. For example, optimizing the composition of the electrolyte, controlling the temperature of the device, and implementing advanced battery management systems can help reduce the impedance and improve the performance [46].

2.8 BAND STRUCTURE OF SOLID AND POLYMER ELECTROLYTE

Electrolytes, being composed of charged particles, do not have band structures in the same way as solid-state materials. Liquid electrolytes, unlike solid-state electrolytes or solid-state materials, do not have well-defined energy bands due to their disordered nature. In a liquid electrolyte, the ions are free to move around and interact with each other in a dynamic environment, making it difficult to describe their electronic properties in terms of energy bands. Instead, the electronic properties of a liquid electrolyte are typically described using concepts such as ion mobility, diffusion, and conductivity, which depend on the concentration and characteristics of the ions, as well as the solvent used. While it is not possible to describe the band structure of a liquid electrolyte, computational methods can be used to study the behavior of ions in a liquid electrolyte. These simulations use classical mechanics to model the ion and solvent dynamics in the electrolyte.

However, solid-state electrolytes are exceptional materials that possess a crystalline structure and can conduct ions like a solid-state material. In solid-state electrolytes, the electronic properties are determined by ions and the crystal lattice interactions. The energy levels of the ions and the lattice can overlap to form energy bands—similar to the energy bands in solid-state materials—as well as determine their conductivity. In general, the band structure of a solid-state electrolyte depends on the specific ions and crystal structure of the material. The band structure can be calculated using density functional theory (DFT) calculations, which use quantum mechanics to predict the electronic properties of a material [47]. However, the band structure of a solid-state electrolyte can be different from that of a solid-state material, due to the presence of charged ions and the interactions between the ions and the lattice [48].

In general, the band structure can be calculated using DFT calculations, which use quantum mechanics to predict the electronic properties of a material. In a solid electrolyte, the energy levels of the ions and the lattice can overlap to form energy bands. These energy bands can determine the conductivity of the solid electrolyte. The electronic properties of a solid electrolyte can also be described using concepts such as effective mass, which describes how electrons or holes move through the material. The band structure of a solid electrolyte can have important implications for its performance as an electrolyte. For example, the presence of a wide bandgap can lead to low electronic conductivity, which is desirable for certain applications such as solid-state batteries. On the other hand, a narrow bandgap can lead to higher electronic conductivity, which is useful for other applications such as solid oxide fuel cells. There are many different types of solid electrolytes, each with their own specific band structures.

Li-ion conducting oxides have a crystalline structure. The band structure of these materials depends on the specific oxide used, but in general, there is a wide bandgap between the valence and conduction bands, which leads to low electronic conductivity [49]. Perovskite oxides are a class of materials with a specific crystal structure that has interesting electronic and ionic properties. For example, some perovskites can conduct both ions and electrons. The band structure of perovskite oxides can be complex and depends on the specific elements and crystal structure of the material [50]. Perovskite-type solid-state electrolytes (SSEs) such as $Li_{3/8}Sr_{7/16}Hf_{1/4}Ta_{3/4}O_3$ (LSTZ) have shown promising ionic conductivity at room temperature. LSTZ was modified by replacing Zr with Hf to create an air-stable SSE $Li_{3/8}Sr_{7/16}Hf_{1/4}Ta_{3/4}O_3$. Doping modifications were also carried out to create a perovskite material $Li_{0.38}Sr_{0.44}Ta_{0.7}Hf_{0.3}O_{2.95}F_{0.05}$ ($LSTHF_5$), which demonstrated perfect stability to H_2O in a wide pH range [51]. In the aforementioned system Li^+ of $LSTHF_5$ ions are isolated, leading to a reduced occurrence of H^+/Li^+ exchange. Additionally, a polymer (PEO)-perovskite composite improves the chemical and structural stability. Additionally, the ionic conductivity of perovskite-type SSEs is also enhanced. The PEO/LiTFSI/LSTZ-based solid-state electrolyte collaborated with perovskite LSTZ as a filler, demonstrate inhibition of H^+ and stability to an aerobic environment. These findings suggest that perovskite-type SSEs are promising candidates to integrate with solid-state batteries.

Sulfide-based solid electrolytes, such as $Li_{10}GeP_2S_{12}$, possess high ionic conductivity, and their band structure reveals the existence of a wide bandgap between the valence and conduction bands, similar to Li-ion conducting oxides [52–54]. Polymer electrolytes are a type of solid electrolyte that can conduct ions through a polymer matrix. The band structure of polymer electrolytes can be complex and depends on the specific polymer used, as well as the ions and solvents incorporated into the material [55, 56].

Li-ion conducting oxides have a specific band structure that is important for their performance as solid electrolytes. In general, Li-ion conducting oxides have a wide bandgap between the valence and conduction bands, which leads to low electronic conductivity. This is desirable for solid electrolyte applications, as it prevents electronic current flow and improves the safety and stability of the device. The specific band structure of a Li-ion conducting oxide depends on the specific oxide used, as well as other factors such as doping and crystal structure.

Oxide-based electrolytes like perovskites, NASICON and LISICON, have been investigated as solid-state electrolytes. NASICON, which stands for sodium superionic conductor, has the chemical formula $Na_3Zr_2Si_2PO_{12}$. Its crystal structure is characterized by a three-dimensional framework of corner-sharing ZrO_6 octahedra, SiO_4 tetrahedra, and PO_4 tetrahedra. The band structure of NASICON has been studied using DFT calculations, which show that the material has a band gap of around 4 eV. The conduction band of NASICON is primarily made up of Zr d and O p orbitals, while the valence band is dominated by P p orbitals [57]. LISICON, which stands for Li-superionic conductor, has the chemical formula $Li_{10}GeP_2S_{12}$. Its crystal structure is characterized by a framework of GeS_4 and PS_4 tetrahedra interconnected by Li ions. The band structure of LISICON has also been studied using DFT calculations, which show that the material has a band gap of around 3 eV. The conduction band of LISICON is primarily made up of P p and S p orbitals, while the valence band is dominated by s orbitals of Li [58].

Perovskite oxides are solid-state electrolyte materials in batteries and other electrochemical devices due to their high ionic conductivity and stability. The band structure of perovskite oxides can have a significant impact on their performance as solid electrolytes. In perovskite oxide-based electrolytes, the valence band is typically made up of oxygen 2p orbitals, while the conduction band is composed of metal d orbitals. The bandgap in these materials can vary depending on the specific elements used but is typically in the range of 2–4 electron volts (eV).

The ionic conductivity in perovskite oxide-based electrolytes is thought to be related to the number of mobile charge carriers in the material, which can be affected by the band structure. In particular, the number of oxygen vacancies, which act as mobile charge carriers, can be influenced by the position of the Fermi level within the bandgap. For example, the perovskite oxide $LiTi_2(PO_4)_3$ has been shown to have a high ionic conductivity due to the presence of oxygen vacancies in the material. The band structure of this material has been studied using computational methods, and it was found that the position of the Fermi level plays an important role in determining the number of oxygen vacancies and the ionic conductivity.

Other perovskite oxide-based electrolytes, such as $Li_{3x}La_{2/3-x}TiO_3$ and $LiAlTiO_4$, have also been studied for their high ionic conductivity and stability. The band structure of these materials has been investigated using various physicochemical techniques like XPS and synchrotron radiation. Further research into the relationship between the band structure and ionic conductivity in these materials could lead to the development of more efficient and stable solid-state electrolytes for use in next-generation batteries and other electrochemical devices.

Sulfide containing solid electrolytes are interesting alternatives to oxide-based solid electrolytes for use in Li-ion batteries and other electrochemical devices. The band structure of sulfide-based solid electrolytes can be complex and depends on the specific elements and crystal structure of the material. In general, sulfide-based solid electrolytes have a similar band structure to other solid-state materials, with a valence band and a conduction band separated by a bandgap. The valence band is typically composed of sulfur p orbitals, while the conduction band is made up of metal d orbitals. The bandgap in these materials can vary depending on the specific elements used but is typically in the range of 2–3 electron volts (eV).

Sulfide-based solid electrolytes are known for their high ionic conductivity, which is thought to be related to the presence of mobile ions such as Li or Na. The band structure can affect the number of mobile ions in the material, and therefore impact its ionic conductivity. For example, the sulfide-based solid electrolyte $Li_{10}GeP_2S_{12}$ (LGPS) has been shown to have a high ionic conductivity due to the presence of Li^+ ions that are able to move through the material [59]. The band structure of LGPS has been studied using computational methods such as density functional theory (DFT), and it was found that the position of the Fermi level within the bandgap plays an important role in determining the number of mobile ions and the ionic conductivity.

Other sulfide-based solid electrolytes, such as undoped as well as doped Na_3PS_4 [52, 60] and $Li_7P_3S_{11}$ [61], have also been studied for their high ionic conductivity and stability. The band structure of these materials has been explored with the help of XRD and DFT calculations. Further research into the relationship between the band structure and ionic conductivity in these materials could lead to the development of more efficient and stable solid-state electrolytes for use in next-generation batteries and other electrochemical devices.

Polymer electrolytes possess unique properties such as flexibility, low toxicity, and compatibility with a wide range of electrode materials. The band structure of polymer electrolytes is different from that of conventional crystalline solid electrolytes, as they are amorphous and lack a well-defined crystal structure. The band structure of polymer electrolytes can vary depending on the specific polymer used and the nature of the ions being transported. In general, the valence band in polymer electrolytes is made up of the highest occupied molecular orbitals (HOMOs) of the polymer chain, while the conduction band is composed of the lowest unoccupied molecular orbitals (LUMOs).

The bandgap in polymer electrolytes is typically in the range of 2–4 electron volts (eV), depending on the specific polymer used. The ionic conductivity in polymer electrolytes is related to the ability of the ions to move through the polymer matrix. The band structure can affect the number of mobile ions in the material, and therefore impact its ionic conductivity. For example, polyethylene oxide (PEO) is a commonly used polymer electrolyte due to its high ionic conductivity and compatibility with a wide range of electrode materials [55, 62, 63]. The band structure of PEO has been studied using computational methods, and it was found that the position of the Fermi level within the bandgap plays an important role in determining the number of mobile ions and the ionic conductivity.

Other polymer electrolytes, such as poly(vinylidene fluoride-co-hexafluoropropylene) (PVDF-HFP) and poly(ethylene oxide)-based block copolymers, have also been studied for their high ionic conductivity and stability [55, 63]. The band structure of these materials has been explored using a combination of experimental techniques like XPS and DFT calculations.

2.9 INFLUENCE OF ELECTROLYTES ON THE CHARGE/DISCHARGE CURVES

Figure 2.3(a) shows the variation of ionic conductivity of the $LiPF_6$ electrolyte system for a range of concentrations from 0.06 to 4 M using molecular dynamics simulations though density functional theory, demonstrating 1 M electrolyte concentration is investigated for Li-ion batteries [64]. Figure 2.3(b) shows >1 mol L^{-1} electrolyte concentration improves charge carrier density and limits Li-ion depletion, thereby increasing the energy density of thick electrode-based batteries and ameliorated current densities [65]. Further, Figure 2.4(a) & (b) show that with an increase in electrolyte concentration, the galvanostatic charge/discharge profiles of the cells become affected for thin and thick electrode cases, respectively. From these observations, it is clear that the average voltage increases while charging and decreases while discharging with an increase in the electrolyte concentration [65].

2.10 CONCLUSIONS AND OUTLOOK

To summarize, here are the most recent advancements in the study and application of electrochemical energy storage (EES) systems. A water-based electrolyte system was employed in the world's first battery and is currently regularly used in a variety of EES devices that require extremely conductive electrolytes and managing security. The latest developments in aqueous electrolytes as well as non-aqueous electrolytes were discussed. When high specific energy from a device is needed in real-world applications, the working voltage constraint of aqueous EES devices is still a disadvantage. A wider operating potential window may be accessible with organic electrolytes than with aqueous electrolytes. Organic electrolytes still face safety concerns, however—because of their high working voltage and high specific energy, they are frequently used in portable device batteries. Ionic liquids' acceptable conductivity and lack of safety concerns are due to their inflammability and nearly nonexistent volatility. Ionic liquid electrolytes have also shown promise for use in EES devices, according to recent studies.

This chapter will draw the attention of researchers to electrolyte materials to comprehend the mechanism of the interaction of materials existing at the electrode and electrolyte interface, and to improve the output of EES devices via understanding the mechanism behind it. Additional investigations are necessary to identify the appropriate electrolyte to EES performance with desired cyclic stability and safety. Moreover, it is crucial to have a thorough grasp of and modulate the physicochemical features of these interphases. Only with careful consideration in the choice of solvents, salts, and/or additions can the perfect interphase form.

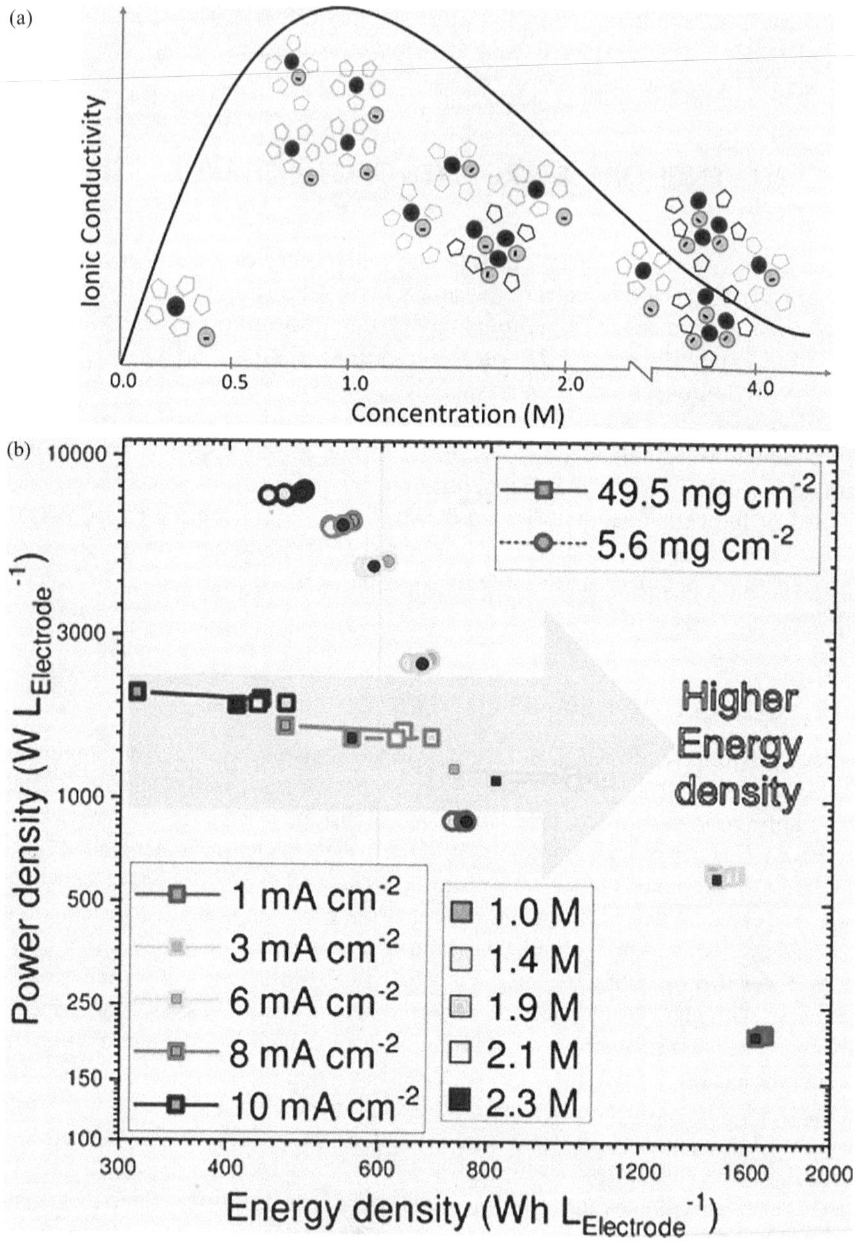

FIGURE 2.3 (a) Variation of ionic conductivity of $LiPF_6$ electrolyte system for a range of concentrations; and (b) effect of electrolyte concentration >1 mol L^{-1} on the energy density and power density.

Source: (a) Reproduced with permission from Ref. [64] Copyright 2018 American Chemical Society. (b) Reproduced with permission from Ref. [65]. Copyright 2020 Wiley.

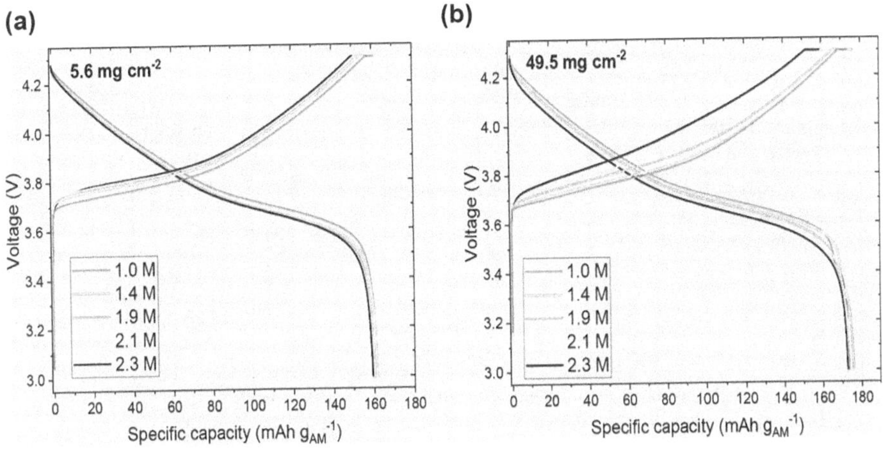

FIGURE 2.4 Charge/discharge curves of (a) the thin and (b) the thick NCM622 cathodes at five different $LiPF_6$ salt concentrations in the ethylene carbonate (EC)/ethyl methyl carbonate (EMC)-based electrolyte at 1 mA cm^{-2}.

Source: Reproduced with permission from Ref. [65]. Copyright 2020 Wiley.

REFERENCES

1. M. Chen, Y. Zhang, G. Xing, S.-L. Chou and Y. Tang, *Energy Environ. Sci.*, 2021, 14, 3323–3351.
2. K. Chen, Q. Wang, Z. Niu and J. Chen, *J. Energy Chem.*, 2018, 27, 12–24.
3. A. Tyagi, K. M. Tripathi and R. K. Gupta, *J. Mater. Chem. A*, 2015, 3, 22507–22541.
4. Y. R. Jeong, G. Lee, H. Park and J. S. Ha, *Acc. Chem. Res.*, 2019, 52, 91–99.
5. W.-J. Song, S. Lee, G. Song, H. B. Son, D.-Y. Han, I. Jeong, Y. Bang and S. Park, *Energy Storage Mater.*, 2020, 30, 260–286.
6. M. Liao, L. Ye, Y. Zhang, T. Chen and H. Peng, *Adv. Electron. Mater.*, 2019, 5, 1800456.
7. H. Yang and N. Wu, *Energy Sci. Eng.*, 2022, 10, 1643–1671.
8. E. Logan and J. Dahn, *Trends Chem.*, 2020, 2, 354–366.
9. O. Borodin, *Curr. Opin. Electrochem.*, 2019, 13, 86–93.
10. J. Kalhoff, G. G. Eshetu, D. Bresser and S. Passerini, *ChemSusChem*, 2015, 8, 2154–2175.
11. F. Mo, G. Liang, D. Wang, Z. Tang, H. Li and C. Zhi, *EcoMat*, 2019, 1, https://doi.org/10.1002/eom2.12008
12. S. Kumar, M. Goswami, N. Singh, N. Sathish, M. V. Reddy and S. Kumar, *J. Energy Storage*, 2022, 55, 105522.
13. C. Y. Chan, Z. Wang, H. Jia, P. F. Ng, L. Chow and B. Fei, *J. Mater. Chem. A*, 2021, 9, 2043–2069.
14. T. Liang, R. Hou, Q. Dou, H. Zhang and X. Yan, *Adv. Funct. Mater.*, 2021, 31, 2006749.
15. J. Tan and J. Liu, *Energy Environ. Mater.*, 2021, 4, 302–306.
16. T. Ye, L. Li and Y. Zhang, *Adv. Funct. Mater.*, 2020, 30, 2000077.
17. C. S. Rustomji, Y. Yang, T. K. Kim, J. Mac, Y. J. Kim, E. Caldwell, H. Chung and Y. S. Meng, *Science*, 2017, 356, https://doi.org/10.1126/science.aal4263
18. L. Wang, Y. Ye, N. Chen, Y. Huang, L. Li, F. Wu and R. Chen, *Adv. Funct. Mater.*, 2018, 28, 1800919.

19. R. Demir-Cakan, M. R. Palacin and L. Croguennec, *J. Mater. Chem. A*, 2019, 7, 20519–20539.
20. A. von Wald Cresce, K. Xu, *Carbon Energy*, 2021, 3, 721–751.
21. S. Tan, Y. J. Ji, Z. R. Zhang and Y. Yang, *ChemPhysChem*, 2014, 15, 1956–1969.
22. Q. Li, S. Jiao, L. Luo, M. S. Ding, J. Zheng, S. S. Cartmell, C.-M. Wang, K. Xu, J.-G. Zhang and W. Xu, *ACS Appl. Mater. Interfaces*, 2017, 9, 18826–18835.
23. C. Arbizzani, G. Gabrielli and M. Mastragostino, *J. Power Sources*, 2011, 196, 4801–4805.
24. K. M. Diederichsen, E. J. McShane and B. D. McCloskey, *ACS Energy Lett.*, 2017, 2, 2563–2575.
25. N. Madria, T. Arunkumar, N. G. Nair, A. Vadapalli, Y.-W. Huang, S. C. Jones and V. P. Reddy, *J. Power Sources*, 2013, 234, 277–284.
26. G. Park, H. Nakamura, Y. Lee and M. Yoshio, *J. Power Sources*, 2009, 189, 602–606.
27. A. Tornheim, M. He, C.-C. Su and Z. Zhang, *J. Electrochem. Soc.*, 2017, 164, A6366.
28. J. Ming, Z. Cao, Y. Wu, W. Wahyudi, W. Wang, X. Guo, L. Cavallo, J.-Y. Hwang, A. Shamim and L.-J. Li, *ACS Energy Lett.*, 2019, 4, 2613–2622.
29. W. Zhao, Y. Ji, Z. Zhang, M. Lin, Z. Wu, X. Zheng, Q. Li and Y. Yang, *Curr. Opin. Electrochem.*, 2017, 6, 84–91.
30. J. A. Mennel and D. Chidambaram, *Front. Energy*, 2022, 17, 1–29.
31. Q. Nian, T. Sun, S. Liu, H. Du, X. Ren and Z. Tao, *Chem. Eng. J.*, 2021, 423, 130253.
32. Q. Zhao, N. W. Utomo, A. L. Kocen, S. Jin, Y. Deng, V. X. Zhu, S. Moganty, G. W. Coates and L. A. Archer, *Angew. Chem. Int. Ed.*, 2022, 134, e202116214.
33. W. Li, H. Yao, K. Yan, G. Zheng, Z. Liang, Y.-M. Chiang and Y. Cui, *Nat. Commun.*, 2015, 6, 7436.
34. V. Aravindan, Y. L. Cheah, W. C. Ling and S. Madhavi, *J. Electrochem. Soc.*, 2012, 159, A1435.
35. Y. Sasaki, *Electrochem.*, 2008, 76, 2–15.
36. X. Zuo, C. Fan, J. Liu, X. Xiao, J. Wu and J. Nan, *J. Electrochem. Soc.*, 2013, 160, A1199.
37. D. Qu, G. Wang, J. Kafle, J. Harris, L. Crain, Z. Jin and D. Zheng, *Small Methods*, 2018, 2, 1700342.
38. F. Feng, R. Yang, J. Meng, Y. Xie, Z. Zhang, Y. Chai and L. Mou, *Energy*, 2022, 242, 122880.
39. F. Feng, R. Yang, J. Meng, Y. Xie, Z. Zhang, Y. Chai and L. Mou, *Energy*, 2022, 243, 123091.
40. L. Suo, Y.-S. Hu, H. Li, M. Armand and L. Chen, *Nat. Commun.*, 2013, 4, 1481.
41. A. Oz, S. Hershkovitz, N. Belman, E. Tal-Gutelmacher and Y. Tsur, *Solid State Ionics*, 2016, 288, 311–314.
42. J. Zhong, L.-Q. Fan, X. Wu, J.-H. Wu, G.-J. Liu, J.-M. Lin, M.-L. Huang and Y.-L. Wei, *Electrochim. Acta*, 2015, 166, 150–156.
43. P. Vadhva, J. Hu, M. J. Johnson, R. Stocker, M. Braglia, D. J. L. Brett and A. J. E. Rettie, *ChemElectroChem*, 2021, 8, 1930–1947.
44. P. Lyu, X. Liu, J. Qu, J. Zhao, Y. Huo, Z. Qu and Z. Rao, *Energy Storage Mater.*, 2020, 31, 195–220.
45. X. Han, L. Lu, Y. Zheng, X. Feng, Z. Li, J. Li and M. Ouyang, *ETransportation*, 2019, 1, 100005.
46. Z. Zhang, G. Zhang and L. Chao, *Energy Storage Mater.*, 2021, 41, 631–641.
47. T. A. Pham, *Int. J. Quantum Chem.*, 2019, 119, https://doi.org/10.1002/qua.25795
48. K. Leung, *J. Phys. Chem. C*, 2013, 117, 1539–1547.

49. Y. Wang, W. D. Richards, S. P. Ong, L. J. Miara, J. C. Kim, Y. Mo and G. Ceder, *Nat. Mater.*, 2015, 14, 1026–1031.
50. Y. Inaguma, C. Liquan, M. Itoh, T. Nakamura, T. Uchida, H. Ikuta and M. Wakihara, *Solid State Commun.*, 1993, 86, 689–693.
51. X. Chen, Z. Guan, F. Chu, Z. Xue, F. Wu, Y. Yu, *InfoMat*, 2022, 4, https://doi.org/10.1002/inf2.12248
52. H. Li, Z. Zhang, B. Liu, J. Liu, B. Chen and J. Zhou, *Comput. Mater. Sci.*, 2023, 216, 111848.
53. Y.-S. Choi and J.-C. Lee, *J. Power Sources*, 2019, 415, 189–196.
54. M. Ribes, B. Barrau and J. L. Souquet, *J. Non-Cryst. Solids*, 1980, 38–39, 271–276.
55. S. Gupta, A. K. Gupta, B. K. Pandey and R. K. Yadav, *Comput. Theor. Chem.*, 2022, 1211, 113667.
56. M. Unge, H. Gudla, C. Zhang and D. Brandell, *Phys. Chem. Chem. Phys.*, 2020, 22, 7680–7684.
57. K. M. Bui, V. A. Dinh, S. Okada and T. Ohno, *Phys. Chem. Chem. Phys.*, 2016, 18, 27226–27231.
58. Y. Mo, S. P. Ong and G. Ceder, *Chem. Mater.*, 2012, 24, 15–17.
59. F. Du, X. Ren, J. Yang, J. Liu and W. Zhang, *J. Phys. Chem. C*, 2014, 118, 10590–10595.
60. H. Huang, H.-H. Wu, X. Wang, B. Huang and T.-Y. Zhang, *Phys. Chem. Chem. Phys.*, 2018, 20, 20525–20533.
61. D. Chang, K. Oh, S. J. Kim and K. Kang, *Chem. Mater.*, 2018, 30, 8764–8770.
62. S. Xue, Y. Liu, Y. Li, D. Teeters, D. W. Crunkleton and S. Wang, *Electrochim. Acta*, 2017, 235, 122–128.
63. S. Gupta, A. K. Gupta and B. Pandey, *Polym. Bull.*, 2022, 79, 4999–5018.
64. B. Ravikumar, M. Mynam and B. Rai, *J. Phys. Chem. C*, 2018, 122, 8173–8181.
65. L. S. Kremer, T. Danner, S. Hein, A. Hoffmann, B. Prifling, V. Schmidt, A. Latz and M. Wohlfahrt-Mehrens, *Batteries & Supercaps*, 2020, 3, 1172–1182.

Section I

Li-ion Batteries

3 Electrolytes and Additives for Li-ion Batteries

Prasanna Sekar, Ananthakumar Ramadoss, and Saravanakumar Balasubramaniam

3.1 INTRODUCTION

Batteries have recently assumed center stage as energy storage devices in electric and hybrid vehicles. Electric vehicle batteries should have high energy densities with appreciable capacity retention. Most of the current research is focused on improving the performance of existing batteries, while parallel efforts are being put forth to develop new generation batteries as well.

Today, Lithium-ion batteries (LIBs) stand out due to their high energy densities (>240 Wh kg^{-1}) and high nominal voltage (~3.6 V), making them the preferred power sources for a variety of devices such as smartphones, laptops, and digital cameras [1]. As described in Chapter 1, the most important components of a Li-ion battery are its active electrode materials, i.e., cathode and anode, and the electrolyte. These three components dictate the overall performance of an LIB and therefore, one should select these materials carefully while enhancing the performance of a battery depending on its applications. The following sections provide a detailed discussion on these components while emphasizing the electrolyte as per the theme of this book. However, the reader should refer to the relevant books and research literature to learn about the electrode materials of their interest, which is beyond the scope of this book.

3.2 ANODE

Graphite and other carbonaceous materials are generally used as anode materials for hosting lithium ions in LIBs accounting for as much as 372 mAh g^{-1} for a maximum stoichiometry of LiC_6 [2]. The lithium storage capacity of carbons depends on various factors such as their crystallinity, morphology, microstructure, and texture [3]. Graphite and meso-carbon microbeads (MCMB) are popular as host materials because of their strong cycling ability and high reversible capacity [4]. However, they are not without problems, which include structural and surface deterioration, capacity fades upon cycling, and exfoliation of graphitic layers owing to the co-intercalation of solvents, especially propylene carbonate (PC) [5]. This has led to the exploration of alternative anode materials such as Si [6], Sn [7], SnO_2/C [8], and Sn/C [9, 10]. In parallel, strategies such as the use of nano architecture for anode hosts are being adopted to improve battery performance [11].

DOI: 10.1201/9781003441182-4

3.3 CATHODE

Cathode materials used in LIBs can be classified into three types based on their structure: layered $LiMO_2$ (M = Mn, Ni, Co), spinel LiM_2O_4 (M = Mn), and olivine $LiMPO_4$ (M = Mn, Fe, Ni, Co) structures. The layered $LiCoO_2$ material has a theoretical capacity of 274 mAh g^{-1} with a working potential of 3.6 V vs. Li^+/Li and excellent cycling ability. However, practical capacities of only 140 mAh g^{-1} can be realized due to the structural instability of highly delithiated forms of $LiCoO_2$ [12]. Higher capacities may be realized by partially or entirely substituting Co with other transition metals, such as Ni and Mn. The spinel $LiMn_2O_4$ material exhibits a working potential of 4.1 V vs. Li^+/Li and improved safety, but it has a moderate theoretical capacity of 148 mAh g^{-1}. The olivine-type $LiFePO_4$ provides an extended cycle life with a working voltage of 3.5 V and a comparable theoretical capacity of 170 mAh g^{-1} [13]. Oxidation of electrolytes at delithiated cathodes is a major issue that can result in capacity fade and poor cyclability.

3.4 ELECTROLYTE

Electrolytes are another vital component in LIBs that determine the rate at which energy is delivered by the device. The electrolyte may be in liquid or solid form (Table 3.1). Liquid electrolytes contain one or more lithium salts ($LiClO_4$, $LiPF_6$, $LiBF_4$, $LiAsF_6$) dissolved in a single solvent or a mixture of solvents. The solvent may be aqueous or non-aqueous. Polymeric and ionic liquid-based electrolytes are also being investigated for LIBs [23, 24]. Electrolytes should remain stable when in contact with other cell constituents throughout the entire voltage range of cell operation. However, they do interact with anode and cathode materials, which can result in the formation of surface passive layers on the electrodes. One such layer is the solid electrolyte interphase (SEI), which is often referred to as the cathode electrolyte interphase (CEI) in the case of the cathode. The SEI prevents further degradation of the electrolyte while allowing the transport of lithium ions [25]. The nature of the SEI can be modified by the utilization of electrolyte additives, which can lead to enhanced battery performance. This chapter describes the different types of electrolytes. However, solid-state electrolytes are discussed in a separate chapter.

3.4.1 Non-Aqueous Electrolytes

Non-aqueous solvent-based electrolytes are preferred for use in lithium-based batteries due to their higher thermodynamic stability windows compared to water-based electrolytes. Popular solvents include linear, cyclic, or fluorinated carbonates such as ethylene carbonate (EC), propylene carbonate (PC), diethyl carbonate (DEC), ethyl methyl carbonate (EMC), and dimethyl carbonate (DMC). These solvents can be used individually or in blends [14].

Criteria for solvent selection include a high dielectric constant (high solubility of salt), low viscosity (high ion transport), low flammability, a wide operating temperature

TABLE 3.1
Electrolytes for Li-Ion Batteries and Their Performance

Electrolyte	Composition	Cell	σ (mS cm^{-1})	Cell Voltage (V)	Initial Capacity (mAh g^{-1})	Energy Density (Wh kg^{-1})	Retention (%)	Temp. (°C)	Ref.
Organic electrolyte	1 M $LiPF_6$ in EC/DMC/DEC (1:1:1 v/v)	Li/$LiNi_{0.5}M_{1.5}O_4$	—	4.5	—	—	—	—	[14]
	1.2 M $LiPF_6$ in F-AEC/F-EMC/F-EPE (2/6/2)	$Li_4Ti_5O_{12}$/$LiNi_{0.5}M_{1.5}O_4$	~1	5	~2.1		~99 (80 cycles)	55	[15]
	1 M $LiPF_6$ EC/DEC (1:1 v/v)	Li/$LiFePO_4$	—	4.5	153	—	~99 (100 cycles)	50	[16]
Aqueous electrolyte	9 M $LiNO_3$	LiV_3O_8/$LiFePO_4$	—	1	90	—	~99 (100 cycles)	—	[17]
	21 m LiTFSI	Mo_6S_8/$LiMn_2O_4$	~10	2.3	47	84	68 (1000 cycles)	RT	[18]
	2 m LiTFSI-94%PEG-6%H_2O	$Li_4Ti_5O_{12}$/$LiMn_2O_4$	0.8	3.2	50	110	82.5 (300 cycles)	RT	[19]
Ionic liquid	INF-Li_{15}	Li/$LiFePO_4$	0.09	4.2	~141.2	—	96 (50 cycles)	RT	[20]
	1 M LiTFSI + BMP-TFSI	Li/$LiFePO_4$	0.6	4.5	91	—	72 (100 cycles)	25	[16]
	1 g(2o1) (2o1)-TFSA	Li/$LiFePO_4$	0.57	4.1	76	—	120 mAh g^{-1} (100th cycle)	25	[21]
	1 g(2o1) (2o2)-TFSA	Li/$LiFePO_4$	0.54	4.3	67	—	107 mAh g^{-1}(100th cycle)	25	[21]
Ionogel	1 m LiTFSI + [EMI][TFSI]-TBOT (1.25 mole ratio)	Li/$LiFePO_4$	2.8	4.2	146.2	—	150.9 mAh g^{-1} (200th cycle)	RT	[22]

Note: Ethyl propyl ether (EPE), fluorinated cyclic carbonate (F-AEC), fluorinated linear carbonate (F-EMC), fluorinated ether (F-EPE), and Ionanofluid (INF) were prepared by dispersing 20 wt.% of SnO_2-NHIF in [DMim][TFSI], butylmethylpyrrolidiniumebis(trifluoromethylsulfonyl)imide (BMP-TFSI), 1-ethyl-3-methylimidazolium bis(trifluoromethylsulfonyl)imide ([EMI][TFSI]), and Tetrabutyl titanate (TBOT).

range, high flash points, and compatibility with electrode materials (for safety reasons). Particularly, EC exhibits superior chemical stability and high permittivity, and it forms a stable passivation layer. DMC is a good choice for low-temperature applications (high conductivity at low temperatures in DMC-based solutions).

Because no single solvent has all the desired properties, solvent mixtures have become the standard practice. Popular electrolytes consist of mixtures of carbonate-based solvents, primarily with $LiPF_6$ as the salt. However, they are unsatisfactory at high voltages (> 4.3 V), especially when compared to the carbon anode. Lucht et al. [15] examined the stability of a standard electrolyte (1 M $LiPF_6$ in EC/DMC/DEC, 1:1:1) between 4.0 and 5.3 V vs. Li^+/Li in cells with $LiNi_{0.5}Mn_{1.5}O_4$ and found that the electrolyte remained stable up to 4.5 V. In a similar study, it was discovered that a 1.2 M $LiPF_6$ in EC/EMC (3/7 wt.%) electrolyte remained stable up to 4.9 V [26].

As the demand for lithium-ion batteries (LIBs) increases in various industries such as personal electronics and electric vehicles, there is a growing need for LIBs that offer higher power, higher capacity, faster-charging rates, longer lifespan, and improved safety. The key to such devices lies in identifying stable electrolytes that can withstand high voltages. Two strategies may be adopted to achieve this objective: (i) using electrolytes with wider thermodynamic windows, and (ii) modifying SEIs on the electrodes.

Fluorinated carbonates are of particular interest here due to their high oxidation potentials. The strong electron-withdrawing nature of fluorine makes it a good alternative to conventional carbonate-based electrolytes. Zhang et al. [27] investigated the high-voltage stability of fluorinated electrolytes and found that a 1.2M $LiPF_6$ in F-AEC/F-EMC/F-EPE (2/6/2) electrolyte exhibited better stability and a wider voltage window compared to the conventional EC/EMC electrolyte (Figure 3.1a). The abbreviations used in the study are as follows: EPE stands for ethyl propyl ether, F-AEC stands for fluorinated cyclic carbonate, F-EMC stands for fluorinated linear carbonate, and F-EPE stands for fluorinated ether. The fluorinated electrolytes also exhibited lower leakage currents than the Gen2 electrolyte at a potential of 5.7 V. The authors concluded that F-EPE significantly improved the electrochemical stability at high voltages. The only significant distinction between E1 and Gen2 is the presence of F-EPE in the former. While the leakage current somewhat increased in the EMC-based electrolyte, i.e., E1 to E4, the currents remained relatively low in the case of E5 and E6. Furthermore, a $Li_4Ti_5O_{12}$/$LiNi_{0.5}Mn_{0.5}O_4$ cell showed significantly better stability in E5 and E6 than in Gen2 after 80 cycles (Figure 3.1b). In contrast, a cell with Gen2 electrolyte showed a higher capacity loss of 70% [27].

The high dielectric constant, low flammability, and outstanding anodic stability, along with a range of cathodes, make alkyl sulfones a good option for high-voltage LIB applications. Their high viscosity (and therefore, poor ion mobility) is a constraint. Moreover, they poorly wet electrodes and separators, and show an inability to form an SEI on the graphite anode [28]. Various types of sulfones (linear, branched, and functionalized) have been investigated for use in LIBs. Sulfoxides have also been explored as solvents because of their inherent stability against oxidation [29].

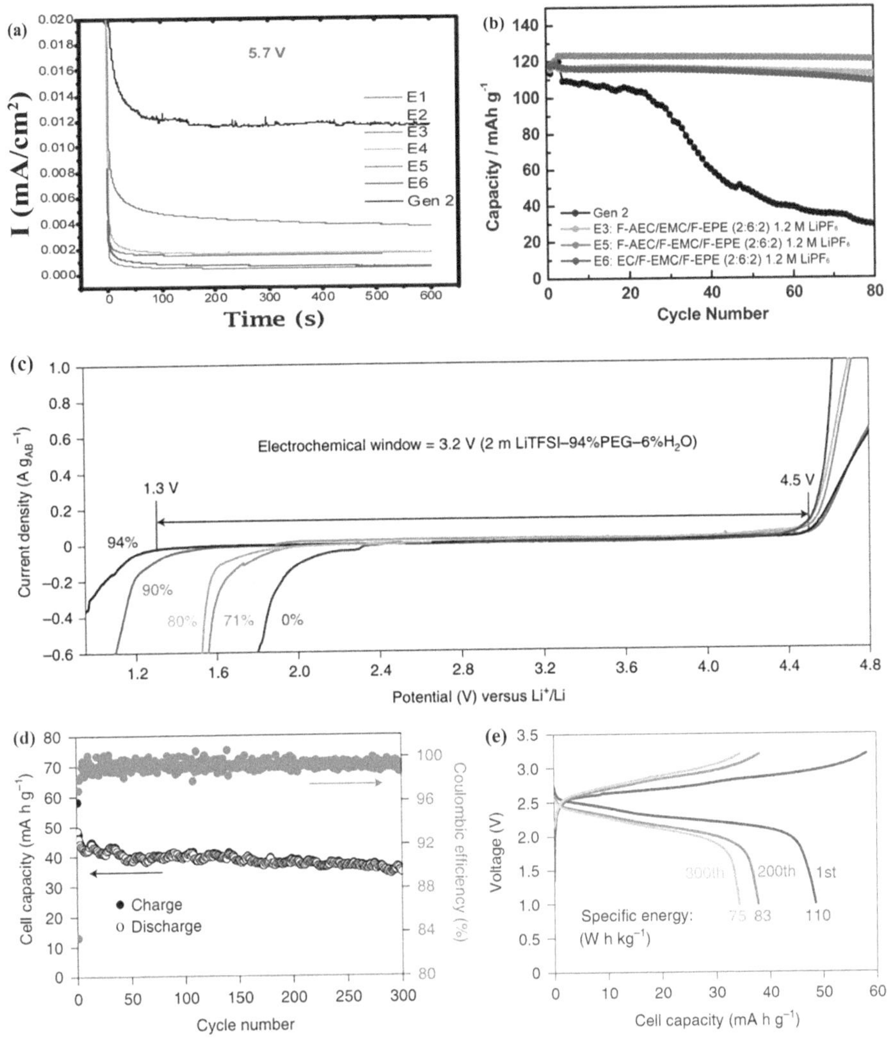

FIGURE 3.1 (a) Electrochemical floating test (3-electrode system) of Gen2 (1.2 M LiPF6 EC/EMC (3/7)) and fluorinated electrolytes E1 to E6 at 5.7 V; (b) LTO/LNMO full cell capacity retention study with Gen2 (baseline) and fluorinated electrolytes E3, E5, and E6 at 55 °C; (c) electrochemical stability windows for molecular crowding electrolyte (2 M LiTFSI–94%PEG–6% H_2O) determined by LSV tests on AB-coated aluminum foil at 0.2 mV s^{-1}; (d) the voltage profiles cell at different cycle intervals (1st, 200th, and 300th) at 1 C; and (e) cycling stability and CE of the L-LTO/LMO cell at 1 C; (f) comparison of isothermal (T = 303 K) conductivity and viscosity; (g) cycling performance for Li/INF-Li_{15}/$LiFePO_4$ at C/5 rate (discharge capacity and Coulombic efficiency); and (h) Li/INF-Li_{10}/$LiFePO_4$ cell at different C-rates and corresponding rate capability. Corresponding voltage profiles are shown in the insets.

Source: (a) and (b) Reproduced with permission from Ref. [27]. Copyright 2013 Royal Society of Chemistry. (c)–(e) Reproduced with permission from Ref. [19]. Copyright 2020 Nature Portfolio. (f)–(h) Reproduced with permission from Ref. [20]. Copyright 2023 Elsevier.

(Continued)

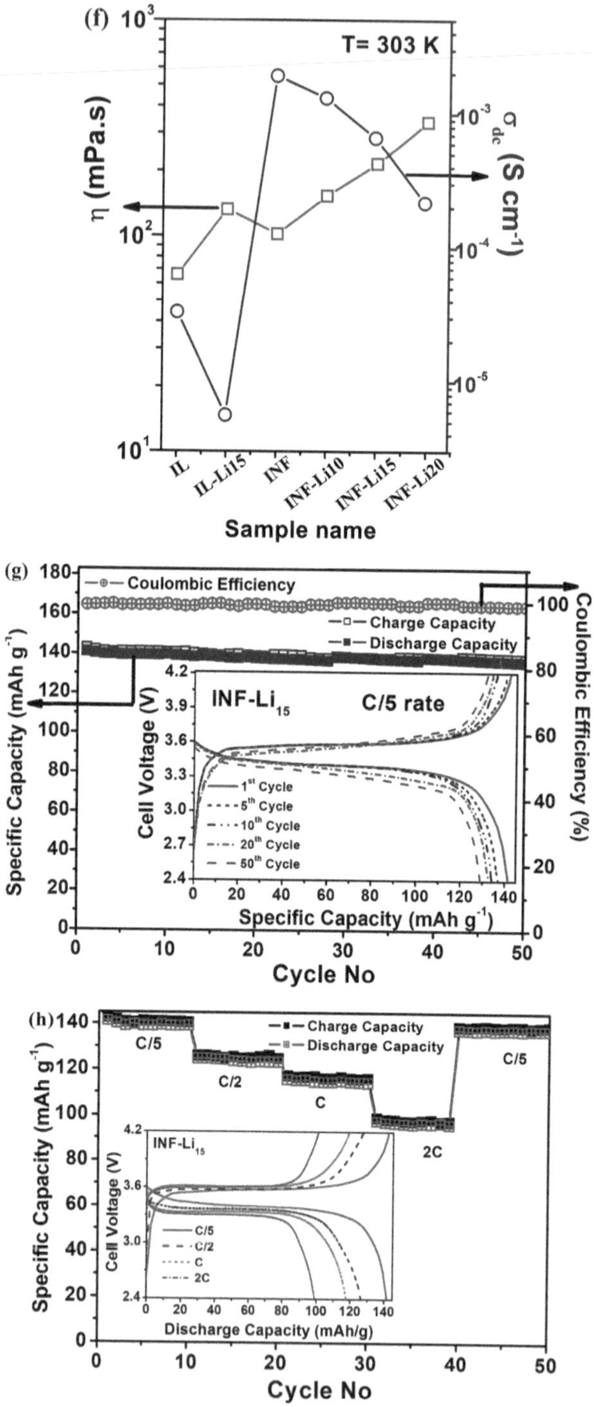

FIGURE 3.1 (Continued)

Due to their moderate dielectric constant and low viscosity, nitrile-based solvents are attractive solvents. Alkoxy-substituted nitriles support high-rate capabilities in $Li_4Ti_5O_{12}$ systems [30]. Furthermore, the stability of high-voltage cathodes was found to improve in electrolytes based on dinitrile [31]. However, the cycling stability of these solvents was not up to par, which hinders their applicability. Phosphorus-containing solvents are also used, especially as flame retardants [32].

As for the lithium salts used in LIB applications, they should meet the following criteria: high solubility, easy dissociation, high ionic conductivity, good redox stability, and compatibility with the solvent. A number of salts that qualify are $LiClO_4$, $LiPF_6$, $LiBF_4$, $LiAsF_6$, LiBOB, LiDFOB, LiFSI, LiTFSI, $LiPF_2O_2$, and LTBP [33]. However, each salt has its merits and demerits, which restrict its application. The best compromise is $LiPF_6$, which has high conductivity and good electrochemical stability, although it exhibits poor thermal stability. It is also incompatible with several recently introduced electrode-active materials. A double-salt approach has been introduced to overcome the shortcomings of $LiPF_6$, significantly improving both performance and safety [34]. Understanding the interplay of various factors is critical in designing electrolytes.

3.4.2 Aqueous Electrolytes

Aqueous electrolytes are desirable alternatives to organic ones because they have higher ionic conductivity and greater solubility for salts. Additionally, they are inexpensive and environmentally friendly. Dahn et al. [17] demonstrated the first aqueous-based LIB ($VO_2(B)/LiMn_2O_4$) in a 5M $LiNO_3$ aqueous electrolyte, which exhibited good energy density, safety, and low cost. Following this, several aqueous systems with lithium nitrate or sulfate were studied. Zhao et al. [18] showed that a $LiV_3O_8/LiFePO_4$ cell with a 9 M $LiNO_3$ aqueous electrolyte delivered a capacity of 60 mAh g^{-1} at 50 C after 500 cycles. However, the narrow working potential (1.23 V), poor cycle stability, and restricted selection of active materials hinder the practicality of aqueous electrolytes.

Suo et al. [35] innovatively introduced an ultra-concentrated (water-in-salt) aqueous electrolyte, which extends the electrochemical stability window to over 3 V due to the reduced water activity and the *in-situ* formation of a passive layer on the anode. Their 21 m LiTFSI-in-water electrolyte exhibited electrochemical stability beyond 3 V. In a full cell, it functioned up to 2.3 V, achieving a Coulombic efficiency (CE) of 100% for over 1000 cycles. However, the higher concentration of salt in electrolytes leads to increased toxicity and higher costs. Xie et al. [19] introduced a new concept-an electrolyte with "molecular crowding", which demonstrated an electrochemical stability window of 3.2 V. This was achieved by reducing the water activity at a low salt concentration (2 m) through the addition of a "crowding" agent. The inclusion of a crowding agent, such as PEG (polyethylene glycol), resulted in a broader electrochemical stability window of 1.3–4.5 V (vs. Li^+/Li) (Figure 3.1c) in full cells containing $Li_4Ti_5O_{12}$ and $LiMn_2O_4$. The full cell delivered an initial specific energy of 110 Wh kg^{-1} and retained 75 Wh kg^{-1} even after 300 cycles with Coulombic efficiencies (CEs) of ~99% at a 1C rate (Figure 3.1d). Furthermore, the mass fraction of water decomposition was estimated to be lower than 0.02 wt.% per cycle (Figure 3.1e). There is, however, a lack of fundamental understanding of water-in-salt systems.

3.4.3 Ionic Liquid Electrolytes

Ionic liquid (IL)-based electrolytes have garnered significant attention due to their high flame retardancy, high ionic conductivity, low vapor pressure, and wide operational voltage range compared to organic electrolytes. Ionic liquids (ILs) consist of organic or inorganic cations and anions in a molten state. Cations such as imidazolium [36], chain quaternary ammonium [37], cyclic quaternary ammonium (pyrrolidinium [38] and piperidinium [39]), anions (imide-type anions such as bis(trifluoromethane sulfonyl)imide ($TFSI^-$)) [40], halides (PF_6^-, BF_4^-) [41], and borates [37]) are commonly found in ILs. The physicochemical properties of the ILs are determined by Coulombic, hydrogen bonding, and van der Waals interactions between the constituted ions. The functional properties of ILs can be adjusted by using different combinations of ions and altering the ratio of the ions. Mostly, the Li-ion conducting ILs have higher melting temperatures due to their small ionic radius, which is not compatible with this application.

On the other hand, non-lithium-based organic cations have a lower melting temperature than Li-ion conducting ILs due to their larger organic cation size. However, this type of ILs exhibits very low Li-ion conductivity, which can be increased through the addition of a small quantity of Li salts (> 0.5 M) to the ILs. ILs may be classified as molecular ILs and polymeric ILs. The former consists of large-sized low-symmetric organic cations and small-sized inorganic anions. The latter have cations and anions in their polymeric backbone. Several ILs are stable over a wide voltage window of 6 V, making them highly appealing for next-generation LIBs.

3.4.3.1 Imidazolium-Based ILs

Imidazolium cations in ILs are attractive due to their high ionic conductivity and low viscosity. Imidazolium-based ILs such as 1-ethyl-2,3-trimethylene imidazolium bis(trifluoromethanesulfonyl)imide ([ETMIm][TFSI]), 1-ethyl-3-methylimidazolium bis(trifluoromethanesulfonyl)imide ([EMIm][TFSI]), 1,2-dimethyl-3-butylimidazolium bis(trifluoromethanesulfonyl)imide ([DMBIm][TFSI]), and 1-allyl-3-vinylimidazolium bis(trifluoromethanesulfonyl)imide ([AVIm][TFSI]) are commonly used as solvents or additives in organic electrolytes. Such ILs form effective passivation layers on the anode and cathode [42]. However, the presence of acidic hydrogen groups in ILs can make the SEIs unstable [43]. Additionally, the high crystalline nature of imidazolium-based ILs and their often insufficient conductivity limit their applicability in high-concentration electrolytes. The low lithium concentration in ILs can create a concentration gradient, resulting in strong polarization and low Li^+ transference numbers. The incorporation of functionalized nanoparticles in the ILs can help control crystallinity, allowing for high salt concentrations.

Recently, Deb et al. [20] reported the SnO_2 nanoparticle grafted 1-methyl-(3-trimethoxysilylpropyl) imidazolium bis(trifluoromethanesulfonyl)imide [m((MeO)$_3$Sip)im][TFSI]) ionanofluid-based ILs with various amount of LiTFSI salt. The ionanofluid electrolyte containing 15 wt.% LiTFSI (INF-Li_{15}) exhibited moderate ion conductivity and high Li^+ ion transportability (Figure 3.1f). The INF-Li_{15} electrolytes in a Li/$LiFePO_4$ cell demonstrated good cycling performance (Figure 3.1g), retaining over 96% of its capacity with a stable CE (99%) at a C/5 rate after 50 cycles. This performance is significantly better than that of a cell with a similar IL electrolyte (81%). INF-Li_{15} also supported a higher rate capability than other electrolytes (Figure 3.1h).

3.4.3.2 Chain Quaternary Ammonium-Based ILs

Quaternary ammonium-based ILs have electrochemical stability windows above 5 V (vs. Li^+/Li) and lower cathodic limiting potential compared to imidazolium-based ILs. However, their higher viscosity and low ionic conductivity restrict their application in LIBs. The properties of these ILs have been sought to be modified by introducing nitrile, ester, and ether groups in the cations. N, N-diethyl-N-methyl-N-(2-methoxyethyl) ammonium TFSA is the most common chain quaternary ammonium-based IL [44]. Alternatively, the ILs could be modified by adding organic solvents. This can lower their viscosity, improve thermal stability, and widen the electrochemical potential window [45].

3.4.3.3 Pyrrolidinium and Piperidinium-Based ILs

Cyclic quaternary ammonium ions, such as pyrrolidinium and piperidinium, are also being investigated for use in energy storage devices. Compared to piperidinium-based ILs, those with pyrrolidinium exhibited lower viscosity due to their smaller molecular size, specifically their five-member ring molecular structure. N-propyl-N-methylpyrrolidinium bis(trifluoromethanesulfonyl)imide is a commonly used pyrrolidinium-based ionic liquid (IL) because of its favorable viscosity and conductivity, non-flammability, wide electrochemical stability window (5.76 V vs. Li^+/Li), and high thermal stability (>400 °C) [16]. Moreover, an improvement in ionic conductivity was observed with the inclusion of organic solvents such as PC, VC (vinylene carbonate), and γ-BL (gamma-butyrolactone). In particular, the addition of γ-BL led to a significant improvement in cell performance [46].

3.4.3.4 Other ILs

Some ILs are used as solvents in electrolytes to enhance the stability and safety of LIBs. The pyrazolium moiety has a ring structure similar to imidazolium, but it has two nitrogen atoms at different positions [47]. Examples of pyridinium-based ILs include 1-butyl-3-methyl pyridinium hydrogen sulfate, 1-butyl-3-methyl pyridinium ethyl sulfate, 1-butyl-3-methyl pyridinium tetrafluoroborate, 1-butyl-3-methyl pyridinium methylsulfate, and 1-butyl-3-methyl pyridinium tetrafluoroborate [21, 48]. Guanidinium-based ILs have also been explored [22]. Recently, ionogels have gained attention due to their high Li+ conductivity resulting from the efficient dissociation of lithium salt [49]. Phosphonium-based ILs are also of interest due to their low viscosity, wide electrochemical window, high anodic stability, and enhanced thermal stability [50]. Often, the viscosity and ionic conductivity of ILs are modified by blending them with other ILs or carbonate solvents. This blending process synergistically improves the performance and rate capability of LIBs [51].

3.5 POLYMER ELECTROLYTES

Polymer electrolytes can address the challenges associated with liquid electrolytes, especially in terms of the mechanical strength and safety of LIBs. Polymer electrolytes are classified as dry and gel polymer electrolytes. Commonly used polymers include PEO (polyethylene oxide), PMA (polymethacrylate), PEG, PVdF-HFP (polyvinylidene fluoride-hexafluoropropylene copolymer), and polyhedral oligomeric silsesquioxane.

3.6 ADDITIVES

Advances in materials and manufacturing technologies continue to improve the performance of LIBs. Current electrolytes are unsatisfactory in terms of capacity retention, thermal stability, and flammability. Several additives minimize the drawbacks of conventional electrolytes without compromising bulk properties. Additives are specifically used to enhance the safety and performance of LIBs. They provide important functions such as forming stable SEIs on graphite, minimizing irreversible capacity and gas production, enhancing flame retardancy and thermal stability of electrolytes, protecting the cathode from dissolution and overcharging, and improving other properties such as ionic conductivity, viscosity, and wettability [52]. Common electrolyte additives include alkyl phosphates, alkyl phosphonates, phosphazenes, hydrofluoroethers, and fluorinated esters [53].

3.6.1 Selection of Electrolyte Additives

Electrolyte additives play a significant part in improving the performance and safety of LIBs in a cost-effective manner. They provide controllable functionality for battery behavior. Functionalities may be grouped under film formation, voltage enhancement, performance enhancement, and safety enhancement (Figure 3.2). The selection also depends on the electrode materials. Table 3.2 lists several common additives and their associated benefits.

3.6.2 Anode Additives

Electrolyte stability against the charged anode is a crucial aspect of battery performance. Stable SEIs inhibit the further decomposition of electrolytes during cycling. Additives may not only help in forming thin, stable SEIs but also in reducing them at high potentials [64, 65]. Mostly, electrolyte additives are chosen based on having lower LUMO levels (lowest unoccupied molecular orbital) than the solvent. This helps to shift electrolyte reduction to more negative potentials [65].

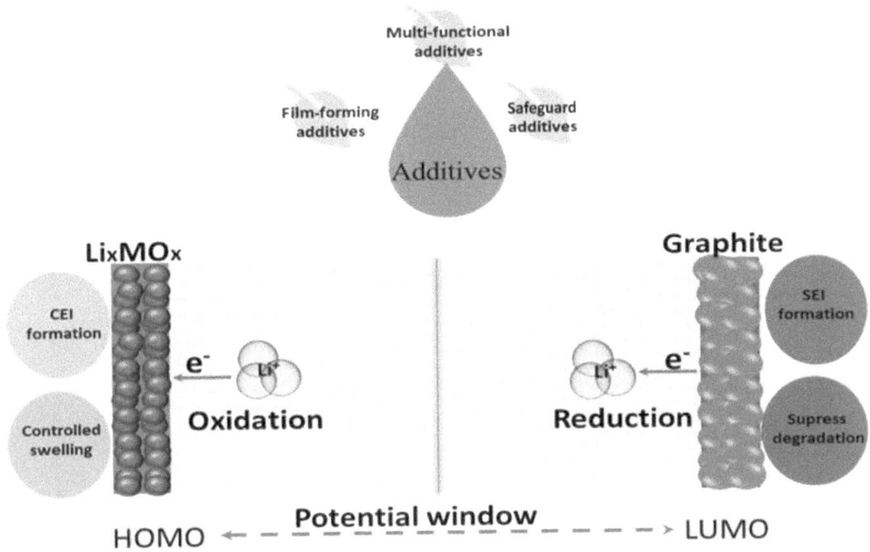

FIGURE 3.2 Schematic diagrams showing the benefits of additives in lithium-ion batteries.

3.6.2.1 Carbon Materials

The carbon materials, mainly graphite, are widely used as anode materials in commercial LIBs due to their high cycling stability and excellent reversibility. The catalytic nature of carbons can cause the electrolyte to decompose, resulting in the formation of a thin layer ($ROCO_2Li$, $(ROCO_2Li)_2$ and Li_2CO_3) on the anode. This process also generates H_2O, O_2, and CO_2. This reduces the capacity and lifespan of the cell and increases its internal pressure.

Any exfoliation of graphite caused by the co-intercalation of electrolytes and cathode dissolution (e.g., Mn^{2+} from $LiMn_2O_4$) can also result in capacity decay [66, 67]. Such drawbacks can be resolved by using additives that form a solid electrolyte interface (SEI). The composition and thickness of SEIs depend on the additives. SEI-forming additives in carbon anodes include unsaturated carbons, sulfur, and halogen-containing compounds [54]. Huang et al. [68] investigated the use of ammonium perfluorocaprylate (1 wt.%) as an electrolyte additive. It forms a uniform and stable SEI layer on the anode at a high potential, which limits electrolyte decomposition and reduces charge transfer resistance.

Another way, the electrolyte decomposes due to the reaction of alkyl carbonate solvents with alkoxide anion impurities. VC minimizes these issues by trapping alkoxide anions. It results in a thin polymeric layer on the anode that prevents electrolyte decomposition and reduces charge transfer resistance. Xu et al. [55] studies show that the addition of vinyl ethylene carbonate (VEC) serves a similar purpose by forming a passive layer on the cathode surface. The film-forming ability of VEC on a graphite anode was confirmed in a three-electrode system through FE-SEM studies.

Figure 3.3a shows the surface morphologies of the electrodes after 10 cycles in VEC and VEC-free electrolytes. The electrode cycled in the VEC-free electrolyte shows the formation of a dense SEI film. Volume changes in the graphite anode can

TABLE 3.2
Electrolytes' and Additives' Role for Li-Ion Batteries and Their Performance

Electrolyte	Additive	Additive (%)	Role	Electrode Materials	Initial Capacity (mAh g^{-1})	Retention (%)	Ref.
1 M $LiPF_6$ in EC/DEC (1:1 v/v)	Ammonium perfluorocaprylate (APC)	0.70 wt.% APC	SEI layer	MCMB	320	74% (30 cycles)	[54]
1 M $LiPF_6$ in EC/DEC/DMC (1:1:1 v/v)	Vinylethylene carbonate	1.0 vol.% VEC	SEI layer	Graphite	405.0	91% (30 cycles)	[55]
1 M $LiPF_6$ in EC/PC/EMC (0.14/0.18/0.68 wt.%)	Sodium chloride	1 wt.% Nacl	Uniform SEI layer	MCMB	304.8	87.2% (52 cycles)	[56]
1 M $LiPF_6$ in EC/DEC (1:1 v/v)	Vinylene carbonate	3 wt.% VC	SEI and rate capability	Si nanowire	2803	79.81% (250 cycles)	[57]
1 M $LiPF_6$ in EC/DEC (1:1 v/v)	Fluoroethylene carbonate	3 wt.% FEC	SEI and rate capability	Si nanowire	2996	59.95% (250 cycles)	[57]
1 M $LiPF_6$ in EC/DEC (1:1 v/v)	Vinyl ethylene carbonate	3 wt.% VEC	SEI and rate capability	Si nanowire	2781	49.79% (250 cycles)	[57]
1 M $LiPF_6$ in EC/DEC (1:1 v/v)	Lithium bis(oxalato) borate	3 wt.% LiBOB	SEI and rate capability	Si nanowire	3934	74.94% (250 cycles)	[57]
1 M $LiPF_6$ in EC/DMC (3:7 v/v)	2,2′:5′,2-terthiophene (3TH)	0.1 wt.% 3TH	CEI layer	$LiCoO_2$	159.6	84.8% (100 cycles)	[58]
1 M $LiPF_6$ in EC/DMC (3:7 v/v)	2,2′-bithiophene (2TH)	0.1 wt.% 2TH	CEI layer	$LiCoO_2$	~159	77.4% (100 cycles)	[58]
1 M $LiPF_6$ in EC/DEC (3:7 wt.%)	Triisopropanolamine cyclic borate (TPCB)	0.5 wt.% TPCB	CEI layer with voltage expansion	$LiCoO_2$	205	82.2% (200 cycles)	[59]
1 M $LiPF_6$ in EC/DMC (1:1 v/v)	Lithium difluorophosphate ($LiPO_2F_2$)	0.5 wt.% $LiPO_2F_2$	CEI layer	Nickel-rich NMC	176	85% (500 cycles)	[60]
1.2M $LiPF_6$ in EC/EMC (3:7 v/v)	1,3-benzodioxan (DBBD1) and 1,4-benzodioxan (DBBD2)	0.2 M DBBD1 or DBBD2	Redox shuttle stability (overcharge protection)	MCMB/ $LiFePO_4$	—	—	[61]
2M $LiPF_6$ in EC/EMC (3:7 wt.%)	3,5-di-tert-butyl-1,2-dimethoxybenzene (DBDB)	5 wt.% DBDB	Redox shuttle stability (overcharge protection)	MCMB/ $LiFeO_4$	—	—	[61]
1 M $LiPF_6$ in EC/DMC (1:1 v/v)	Ethoxy (pentafluoro) cyclotriphosphazene (PFPN)	5 vol.% PFPN	Flame retardant	Li/$LiCoO_2$	150.7	99.1% (30 cycles)	[62]
1 M $LiPF_6$ in EC/DMC (1:1 wt.%)	(Phenoxy) pentafluorocyclo-triphosphazene (FPPN)	5 wt.% FPPN	Flame retardant	MCMB/ NMC111	—	—	[63]

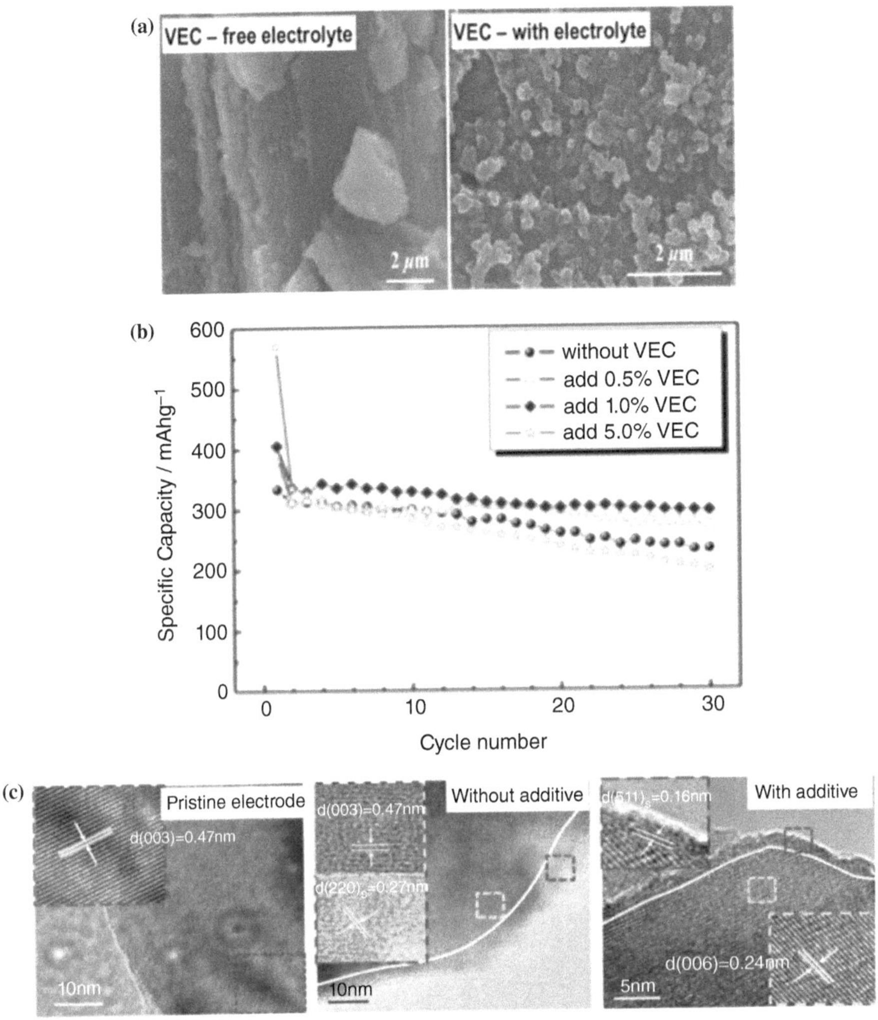

FIGURE 3.3 (a) SEM images of graphite electrode after CV cycles in VEC-free electrolyte and VEC-containing electrolyte; (b) cycle performance of graphite electrode using 1 mol $LiPF_6$ in EC: DEC: DMC with or without VEC; (c) (layered cathode materials) HR-TEM images of pristine LRNMC particle and cycled LRNMC electrodes for 500 cycles at 3C, without additive (LP30); with additive (LP30 with 0.5 wt.% $LiPO_2F_2$), (d) cycling performance up to 500 cycles at 3C, (e) voltage profiles of MCMB/$LiFePO_4$ cells containing 0.2 M DBBD1 and DBBD2 during the overcharge tests (C/5 charging rate, time axis breaks between 150 and 300 h), (f) digital photos of flammability testing of the electrolytes with different PFPN contents.

Source: (a) and (b) Reproduced with permission from Ref. [55]. Copyright 2013 The Electrochemical Society. (c) and (d) Reproduced with permission from Ref. [69]. Copyright 2015 The Electrochemical Society. (e) Reproduced with permission from Ref. [61]. Copyright 2010 Elsevier. (f) Reproduced with permission from Ref. [62]. Copyright 2014 Elsevier.

(Continued)

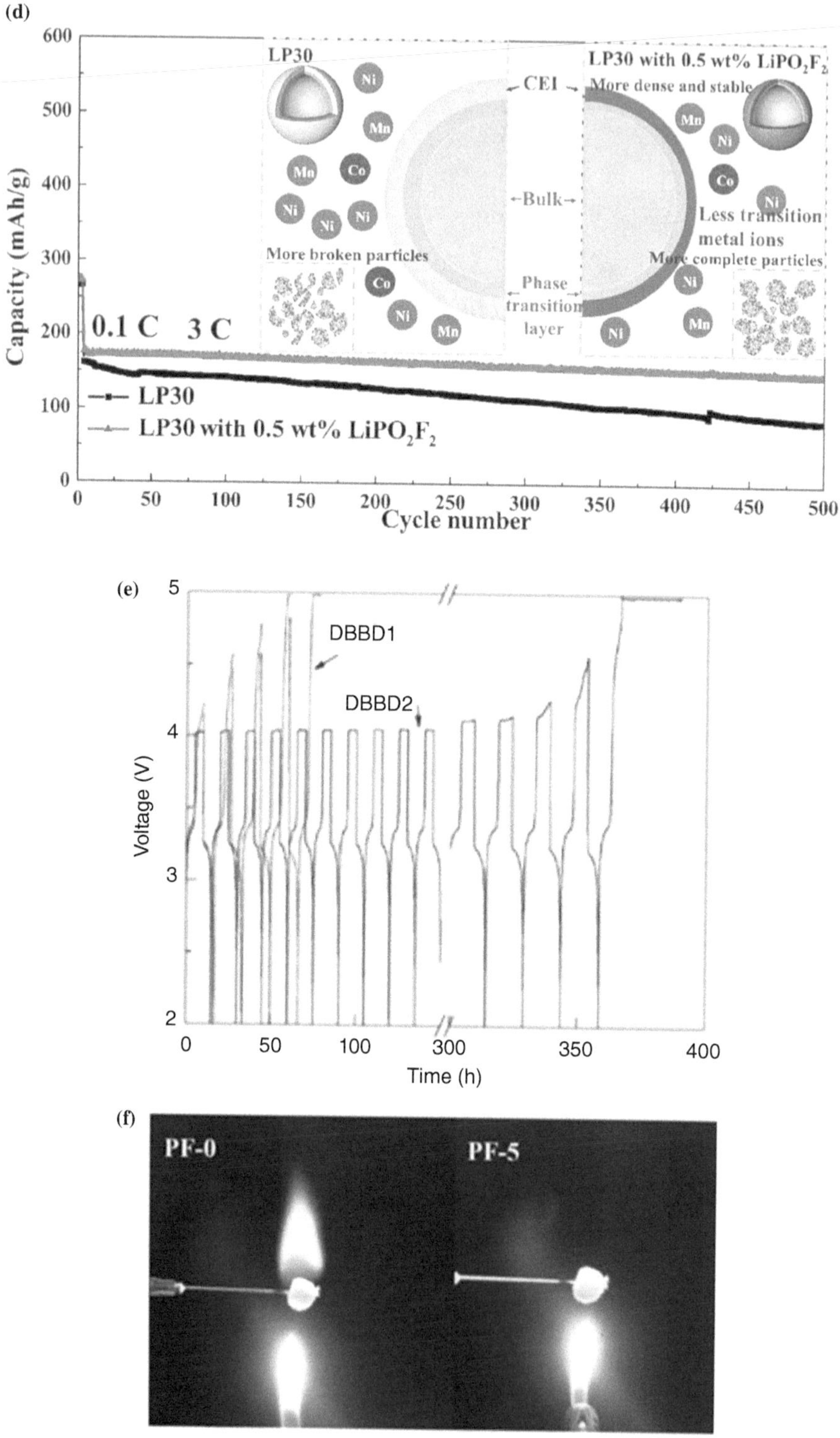

FIGURE 3.3 (Continued)

result in inconsistent SEI layers. Cell stability at various concentrations of VEC (0, 0.5, 1, and 5 wt.%) is depicted in Figure 3.3b. The performance of the cell is better with lower VEC concentrations (0.5% and 1%) than without VEC.

Additionally, increasing the content of VEC leads to inferior cell performance, even worse than the cell without VEC. Wang et al. [70] investigated a series of additives based on phenylene dimaleimide with maleimide groups in the ortho, meta, and para positions. All the maleimide-based additives led to improved performance due to the formation of stable SEIs. Glycolide derivatives were also introduced for high-temperature applications. 3-methyl glycolide additive performed well at high temperatures (Tonset 120 °C) [71].

Graphite exfoliation in PC can be prevented by using various additives. These include olefinic compounds [72], butyrolactone derivatives [73], dimethyl 2-oxo-1,3 -dioxolane-4,5-dicarboxylate [74], and quaternary ammonium ILs such as trimethylhexyl ammonium bis(trifluoromethane sulfonyl)imide [75], sulfur-containing compounds [56], and inorganic additives [76, 77].

3.6.2.2 High-Energy Anode Materials

Anodes based on silicon (Si), germanium (Ge), tin (Sn), and antimony (Sb) are attractive due to their high lithium-insertion capacities. However, they suffer from large volume changes, which leads to unstable SEI layers. Due to significant volume changes and accompanying stress, these active materials disintegrate. This issue can be addressed by forming a stable and flexible SEI with good mechanical robustness [78].

In the case of Si, impurities such as O_2 or H_2O in the electrolyte generate layers that contain silanol (Si-OH) and silicon oxide (Si-O-Si). These impurities reduce conductivity and the overall life cycle of the cell. VC and FEC additives are primarily used for Si anodes. FEC is primarily used as a co-solvent and forms an SEI layer with strong bonding energy, consisting of -Si-F and LiF [79]. VC forms a polycarbonate layer on Si, which prevents electrolyte decomposition and electrode rupture [57].

Succinic anhydride was used as an additive for the Si anode, which helps increase the formation of a stable Li_2CO_3 layer and prevents electrolyte decomposition [80]. Studies have shown the improved high-temperature performance of Si anodes with VC and FEC additives [81]. The improved performance is due to the formation of a secondary polycarbonate layer over the primary SEI layer.

$Li_4Ti_5O_{12}$, as an anode, exhibits a high operating potential, excellent cycle life, high rate performance, and safety [82]. Due to the nobler potentials at which $Li_4Ti_5O_{12}$ works, it is hardly covered with an SEI. Various film-forming additives have been tested with $Li_4Ti_5O_{12}$ anodes, including p-toluenesulfonyl isocyanate [83], methylene methane disulfonate [56], and glutaric anhydride [84], to generate artificial SEI layers.

3.6.3 Cathode Additives

Failure of LIBs can mainly be attributed to the degradation of cathodes, which can occur due to the dissolution of transition metals, loss of conductivity caused by cathode mass rupture, and secondary reactions at the electrode-electrolyte interface. $LiMn_2O_4$, in particular, is prone to dissolution, which can destroy the SEI on the anode. Forming passive layers is an effective way to control dissolution, and this can be achieved using cathode additives. Electrolyte additives form passive layers prior to the oxidation of the electrolyte, which protects the electrolytes from further oxidation.

3.6.3.1 Layered Cathode Materials

$LiCoO_2$ is the most commonly used cathode material for LIBs due to its higher energy density. But this electrode is thermally unstable, prone to cobalt dissolution, and highly reactive towards electrolytes in its highly charged states [85]. It can be circumvented by using electrolyte additives that can create a CEIs layer. A variety of additives have been investigated, including biphenyl (4.44 V), o-terphenyl (4.45 V), m-terphenyl (4.25 V) [58], thiophene derivatives [86], sulfate additive [87], maleimide-based self-polymerizing branched oligomer [59], tri-isopropanol amine cyclic borate (TPCB) [88], and inorganic additives such as Li_2CO_3 and nano-Al_2O_3 [59]. Layered $LiNi_yMn_yCo_{1-2y}O_2$ (NMCs) have attracted attention due to their excellent cycle stability, thermal stability, high reversible capacity, and ability to operate up to 4.6 V. Similar to $LiCoO_2$, NMCs also encounter challenges related to transition metal dissolution and the evolution of reactive gases [89]. These issues are being addressed through the incorporation of additives.

Polyfluoroalkyl-based additives have been demonstrated to result in improved high-voltage performance [90]. This is because additives oxidize earlier than electrolytes and results in the formation of the CEI layer, which protects the electrolyte from oxidation. Likewise, glymes [91], nitriles [60], and lithium difluorophosphate ($LiPO_2F_2$) [69] have also been used with NMC cathodes. Jiang et al. [69] studied the cycling stability of $Li_{1.14}Ni_{0.133}Co_{0.133}Mn_{0.544}O_2$ with $LiPO_2F_2$ as an additive.

The formation of a CEI on the cathode was examined by HR-TEM after 500 cycles at a rate of 3C (Figure 3.3c). HR-TEM images of the raw cathode showed good crystallinity with a d-spacing of 0.47 nm. The cycled cathode retained the crystallinity but exhibited additional phases with a d-spacing of 0.27 nm and a surface layer of 8-15 nm in thickness. The introduction of additives inhibits the growth of the phase transition layer at the cathode interface, as evidenced by the decreased width of the phase change layer (3 nm) in the HR-TEM images. The formation of a smooth CEI layer and a reduced phase change layer results in improved cell performance (Figure 3.3d). The role of the additive is schematically represented in the inset of Figure 3.3d.

3.6.3.2 Li-Rich Cathodes

Lithium-rich cathodes operate at potentials higher than 4.5 V, have capacities above 250 mAh g^{-1}, and exhibit better thermal stability. But they are plagued with problems such as dissolution, volume change, and the formation of irreversible Li_2O surface layers [92]. These issues could potentially be resolved by incorporating appropriate electrolyte additives that form a CEI on the cathode surface [93]. Compounds containing sulfur, fluorine, boron, and phosphorus are utilized to enhance cell performance and prevent metal dissolution and structural change [94–96].

3.6.3.3 Spinel Cathodes

As a cathode material, $LiMn_2O_4$ exhibits a high specific capacity of 148 mAh g^{-1}, 4.1 V (vs. Li^+Li), which is a cost-effective and environmentally friendly option. Apart from issues such as capacity decay and poor cyclic performance, manganese dissolution occurs due to its incompatibility with carbonate electrolytes. There is also the problem of structural instability brought about by Jahn-Teller distortion, as the Mn^{3+} content increases upon discharge [97]. The material is particularly vulnerable to hydrofluoric acid (HF), which can leach out manganese.

This vulnerability is especially pronounced when HF is formed through the hydrolysis of $LiPF_6$. Organo silicon compounds, such as hexamethyldisilazane and heptamethyldisilazane, have been found to dehydrate the electrolyte and prevent the generation of HF. Likewise, tris(pentafluorophenyl)silane as an electrolyte additive with $LiNi_{0.5}Mn_{1.5}O_4$ forms a passive layer through oxidation before the solvent does. The CEI prevents electrolyte decomposition, which improves CE and capacity retention [98]. Sulfur-containing additives and inorganic sodium phosphate have also been investigated for their ability to suppress HF formation and create robust CEI layers on the cathode [99].

3.6.3.4 Olivine Cathodes

Olivine-type cathodes (e.g., $LiFePO_4$) have gained much interest, especially for large-area devices, due to their affordability, moderate capacities, and high level of intrinsic safety. Dissolution of iron and subsequent decomposition of the electrolyte generate HF and H_2O, which can dissolve the SEI on the anode. Additives that help control or prevent hydrolysis and HF attack include N,N-diethylaminotrimethyl silane [100], triphenylborate and trimethylborate [101], VC, and FEC [102].

3.6.4 Multifunctional Additives

3.6.4.1 Redox Shuttles and Overcharging

Redox shuttle additives have been used to protect batteries from overcharging. Overcharging provokes unwanted chemical reactions, which degrade the electrode and electrolyte. The unintended reactions generate gases such as O_2, CO_2, and CO, which contribute to a rapid rise in temperature, leading to both explosion and fire.

Excessive lithium deintercalation could cause the cathode structure to collapse. On the anode side, the formation of lithium dendrites and the dissolution of copper can lead to short-circuiting, which can eventually result in thermal runaway. Shuttling compounds exhibit highly reversible redox behavior and slightly higher oxidation potentials than the cathode [62]. In their overcharged condition, the cations (RS^+) in the additive molecule (RS) are oxidized at the cathode, which limits the oxidation of the cathode and, consequently, the decomposition of the electrolyte. The oxidized cations subsequently migrate to the anode, where they are reduced back to RS (Equations (3.1) and (3.2)).

$$\text{Cathode: RS} \rightarrow \text{RS}^+ + ne^- \tag{3.1}$$

$$\text{Anode: RS}^+ + ne^- \rightarrow \text{RS} \tag{3.2}$$

It was Behl et al. [103] who first demonstrated this effect with an I^-/I^{3-} couple ~3.25 V (vs. Li^+/Li) in a $LiAsF_6$/THF electrolyte. Subsequently, the Br^-/Br^{3-} couple (~3.78 V) was investigated for 3 V LIBs [104]. However, the high reactivity and volatility of the halogens hinder their application in commercial LIBs. Golovin et al. [105] examined derivatives of ferrocene, including carbamyl ferrocene, carbomethoxyferrocene, methoxymethyl ferrocene, and (dimethyl aminomethyl) ferrocene, with redox potentials ranging from 3.2 to 3.5 V vs. Li^+/Li. Dahn et al. [106] introduced 2,5-di-tert-but yl-1,4-dimethoxybenzene, which operates stably at ~4 V. However, its low solubility in carbonate electrolytes became an issue. This was circumvented by using its isomers (3,5-di-tert-butyl-1,2-dimethoxybenzene [107], 4,6-di-tert-butyl-1,3-benzodio xole (DBBD1), and 5,7-di-tert-butyl-1,4-benzodioxin (DBBD2) [61].

The bi-functional redox shuttle, 4-tert-butyl-1,2-dimethoxybenzene, was found to be suitable for operation at 4.1–4.2 V [108]. The overcharging performance of DBBD1 and DBBD2 was tested in MCMB/LiFePO4 cells with 0.2 M of additives, cycling at C/5 with a 100% overcharge (Figure 3.3e). 1,4-benzodioxan (DBBD2) exhibited slightly better stability due to the larger number of CH_2 groups [61]. The bifunctional redox shuttle, 2-(pentafluorophenyl)-tetrafluoro-1,3,2-benzodioxaborol e, was introduced for the high-voltage battery. This redox shuttle allows for the potential to be shuttled over 4.2 V [109].

3.6.4.2 Flame Retardants

Safety is a major concern with LIBs. An effective method for controlling the flammability of liquid electrolytes is by adding flame-retardant additives to them. Flame retardants are chemical compounds that inhibit combustion reactions by scavenging highly reactive radicals. Phosphorus and halogen-containing organic compounds undergo rapid reactions with free radicals (•OH and H•). Halogen-based components, however, are not ideal for this application due to their toxicity to the environment and humans. Prominently, phosphorus-based compounds such as trimethyl phosphate [110], triphenyl phosphate [111], ethylene ethyl phosphate [112], tris(2,2,2-trifluoroethyl)phosphate [113], tris(2,2,2-trifluoroethyl)phosphite [114], bis(2,2,2-trifluoroethyl)methyl phosphonate [115], (ethoxy)pentafluorocyclotriphosphazene [62], and (phenoxy)pentafluorocyclotri-phosphazene [63] are widely used flame retardants in LIBs. The flame retardancy of ethoxy(pentafluoro)cyclotriphosphazene is illustrated in Figure 3.3f. The additive has a self-extinguishing time of 12.38 s g^{-1} and a critical oxygen index of 22.9 [62].

3.7 SUMMARY AND FUTURE PROSPECTIVE

The performance of lithium-ion batteries is continuously improving. However, high-power applications still demand improvements, particularly in terms of power capability and safety. In this chapter we have outlined the problems, challenges, and perspectives regarding liquid electrolytes and additives for LIBs. We specifically draw attention to their design and selection. The ionic conductivity, viscosity, liquidus range, and wettability are critical factors for electrolytes. They reflect on the ion transport, rate capability, and cycle stability of LIBs. SEIs are crucial determinants of performance parameters such as service life, life cycle, capacity, and safety. Thus, the selection of appropriate combinations of electrolytes and additives is crucial for improving the performance and safety of LIBs. A summary of electrolytes and additives used in the LIBs is presented in Figure 3.4.

A deep understanding of the constituents of electrolytes and additives, their interaction with active materials, as well as failure mechanisms, is crucial for the

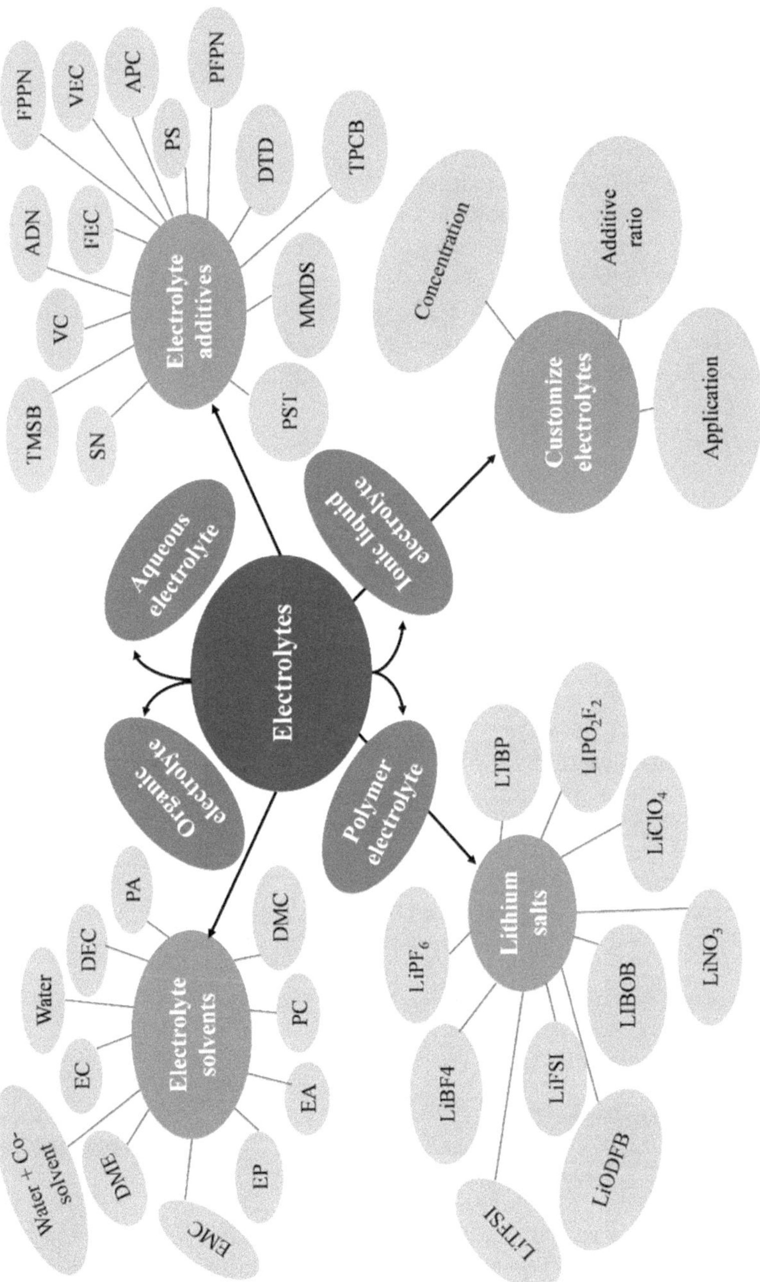

FIGURE 3.4 Summary of the electrolyte components in LIBs.

development of stable electrolytes and multifunctional additives. This requires the use of advanced analytical and characterization techniques, particularly to study SEIs and mechanical and ion transport properties. This must be backed by theoretical modeling and validation.

ACKNOWLEDGEMENTS

BS and PS acknowledge the Start-up Research Grand (SRG)-Science and Engineering Research Board (Grant No. SRG/2020/001801) for the financial support. BS acknowledges Technology Development Transfer (TDT) (DST/TDT/AM/2022/169G), Department of Science & Technology, Ministry of Science & Technology, New Delhi for the financial support. RA thanks the Technology Mission Division, Department of Science & Technology, Ministry of Science & Technology, New Delhi for the financial support under the scheme of IC-MAP-2021, Project title: DST-Storage MAP (Ref. No. DST/TMD/IC-MAP/2K20/01).

REFERENCES

1. R. Kumar, K. Aifantis, P. Hu, *Rechargeable Ion Batteries: Materials, Design and Applications of Li-Ion Cells and Beyond*, 2023, 4, 83–103, Wiley-VCH GmbH.
2. G. Wang, M. Yu and X. Feng, *Chem. Soc. Rev.*, 2021, 50(4), 2388–443.
3. M. Endo, C. Kim, K. Nishimura, T. Fujino and K. Miyashita, *Carbon*, 2000, 38(2), 183–97.
4. S. Prakash, R. Kumar, A. Gupta, A. Chaudhary, V.K. Chandaliya, P.S. Dash, P. Gurunathan, K. Ramesh, S. Kumari and S.R. Dhakate, *Mater. Chem. Phys.*, 2022, 1, 281, 125836.
5. W. Cai, Y.X. Yao, G.L. Zhu, C. Yan, L.L. Jiang, C. He, J.Q. Huang and Q. Zhang, *Chem. Soc. Rev.*, 2020, 49(12), 3806–33.
6. Q. Shi, J. Zhou, S. Ullah, X. Yang, K. Tokarska, B. Trzebicka, H.Q. Ta and M.H. Rümmeli, *Energy Storage Mater.*, 2021, 1, 34, 735–54.
7. B. Wang, B. Luo, X. Li and L. Zhi, *Mater. Today*, 2012, 15(12), 544–52.
8. X. Lan, X. Xiong, J. Liu, B. Yuan, R. Hu and M. Zhu, *Small*, 2022, 18(26), 2201110.
9. S. Levchenko, S. Wei, V. Marangon and J. Hassoun, *Energy Technol.*, 2022, 1, 2200725.
10. H. Zhang, Y. Yang, D. Ren, L. Wang and X. He, *Energy Storage Mater.*, 2021, 1(36), 147–70.
11. M. Nazhipkyzy, A.B. Maltay, K. Askaruly, D.D. Assylkhanova, A.R. Seitkazinova and Z.A. Mansurov, *Nanomaterials*, 2022, 21, 12(20), 3710.
12. J. Qian, L. Liu, J. Yang, S. Li, X. Wang, H.L. Zhuang and Y. Lu, *Nat. Commun.*, 2018, 9(1), 4918.
13. X. Shen, X.Q. Zhang, F. Ding, J.Q. Huang, R. Xu, X. Chen, C. Yan, F.Y. Su, C.M. Chen, X. Liu and Q. Zhang, *Energy Mater. Adv.*, 2021, 2021. https://doi.org/10.34133/2021/1205324
14. K. Xu, *Chem. Rev.*, 2004, 16, 104(10), 4303–418.
15. L. Yang, B. Ravdel and B.L. Lucht, *Electrochem. Solid-State Lett.*, 2010, 13(8), A95.
16. G.B. Appetecchi, M. Montanino, D. Zane, M. Carewska, F. Alessandrini and S. Passerini, *Electrochim. Acta*, 2009, 54, 4, 1325–32.
17. W. Li, J.R. Dahn and D.S. Wainwright, *Science*, 1994, 264(5162), 1115–8.
18. M. Zhao, B. Zhang, G. Huang, H. Zhang and X. Song, *J. Power Sources*, 2013, 232, 181–6.

19. J. Xie, Z. Liang and Y.C. Lu, *Nat. Mater.*, 2020, 19(9), 1006–11.
20. D. Deb and S. Bhattacharya, *J. Mol. Liq.*, 2023, 379, 121645.
21. T. Stettner and A. Balducci, *Energy Storage Mater.*, 2021, 40, 402–14.
22. Y. Jin, S. Fang, L. Yang, S. Hirano and K. Tachibana, *J. Power Sources*, 2011, 196(24), 10658–66.
23. P. Fan, H. Liu, V. Marosz, N.T. Samuels, S.L. Sui, L. Sun, L. Liao, *Adv. Funct. Mater.*, 2021, 31(23), 2101380.
24. A. Balducci, *Top. Curr. Chem.*, 2017, 375, 1–27.
25. C. Fang, T.N. Tran, Y. Zhao and G. Liu, *Electrochim. Acta*, 2021, 10, 399, 139362.
26. L. Hu, Z. Zhang and K. Amine, *J. Power Sources*, 2013, 236, 175–80.
27. Z. Zhang, L. Hu, H. Wu, W. Weng, M. Ko, P.C. Redfern, L.A. Curtiss and K. Amine, *Energy Environ. Sci.*, 2013, 6(6), 1806–10.
28. X.G. Sun and C.A. Angell, *Solid State Ionics*, 2004, 175(1–4), 257–60.
29. K. Xu, *Chem. Rev.*, 2014, 114(23), 11503–618.
30. Q. Wang, P. Pechy, S.M. Zakeeruddin, I. Exnar and M. Grätzel, *J. Power Sources*, 2005, 146(1–2), 813–6.
31. M. Nagahama, N. Hasegawa and S. Okada, *J. Electrochem. Soc.*, 2010, 157(6), A748.
32. S. Dalavi, M. Xu, B. Ravdel, L. Zhou and B.L. Lucht, *J. Electrochem. Soc.*, 2010, 157(10), A1113.
33. Y.K. Liu, C.Z. Zhao, J. Du, X.Q. Zhang, A.B. Chen and Q. Zhang, *Small*, 2023, 19, 2205315.
34. G. Xu, X. Shangguan, S. Dong, X. Zhou and G. Cui, *Angew. Chem. Int. Ed.*, 2020, 59(9), 400–15.
35. L. Suo, O. Borodin, T. Gao, M. Olguin, J. Ho, X. Fan, C. Luo, C. Wang and K. Xu, *Science*, 2015, 350(6263), 938–43.
36. M. Huang, L. Kan, W. Zhao, Y. Wang, Y. Xiong, W. Shan and Z. Lou, *Chem. Eng. J.*, 2021, 421, 127763.
37. M.R. Shimpi, P. Rohlmann, F.U. Shah, S. Glavatskih and O.N. Antzutkin, *Phys. Chem. Chem. Phys.*, (2021), 23, 6190–203
38. F.L. Wu, G.T. Kim, T. Diemant, M. Kuenzel, A.R. Schur, X.P. Gao, B.S. Qin, D. Alwast, Z. Jusys, R.J. Behm, D. Geiger, U. Kaiser and S. Passerini, *Adv. Energy Mater.*, (2020), 10, 2001830.
39. J.M. Lee, *Chem. Eng. J.*, 2011, 172(2–3), 1066–71.
40. K.E. Gunderson-Briggs, T. Ruther, A.S. Best, M. Kar, C. Forsyth, E.I. Izgorodiana, D.R. MacFarlane and A.F. Hollenkamp, *Angew. Chem. Int. Ed.*, 2019, 58(13), 4390–4.
41. T. Rüther, A.I. Bhatt, A.S. Best, K.R. Harris and A.F. Hollenkamp, *Batteries Supercaps*, 2020, 3, 793–827.
42. S. Javadian, P. Salimi, H. Gharibi, A. Fathollahi, E. Kpwsari and J. Kakemam, *J. Iran. Chem. Soc.*, 2019, 16, 2123–34.
43. Z. Wang, Y. Cai, Z. Wang, S. Chen, X. Lu and S. Zhang, *J. Solid State Electrochem.*, 2013, 17, 2839–48.
44. A. Ray and B. Saruhan, *Materials*, 2021, 4(11), 2942.
45. H. Niu, L. Wang, P. Guan, N. Zhang, C. Yan, M. Ding, X. Guo, T. Huang and X. Hu, *J. Energy Storage*, 2021, 40, 102659.
46. N. Wongittharom, C.H. Wang, Y.C. Wang, G.T. Fey, H.Y. Li, T.Y. Wu, T.C. Lee and J.K. Chang, *J. Power Sources*, 2014, 260, 268–75.
47. M. Ghirardello, M. Costantini, A. Vecchi, D. Pazzi, F. Castiglione, A. Mele and A. Marra, *Eur. J. Org. Chem.*, 2022, (21), https://doi.org/10.1002/ejoc.202200100
48. C. Wu, A.D. Visscher and I.D. Gates, *J. Phy. Chem. B*, 2018, 122(26), 6771–80.
49. F. Wu, N. Chen, R. Chen, Q. Zhu, J. Qian and L. Li, *Chem. Mater.*, 2016, 28(3), 848–56.

50. K. Tsunashima, Y. Ono and M. Sugiya, *Electrochim. Acta*, 2011, 56, 4351–5.
51. L. Lombardo, S. Brutti, M.A. Navarra, S. Panero and P. Reale, *J. Power Sources*, 2013, 227, 8–14.
52. A.M. Haregewoin, A.S. Wotango and B. J. Hwang, *Energy Environ. Sci.*, 2016, 9(6), 1955–88.
53. A. Lex-Balducci, W.A. Henderson, S. Passerini, Lithium-ion batteries advanced materials and technologies, in: J. Yuan, X. Liu, H. Zhang (Eds.), *Electrolytes for Lithium Batteries*, 2011, CRC Press.
54. J. Xing, S. Bliznakov, L. Bonville, M. Oljaca and R. Maric, *Electrochem. Energy Rev.*, 2022, 5(4), 14.
55. S.D. Xu, Q.C. Zhuang, J. Wang, Y.Q. Xu and Y.B. Zhu, *Int. J. Electrochem. Sci.*, 2013, 8(6), 8058–76.
56. B. Tong, Z. Song, H. Wan, W. Feng, M. Armand, J. Liu, H. Zhang, Z. Zhou, *InfoMat*, 2021, 3(12), 1364–92.
57. H. Shobukawa, J. Alvarado, Y. Yang and Y.S. Meng, *J. Power Sources*, 2017, 359, 173–81.
58. K. Abe, T. Takaya, H. Yoshitake, Y. Ushigoe, M. Yoshio and H. Wang, *Electrochem. Solid-State Lett.*, 2004, 7(12), A462.
59. Y.H. Li, M.L. Lee, F.M. Wang, C.R. Yang, P.P. Chu and J.P. Pan, *Electrochim. Acta*, 2012, 85, 72–7.
60. G.Y. Kim and J.R. Dahn, *J. Electrochem. Soc.*, 2015, 162(3), A437.
61. W. Weng, Z. Zhang, P.C. Redfern, L.A. Curtiss and K. Amine, *J. Power Sources*, 2011, 196, 1530.
62. X. Li, W. Li, L. Chen, Y. Lu, Y. Su, L. Bao, J. Wang, R. Chen, S. Chen and F. Wu, *J. Power Sources*, 2018, 378, 707.
63. T. Dagger, V. Meier, S. Hildebrand, D. Brüggemann, M. Winter and F.M. Schappacher, *Energy Technol.*, 2018, 6, 2001.
64. P.U. Nzereogu, A.D. Omah, F.I. Ezema, E.I. Iwuoha and A.C. Nwanya, *Appl. Surf. Sci. Adv.*, 2022, 9, 100233.
65. A. Wang, S. Kadam, H. Li, S. Shi and Y. Qi, *NPJ Comput. Mater.*, 2018, 4(1), 15.
66. G.M. Hobold, A. Khurram and B.M. Gallant, *Chem. Mater.*, 2020, 32(6), 2341–52.
67. S.J. An, J. Li, C. Daniel, D. Mohanty, S. Nagpure and D.L. Wood III, *Carbon*, 2016, 105, 52–76.
68. Q. Huang, T. Wang, M. Yan and Z. Jiang, *J. Electrochem. Soc.*, 2006, 153(12), A2363.
69. B. Jiang, J. Li, B. Luo, Q. Yan, H. Li, L. Liu, L. Chu, Y. Li, Q. Zhang and M. Li, *J. Energy Chem.*, 2021, 60, 564–71.
70. F.M. Wang, M.H. Yu, C.S. Cheng, S.A. Pradanawati, S.C. Lo and J. Rick, *J. Power Sources*, 2013, 231, 18–22.
71. J. Jeon, S. Yoon, T. Park, J.J. Cho, S. Kang, Y.K. Han and H. Lee, *J. Mater. Chem.*, 2012, 22(39), 21003–8.
72. K. Abe, H. Yoshitake, T. Kitakura, T. Hattori, H. Wang and M. Yoshio, *Electrochim. Acta*, 2004, 49(26), 4613–22.
73. Y. Matsuo, K. Fumita, T. Fukutsuka, Y. Sugie, H. Koyama and K. Inoue, *J. Power Sources*, 2003, 119, 373–7.
74. M. Khasanov, E. Pazhetnov, W.C. Shin, *J. Electrochem. Soc.*, 2015, 162(9), A1892.
75. X. Wang, M. Salari, D.E. Jiang, J. Chapman Varela, B. Anasori, D.J. Wesolowski, S. Dai, M.W. Grinstaff and Y. Gogotsi, *Nat. Rev. Mater.*, 2020, 5(11), 787–808.
76. B. Wu, Y. Ren, D. Mu, C. Zhang, X. Liu, G. Yang and F. Wu, *Int. J. Electrochem. Soc.*, 2013, 8, 670–7.

77. S. Komaba, B. Kaplan, T. Ohtsuka, Y. Kataoka, N. Kumagai and H. Groult, *J. Power Sources*, 2003, 119, 378–82.
78. S.H. Lee, J.S. Kang and D. Kim, *Materials* 2018, 11(12), 2557.
79. A. Casimir, H. Zhang, O. Ogoke, J.C. Amine, J. Lu and G. Wu, *Nano Energy*, 2016, 27, 359–76.
80. G. Xu, X. Shangguan, S. Dong, X. Zhou and G. Cui, *Angew. Chem. Int. Ed.*, 2020, 59(9), 3400–15.
81. H. Zhao, X. Yu, J. Li, B. Li, H. Shaoo, L. Li and Y. Deng, *J. Mater. Chem. A*, 2019, 7(15), 8700–22.
82. Y. Zhang, J. Huang, N. Saito, Z. Zhang, L. Yang, S.I. Hirano, *Energy Storage Mater.*, 2023, 55, 364–87.
83. H. Cavers, P. Molaiyan, M. Abdollahifar, U. Lassi and A. Kwade, *Adv. Energy Mater.*, 2022, 12(23), 2200147.
84. Z. Wang, N. Dupré, L. Lajaunie, P. Moreau, J.F. Martin, L. Boutafa, S. Patoux and D. Guyomard, *J. Power Sources*, 2012, 215, 170–8.
85. Y. Lyu, X. Wu, K. Wang, Z. Feng, T. Cheng, Y. Liu, M. Wang, R. Chen, L. Xu, J. Zhou and Y. Lu, *Adv. Energy Mater.*, 2021, 11(2), 2000982.
86. L. Xia, Y. Xia and Z. Liu, *Electrochim. Acta*, 2015, 151, 429–36.
87. X.Q. Liao, F. Li, C.M. Zhang, Z.L. Yin, G.C. Liu and J.G. Yu, *Nanomaterials*, 2021, 11(3), 609.
88. Y. Zou, Y. Cheng, J. Lin, Y. Xiao, F. Ren, K. Zhou, M.S. Wang, D.Y. Wu, Y. Yang and J. Zheng, *J. Power Sources*, 2022, 532, 231372.
89. B. Jeevanantham and M.K. Shobana, *J. Energy Storage*, 2022, 54, 105353.
90. Y. Zhu, M.D. Casselman, Y. Li and A. Wei, *J. Power Sources*, 2014, 246, 184–91.
91. F. Chrétien, J. Jones, C. Damas, D. Lemordant, P. Willmann and M. Anouti, *J. Power Sources*, 2014, 248, 969–77.
92. W. Zuo, M. Luo, X. Liu, J. Wu, H. Liu, J. Li, M. Winter, R. Fu, W. Yang and Y. Yang, *Energy Environ. Sci.*, 2020, 13(12), 4450–97.
93. J. Pires, L. Timperman, A. Castets, J.S. Peña, E. Dumont, S. Levasseur, R. Dedryvère, C. Tessier and M. Anouti, *RSC Adv.*, 2015, 5(52), 42088–94.
94. N. Xu, J. Shi, G. Liu, X. Yang, J. Zheng, Z. Zhang and Y. Yang, *J. Power Sources Adv.*, 2021, 7, 100043.
95. J. Zhang, J. Wang, J. Yang and Y. NuLi, *Electrochim. Acta*, 2014, 117, 99–104.
96. Z. Zhou, Y. Ma, L. Wang, P. Zuo, X. Cheng, C. Du, G. Yin and Y. Gao, *Electrochim. Acta*, 2016, 216, 44–50.
97. Z.I. Radzi, K.H. Arifin, M.Z. Kufian, V. Balakrishnan, S.R. Raihan, N. Abd Rahim and R. Subramaniam, *J. Electroanal. Chem.*, 2022, 920, 116623.
98. T.J. Lee, J.B. Lee, T. Yoon, D. Kim, O.B. Chae, J. Jung, J. Soon, J.H. Ryu, J.J. Kim and S.M. Oh, *J. Electrochem. Soc.*, 2016, 163(6), A898.
99. M. Jo, S.H. Park and H. Lee, *Materials*, 2021, 14(16), 4670.
100. M. Cuisinier, N. Dupré and D. Guyomard, *Electrochem. Commun.*, 2014, 38, 138–41.
101. C.C. Chang, K.Y. Lee, H.Y. Lee, Y.H. Su and L.J. Her, *J. Power Sources*, 2012, 217, 524–9.
102. E.R. Logan, H. Hebecker, A. Eldesoky, A. Luscombe, M.B. Johnson and J.R. Dahn, *J. Electrochem. Soc.*, 2020, 167(13), 130543.
103. W.K. Behl and D.T. Chin, *J. Electrochem. Soc.*, 1988, 135, 16.
104. W.K. Behl, *J. Electrochem. Soc.*, 1989, 136, 2305.
105. M.N.W. Golovin, P. David, J.T. Dudley, D. Holonko and S. Woo, *J. Electrochem. Soc.*, 1992, 139, 5.
106. J.R. Dahn, J. Jiang, L.M. Moshurchak, M.D. Fleischauer, C. Buhrmester and L.J. Krause, *J. Electrochem. Soc.*, 2005, 152, A1283.

107. Z. Zhang, L. Zhang, J.A. Schlueter, P.C. Redfern, L. Curtiss and K. Amine, *J. Power Sources*, 2010, 195, 4957.
108. G.B. Han, M.H. Ryou, K.Y. Cho, Y.M. Lee and J.K. Park, *J. Power Sources*, 2010, 195(11), 3709–14.
109. Z. Chen, Q. Wang and K. Amine, *J. Electrochem. Soc.*, 2006, 153, A2215.
110. X.M. Wang, E. Yasukawa and S. Kasuya, *J. Electrochem. Soc.*, 2001, 148, A1058.
111. Y.E. Hyung, D.R. Vissers and K. Amine, *J. Power Sources*, 2003, 119, 383.
112. D. Gao, J.B. Xu, M. Lin, Q. Xu, C.F. Ma and H.F. Xiang, *RSC Adv.*, 2015, 5, 17566.
113. M.S. Ding, K. Xu and T.R. Jow, *J. Electrochem. Soc.*, 2002, 149, A1489.
114. S.S. Zhang, K. Xu and T.R. Jow, *J. Power Sources*, 2003, 113, 166.
115. Z. Zeng, X. Jiang, B. Wu, L. Xiao, X. Ai, H. Yang and Y. Cao, *Electrochim. Acta*, 2014, 129, 300.

4 Designing a Battery's Electrolyte through Simulation

Anoop Kumar Kushwaha and Saroj Kumar Nayak

4.1 INTRODUCTION

Electrolyte is a crucial component of batteries, impacting their efficiency, performance, and safety. Electrolyte acts as an ionic conductor and an electronic insulator. It controls how much and how quickly the battery can be discharged or charged, whereas electrode materials determine energy density. In general, faradaic reactions occur at the electrodes, while the electrolyte remains an inert element. However, electrolytes interact directly with both electrodes and may even participate in the cell reactions as in the case of the lead-acid battery. The active materials are strongly oxidizing/reducing and, therefore, electrolyte should be electrochemically stable against both electrodes. Put differently, the success of a battery depends on the electrolyte's stability and compatibility with the electrodes (i.e., electrolyte-electrode interface) [1–3].

Recent years have witnessed a huge rise in demand for high-voltage and high-capacity batteries for application in electric vehicles and smart grids. In response to this demand, new and efficient electrode materials are being explored, which in turn drives the development of compatible electrolytes [4–6]. Electrolytes may be classified into two types: solid-state and liquid electrolytes. The latter may be further classified as aqueous, non-aqueous, and ionic liquids. The chemistry of liquid electrolytes differs from that of solid electrolytes [7]. Salts [8], solvents [9], and additives [10] constitute electrolytes.

Efficient electrolytes must satisfy several criteria. They must (i) be good conductors of ions and an insulator of electrons, allowing easy ion transport and limiting self-discharge; (ii) have a wide electrochemical window to avoid degradation within the operating potentials of the electrodes; (iii) be inactive towards other battery components (electrodes, separator, casing, binders); (iv) be resistant to thermal, mechanical, and electrical stress; and (v) be composed of non-hazardous and environment-friendly materials.

Conventional methods for electrolyte discovery and optimization rely primarily on chemical intuition and trial-and-error experimentation. They, however, can be highly expensive and time-consuming. Modern frameworks relying on atomistic simulation (based on quantum and classical theories) can predict and design efficient electrolytes at a low cost and in a short time. The simulation is effective because it can mimic real electrolytes. These techniques require fundamental knowledge, at the

DOI: 10.1201/9781003441182-5

atomic scale, of the structure and properties (electronic, physical, and chemical) of constituent electrolyte materials. In this chapter, we discuss parameters estimated by simulation (parameters required for designing electrolytes) and computational methodologies. Prior to overview, we discuss electrolyte structure that is required for modeling [11].

4.1.1 Electrolyte Solvation Structures

The transport of cations in electrolytes provides estimation of the rate at which energy is transferred and accumulated within the electrodes. Cation transport is controlled by a two-step process: (i) formation of a solvation shell around the cation, and (ii) migration of the solvation shell. The solvation structure strongly affects the electrochemical properties (including ion transport and redox potential) of electrolytes. Particularly, cation-solvent and cation-anion interactions control these properties. The first-solvation shell contains cation-linked solvent molecules, while the second-solvation shell interacts with them. During ionic transport between electrodes, the solvation number (the number of solvent molecules surrounding the cation) changes. Solvation numbers are generally lower at the interface region and higher in the bulk electrolyte. For example, Li^+ interacts with one or two ethylene carbonate (EC) molecules on the interface, but it is coordinated by three to six EC molecules in the bulk, the preferred solvation number being four [12].

Figure 4.1a depicts a schematic diagram of an LIB that shows the solvation shells in the electrolyte [13, 15]. IR and Raman spectroscopy validate the formation of solvation shells but cannot explain it at the atomic level. A theoretical simulation based on density functional theory (DFT) explains how ions and solvents interact at the atomic level within the solvation shell, whereas *ab initio* molecular dynamics (AIMD) shows how a solvation shell moves inside the electrolyte. The exchange dynamics (how long a solvation shell resides) of a solvation shell can be studied by molecular dynamics (MD) simulations. The exchange is dynamic while the rigidity of the solvation shell determines the ease with which the structure can break. Thus, the mobility of ions depends on the rigidity of the solvation shell. Furthermore, due to weak bonding with the cation, fast exchange dynamics invoke poor rigidity of the solvation shell [11, 16–18].

4.1.2 Electrode-Electrolyte Interfacial Structures

Most of the chemical and electrochemical reactions, such as solvation shell formation, electrolyte degradation, and (de)solvation of cations, take place at the electrode-electrolyte interface. Figure 4.1b depicts cathode-electrolyte (CEI) and solid-electrolyte (SEI) interphases. The cations (e.g., Li^+) are desolvated on the SEI and solvated on the CEI. In general, the interphases permit cations to pass while blocking electrons, allowing the electrochemical reaction to continue while stopping electrolyte decomposition. The interphases strongly impact the battery's rate performance, lifetime, cycle life, and safety [19]. A prior knowledge of the interphases facilitates the design of high-performance, robust, durable, and efficient batteries. A microscopic understanding of electrode-electrolyte interfaces offers insight into the thermodynamics and kinetics of interfacial reactions.

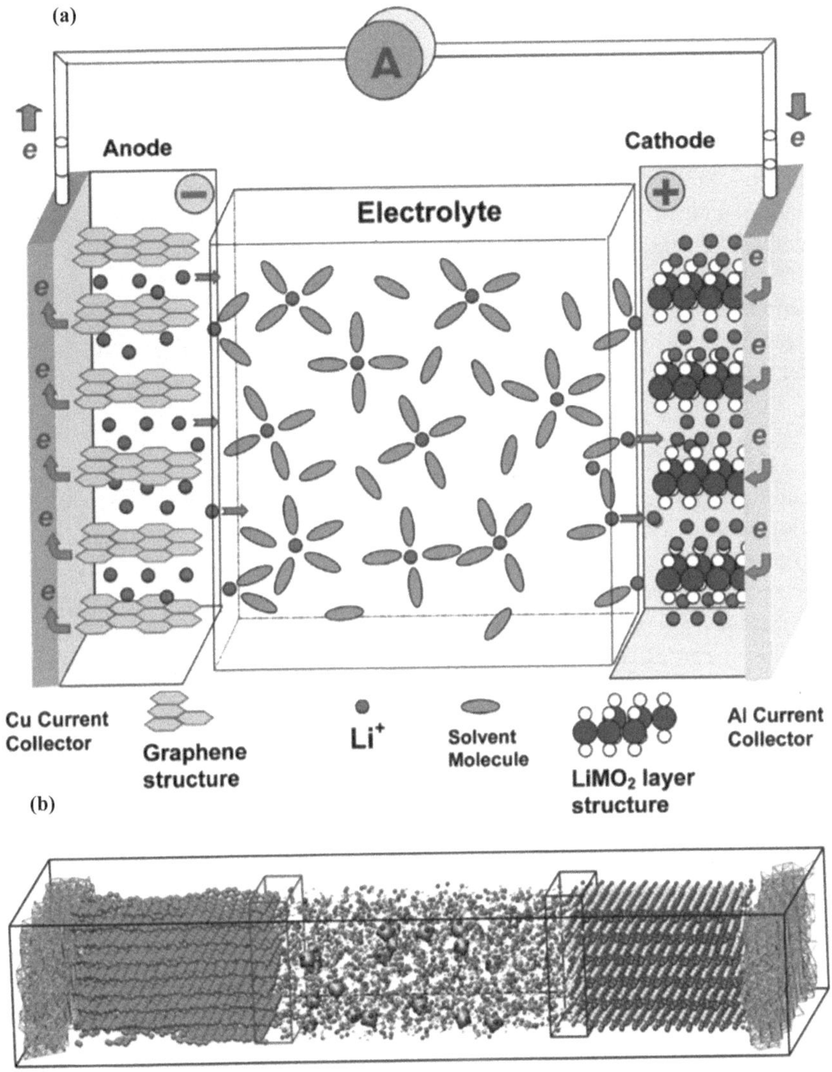

FIGURE 4.1 (a) Schematic diagram and (b) atomic-scale representation of LIB with associated electrolyte-electrode interfaces. During charging, oxidation and reduction take place in right and left boxes, respectively.

Source: Figures are reproduced from permission of Ref [13] Copyright (2004) American Chemical Society and from Ref [14] Copyright (2017) American Chemical Society.

Experimental characterization and mathematical modeling of complex interfacial structures are still challenging. For example, to comprehend the SEI, we must study each stage of its formation using simulation. Generally, SEI is formed in three steps: (i) reduction of solvents/anion, (ii) growth of the reaction product into SEI, and (iii) deposition of SEI layers. The first step could not be observed through experiments, whereas DFT, AIMD, and MD simulations effectively explain electrolyte reduction on electrodes. The second step, i.e., the growth of the reaction product to form SEI, is rarely studied due to the complexity of both experiment and simulation. However, *in-situ* experiments have explored the compositional and morphological evolution during SEI growth, providing limited insight into SEI structure creation.

In recent years, the multi-scale approaches, for example, Monte Carlo (MC) [20, 21] and Red Moon (RM) [22, 23], have evolved to bridge the gap between simulation (reduction of electrolyte) and experiment (morphology of SEI) [24]. At the interface, the electrolyte's thermodynamic stability is governed by the electrode's Fermi level and the electrolyte's orbital energies [i.e., highest occupied (HOMO) and lowest unoccupied molecular orbital (LUMO)]. More precisely, the electrolyte is reduced if its LUMO is lower compared to anode's Fermi level and is oxidized if its HOMO is greater than the cathode's Fermi level.

4.1.3 Atomic-Scale Electrolyte Structure

Conventionally, electrolytes have been developed by trial and error. With the advancement of computational facilities and robust algorithms, electrolytes are now designed by simulation. For the electrolyte, simulations efficiently obtain a structure-property relation, strongly correlatable with experiments. Their prediction efficiency, however, is highly dependent on the atomic structure and electrochemical properties of electrolytes. Furthermore, because of a lack of knowledge of structure-property-performance relations, new strategies are found wanting in designing desirable electrolytes.

The relationship between battery performance and electrolyte properties is far more complex than the relationship between structures and properties. Because battery performance is affected by factors other than electrolyte properties, atomic-scale simulations of electrolytes only help to compare their performances and assist in selecting and anticipating efficient ones. Certain parameters such as redox potentials, electronegativity, (de)solvation energy, ionic diffusivity, and conductivity are used to predict efficient electrolytes.

4.2 A BRIEF OVERVIEW OF THEORETICAL METHODOLOGY

Several computational modeling and theoretical strategies have been established to capture specific properties of electrolytes over a wide time scale and length range. Figure 4.2 depicts the theoretical model for calculations and various stages of reactions in the electrolytes. DFT, AIMD, MD, and solvation models are widely used to study ionic interactions and electrode-electrolyte interfaces at the atomic scale. These methodologies are summarized below:

4.2.1 Density Functional Theory (DFT)

DFT is a well-established and important technique for designing and developing efficient electrolytes. Within the DFT framework, the electronic properties of molecular structures are investigated through solving a many-body Schrödinger equation.

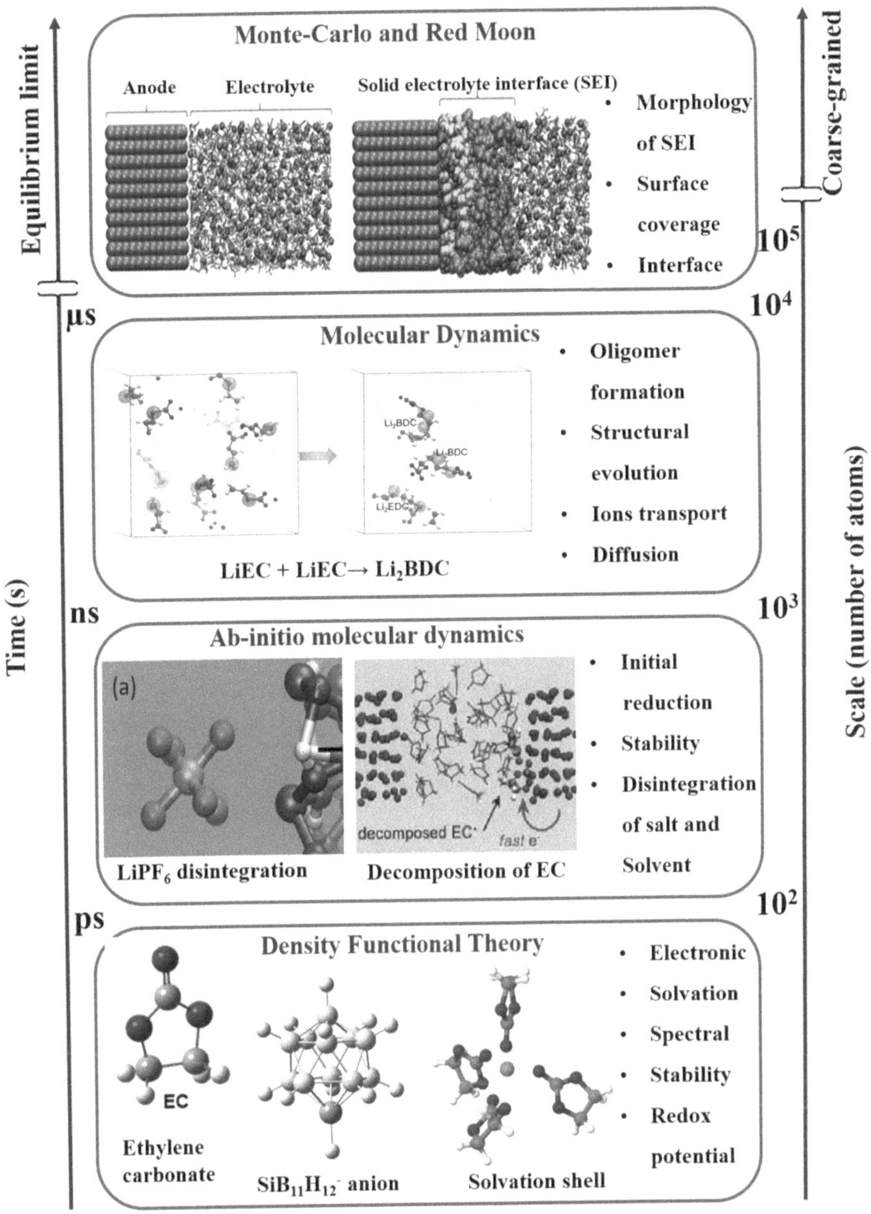

FIGURE 4.2 Overview of theoretical simulation methods for analyzing the electrolyte and electrode-electrolyte interfaces. The applications of various theoretical techniques such as DFT [25, 26], AIMD [27, 28], MD [29], Red-Moon (RD) [27], and Monte-Carlo (MC) methods in the context of electrolyte development are also shown.

In this formulism, the Hamiltonian is operated on the wave function to obtain energy eigenvalues.

A few approximations simplify the Schrödinger equation and reduce computational cost without impacting accuracy. For example, Born-Oppenheimer approximation neglects nuclei's kinetic energy and nuclei-nuclei repulsion terms to simplify the molecular Hamiltonian. Moreover, the electron-nuclei Coulombic interaction was changed to external potential to incorporate all environmental effects. In this formulism, electron density is a key parameter rather than a wave function.

The Hohenberg-Kohn theorems and Kohn-Sham method transform the many-body problem to many-one-body by including an exchange-correlation term. This term is determined by certain approximation methods, e.g., local density approximation (LDA), generalized gradient approximation (GGA), and hybrid functional. The computational cost and accuracy of calculations are balanced by Jacob's ladder. To enhance accuracy and reduce cost, the electrons are divided into inner core and valence electrons. The inner shell electrons are strongly bound, whereas the electrons in the valence shell carry all chemical properties. Therefore, inner shell electrons are ignored by reducing an atom to an ionic core where they interact with the valence electrons.

The valence shell electrons are defined by the basis set derived from Slater- and Gaussian-type functions. The potential felt by the valence electron is introduced by the pseudopotential. Several books and articles can assist in the understanding of DFT [30–35] in the context of electrolyte development [36]. This method is used to explore electrolyte reaction pathways, to understand the additive's role in SEI formation, and to design novel electrolytes where no information is available through experiments.

4.2.2 Solvent Models (SM)

Most reactions in aqueous electrolytes such as salt-solvent interactions and electrolyte-electrode interactions occur in the fluid medium. This medium could be incorporated into DFT through a suitable solvent model. These models generate a realistic aqueous electrolyte medium, which improves prediction and understanding of chemical reactions in the solvated phase. The solvent models are classified into two types based on their physical appearance, namely implicit and explicit.

The explicit solvent model specifically considers solvent molecules, whereas continuum solvent media are used in the implicit solvent model. The explicit solvent model provided a realistic description of salt-solvent or cation-solvent interactions and was thus employed to study the formation of solvation-shell around various cations (Li^+, Na^+). Although this model gives details of solvation structure, its application is limited due to the high computational cost.

In the implicit solvent model, a homogeneous polarizable medium depicts the continuum solvent distribution. Because this model employs a thermally averaged isotropic solvent, only a few parameters are required to accurately represent it. In general, the dielectric constant (ε) defines the continuum of solvent media. Water, EC, and acetonitrile—commonly used to define solvent media—have dielectric constants of ~78, ~80, and ~35, respectively. Because it does not use molecular coordinates, this solvent model is computationally less expensive than the explicit solvent model.

Various implicit solvent models—e.g., polarized continuum model (PCM), conductor-like PCM (C-PCM), and universal solvent model (SMD)—have been developed in recent years. In the PCM, the cavity-embedded solute is referred to as polarizable charge distribution, whereas dielectric media, representing the solvent, is polarized by the solute. The charge distributions and geometrical shapes of the solute are affected by the induced charge on the solvent. The induced charge could be obtained by solving the non-homogeneous Poisson equation using the self-consistent method, which provides the solvent (or Onsager) reaction field. These reaction fields are mathematically defined by effective potential.

Because PCM includes only electrostatic solute-solvent interactions, an integral equation formalism (IEF) version of PCM (IEF-PCM) was developed to include other interaction terms. Solvation models provide more accurate energies (including free energy) to compute associated properties by embedding (or coupling) with DFT. We would like to recommend a few books and articles that can help to understand the solvation models [37–40]. Several studies have investigated electrolytes using the solvation model [41–43]. For example, Shao et al. [44, 45] found a high electrochemical window for sulfone-based high-voltage electrolytes using the PCM model, which was later confirmed by experiments.

4.2.3 *Ab Initio* Molecular Dynamics (AIMD)

AIMD, based on quantum mechanical theory (especially DFT), is widely used to simulate the motion (dynamics) of atoms and molecules. In AIMD simulation, atomic interactions are calculated directly from electronic structure (as determined by DFT), rather than using empirical potentials or force fields. DFT is used in the AIMD framework to calculate a system's total energy at each time step. Forces acting on each atom are determined through energy gradient. The atom's trajectory is then predicted using equations of motion.

With the inclusion of the polarization effect, the AIMD efficiently determines bond breaking/forming mechanisms. Thus, AIMD simulations are widely employed to investigate structural, electronic, thermal, physical, chemical, and dynamical properties of materials. AIMD is especially useful for systems that cannot be described using empirical potentials (or force fields) such as strongly electron-correlated systems or systems with many degrees of freedom. AIMD simulations are computationally demanding and require high computational resources.

The Car-Parrinello molecular dynamics (CPMD) method is a popular method for studying AIMD. In this method, the core and valence electrons are described by a pseudopotential and a plane wave basis set. The CPMD employs fictional dynamics to hold electrons near to the ground states, thereby reducing self-consistent iterative minimization in every step. References [46] and [47] are recommended to understand AIMD. In recent years, AIMD has been used to study the structural stability of electrolytes at high temperatures and during the formation of SEI [11, 48].

4.2.4 Molecular Dynamics

The physical movement of atoms (or molecules) is studied using a molecular dynamics (MD) simulation based on Newton's equation of motion. This technique allows atoms/molecules to interact for a certain amount of time, resulting in a dynamical evolution of the system. The numerical solution of Newton's equation of motion determines the trajectories for interacting atoms/molecules in a system. This interaction between atoms is simulated by using an appropriate interatomic potential or force field that carries information about the acting force between atoms and the potential energies of the system.

MD simulation uses numerical methods to determine the dynamical properties of complex-molecular systems that cannot be achieved analytically. To begin MD simulation, the initial and boundary conditions, interatomic potential or force field, and integrator/ensemble must be specified. Isolated (IBC) and periodic (PBC) boundary conditions are used in clusters/molecules and solid systems, respectively. Newton's equations of motion, which are second-order ordinary differential equations, are solved by providing initial conditions such as atom positions and velocities. The interatomic potentials—such as pair potential (e.g., Lennard-Jones potential), repulsive potential, and Stillinger-Weber potential—are used in MD simulations to define the interaction between atoms with force-field incorporation. Force fields are a collection of parameters used to describe physical interactions between atoms through a given energy expression.

The ensemble and integrator are grouped together because several methods based on time-integration create an appropriate ensemble distribution. These approaches modify the equation of motion to generate a specific dynamic trajectory that creates a cloud in phase space with appropriate density distribution. As a result, the time-average of the single-point operator approaches the thermodynamic average in one such trajectory. Although MD simulation has many applications, the main issue is accuracy (error caused by the interatomic potential) and efficiency. To assist the reader in understanding MD simulations, we recommend references [49] and [50]. MD simulation is employed in electrolytes to determine the cation's dynamics and transport, salt concentration, and interfacial evolution. Monte-Carlo [51] and Red-Moon [23] methods are now used to simulate the morphology of electrode-electrolyte interfaces and formation of SEI [24].

4.3 PARAMETERS AND KEY PROPERTIES

Simulation is used to design and predict efficient electrolytes by investigating various parameters and properties. For example, structural stability could be investigated by analyzing solvation energy, redox potential, electronic structure, and the energy fluctuation of the total system. Here, we discuss key properties and parameters that provide a qualitative roadmap for developing efficient electrolytes.

4.3.1 Electronic Structure

4.3.1.1 Electronic and Orbital Energies

The investigation of electronic properties [e.g., electron affinity (EA), ionization potential (IP), vertical detachment energy (VDE), electronegativity, charge transfer, and electron density distribution] is the primary and essential step for distinguishing an electrolyte molecule from a sea of molecules. Electronic properties are crucial in determining stability and electrolytic performance. For example, a high VDE [$=E$(neutral, at anion geometry) – E(anion)] and a high EA ($=E$(neutral) – E(anion)) of $LiCB_{11}H_{12}$ (salt) indicate a strong ability to bind an extra electron, resulting in robust electronic stability. The difference between EA and VDE measures the cluster's relaxation when an electron is photo-detached from an anion. A small difference between EA and VDE (~0.5 eV) shows a quick relaxation of structure ($LiCB_{11}H_{12}$) without any significant structural distortion during an electron's photo-detachment. The structure of $B_{11}H_{12}^{2-}$ is given in Figure 4.3a.

A high value for second electron affinity [SEA = E(nion) − E(dianion)] indicates strong structural and electronic stability, making $B_{12}(CN_{12})^{2-}$ an appropriate electrolyte for lithium- and magnesium-ion batteries [52]. The third electron affinity [TEA = E(dianion) − E(trianion)) and VDE (=E(dianion, at trianion geometry) − E(trianion)] is found to be high for $BeB_{11}(CN_{12})$, indicating that it could be used as a halogen-free electrolyte in aluminum-ion batteries [53]. These findings indicate that determining EA and VDE is a critical step in identifying an electrolyte molecule. A chemically stable and less-reactive molecule has a high electronegativity [$\chi = ((-E_{HOMO}) + (-E_{LUMO}))/2$] and a large HOMO–LUMO gap (E_{gap}).

The orbital energies, particularly the HOMO (E_{HOMO}) and LUMO (E_{LUMO}) energies, as well as the HOMO–LUMO energy gap (ΔE_{gap}), determine the reactivity, structural stability, and other properties of salts and solvents at the atomic level. An efficient electrolyte (both salt and solvent) should be chemically/structurally stable, avoid unwanted chemical reactions, and be compatible with electrodes, which can be determined through orbital energy analysis. For example, an eligible electrolyte should have a broad potential window with HOMO lower than the cathode's Fermi level (μ_C) and LUMO higher than the anode's Fermi level (μ_A). Figure 4.3c depicts a schematic diagram of the HOMO and LUMO distributions. A stable SEI mitigates these issues by preventing electron movement between electrode and electrolyte, although it reduces the battery's performance. Thus, the electrolyte should be compatible with both the anode and the cathode. Orbital energies calculated using DFT-based simulations can be used to examine this compatibility.

4.3.1.2 Electrochemical Stability Windows

The electrochemical stability window (ESW), directly connected to orbital energies, is the voltage span in which electrolytes are neither reduced nor oxidized. The ESW ($V_{AL} - V_{CL}$) of an electrolyte is determined by its reduction potential (or anodic limiting potential, V_{AL}) and oxidation potential (cathodic limiting potential, V_{CL}). The V_{AL} (=E_{HOMO}/e) and V_{CL} (=E_{LUMO}/e) are ascertained through HOMO and LUMO, respectively, for single electron transfer. A large ESW indicates that the electrolyte is chemically stable at the high voltage. For example, $CB_{11}H_{12}^-$ (9.82 V)- and $SiB_{11}H_{12}^-$ (9.84 V)-based electrolytes possess high ESWs larger than those of commercial electrolytes such as $TFSI^-$ (~4.5 V), TFO^- (~4.1 V), BF_4^- (6.1 V), and PF_6^- (~7.1 V) that have been applied in high-voltage batteries [25, 54, 55].

Thus, investigating ESW is also an essential step in designing and predicting high-voltage electrolytes. The origin of robust stability of an electrolyte could also be determined by orbital energies and their molecular orbital distribution. For example, low-lying E_{HOMO}, which arises due to weak orbital interactions between bonding and antibonding MOs, creates a large HOMO–LUMO gap in $B_{11}(BO)_{12}^{2-}$ in comparison to a $B_{12}(CN)_{12}^{2-}$ ion-based salt in a magnesium-ion battery [56]. Figure 4.3c is a schematic of the HOMO and LUMO distributions for $B_{12}(CN)_{12}^{2-}$.

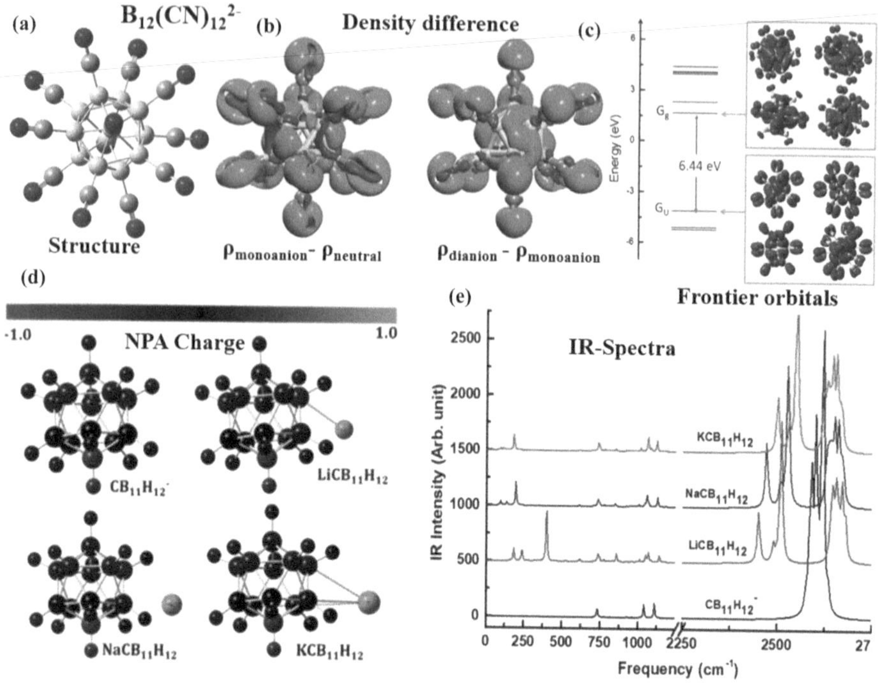

FIGURE 4.3 (a) Optimized structure; (b) electron density difference; (c) frontier orbitals and orbital energy levels of $B_{12}(CN)_{12}^{2-}$ dianions; (d) mapping of NPA charge; and (e) IR spectra of $MCB_{11}H_{12}$ (M = Li, Na, K).

Source: (a)–(c) Reproduced with permission from [52]. Copyright 2016 Wiley-VCH. (d) and (e) Reproduced with permission from [25]. Copyright 2022 American Chemical Society.

4.3.1.3 Electron and Charge Density Distributions

The atomic-scale analysis of electron/charge density distribution determines the ability of binding electrons and the chemical and structural stability of salts and solvents. The electron density difference between different charge states of electrolyte molecules (salts and solvents) is represented by the electron deformation density (EDD), which is used to analyze the excess electron distribution. For example, the EDD of $B_{12}(CN_{12})^{2-}$ anion (a constituent of $MgB_{12}(CN_{12})^{2-}$ salt) is calculated by the density difference of the neutral state to monoanion and the monoanion to dianion, which reveals that excess electrons were mainly distributed on the ligand (CN) [52]. Figure 4.3b depicts the EDD of $B_{12}(CN_{12})^{2-}$.

To be more precise, each –CN pair gains 0.08 and 0.12 electrons, whereas cage-B gains only 0.02 and 0.03 electrons by adding the first and second electrons. This arises due to the high electronegativity of –CN. Furthermore, charge/electron distribution (shown in Figure 4.3d) has been used to analyze borane-based salts such as $LiCB_{11}H_{12}$ and $LiSiB_{11}H_{12}$ [25]. In this case, C (~−0.66) and Si (~+1.05) carry negative and positive charges, respectively, altering charge distributions on the B-cage even though the B atoms are linked to Li^+. The more negatively charged

B-atom of $LiSiB_{11}H_{12}$ binds Li^+ more strongly than $LiCB_{11}H_{12}$. Thus, the charge density distribution calculated by DFT determines the bonding nature and strength of the electrolyte.

The bonding nature and strength in an electrolyte are more precisely investigated by analyzing parameters (e.g., electron density (ρ), Laplacian of electron density ($\nabla 2\rho$), energy density $Ed(r)$, and Lagrangian kinetic energy $G(r)$, etc.) at bond critical points (BCPs). These properties are calculated through the quantum theory of atom in molecule (QTAIM) method, which is embedded in Multiwfn [57]. Boggs criteria ($\rho > 0.1$ and $\nabla^2\rho < 0$ at BCPs) determine the existence of covalent bonding. Using this criteria, Ghanty et al. [58] observed the covalent bond between the B_{12}-cage and ligands (X = –C≡H, –C≡C–CN, –C≡C–BO) in the formation of $B_{12}X_{12}$ [58]. This method also indicates the covalent bond between B–B atoms inside the B_{12}-cage. The mapping of $\nabla^2\rho$ and the electron localization function (ELF) also captures the bonding nature of the salt and solvent.

4.3.1.4 Vibrational Spectra

Vibrational spectra, such as IR and Raman, are generally used to analyze doping effects in an electrolyte at the atomic scale. For example, fluorination of a carbonate-based solvent shows a blue shift (i.e., towards higher frequencies) in the stretching frequency associated with the most affected C=O bond [59]. Thus, fluorination enhances the bonding strength of C=O while decreasing the associated bond length, resulting in a reduction in the bonding strength between the fluorinated carbonate and Li^+, which is required for efficient solvent.

Similar to the solvent molecules, the IR spectra (Figure 4.3e) of salts such as $B_{12}H_{12}$, $CB_{11}H_{12}$, and $SiB_{11}H_{12}$ shows that the peak associated with the B–H bond is the most affected (split) by the coordinating Li^+ cation. A red shift in M^+–B bond is also observed by its movement toward Li (397.8) → Na (195.95) → K (181.33) in $MCB_{11}H_{12}$ salt (M = Li, Na, K). The red shift represents an enhancement in bond length and a reduction in binding energy, which was verified by structural analysis. Thus, vibrational spectral analysis captures changes in the electronic properties after doping in salt and solvent.

4.3.1.5 Band Structure and Density of State

In solid-state electrolytes, electronic properties are studied by analyzing the band structure and density of states (DOS) [including projected DOS (PDOS) and localized DOS (LDOS)]. The band structure determines the energy levels of electrolytes that can be occupied by electrons, and their energy band gap is directly associated with electronic conductivity. The number of electron states per unit volume per unit of energy is analyzed by DOS. For example, Li_2PO_2N solid-state electrolytes based on lithium phosphorous oxy-nitride (LiPON; composition: $Li_xPO_yN_z$, where $x = 2y + 3z - 5$) have been reported as efficient electrolytes [60]. Their band structure analysis found a flat nature to the valence bond, composed primarily of 2p-π state of N and spatial separation forming non-dispersive bands. The band gap of Li_2PO_2N is >6 eV, which makes it a highly stable, solid-state electrolyte with high ionic conductivity.

4.3.2 Solvation and Dissociation

4.3.2.1 Dissociation Energy of the Electrolyte Salt

The dissociation energy [$E_d = E_{AM} - (E_{A-} + E_{M+})$] is the energy released when an anion (A^-) and an alkali cation ($M^+ = Li^+, Na^+, K^+$) separate from a salt (AM). This is an important property to simulate in order to find a suitable salt. Dissociation energy should be small for efficient salts so that the anion and the alkali cation can be easily detached. For example, $LiPF_6$ (5.73 eV), used in commercial LIBs, has a lower dissociation energy than $LiBF_4$ (6.08), $LiBH_4$ (6.62 eV), $LiNO_3$ (6.53 eV), and $LiBO_2$ (6.36 eV) [61]. However, the dissociation energies of $LiCB_{11}H_{12}$ (5.09 eV) and $LiSiB_{11}H_{12}$ (5.17 eV) are lower than that of $LiPF_6$, which suggests that borane-based electrolytes are potential candidates for salts in LIBs [25].

Similar to the salt, the binding between the solvent molecule and the alkali cation should be weaker. For example, EC has a binding energy with Li^+ of ~2.20 eV, which can be reduced to ~0.6 eV by fluorination [59]. This is one of the major reasons for using fluorinated EC as a solvent instead of EC. Generally, fluorination reduces the interaction between cations and carbonate-based solvents.

4.3.2.2 Solvation and De-Solvation Energy

The cations in the electrolyte travel between the electrodes as solvated ions. The first solvation-shell refers to the solvent molecules directly linked with the cation. The molecules that directly interact with the first solvation-shell are classified as the second solvation shell, and so on. During transport between electrodes, the solvation numbers of M^+ change and become smaller at the electrolyte-electrode interface. For example, Li^+ ions in general are surrounded by four EC molecules in the first solvation shell, which gets reduced to one or two at the interface. The cations undergo a specific type of interaction with the concentric shell during the solvation process. The type and strength of this interaction is investigated with DFT-based simulation, which gives vital information on reactivity, solubility, viscosity, density, and solvation and de-solvation energies. The solvation (ΔE_{solv}) and de-solvation (ΔE_{desolv}) energies characterize the interaction and disintegration of the cation with the solvent molecule in the solvation shell. These energies for Li^+ surrounded by different numbers (n) of solvent molecules (S) are calculated by using the following equations:

$$\Delta E_{solv} = E_{total}\left[Li^+(S)_n\right] - E_{total}\left[Li^+\right] - n \times E_{total}\left[(S)_n\right]$$

$$\Delta E_{desolv} = E_{total}\left[Li(S)_n\right] - E_{total}\left[Li\right] - n \times E_{total}\left[Li(S)_{n+1}\right]$$

An efficient solvent should have high solvation energy and low de-solvation energy. Based on this, EC is a superior solvent than propylene carbonate [62]. However, fluorinated EC shows a contrasting behavior as supported by experiments [26]. For a realistic approach, MD simulation is used to understand the solvation of cations in pristine solvent and in mixtures of solvents. This method provides the solvation behavior as well as the diffusivity of cations [16, 63].

4.3.2.3 Gibbs Free Energy and Redox Potential

The redox potential determines whether a solvent molecule or salt donates or accepts an electron in electrolytic solutions. A solution with a high oxidation potential will gain electrons from the solute, whereas the solution with a small reduction potential will lose electrons to the solute. The redox potential of the electrolyte molecule in the solvent is calculated by the Born-Haber thermodynamic cycle. DFT-based simulation and a solvation model provide the Gibbs free energy (ΔG) of the electrolyte molecule. In order to get oxidation potential on the Li^+/Li scale, the absolute oxidation potential is subtracted from 1.37 eV, as given below:

$$E_{ox}\left(vs.\frac{Li}{Li^+}\right)=\frac{\Delta G_{ox}}{F}-1.37=\left[\Delta G_e+\Delta G_{sol}\left(A^+\right)-\Delta G_{sol}\left(A^+\right)\right]/F-1.37$$

Investigations on oxidation potentials have shown that $B_{12}(C{\equiv}C{-}BO)_{12}^{2-}$ and $B_{12}(C{\equiv}C{-}CN)_{12}^{2-}$ could be effective anions in Li- and Mg-ion batteries [58]. The same methodology also found that borane ($B_{12}H_{12}$) and its modified version can be highly efficient high-voltage electrolytes in alkali and alkaline-earth metal ion batteries.

4.3.2.4 Coordination of Cation with Solvent Molecule

The interaction strength between alkali cations and solvent molecules is also a crucial parameter for designing electrolytes. To determine the composition of the lithium solvation shell, the binding energy of Li^+-solvent is required. The binding energy should neither be high nor low and must be in a compatible range. Analysis of binding energy aids in the selection of solvents. Doped or engineered molecules could be easily examined for their application as solvents by analyzing their binding energy. For example, Kushwaha et al. [59] have compared the binding energies of Li^+ ions with pristine and fluorinated carbonates and found that the latter are superior to the pristine carbonate, which is in good agreement with experiments.

4.3.3 Ion/Molecule Transport Kinetics (Diffusivity and Conductivity)

Ion transport, a key property of the electrolyte, is used to study the electrolyte's ionic conductive and electronic insulative behavior. Low-temperature and fast-charging batteries require electrolytes with excellent ion transport. Ion transport determines the ionic diffusivity, conductivity, and transference number of the electrolyte. The study of the ionic diffusion process occurring at the atomic scale within the battery is extremely instructive. The ionic diffusivity is crucial to the performance of the electrode-electrolyte and is strongly related to the battery's efficiency in terms of cycling performance and structural stability. Generally, experimental studies measure ion transport by analyzing various parameters; however, macroscopic properties and micro-mechanisms do not correlate. To unveil the limitations and rationally design an advanced electrolyte, a deep understanding of ion transport is needed. Simulations based on DFT and MD are widely employed to understand ion transport mechanisms in batteries. Ion transport mechanisms are different in liquid and solid-state electrolytes.

4.3.3.1 Ionic Diffusion in Liquid Electrolytes

The electric field applies a driving force for the movement of ions. Generally, transport mechanisms in liquid electrolytes are categorized as vehicular and structural diffusion. Vehicular diffusion implies the concerted diffusion of molecules/ions in the form of a solvation shell, whereas in the structural (or ion-hopping) diffusion, ions jump from one solvation shell to another.

The two diffusion modes are mixed in practical situations. Figure 4.4 illustrates the two modes of diffusion. Ion transport processes in liquid electrolytes involve both de-solvation and re-solvation processes. It is not simple to develop an exact diffusion mechanism with the inclusion of the two diffusion mechanisms. However, MD simulations can qualitatively capture ionic diffusion in a liquid electrolyte.

For example, Kankanamge et al. [66] used MD simulation and IR spectral analysis to study ionic diffusion in an electrolyte of LiTFSI in a carbonate solvent. They showed that coordination with and de-coordination from Li^+ ions occur in the time scale of picoseconds, which corresponds to the structural diffusion mechanism. However, Borodin et al. [67] showed that vehicular diffusion occurred with the same salt in another solvent. They found that the movement of water-coordinated Li^+ ions was faster than that of TFSI-coordinated Li^+ ions. The solvent-assisted diffusion mechanism explains the faster Li^+ diffusion in a cyanogen-containing electrolyte, as studied by AIMD. In this mechanism, Li^+ is separated from the salt and diffused with the solvent molecule (see Figure 4.4) without following the Stokes-Einstein equation [65]. For a detailed understanding of ionic diffusion in liquid electrolytes, the reader is directed to reference [11].

4.3.3.2 Ionic Diffusion in Solid-State Electrolytes

The slower lithium diffusion in solid-state electrolytes leads to slower charging and discharging compared to those in liquid electrolytes. Moreover, poor contact between the electrode and solid-state electrolyte due to rigidity restricts large active sites and increases interfacial resistance. Hence, ionic diffusion must be studied to design novel, high-performance solid electrolytes. DFT-based simulation is commonly used to investigate the migration-path, barrier energy, migration energy, and ionic conductivity (σ) in solid electrolytes at the atomic scale. Fast ion conductors in LIBs are composed of disordered (consisting of movable ions) and rigid (consisting of immobile ions) sublattices. Fast ion conductors possess disordered ion sublattices, whereas rigid ones provide continuous ionic migration pathways with low potential barriers. Thus, the ionic conductivity in solid electrolytes depends not only on the crystal structure but also on other factors. The conductivity is determined by the concentration of mobile ion sites (n_c), mobile ion charge (Z_e), and migration activation energy (E_a), according to the following equation:

$$\sigma = n_c Z_e \mu = \frac{A}{T} n_c \exp\left(-\frac{E_a}{k_B T}\right)$$

where T is the absolute temperature, k_B is the Boltzmann constant, and A is a crystal structure constant.

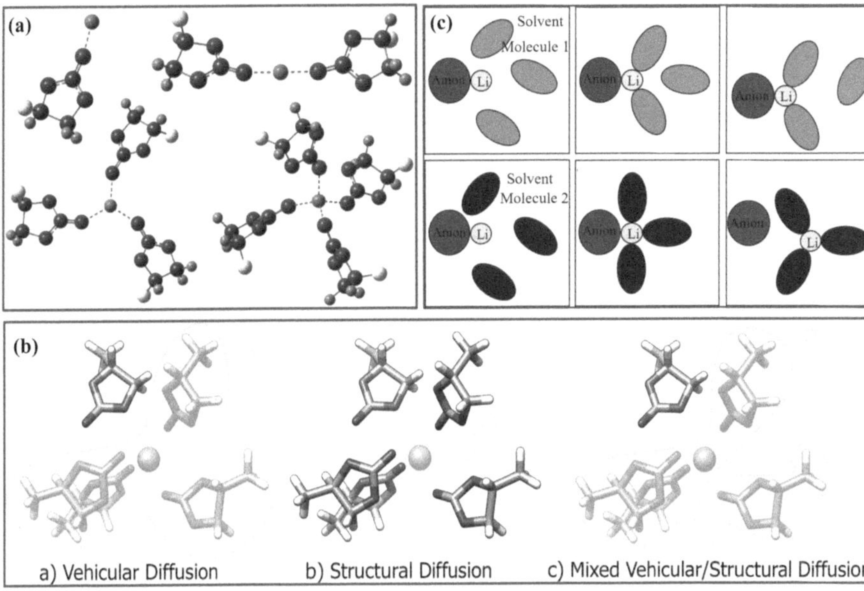

FIGURE 4.4 (a) Solvation shell of ethylene carbonate (EC) around Li^+ cation; (b) schematic representation of structural, vehicular, and mix diffusion mechanisms; (c) schematic representation of (bottom) solvent-assisted Li^+ diffusion and (top) Li^+-salt diffusion in different solvents.

Source: (a) Reproduced with permission from [26]. Copyright 2021 Springer. (b) Reproduced with permission from [64]. Copyright 2019 American Chemical Society. (c) Reproduced with permission from [65]. Copyright 2016 American Chemical Society.

This equation provides the fundamental concept for improving ionic conductivity. For example, potential wells (migration-activation energy) formed by interacting ions and electrons can hinder lithium-ion migration. Furthermore, multi-ion coordinated transport and crystal structure opening can effectively reduce the migration energy barrier. The ionic migration path is investigated by the transition state theory, which states that converting a reactant into product molecules necessitates a transition state and activation energy—and that the reaction always takes the path with the lowest energy. DFT-based nudged elastic band (NEB) calculation is commonly used in solid electrolytes to determine the transition state with the lowest energy path and saddle points [68]. AIMD simulation is also used to estimate the ionic migration and conductivity of solid electrolytes. With a given ionic charge (q), number of ions (N), and self-diffusion coefficient (D_s), σ may be calculated by the Nernst-Einstein relationship:

$$\sigma = \frac{Nq^2D_s}{k_B T}$$

The DFT-based investigation found that sulfide-, oxide-, and halide-based solid-state electrolytes exhibit high ionic conductivity. AIMD simulation has outstanding advantages in the calculation of diffusivity, stability, and the electrochemical window of solid electrolytes.

4.4 DESIGNING ELECTROLYTE AND FUTURE OUTLOOK

Designing efficient electrolytes through simulation requires an in-depth understanding of the electronic, physical, chemical, and transport properties of atomic/molecular systems. Simulation provides numerical values or analytical data on these parameters, which can be used for electrolyte evaluation. Substitution of atoms/molecules in existing electrolytes is a useful strategy for designing new electrolytes. A comparative analysis of these properties would reveal the relative performance of electrolytes.

For example, stability is a most important and necessary factor for electrolytes, which must first be investigated. The HOMO–LUMO gap, electron affinity, VDE, and ESW can all be used to measure stability. For high-voltage electrolytes, the potential windows and anodic and cathodic limits should be investigated. For better cation conduction, it is imperative to evaluate the surface area and molecular volume, both of which must have high values.

It is necessary to investigate the dissociation energy between the cation and anion, which must be low. The salt should be non-reactive, which can be determined by a large HOMO–LUMO gap. Simulation techniques have been extensively used to study the effect of substitution in electrolytes on performance in terms of voltage range, stability, and rapid charging.

For example, the substitution of H with CN in the $B_{12}H_{12}^{2-}$ anion increases second electron affinity by more than six times, indicating that $B_{12}(CN)_{12}^{2-}$ anion is a stabler electrolyte for Mg-ion batteries. By replacing the H in the $CB_{11}H_{12}^{-}$ anion with CN or BO, the potential gaps were found to be larger. Therefore, high-voltage electrolytes (more than 5 volts) employ $CB_{11}(CN)_{12}^{-}/CB_{11}(BO)_{12}^{-}$ anions. This approach was used to design salts such as $MgB_{12}(CN)_{12}$, $CaB_{12}(CN)_{12}$, $LiCB_{11}(CN)_{12}$, $NaCB_{11}(CN)_{12}$, $KCB_{11}(CN)_{12}$, $LiCB_{11}(BO)_{12}$, $LiCB_{11}(SCN)_{12}$, and $LiCB_{11}(OBO)_{12}$ for alkali and alkaline-earth metal-ion batteries.

Thus, novel electrolytes could be designed by manipulating ligands and substitutions. Solvation/de-solvation, like electrical characteristics, can be modified by engineering the solvent molecule. Furthermore, atomic-scale engineering is especially helpful for improving electrolyte transport characteristics. The study of these properties at the atomic level by simulation needs to be explored in order to design efficient and resilient electrolytes.

4.5 CHALLENGES AND LIMITATIONS

Although theoretical modeling and simulations have been successful in developing efficient electrolytes, there are still many unresolved issues. For example, current theoretical simulations demonstrate a material's intrinsic properties without considering the complex real environment. To increase accuracy, practical reaction conditions (such as charge surfaces and solvents) should be included in the development of new algorithms and theories. Further, most electrolyte systems are simulated at the nanoscale level. For example, to reduce computational cost and complexity, generally simulation is performed for the surface and edge of electrolyte or electrode systems. For mesoscale simulation, the DFT-based kinetic Monte-Carlo approach can be applied. Machine learning is expected to lead to a revolution in theoretical research methodology.

4.6 SUMMARY

The increasing demand for electric vehicles and smart grids requires the development of highly efficient battery systems. Alkali metal-ion batteries—such as lithium-, sodium-, and potassium-ion batteries—are among potential candidates. Lithium-ion batteries are widely used for various applications, although others need an elevation in their efficiency and performance. Generally, the efficiency of the battery is enhanced by increasing capacity or operating voltage or both. The electrolyte is a pivotal component but could decompose or become unstable at high voltages. It should also be non-toxic, non-corrosive, highly stable, and safe. Today, the discovery and optimization of electrolytes mostly rely on chemical intuition and trial-and-error methods. These conventional methods are expensive and time-consuming. Modern-day frameworks (based on quantum and classical theories) for atomistic simulation can be employed to predict and design efficient high-voltage electrolytes at reduced costs and duration. They are effective at mimicking real electrolytes.

REFERENCES

1. M. Winter, B. Barnett and K. Xu, *Chem. Rev.*, 2018, 118(23), 11433–11456.
2. M. Gauthier, T. J. Carney, A. Grimaud, L. Giordano, N. Pour, H. H. Chang, D. P. Fenning, S. F. Lux, O. Paschos and C. Bauer, *J. Phys. Chem. Lett.*, 2015, 6(22), 4653–4672.
3. S. K. Heiskanen, J. Kim and B. L. Lucht, *Joule*, 2019, 3(10), 2322–2333.
4. F. Q. Liu, W. P. Wang, Y. X. Yin, S. F. Zhang, J. L. Shi, L. Wang, X. D. Zhang, Y. Zheng, J. J. Zhou and L. Li, *Sci. Adv.*, 2018, 4(10), https://doi.org/10.1126/sciadv.aat5383
5. H. Sun, P. Liang, G. Zhu, W. H. Hung, Y.-Y. Li, H. C. Tai, C. L. Huang, J. Li, Y. Meng and M. A. Angell, *Proc. Natl. Acad. Sci.*, 2020, 117(45), 27847–27853.
6. X. He, D. Bresser, S. Passerini, F. Baakes, U. Krewer, J. Lopez, C. T. Mallia, Y. Shao-Horn, I. Cekic-Laskovic and S. Wiemers-Meyer, *Nat. Rev. Mater.*, 2021, 6(11), 1036–1052.
7. A. Banerjee, X. Wang, C. Fang, E. A. Wu and Y. S. Meng, *Chem. Rev.*, 2020, 120(14), 6878–6933.
8. G. Xu, X. Shangguan, S. Dong, X. Zhou and G. Cui, *Angew. Chem. Int. Ed.*, 2020, 59(9), 3400–3415.
9. K. Xu, *Chem. Rev.*, 2014, 114(23), 11503–11618.
10. H. Zhang, G. G. Eshetu, X. Judez, C. Li, L. M. Rodriguez-Martínez and M. Armand, *Angew. Chemie Int. Ed.*, 2018, 57(46), 15002–15027.
11. N. Yao, X. Chen, Z.-H. Fu and Q. Zhang, *Chem. Rev.*, 2022, 122(12), 10970–11021.
12. W. Ding, X. Lei and C. Ouyang, *Int. J. Quantum Chem.*, 2016, 116(2), 97–102.
13. K. Xu, *Chem. Rev.*, 2004, 104, 4303–4418.
14. V. Ponce, D. E. Galvez-Aranda, J. M. Seminario, *J. Phys. Chem. C*, 2017, 121(23), 12959–12971.
15. Y. S. Meng, V. Srinivasan and K. Xu, *Science*, 2022, 378(6624), https://doi.org/10.1126/science.abq3750

16. M. T. Ong, O. Verners, E. W. Draeger, A. C. T. Van Duin, V. Lordi and J. E. Pask, *J. Phys. Chem. B*, 2015, 119(4), 1535–1545.
17. C. Liang, K. Kwak and M. Cho, *J. Phys. Chem. Lett.*, 2017, 8(23), 5779–5784.
18. S. Han, Structure and Dynamics in the Lithium Solvation Shell of Nonaqueous Electrolytes. *Sci. Rep.*, 2019, 9(1), 5555.
19. C. Yan, R. Xu, Y. Xiao, J. Ding, L. Xu, B. Li and J. Huang, *Adv. Funct. Mater.*, 2020, 30(23), 1909887.
20. J. Hammersley, *Monte Carlo Methods*; Springer Science & Business Media, 2013.
21. R. Y. Rubinstein and D. P. Kroese, *Simulation and the Monte Carlo Method*; John Wiley & Sons, 2016.
22. Y. Suzuki and M. A. Nagaoka, *J. Chem. Phys.*, 2017, 146(20), 204102.
23. T. Fujie, N. Takenaka, Y. Suzuki and M. Nagaoka, *J. Chem. Phys.*, 2018, 149(4), 44113.
24. N. Takenaka, A. Bouibes, Y. Yamada, M. Nagaoka and A. Yamada, *Adv. Mater.*, 2021, 33(37), 2100574.
25. A. K. Kushwaha, S. S. Jena, P. Jena and S. K. Nayak, *J. Phys. Chem. C*, 2022, 126(11), 5112–5121.
26. A. K. Kushwaha, S. S. Jena, M. R. Sahoo and S. K. Nayak, *J. Electron. Mater.*, 2021, 50(4), 1807–1816.
27. K. Leung, Y. Qi, K. R. Zavadil, Y. S. Jung, A. C. Dillon, A. S. Cavanagh, S. H. Lee and S. M. George, *J. Am. Chem. Soc.*, 2011, 133(37), 14741–14754.
28. P. Ganesh, P. R. C. Kent and D. Jiang, *J. Phys. Chem. C*, 2012, 116(46), 24476–24481.
29. M. M. Islam and A. C. T. Van Duin, *J. Phys. Chem. C*, 2016, 120(48), 27128–27134.
30. A. Szabo and N. S. Ostlund, *Modern Quantum Chemistry: Introduction to Advanced Electronic Structure Theory*; Courier Corporation, 2012.
31. R. G. Parr, Density Functional Theory of Atoms and Molecules. (Kenichi Fukui and Bernard Pullman, Eds.), in *Horizons of Quantum Chemistry*; Springer, 1980; pp. 5–15.
32. P. Geerlings, F. De Proft and W. Langenaeker, *Chem. Rev.*, 2003, 103(5), 1793–1874.
33. D. Chakraborty and P. K. Chattaraj, *Chem. Sci.*, 2021, 12(18), 6264–6279.
34. W. Kohn, A. D. Becke and R. G. Parr, *J. Phys. Chem.*, 1996, 100(31), 12974–12980.
35. A. J. Cohen, P. Mori-Sánchez and W. Yang, *Science*, 2008, 321(5890), 792–794.
36. W. Chen, Y. Li, D. Feng, C. Lv, H. Li, S. Zhou, Q. Jiang, J. Yang, Z. Gao and Y. He, *J. Power Sources*, 2023, 561, 232720.
37. B. Mennucci and J. Tomasi, *J. Chem. Phys.*, 1997, 106(12), 5151–5158.
38. J. Tomasi, B. Mennucci and R. Cammi, *Chem. Rev.*, 2005, 105(8), 2999–3094.
39. M. Cossi, V. Barone, R. Cammi and J. Tomasi, *Chem. Phys. Lett.*, 1996, 255(4–6), 327–335.
40. B. Mennucci and R. Cammi, *Continuum Solvation Models in Chemical Physics: From Theory to Applications*; John Wiley & Sons, 2008.
41. A. K. Kushwaha and S. K. Nayak, *Phys. E Low-dimensional Syst. Nanostruct.*, 2018, 97, 368–374.
42. A. K. Kushwaha, M. R. Sahoo and S. Nayak, *AIP Conference Proceedings*, 2018, 1951, 20014; AIP Publishing.
43. A. K. Kushwaha, M. R. Sahoo, J. Nanda and S. K. Nayak, *J. Clust. Sci.*, 2017, 28(5), 2779–2793.
44. N. Shao, X. G. Sun, S. Dai and D. Jiang, *J. Phys. Chem. B*, 2011, 115(42), 12120–12125.
45. E. Björklund, M. Göttlinger, K. Edström, R. Younesi and D. Brandell, *Batter. Supercaps*, 2020, 3(2), 201–207.
46. D. Marx and J. Hutter, *Ab Initio Molecular Dynamics: Basic Theory and Advanced Methods*; Cambridge University Press, 2009.
47. M. E. Tuckerman, *J. Phys. Condens. Matter.*, 2002, 14(50), R1297.

48. K. Leung and J. L. Budzien, *Phys. Chem. Chem. Phys.*, 2010, 12(25), 6583–6586.
49. D. Frenkel and B. Smit, *Understanding Molecular Simulation: From Algorithms to Applications*; Elsevier, 2001; Vol. 1.
50. D. C. Rapaport and D. C. R. Rapaport, *The Art of Molecular Dynamics Simulation*; Cambridge University Press, 2004.
51. M. H. Kalos and P. A. Whitlock, *Monte Carlo Methods*; John Wiley & Sons, 2009.
52. H. Zhao, J. Zhou and P. Jena, *Angew. Chem. Int. Ed.*, 2016, 128(11), 3768–3772.
53. T. Zhao, J. Zhou, Q. Wang and P. Jena, *Angew. Chem. Int. Ed.*, 2017, 129(43), 13606–13610.
54. M. Zhong, J. Zhou, H. Fang and P. Jena, *Phys. Chem. Chem. Phys.*, 2017, 19(27), 17937–17943.
55. Q. Li, J. Chen, L. Fan, X. Kong and Y. Lu, Progress in Electrolytes for Rechargeable Li-Based Batteries and Beyond. *Green Energy Environ.*, 2016, 1(1), 18–42.
56. J. Moon, H. Baek and J. Kim, *Chem. Phys. Lett.*, 2018, 698, 72–76.
57. T. Lu and F. Chen, *J. Comput. Chem.*, 2012, 33(5), 580–592.
58. M. Joshi and T. K. Ghanty, *J. Phys. Chem. C*, 2018, 122(49), 27947–27954.
59. A. K. Kushwaha, M. R. Sahoo and S. K. Nayak, *ChemistrySelect*, 2019, 4(4), 1251–1258.
60. K. Senevirathne, C. S. Day, M. D. Gross, A. Lachgar and N. A. W. Holzwarth, *Solid State Ion.*, 2013, 233, 95–101.
61. A. K. Srivastava and N. Misra, *Polyhedron*, 2016, 117, 422–426.
62. M. D. Bhatt, M. Cho and K. Cho, *J. Solid State Electrochem.*, 2012, 16(2), 435–441.
63. N. Kumar and J. M. Seminario, *J. Phys. Chem. C*, 2016, 120(30), 16322–16332.
64. J. Self, K. D. Fong and K. A. Persson, *ACS Energy Lett.*, 2019, 4(12), 2843–2849.
65. Z. K. Tang, J. S. Tse and L. M. Liu, *J. Phys. Chem. Lett.*, 2016, 7(22), 4795–4801.
66. S. R. Galle Kankanamge and D. G. Kuroda, *J. Phys. Chem. B*, 2020, 124(10), 1965–1977.
67. O. Borodin, L. Suo, M. Gobet, X. Ren, F. Wang, A. Faraone, J. Peng, M. Olguin, M. Schroeder and M. S. Ding, *ACS Nano*, 2017, 11(10), 10462–10471.
68. A. Urban, D. H. Seo and G. Ceder, *NPJ Comput. Mater.*, 2016, 2(1), 1–13.

5 Solid-State Electrolytes for Batteries

R. Meera Naachiyar, S. Aafrin Hazaana, N. Muniraj Vignesh, and S. Selvasekarapandian

5.1 OVERVIEW OF SOLID ELECTROLYTE

Over the past 30 years, the global energy needs of the world have increased by 125%. However, the persistent usage of fossil fuels leads to abiding environmental pollution, global warming, ozone layer depletion, etc. In order to balance the environmental pollution and energy crisis, the global communities are moving towards zero carbon emission energy storage devices. Batteries and supercapacitors are the commonly used energy storage devices that rule the modern world. A battery is an electrochemical energy storage device composed of an anode, cathode, and electrolyte, whereas electrolyte is the medium in which ions are transported between electrodes. Electrolytes in various forms such as liquid, gel, and solid have become standard in numerous fields of applications [1].

The evolution of solid-state electrolytes was initiated by the discovery of M. Faraday's inorganic solid-state electrolyte in the 19th century. The inorganic solid-state electrolytes initially used were silver sulphide (Ag_2S) and lead (II) fluoride (PbF_2). Various kinds in solid-state electrolytes such as phosphate, borohydride, chalcogenide, and metal-organic framework [2–5] based solid ion conductors have been utilized for more than several decades. In the 1970s, Peter V Wright [6] discovered the first solid-state ion-conducting polymeric material, Poly ethylene oxide (PEO), incorporated with sodium and potassium salts. As termed by Japanese scientist Takehiko Takahashi, the study of solid-state ion conducting materials is **Solid-State Ionics**.

The importance of solid-state electrolytes in electrochemical energy storage devices, i.e., batteries, increases every day due to its assured safety, high mechanical stability, and desirable properties. Solid-state electrolytes are classified into Solid Inorganic/Ceramic electrolytes and Polymer electrolytes. Table 5.1 lists different types of membranes and their properties.

In this chapter, the technological innovations and key features of solid-state electrolytes as an ion conducting medium in batteries will be discussed highlighting their properties for commercial interest in daily applications. The importance of solid-state electrolytes in the present and future needs of energy storage devices will be described comprehensively by comparing the properties of different solid-state ionic conductors.

DOI: 10.1201/9781003441182-6

TABLE 5.1
Different Types of Solid-State Electrolytes and Their Properties

Types	Advantages	Drawbacks	Ionic Conductivity	Device
Ceramic electrolyte	High electrochemical stability, high thermal stability, and dendrite suppression capability [7]	Brittle, fragile, high cost	Ranges between 10^{-4} and 10^{-3} S cm^{-1}	Solid oxide fuel cell (SOFC), lithium-ion battery (LIB)
	Possible applicability in high temperature operating devices	Poor interfacial compatibility with electrodes		
Solid polymer electrolytes	Flexibility, enhanced interfacial compatibility [8]	Low ionic conductivity at room temperature	Ranges from 10^{-4} to 10^{-1} S cm^{-1}	Solid-state batteries, fuel cell, supercapacitor, solar cells, etc.
Composite electrolyte	Ideal mechanical strength, affordable electrochemical window, better interfacial contact with electrode [9]	Complex fabrication process	Ranges between 10^{-4} and 10^{-5} S cm^{-1}	Lithium-ion battery (LIB) and proton-exchange membrane fuel cell (PEMFC)
	Improved ionic transference number and room temperature ionic conductivity, reduced chance of dendrite formation			

Ultimately, the influence of solid-state electrolytes as a revolutionary replacement for existing liquid electrolytes in high-energy devices will be outlined.

5.2 SOLID-STATE ELECTROLYTE

The functional properties required for better solid-state electrolytes are:

1. High room temperature ionic conductivity in the order of >10^{-4} S cm^{-1}.
2. Low interfacial resistance between solid-solid electrode-electrolyte interfaces.
3. Optimized stability with anode and cathode.
4. Wide electrochemical window.
5. High density to overcome dendrite formation.

Two main classifications in solid-state electrolytes are ceramic-based solid-state electrolytes and polymer-based electrolytes.

5.3 CERAMIC-BASED SOLID-STATE ELECTROLYTES

The emergence of Solid Ceramic electrolytes with high ionic conduction at high temperatures is by the movement of ionic point defects, where the formation and movement of ionic point defects in solids require high energy [7]. The observation of better ionic conductivity in solid phase materials like Fluoride, Oxide-ion conductors [10], β – Alumnus [11], Silver & Copper ion conductors [12], and NASICON [13] results in the evolution of solid-state electrolytes, which greatly stabilizes the performance of energy devices. Several oxides and phosphate-based solid ceramic electrolytes promise their application in energy conversion devices such as solid oxide fuel cells at high temperatures [14].

Ceramic electrolytes can be divided into three categories: NASICON-type, Perovskite-type, and Garnet-type. The crystalline data of Sodium Super Ionic Conductors (NASICON) structured $NaA_2^{IV}(PO_4)_3$, where A^{IV} = Ge, Ti, and Zr, was interpreted by Ove Hagman et al. in 1968. In 1976, the NASICON-type ceramic electrolytes were proposed by Hong and Goodenough as a class of structurally isomorphous 3D framework compounds with high ionic conductivity. Low thermal expansion behavior, large surface area, and ability to accommodate ions in the crystalline sites make NASICON an interesting commercial alternate for liquid electrolytes in battery applications. Also, LISICON and several other crystalline-based solid electrolytes possess essentially the same NASICON crystalline structure with high-temperature superionic conduction [8]. The surface morphology of the ceramic electrolyte Na_2PS_4 is shown in Figure 5.5 to illustrate the surface nature of the ceramic electrolytes.

Ideal perovskite-type solids exhibit an ABO_3 structure where A and B are six and 12 oxygen-coordinated sites, respectively. Belous et al. reported the stable perovskite-structured Li ion conductor in 1984. Further, research is focused on the area of perovskite-type super ionic conductors resulting from the innovation of solid perovskite electrolytes for battery applications. Stramare et al. reported the super ionic conduction at room temperature in the order of 10^{-3} S/cm for $Li_{3x}La_{(2/3)-x}TiO_3$ ($x = 0.11$), which is in the super ion conducting regime. Increased chemical stability, wide electrochemical window, and high ionic conductivity make perovskite-type electrolytes a possible substitute in electrochemical applications [9].

Garnet structures are introduced to reduce the interfacial resistance experienced because of the solid/solid electrode-electrolyte interface in batteries. Li-ion conducting garnet structured solid electrolytes such as $Li_7La_3Zr_2O_{12}$ possess ultralow interfacial resistance in all-solid-state Li-ion batteries. To reduce and neglect the interfacial resistance, various methods such as surface modification strategy [15], usage of separator [16], and different coating methods such as chemical coating and atomic layer deposition have been carried out [10]. A bird's-eye view of Li-stuffed garnet-type $Li_7La_3Zr_2O_{12}$ ceramic electrolytes for advanced all-solid-state Li batteries [11] and Metastable properties of a garnet-type $Li_5La_3Bi_2O_{12}$ solid electrolyte towards low-temperature pressure-driven densification [12] are works reported on Li-ion conducting garnet type electrolytes. Table 5.2 shows some of the ceramic-based solid electrolytes used in batteries.

TABLE 5.2
Some of the Major Results Obtained Using Solid Ceramic-Based Electrolytes in Batteries

S. No	Ceramic Electrolyte Composition	Highest Ionic Conductivity (S/cm)	Discharge Capacity (mAh g^{-1})	Ref.
1	$Li_5La_3Ta_2O_{12}$	1.3×10^{-4}	83	[19]
2	NASICON/graphene	0.6×10^{-3}	108	[20]
3	Na_3PS_4	2×10^{-4}	90	[21]
4	Honeycomb structured $Li_{0.35}La_{0.55}TiO_3$	1.1×10^{-3}	73 (mAh cm^{-2})	[22]
5	$Na_3Zr_2Si_2PO_{12}$	1.93×10^{-4}	62.23	[23]
6	$Li_{6.4}La_3Zr_{1.4}Ta_{0.6}O_{12}$	5×10^{-4}	685	[24]
7	$Li_{0.34}La_{0.56}TiO_3$	2×10^{-5}	145	[25]
8	$Li_7P_3S_{11}$	1.1×10^{-2}	140	[26]
9	$Na_{3.06}P_{0.94}Si_{0.06}S_4$	7.4×10^{-4}	90	[26]
10	$Li_{1.3}Al_{0.3}Ti_{1.7}(PO_4)_3$	0.13×10^{-3}	160	[27]
11	80 Li_2S·20 P_2S_5	1.3×10^{-3}	400	[9]
12	Cubic Na_3PS_4	4.6×10^{-4}	130	[28]
13	$Li_7La_3Zr_2O_{12}$	1.6×10^{-3}	150	[29]
14	$Li_7P_3S_{11}$	9.7×10^{-4}	—	[30]
15	Li_2S–P_2S_5	~10^{-3}	100	[31]
16	78 Li_2S·22 P_2S_5	1.7×10^{-2}	480	[32]
17	Li_3BO_3 doped Li_2S-P_2S_5	1.03×10^{-3}	156	[33]

5.4 POLYMER-BASED SOLID ELECTROLYTES

Depending on the phase of the material, the polymer electrolytes were classified into Gel polymer electrolyte, Solid polymer electrolyte, and Composite polymer electrolyte. The interest towards the development of a solid polymer ion-conducting medium is due to its peculiar properties like flexibility, amorphous nature, and ambient temperature operation. The polymers are chosen as an ion-conducting medium on the detailed analysis of its chemical structure, which contains a considerable amount of polar groups. These polar groups (–OH, –NH) can attract the cation of any ionic salts. The movement of cations from one polar site to another paves the way for high ionic motion in an electrolyte [17, 18]. Figure 5.5 depicts the surface morphological nature of Solid polymer electrolytes based on Agar-Agar with LiCl.

The prime invention of ionic conduction in the solid polymer medium using PEO [34] ascribes the applicability of polymers as a host material. The mechanical and electrochemical stability of Solid polymer electrolytes is predominantly based on the chemical, physical, and dielectric properties of the polymeric materials. Based on their origin, Solid polymer electrolytes can be further divided into Synthetic polymer electrolytes and Biopolymer electrolytes.

In this way of research, the trend of bio-based materials in solid-state electrolytes ignites the employment of chemically unprocessed naturally obtained biomaterials like leaves, seeds, flowers, barks, and roots utilized as such, as ion conducting membrane (Electrolyte) [52]. The battery performance of Moringa Oleifera gum and seed-based membranes as a proton conducting medium stimulates the applicability of biomaterial in various electrochemical devices [51, 52]. Composite polymer electrolyte ascribes the combination of both ceramic and polymer components together. Through proper management of composition and adequate fabrication methods, the acquired solid-state ceramic-polymer electrolyte may provide the expected performance with enhanced properties. Table 5.3 lists various polymer-based solid electrolytes and their performance.

TABLE 5.3
List of Several Polymer-Based Solid Electrolytes

S. No	Polymer	Additive	σ^k (S cm^{-1})	Ref.
1	Polyvinyl alcohol (PVA)	Piperidinium (plasticizer)/ LiTFSI[d] (ionic salt)	1.2×10^{-3} (RT)	[35]
2	Carboxymethyl cellulose	Oleic acid	2.11×10^{-5} (RT)	[36]
3	PEO[h]	KTFSI	3.54×10^{-5} (65 °C)	[37]
4	PVA[a]	$LiFePO_4$	1.18×10^{-5} (RT)	[38]
5	PPC[g]	LiTFSI	3.0×10^{-4} (RT)	[39]
6	Agar–Agar	LiCl	3.12×10^{-2} (RT)	[40]
7	Gellan gum	NH_4HCO_2	5.62×10^{-3} (RT)	[41]
8	Gellan gum	NH_4SCN	1.41×10^{-2} (RT)	[42]
9	Gellan gum	LiCl	4.08×10^{-3} (RT)	[8]
10	Guar gum	NH_4SCN	4.91×10^{-3} (RT)	[43]
11	Corn starch	Na_3PO_4	7.27×10^{-6} (RT)	[44]
12	Chitosan	NH_4SCN	8.57×10^{-4} (RT)	[45]
13	70 wt.% of chitosan and 30 wt.% of methyl cellulose	$MgCl_2$	2.75×10^{-3} (RT)	[46]
14	Polyvinyl alcohol (PVA): Chitosan	NH_4SCN	3.5×10^{-3} (RT)	[47]
15	Carboxylmethyl cellulose (CMC)[c]	NH_4F	2.67×10^{-7} (RT)	[48]
16	Dextran: Chitosan	$Mg(CH_3COO)_2$:Ni	7.71×10^{-6} (RT)	[49]
17	Pectin	LiTFSI	3.84×10^{-4} (RT)	[50]
18	Moringa Oleifera gum	NH_4NO_3	2.66×10^{-3} (RT)	[51]
19	Moringa Oleifera seed	NH_4SCN	1.83×10^{-2} (RT)	[52]
20	PVA/PAN[b]	LATP[e]/LiTFSI/SN[f]	1.13×10^{-4} (25 °C)	[53]
21	PVA – Chitosan	NH_4I	1.77×10^{-6} (RT)	[54]
22	PEO – SiO_2[j]	NH_4HSO_4	2.5×10^{-4} (RT)	[55]
23	PVDF	$LiClO_4$	7.0×10^{-4} (RT)	[56]
24	Sodium Alginate – GQD[i]	NH_4HCO_2	2.01×10^{-2} (RT)	[57]

a PVA – Polyvinyl alcohol.
b PAN – Polyacrylonitrile.
c CMC – Carboxymethyl cellulose.
d LiTFSI – Lithium bis(trifluoromethanesulfonyl)imide.
e LATP – $Li_{1.4}Al_{0.4}Ti_{1.6}(PO_4)_3$ (NASICON-type structure).
f SN – Succinonitrile.
g PPC – Polypropylene carbonate.
h PEO – Poly (ethylene oxide).
i GQD – Graphene quantum dot.
j SiO_2 – Silicon dioxide.
k σ – Ionic conductivity.

5.5 PREPARATION TECHNIQUES

Solid Electrolytes can be prepared using various techniques. The techniques involved in the preparation of ceramic-based electrolytes are the Dry pressing method, Colloidal process, Pulsed laser deposition, Chemical vapor deposition, Sputter deposition, sol-gel deposition, melt quenching, mechanical milling, and wet chemical synthesis. Among these, the Dry pressing method is one of the most commonly used to produce high-quality ceramic components, where the dry ceramic powders (moisture-free material—less than 3%) are compressed firmly using the die/punch arrangement. The Colloidal process is a versatile technique used to shape and tune ceramic-based components. This method can mold ceramics with defined pore size, geometry, and surface area. Sputter deposition, which is a physical vapor deposition technique, involves the phenomenon of sputtering to deposit and achieve thin film-based materials.

Pulsed laser deposition and chemical vapor deposition methods are other important techniques utilized for the deposition of materials to obtain the solid form of electrolytes. Sol-gel deposition is another flexible and simple method for synthesizing porous, ceramic, and glass-like materials in the form of thin films. This method involves various processes in chronological order as hydrolysis and poly-condensation, gelation, aging, drying, densification, and crystallization (refer Figure 5.1). This extended processing technique results in high-quality components with superior homogeneity and purity. Polymer-based solid electrolytes were synthesized using common techniques such as sol-gel, hot-pressing, phase inversion method, electrospinning, solution casting technique, etc. (refer Figure 5.2).

The phase inversion method is a chemical phenomenon to form filtration membranes using polymers. It controls the polymer transformation from the liquid phase to the solid phase. Phase inversion is highly influenced by the nature of the polymer and the solvent used to dissolve the polymer. The electro-spinning technique is a fiber production technique that uses electric force to draw charged threads of polymer solutions or polymer melts up to fiber diameters in the range of a hundred nanometers. This method exhibits the characteristics of both solution dry spinning of fibers and electro-spraying. The solution casting method is the most compatible, popular, and safest technique to prepare solid polymer electrolytes. Due to its ease of systemization and less preparation time, it plays a significant role in polymer electrolyte preparation compared to other synthesis methods.

5.6 MECHANISM INVOLVED

The conduction in solid substances can be classified into two broad categories: Electron-limited conduction mechanism and Bulk-limited conduction mechanism. Electron-limited conductions mainly deal with electron conduction, whereas bulk-limited conduction defines the flow of ions in solid species. In Solid Electrolytes, the motion of ions from one side to another is generally defined with bulk-limited conduction models where some of the important conduction models are the Hopping conduction model, Poole-Frenkel conduction model, Space-charge limited conduction model, and Grain boundary limited conduction model [58]. Present research

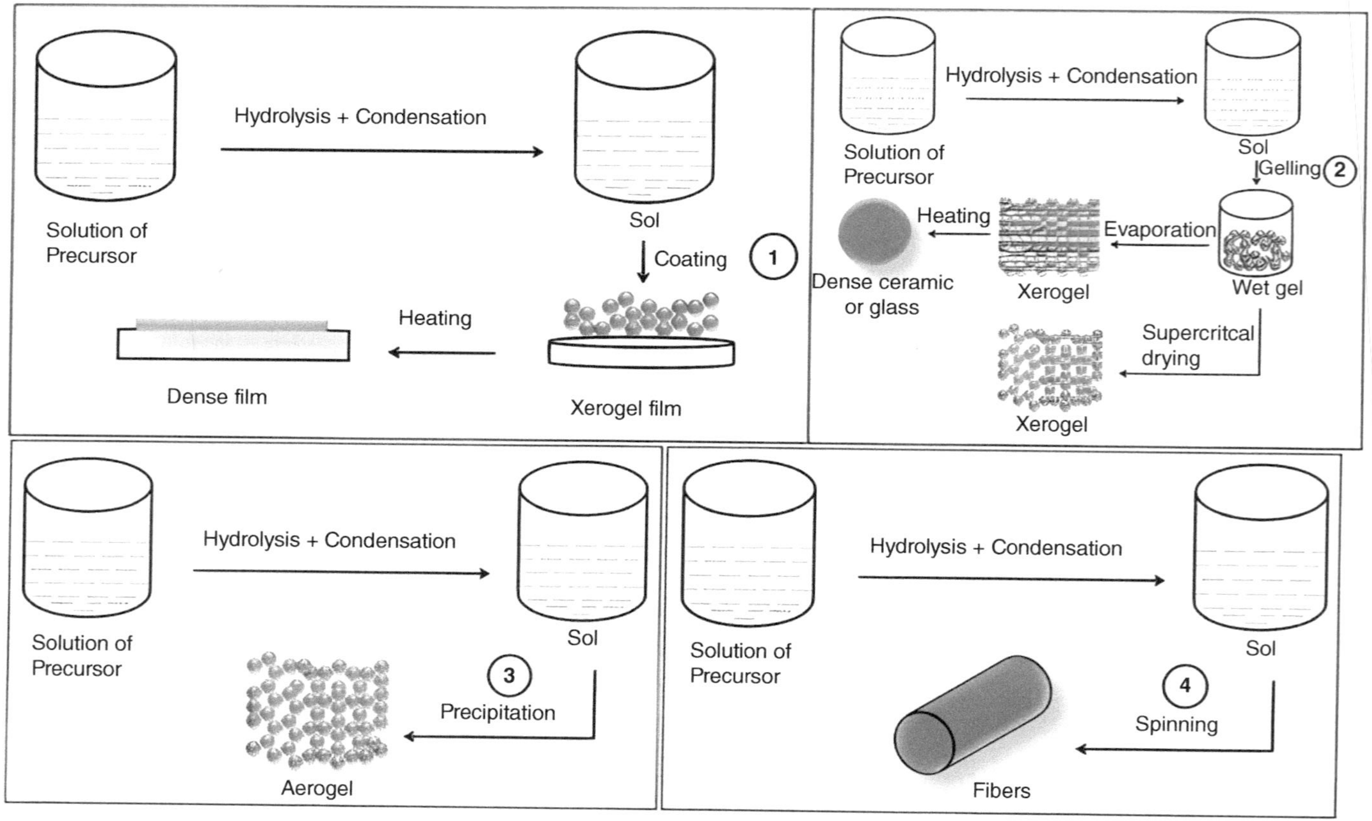

FIGURE 5.1 Schematic diagram of the sol-gel method.

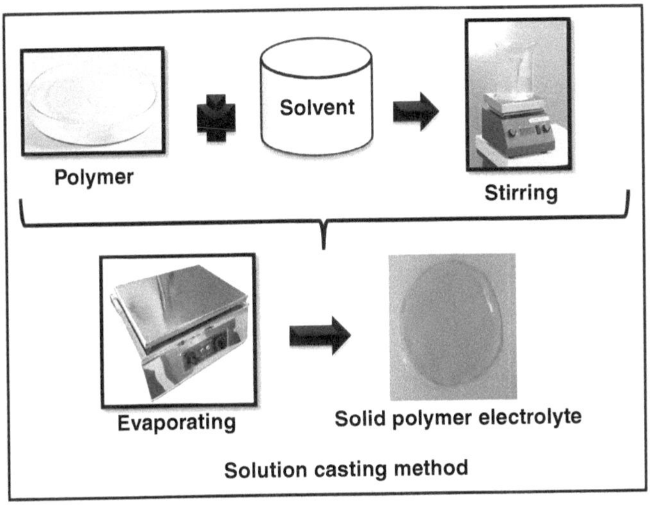

FIGURE 5.2 Schematic diagram of the solution casting method.

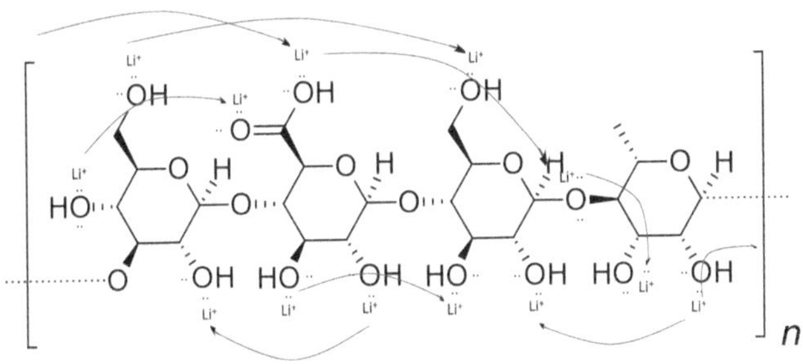

FIGURE 5.3 Graphical representation of hopping conduction.

enhances the properties of Solid Electrolytes, which need a basic understanding of ion conducting mechanisms with clear experimental and theoretical evidence.

The conduction in Solid Polymer Electrolytes like Solid Electrolytes is generally idealized by the hopping (jumping) of ions from one site to another. In principle, through Coulombic attraction, the positive ions in Solid Polymer Electrolytes are attached to the negative polar sites. These negative sites can be assumed as trapping locations. To jump from one trap site to another, ions need some amount of energy, which is termed activation energy. Figure 5.3 shows the graphical representation of hopping conduction [59]. The polarization of ions by the applied voltage is one of the important characteristics of solid electrolytes for energy applications. The response of the solid electrolyte to the applied voltage (which may be due to the presence of an anode and cathode in the battery) results in the evolution of interfacial resistance between electrode-electrolyte interfaces. This interfacial resistance is theoretically represented by the addition of the Constant Phase Element in the equivalent circuit analysis.

In 1971, Rice and Roth formalized a valid theoretical ionic conduction using the hopping mechanism. According to Rice and Roth [60], the average time spent by the ion at the trapping site or relaxation time (t_0) can be given by

$$t_0 = \frac{1}{v_0}$$

where v_0 is the ionic oscillator frequency. The self-diffusion of the solid electrolyte is given by

$$D_s(T) = \frac{1}{3} v_0 l_0 e^{-\varepsilon_0 / k_B T}$$

Here, l_0 is the average distance between two trapping sites, and ε_0 is the dielectric constant. Using the hopping theory, the ionic conductivity of the solid electrolyte can be calculated from the formula

$$\sigma = \frac{2}{3}\left(\frac{(Ze)^2}{kTm_c}\right) n_d E_a \tau \exp\left(\frac{-E_a}{kT}\right)$$

Ideally, activation energy is the measure of the energy difference between two trapping sites. The activation energy E_a plays a vital role in the conduction of ions. The hopping theory exposes the importance of dielectric parameters in ionic conduction and the electrochemical performance of solid electrolytes. Comparatively, the availability of high flexible nature in polymer membrane electrolytes endorses their ionic mobility and electrochemical properties. Due to the high activation energy required for conduction in ceramic electrolytes (the energy between two trapping sites in ceramic electrolytes is very high), it needs an external energy source in the form of heat. So, ceramic electrolytes are only applicable for high-temperature operating energy systems.

5.6.1 Poole-Frenkel Conduction Model

Apart from dielectric performances, the response of electrically insulating materials to the applied electric field can be explained by using the Poole-Frenkel conduction theory, where solid electrolytes are identified as electrically insulating materials [61]. In this model, the ions are assumed to be bound to the charged sites. The potential of the ion bound with the charged site is given by

$$\varnothing = -\frac{e^2}{4\pi\epsilon_0\epsilon_r\gamma} + e(\boldsymbol{F}.\boldsymbol{r})$$

Here, $\boldsymbol{F}$ and $\boldsymbol{r}$ are the applied field and distance between the ion and charged site. To flow from one charged site to another, the ion must overcome this potential energy that is offered in the form of barrier potential by the charged site. This barrier potential increases with the increase in distance between the ion and charged site, and the probability of the ion overcoming this barrier potential increases with the increase in strength of the applied electric field [62]. This allows us to express the ionic conduction in solid insulators following Ohm's law in a low electric field and obeys the Poole-Frenkel equation in a high electric field. The Poole-Frenkel equation is given by

$$\sigma = e\mu K\left(-\frac{E_0}{2kT}\right)$$

In the Poole-Frenkel model the energy of the barrier site is expressed in the order of k and T. The factor E_0 in the above equation is the strength of the applied electric field, and K is the constant related with the number density n of the carrier species. It is obvious that the energy required to overcome the barrier potential in the direction opposite to the applied field is higher than the direction along with the field. Accordingly, the internal resistance of the ceramic electrolytes is high, as compared with the polymer-based membrane electrolytes. The ionic mobility in ceramic electrolytes is limited by the rigid behavior of the barrier sites where mobility is defined as a field independent parameter.

5.6.2 Space-Charge Limited Conduction Model

With the Space-charge limited conduction model [15], the charge transport properties of solid electrolytes can be defined by using the delocalization and localization of charge carriers. The ionic conduction through this model is interpreted by the concentration of trap sites available in the solid electrolyte. In general, there are two types of traps in a solid species: a shallow trap and a deep trap. The traps located near to the Lowest Occupied Molecular Orbital (LUMO) region of the solid electrolyte are called shallow traps, and near to Highest Occupied Molecular Orbital (HOMO), region traps are deep traps. Shallow traps are for electron conduction, and deep traps are for conduction of holes or ions. In this model, the conducting ions are considered as the mobile trap sites [63].

In the Space-charge limited conduction model, the mobility of the carrier ions is calculated by sandwiching the solid electrolyte between two electrodes. The system is designed so that the cathode behaves as the blocking electrode. When the strong voltage pulse is applied at the anode end, the conduction of ions occurs from one side of the electrolyte to another, whereas it can be detected using an oscilloscope. And, the mobility of the ions can be expressed using the following formula:

$$\mu = \frac{0.787d^2}{V\tau}$$

The arriving time τ is the measure of time taken by the ion to reach the respective cathode, where 0.787 is the correction factor for space charge-free carrier transit time. At the space charge region, which is nothing but the interfacial region between the electrode and electrolyte, the ions are distributed accordingly due to the presence of a high field by the electrodes. An electrode-electrolyte double layer is formed, and the conduction follows the original Poole-Frenkel equation.

5.6.3 Grain Boundary Limited Conduction Model

The conduction of ions in poly crystalline solid electrolytes is explained using the Grain boundary limited conduction model. In this model, the ionic conductivity is highly influenced by the grain boundaries present between two-grain sites. Several experimental and theoretical results show that the increase in grain size increases the ionic conductivity. The conducting ions have to cross a large amount of energy barrier, which is not possible in many cases. Hence, grain size plays a dominant role in the conduction of ionic charge carriers. The Grain boundary limited conduction model [64] can be numerically calculated from the formula.

$$\sigma_0 = \alpha_{eff} L g_b$$

The average grain size is denoted by L, and α_{eff} runs from 0 to 1, representing the conductivity reduction due to non-uniform grain distribution. The term g_b is the barrier conduction per unit area, which could be found by

$$g_b = \frac{4\pi e^2 m_e}{h^3} \frac{l}{l + \frac{3}{4} w} k_B T \ln\left(1 + \exp\frac{-\phi}{k_B T}\right)$$

where l is the carrier mean free path, w is the barrier width, and φ is the Fermi energy value. In this method, the carrier density n is derived from the Hall Effect, related to the conductivity σ (conveniently measured from impedance technique) by

$$\sigma = ne\mu^*$$

μ^* is the effective mobility. By theory, increasing grain size is achieved by the thermal treatment of solid materials. In solid polymer electrolytes, the thermal treatment results in the extension of polymer chains due to the flexibility of polymeric behavior. Recent research on enhanced solid electrolytes for energy applications concentrates on the thermal and mechanical stability of energy devices. To increase the flexibility nature of the solid electrolytes, many polymers are investigated for their thermal properties.

All the above models suggest that the solid electrolytes with low activation or barrier energy sites designed with commendable flexibility behave as a potential alternate in energy applications instead of liquid electrolytes. The increase in the amorphous nature of the solid electrolytes helps it to increase its flexibility. Also, the amorphous area in a solid electrolyte is comparatively more flexible than the crystalline area of that material. Hence, the increase in ionic conductivity is observed by increasing the amorphous nature of the solid electrolytes.

5.7 CHARACTERIZATION TECHNIQUES

Solid electrolytes undergo various characterization techniques such as structural, vibrational, thermal, electrochemical, and morphological analysis that helps to analyze and justify their properties for solid-state battery applications.

X-ray diffraction is a structural analysis technique that determines the physical properties of metals, polymeric materials, and other solids. It is mainly used as a "fingerprint" characterization of the crystalline/amorphous nature of the material. This helps to understand the structure and nature (i.e., crystalline/amorphous) of the materials [65]. It also involves the study of various parameters such as crystalline percentage, variation in inter-planar spacing, and polycrystalline phase of the material.

The XRD technique also paves the way for understanding several other properties such as stress-strain measurement, phase equilibrium, structural order-disorder transformation, particle size determination, and chemical analysis [66]. The difference in the crystalline and amorphous nature of the material can be easily understood from the nature of peaks (Figure 5.4). The crystalline materials will have a sharp peak with high intensity, whereas the amorphous materials appear to have a broad peak with comparatively low intensity. In solid-state electrolytes, ion conduction occurs mainly in the amorphous region.

Infrared (IR) spectroscopic analysis is an important vibrational technique that helps identify the organic functional groups present in the system and studies the interaction of functional groups with the cation and anion of the salt added to the

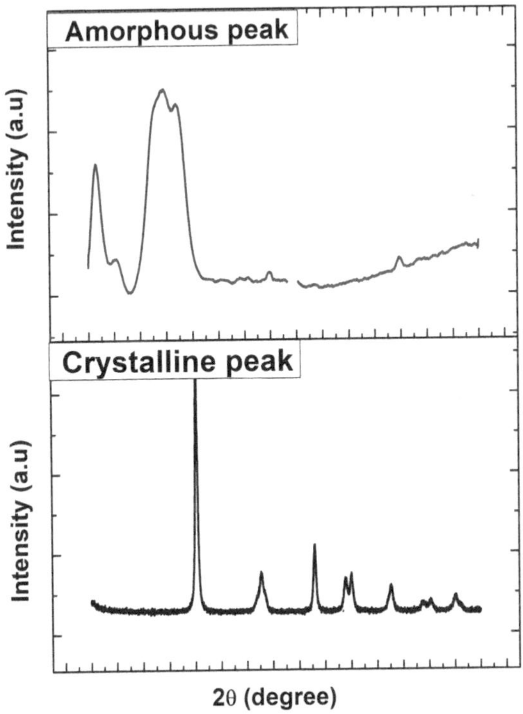

FIGURE 5.4 XRD spectrum.

solid-state electrolyte. This interaction leads to changes in the frequency of the functional groups that can be observed from the FTIR spectrum. The change in frequency due to the interaction leads to the change in bond length in the various molecules of the functional groups. Nuclear Magnetic Resonance Spectroscopy is a powerful method to investigate the structure and ion dynamics (diffusion and conductivity) of solid electrolytes.

Differential Scanning Calorimetric is used to determine the glass transition temperature of solid polymer electrolytes. If the glass transition is very low, then the solid polymer electrolyte will be more flexible in nature. Thermogravimetric analysis is used to study the physical changes of the material, especially the weight loss or gain of the sample due to decomposition, oxidation, or dehydration while heating. This analysis is suitable for all types of solid materials, including organic or inorganic materials. The frequency response of solid electrolytes is analyzed by using Impedance spectroscopy.

This technique reveals the quantitative measure of ionic conductivity, one of the most essential parameters for battery applications. Impedance spectroscopy avails the frequency characteristics of an electrical circuit containing resistance, capacitance, and inductance (LCR circuit) and its voltage versus current responses. The plot between $Z\cos\theta$ versus $Z\sin\theta$ is termed as Nyquist plot, which allows the bulk resistance (R_b) of the solid electrolyte to be measured (refer Figure 5.5).

In general, the impedance (Z) of any circuit is given by $Z = Z' + Z'' = Z\cos\theta + IZ\sin\theta$, where θ represents the phase angle.

The ionic conductivity of the solid electrolyte is evaluated from the formula

$$\sigma = \frac{l}{AR_b}$$

l is the thickness and A is the area.

The Voltammetric analysis includes Linear Sweep Voltammetry (LSV), and Cyclic Voltammetry (CV) is the necessary electrochemical technique used to determine the working potential of the solid-state electrolyte. LSV helps determine the electrochemical window of the solid-state electrolyte. CV is an electrochemical technique that provides the knowledge of thermodynamics of the oxidation and reduction process. Scanning electron microscopy and transmission electron microscopy techniques discuss the study of the surface morphology of solid-state electrolytes. These studies provide information about the porous nature and distribution of anions and cations in solid-state electrolytes. Transference number measurements such as Wagner and Evans provide information about the nature of contributing species, either ions or electrons, and their contribution of particular ions towards ionic conductivity.

5.8 ADVANTAGES AND CHALLENGES WITH SOLID ELECTROLYTES

Electrolytes are one of the major components in batteries. Commercially available high-energy batteries make use of Liquid electrolytes due to their superior ion-conducting properties [67]. Though Liquid electrolytes have immense commercial

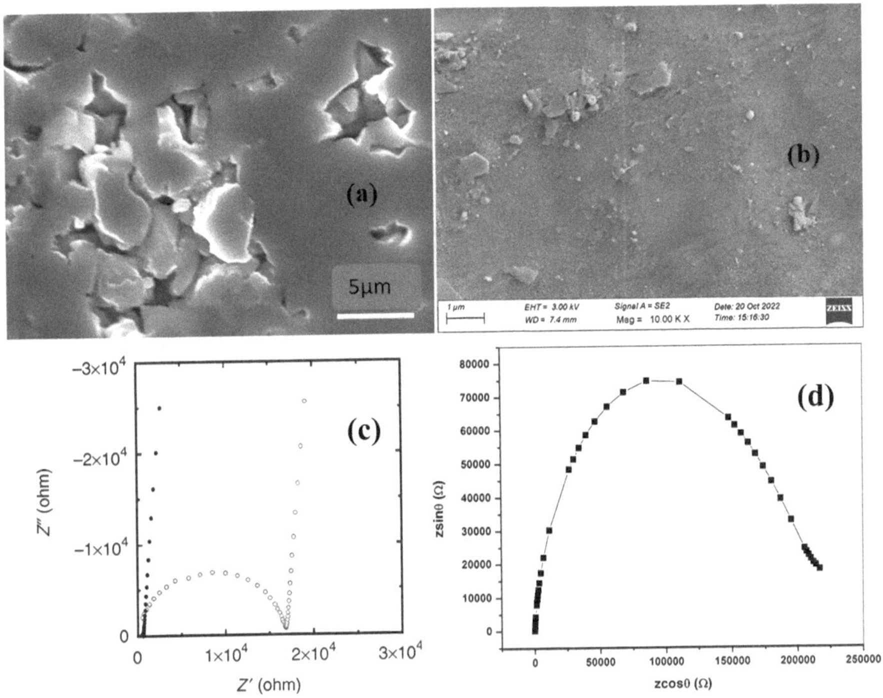

FIGURE 5.5 SEM images of (a) Na_3PS_4 ceramic electrolyte; (b) Agar-Agar with LiCl biopolymer membrane, Impedance plot for (c) Na_3PS_4 ceramic electrolyte; (d) Agar-Agar with LiCl biopolymer membrane [21, 40].

Source: The pictures are published under Common Creative Licence of Nature Publications.

value, they possess various drawbacks such as leakage, low stability, the chance of dendrite formation, short-circuiting, flammability, heavy weight, less portability, etc. [68]. Similar problems have occurred with Gel electrolytes. This leads to the emergence of solid-state electrolytes, which acts as a superior alternative for Liquid electrolytes and Gel electrolytes with remarkable properties such as being compact, safe, good mechanical stability, and versatile nature, providing better interfacial contact between electrodes and electrolyte [69].

The main advantage of Ceramic solid electrolytes is that they can be used in all kinds of Fuel cells, where the electrolytes remain stable for a long time, which provides a better life in Solid Oxide Fuel Cells. The usage of low-temperature operating ceramic electrolytes is unfavorable for energy storage devices like batteries because of their low electrolytic performance (Ionic conductivity). Recent energy research has concentrated on the selection of low-cost, high-ion conducting solid-state electrolytes for better battery performance.

In recent decades, critical research in Solid Electrolytes has identified numerous synthetic polymers like PAN, PVA, PMMA, PVDF, etc. [70–76] as excellent polymeric mediums. These synthetic polymer-based SPEs show relatively high ionic conductivity at room temperature, better mechanical stability, flexibility, and commendable performances in several small-scale batteries. Apart from its good performance, few Solid Electrolytes with the high cost and eco-toxic nature are not feasible for commercial applications and green energy motives.

To overcome the above issues, naturally abundant biopolymers, mainly derived from the chemical processing of biological sources, are extensively examined for their applicability as a polymer medium in Solid Polymer Electrolytes. Bio polymers like cellulose, pectin, seaweeds, gums, and several polysaccharides [77–86] are investigated for their electrochemical performances as Solid Polymer Electrolytes in batteries. Though bio polymers show good flexibility, mechanical stability, and eco-friendly behavior, cost-effective and abundant bio-based materials with better electrochemical properties may extend the commerciality of solid-state batteries to global needs.

Along with the chief material (Ceramic/Polymer), the addition of suitable ionic liquids, nano-fillers, plasticizers, etc., provides effective ion-conducting properties. Here, the evolution of each material is to overcome the issues faced by the other. Lithium-ion batteries are one of the promising candidates for electrochemical density, enhanced cycling stability, and less weight. The evolution of lithium-ion batteries has progressed from Li-metal batteries with Lithium metal as an anode to Li-intercalation compound electrodes to overcome the major issues such as dendrite growth and flammability faced by Li-metal batteries. As a result, new approaches such as electrolyte additives and the usage of advanced separators are utilized to improve the device's performance and safety [16].

In Li-ion batteries, the main requirements associated with the electrolyte include a) high ionic conductivity, b) inertness towards the various species that may be present at the stage of assembly and/or resultant from the electrochemical or side reactions, thus ensuring low or no emissions of pollutant species, c) reasonably low cost, d) being able to stand a relatively wide temperature range warranting safety conditions, and e) chemical and electrochemical stability. It is important to point out that such property does not refer to just intrinsic electrolyte characteristics, but it is also related to its interactions with the electrode materials.

The most typical and widespread electrolyte materials are those based on non-aqueous liquid solvents and include salts and additives. This class of liquid electrolytes is highly polar, thus having very good ionic conductivity properties, with low toxicity. However, these electrolytes are usually flammable, thus posing important safety issues. For that reason, solid electrolytes, including polymers, gels, and glassy matrices, are extensively investigated and utilized. Overall, the main issue associated with solid electrolytes is related to their low ionic conductivities. Working both in the chemistry of the basic polymer and adding other components are the most popular solutions to overcome this problem.

In device fabrication, a well-known problem using solid electrolytes is high interfacial resistance; to overcome this drawback, various methods such as surface modification strategy [16], usage of separator [87], and different coating methods have been employed [88]. The use of liquid electrolytes in lithium batteries results in a high discharge capacity of 2000 mAh^{-1} g^{-1} [89], but the evolving solid-state electrolytes have been inferred with a maximum discharge capacity of 800 mAh^{-1} g^{-1} [90]. Essential steps should be taken to reduce the interfacial resistance and increase the discharge capacity of solid-state electrolytes to utilize the same in future commercial applications of batteries.

5.9 SOLID-STATE ELECTROLYTE FOR BATTERIES

According to the literature review [69, 91, 92], Solid Polymer Electrolytes with advanced advantages act as compatible materials in the fabrication of electrochemical energy storage devices such as batteries because of their acceptable mechanical stability, flexible nature, and excellent interfacial contact between electrodes and electrolytes. This results in a better performing solid-state battery. There are various kinds of primary and secondary batteries such as Lithium-ion batteries, Magnesium-ion batteries, Sodium-ion batteries, Zinc-ion batteries, Proton batteries—as well as Aluminium-air batteries, Zinc-air batteries, etc. As per the survey, numerous research works have been carried out using different solid-form electrolytes in the construction of solid-state batteries and are tabulated in Table 5.4.

5.10 FUTURE SCOPE OF SOLID ELECTROLYTES AND SOLID-STATE ENERGY STORAGE DEVICES

The solid-state battery is one of the most promising solutions for the future generation of batteries with their high thermal stability, making them significantly safer and longer-lasting than traditional electric vehicle batteries. Solid-state batteries employ a solid electrolyte consisting of glass, ceramics, and solid polymers. All-solid-state batteries promise greater energy storage density, increased reliability and wear resistance, fast charging, and, most importantly, improved operational safety. At high temperatures, liquid electrolytes become volatile and flammable. However, solid electrolytes have high thermal stability, which limits the risk of fire or explosion. Solid-state batteries have a higher energy density per unit area due to their compact size [17].

The high thermal stability of solid-state electrolytes classifies it as a significant substitute for conventional LIBs. Solid-state batteries are promoted by their greater energy density, wear resistance, reliability, and improved operating stability. The dendrite formation, particularly in lithium metal batteries, can be eliminated by the employment of solid-state electrolytes in the commercial lithium battery arena. Considerable efforts to increase the ionic conductivity of solid-state electrolytes to $\sim 10^{-3}$ S/cm create tremendous application opportunities. Various research approaches to minimize the interfacial resistance between solid-state electrolyte/electrode interfaces such as flat arrangement of solid-state electrolyte on electrode [98] and introduction of interlayer production [16] validate solid-state electrolytes for near-future commerciality.

Currently, batteries in electric vehicles have numerous shortcomings such as limited drive range, intolerance to hard usage, short operating life, expensive materials, and a complicated packaging process. The application of solid-state batteries in energy vehicles is expected to overcome these shortcomings by simplifying the design and handling of energy storage devices. For considerable commercialization of solid-state batteries in market applications, accurate optimization between electrodes and solid-state electrolytes is necessary. Several crystalline and semi-crystalline materials like $LiCoO_2$, $LiMn_2O_4$, $LiFePO_4$, and $LiFeSO_4$ are extensively examined for their applicability as cathode material in all-solid-state Li-ion batteries. It is expected that development in solid-state Li-ion batteries will reduce 35% lithium usage from present use

TABLE 5.4
Review on the Fabrication of Solid-State Batteries with Solid Electrolytes

S. No	SSE	Electrodes (C[e]/A[f])	Battery	Voltage (V)	Discharge Capacity ($mAh^{-1}\ g^{-1}$)	No. of cycles	Rate	Ref.
1	$Li_5La_3Ta_2O_{12}$	$LiCoO_2$/Li	LIB(S)[a]	—	10.96	30	0.02	[19]
2	Ta-$Li_7La_3Zr_2O_{12}$	$LiFePO_4$/Li	LIB(S)	—	150	100	0.05	[29]
3	Cubic Na_3PS_4	Na–Sn/TiS_2	SIB(S)[b]	1.6	90	10	—	[21]
4	80 Li_2S: 20 P_2S_5	$LiCoO_2$/80 SnS: 20 P_2S_5	LIB(S)	—	400	50	—	[9]
5	80 Li_2S: 20 P_2S_5	CuS/Li–In	SRB(S)[d]	—	650	20	—	[102]
6	PPC/LiTFSI	$LiFePO_4$/Li	LIB(S)	3.35	138	1000	0.05	[39]
7	1 MEC:DMCLiPF_6	NCM/Li foil	LIB(S)	—	185	250	2	[98]
8	Agar-Agar/LiCl	$LiFePO_4$/G	LIB(S)	2.62	—	15	—	[40]
9	Gellan gum/NH_4HCO_2	MnO_2+G/Zn+Zn.SO_4 + G	PIB(S)[c]	2.27	—	11	—	[41]
10	Gellan gum/NH_4SCN	PbO_2+V_2O_5+G/Zn + Zn.SO_4	PIB(P)	1.62	—	—	—	[42]
11	Gellan gum/LiCl	PbO_2 + V_2O_5 + G/Zn + Zn.SO_4	LIB(P)	1.88	—	—	—	[8]
12	Moringa seed/NH_4SCN	PbO_2 + V_2O_5 + G/Zn + Zn.SO_4	PIB(P)	1.68	—	—	—	[52]
13	PEO/$NaClO_4$/TiO_2	$Na_{2/3}Co_{2/3}Mn_{1/3}O_2/Na_{halfcell}$	SIB(S)	—	49.2	25	0.1	[103]

a LIB – Lithium-ion conducting solid-state battery.
b SIB – Na-ion conducting solid-state battery.
c PIB – Proton (H^+) conducting solid-state battery.
d SRB – Sulphur-based rechargeable battery.
e C – Cathode.
f A – Anode.

[99]. Various solid-state Li-ion conducting electrolytes such as Lithium-phosphorous-oxynitride, NASICON-based LAGP/LATP electrolytes, perovskite-type LLTO electrolytes, garnet-type LLZO electrolytes, and polymer-based PEO, PMEEGE, PPC-LiTFSI solid electrolyte are investigated for their applicability in commercial solid-state batteries [100, 101]. Further, solid-state batteries will reduce carbon footprints 24%. Figure 5.6 represents the charge-discharge curves for various solid-state batteries with ceramic and polymer-based electrolytes. The charge-discharge curve ascribes the discharge capacity (mAh g^{-1}) with respect to voltage (V).

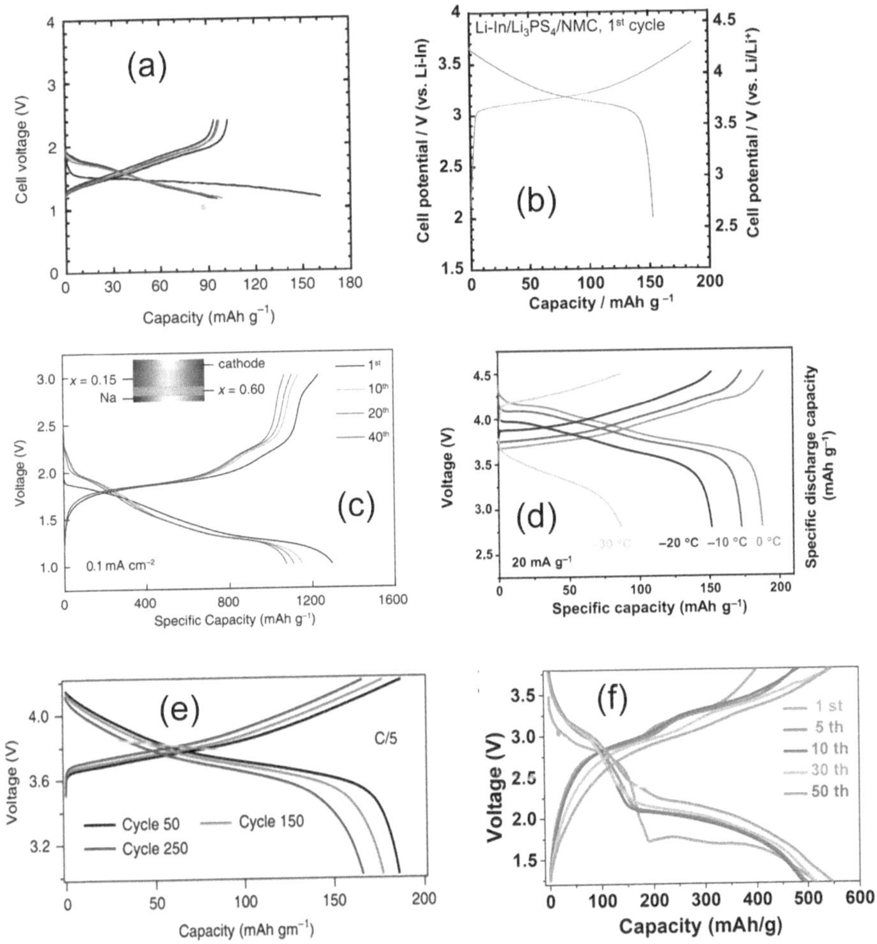

FIGURE 5.6 Typical charge-discharge curves for ceramic and based electrolytes – (a) Na_3PS_4 (CE); (b) Li_3PS_4 (CE); (c) $Na_3\ S_{4-x}O_x$ (where $0 < x \leq 0.60$) (CE); (d) $LiNi_{0.8}CO_{0.1}Mn_{0.1}O_2$ (PE); (e) 1M EC: DMC $LiPF_6$ (PE); (f) FeF_3 (PE) [21, 93–97]. (The pictures are published under Common Creative License of Nature Publications.) Notes: *CE – Ceramic-based electrolytes, *PE – Polymer-based electrolytes, *Na_3PS_4 – Trisodium sulfanylidene (trisulfido) – lambda 5-phosphane, *Li_3PS_4 – Lithium thiophosphate, *EC – Ethylene Carbonate, *DMC – Dimethyl Carbonate, *$LiPF_6$ – Lithium hexafluorophosphate, *FeF_3– Ferric fluoride.

5.11 CONCLUSIONS

The future of human civilization relies heavily on sustainable energy production. The need for green and reliable energy sources is now a global necessity, and society must make every effort to produce more efficient and affordable energy devices. Solid-state electrolytes offer a promising solution for meeting these demands and have the potential to be used as a commercial application in high-performance solid-state batteries.

The current trend in energy research has shed light on the prospects of solid-state electrolytes as an electrolyte in simple and complex energy storage and conversion devices. However, widespread application of solid-state electrolytes in various energy devices, particularly batteries, will require extensive research and expertise.

It is encouraging to note that the evolution of human culture and science has always led to solutions for global challenges such as the current energy crisis. Solid-state electrolytes hold immense promise for resolving this crisis, but much effort and innovation will be needed to fully harness their potential. Therefore, it is imperative to continue research and development in this field to realize the dream of sustainable and reliable energy production for a better future.

REFERENCES

1. P. Knauth and H. L. Tuller, *J. Am. Ceram. Soc.*, 2002, 85(7), 1654–1680.
2. O. Paschos, J. Kunze, U. Stimming and F. Maglia, *J. Phys.: Condens. Matter*, 2011, 23(23), 234110–234117.
3. C. Jing, Z. You, Z. Tengfei, L. Shigang, R. Kun, G. Zaiping, L. Huakun and Y. Xuebin, *Adv. Mater.*, 2019, 31(1), 1803533.
4. R. E. Pino, J. W. Bohl, N. McDonald, B. Wysocki, P. Rozwood, K. A. Campbell, A. Oblea and A. Timilsina, *IEEE/ACM Int. Symposium on Nanoscale Architectures*, 2010, 30(10), 225–229.
5. W. Xue, C. D. Sewell, Q. Zhou and Z. Lin, *Angew. Chem. Int. Ed.*, 2022, 134(34), 202206512–202206519.
6. P. V. Wright, *Electrochim. Acta*, 1998, 43(10–11), 1137–1143.
7. J. W. Fergus, *J. Power Sources*, 2010, 195(15), 4554–4569.
8. S. Aafrin Hazaana, A. Joseph, S. Selvasekarapandian, R. Meera Naachiyar, M. Vengadesh Krishna and N. Muniraj Vignesh, *Polym. Bull.*, 2023, 80, 5291–5311.
9. M. Tatsumisago, F. Mizuno and A. Hayashi, *J. Power Sources*, 2006, 159(1), 193–199.
10. H. Nafe, *Solid State Ion.*, 1984, 13(3), 255–263.
11. F. G. Will and S. P. Mitoff, *J. Electrochem. Soc.*, 1975, 122(4), 457–465.
12. P. Boolchand and W. Bresser, *Nat.*, 2001, 410(6832), 1070–1073.
13. M. Hou, F. Liang, K. Chen, D. Yongnian and X. Dongfeng, *Nanotechnol.*, 2020, 31(13), 132003–132010.
14. A. M. Azad, S. Larose and S. Akbar, *J. Mater. Sci.*, 1994, 29(16), 4135–4151.
15. E. Kuhnert, L. Ladenstein, A. Jodlbauer, C. Slugovc, T. Gregor, H. Martin R. Wilkening and D. Rettenwander, *Cell Rep.:Phys. Sci.*, 2020, 1(10), 100214–100219.
16. L. Wang, D. Liu, T. Huang, Z. Geng and A. Yu, *RSC Adv.*, 2020, 10(17), 10038–10045.
17. M. V. Leena Chandra, S. Karthikeyan, S. Selvasekarapandian, D. Vinoth Pandi, S. Monisha and S. Arulmozhi Packiaseeli, *Ionics*, 2016, 22(12), 2409–2420.
18. T. Maheshwari, K. Tamilarasan, S. Selvasekarapandian, R. Chitra and S. Kiruthika, *J. Solid State Electrochem.*, 2021, 25(3), 755–765.

19. M. Kotobuki and K. Kanamura, *Ceram. Int.*, 2013, 39(6), 6481–6487.
20. E. Matios, H. Wang, C. Wang, X. Hu, X. Lu, J. Luo and W. Li, *ACS Appl. Mater. Interfaces*, 2019, 11(5), 5064–5072.
21. A. Hayashi, K. Noi, A. Sakuda and M. Tatsumisago, *Nat. Commun.*, 2012, 3(1), 856–863.
22. M. Kotobuki, Y. Suzuki, H. Munakata, K. Kanamura, Y. Sato, K. Yamamoto and T. Yoshida, *J. Electrochem. Soc.*, 2011, 56(3), 1023–1029.
23. Y. Ruan, F. Guo, J. Liu, S. Song, N. Jiang and B. Cheng, *Ceram. Int.*, 2019, 45(2), 1770–1776.
24. X. Huang, C. Liu, Y. Lu, T. Xiu, J. Jin, M. E. Badding and Z. Wen, *J. Power Sources*, 2018, 382,190–197.
25. Z. Jiang, S. Wang, X. Chen, W. Yang, X. Yao, X. Hu, Q. Han and H. Wang, *Adv. Mater.*, 2020, 32(6), 1906221–1906228.
26. M. Tatsumisago and A. Hayashi, *Int. J. Appl. Glass Sci.*, 2014, 5(3), 226–235.
27. X. Song, C. Wang, J. Chen, S. Xin, D. Yuan, Y. Wang, K. Dong, L. Yang, G. Wang and H. Zhang, *Adv. Funct. Mater.*, 2022, 32(10), 2108706–2108711.
28. A. Hayashi, K. Noi, N. Tanibata, M. Nagao and M. Tatsumisago, *J. Power Sources*, 2014, 258, 420–423.
29. F. Du, N. Zhao, Y. Li, C. Chen, Z. Liu and X. Guo, *J. Power Sources*, 2015, 300, 24–28.
30. R. C. Xu, X. H. Xia, Z. J. Yao, X. L. Wang, C. D. Gu and J. P. Tu, *Electrochim. Acta.*, 2016, 219, 235–240.
31. F. Mizuno, S. Hama, A. Hayashi, K. Tadanaga, T. Minami and M. Tatsumisago, *Chem. Lett.*, 2002, 31(12), 1244–1245.
32. Y. Zhang, R. Chen, T. Liu, B. Xu, X. Zhang, L. Li, Y. Lin, C. W. Nan and Y. Shen, *ACS Appl. Mater. Interfaces*, 2018, 10(12), 10029–10035.
33. M. Eom, S. Choi, S. Son, L. Choi, C. Park and D. Shin, *J. Power Sources*, 2016, 331, 26–31.
34. W. Gorecki, M. Jeannin, E. Belorizky, C. Roux and M. Armand, *J. Phys.: Condens. Matter*, 1995, 7(34), 6823–6829.
35. V. S. Rangasamy, S. Thayumanasundaram and J. P. Locquet, *Solid State Ion.*, 2019, 333, 76–82.
36. C. M. Nyuk and M. I. N. B. M. Isa, *J. Sustainability Sci. Manage.*, 2017, 66, 42–48.
37. A. D. Khudyshkina, P. A. Morozova, A. J. Butzelaar, M. Hoffmann, M. Wilhelm, P. Theato, S. S. Fedotov and F. Jeschull, *ACS Appl. Polym. Mater.*, 2022, 4(4), 2734–2746.
38. V. M. Mohan, W. Qiu, J. Shen and W. Chen, *J. Polym. Res.*, 2010, 17(1), 143–150.
39. J. Zhang, J. Zhao, L. Yue, Q. Wang, J. Chai, Z. Liu, X. Zhou, H. Li, Y. Guo, G. Cui and L. Chen, *Adv. Energy Mater.*, 2015, 5(24), 1501082–1501087.
40. S. Aafrin Hazaana, A. Joseph, S. Selvasekarapandian, R. Meera Naachiyar, N. Muniraj Vignesh, *J. Solid State Electrochem.*, 2022, 27(3), 539–557.
41. R. Meera Naachiyar, M. Ragam, S. Selvasekarapandian, G. Aristatil, S. Aafrin Hazaana, N. Muniraj Vignesh and M. Vengadesh Krishna, *J. Polym. Res.*, 2022, 29(8), 89–94.
42. R. Meera Naachiyar, M. Ragam, S. Selvasekarapandian, M. Vengadesh Krishna and P. Buvaneshwari, *Ionics*, 2021, 27(8), 3415–3429.
43. K. Venkatesh, I. Jenova, S. Karthikeyan, S. Madeswaran, M. Arivanandhan, D. Joice Sheeba and S. Nithya, *Polym. Bull.*, 2022, 68, 1–23.
44. S. Shahrudin and A. H. Ahmad, *Sci. Lett. (ScL)*, 2016, 10(2), 26–30.
45. S. B. Aziz, M. M. Nofal, R. T. Abdulwahid, M. F. Z. Kadir, J. M. Hadi, M. M. Hessien, W. O. Kareem, E. M. A. Dannoun and S. R. Saeed, *Results Phys.*, 2021, 29, 104770–104778.
46. P. Nayak, Ismayil, V. Cyriac, S. Hegde, G. Sanjeev, M. S. Murari and Y. N. Sudhakar, *J. Non-Cryst. Solids*, 2022, 592, 121741.

47. S. B. Aziz, J. M. Hadi, E. M. A. Dannoun, R. T. Abdulwahid, S. R. Saeed, A. Shahab Marf, W. O. Karim and M. F. Z. Kadir, *Polymers*, 2020, 12(9), 1938–1945.
48. M. A. Ramlli and M. I. N. Isa, *Res. J. Recent Sci.*, 2014, 3(6), 59–66. E-ISSN: 2277, 2502.
49. A. S. F. M. Ahmad, S. B. Aziz, S. R. Saeed, Y. M. Yusof, R. T. Abdulwahid, S. Al-Zangana, W. O. Karim and M. F. Z. Kadir, *Membranes*, 2020, 10(12), 389.
50. R. Shilpa and R. Saratha, *Mater. Today: Proc.*, 2021, 47(9), 819–824.
51. R. Chitra, M. V. Krishna and S. Selvasekarapandian, *Polym. Bull.*, 2022, 79(6), 3555–3572.
52. N. Muniraj Vignesh, S. S. Jayabalakrishnan, S. Selvasekarapandian, S. Aafrin Hazaana, P. Kavitha and M. Vengadesh Krishna, *Ionics*, 2023, 29(6), 1–14.
53. H. K. Tran, Y. S. Wu, W. C. Chien, S. H. Wu, R. Jose, S. J. Lue and C. C. Yang, *ACS Appl. Energy Mater.*, 2020, 3(11), 11024–11035.
54. M. H. Buraidah and A. K. Arof, *J. Non-Cryst. Solids*, 2011, 357(16–17), 3261–3266.
55. G. P. Pandey, S. A. Hashmi and R. C. Agrawal, *Solid State Ion.*, 2008, 179(15–16), 543–549.
56. Y. J. Shen, M. Jaipal Reddy and P. P. Chu, *Solid State Ion.*, 2004, 175(1–4), 747–750.
57. N. Vanitha, C. Shanmugapriya, S. Selvasekarapandian, R. Meera Naachiyar, M. Vengadesh Krishna, S. Aafrin Hazaana, K. Nandhini and R. Mangalam, *Ionics*, 2022, 28(6), 2731–2749.
58. F. C. Chiu, *Adv. Mater. Sci. Eng.*, 2014, (1), 1–18.
59. Z. Raddaoui, S. El Kossi, R. Brahem, A. Bajahzar, A. Valentinovich Trukhanov, A. Leonidovich Kozlovskiy, M. Vladimirovich Zdorovets, J. Dhahri and H. Belmabrouk, *J. Mater. Sci.: Mater. Electron.*, 2021, 32(12), 16113–16125.
60. M. J. Rice and W. L. Roth, *J. Solid State Chem.*, 1972, 4(2), 294–310.
61. H. Schroeder, *J. Appl. Phys.*, 2015, 117(21), 215103–215114.
62. M. Ieda, G. Sawa and S. Kato, *J. Appl. Phys.*, 1971, 42(10), 3737–3740.
63. S. A. Moiz, I. A. Khan, W. A. Younis and K. S. Karimov, *Conduct. Polym.*, 2016, 5, 91–103.
64. M. C. Martin and M. L. Mecartney, *Solid State Ion.*, 2003, 161(1–2), 67–79.
65. J. Bonarski and W. Olek, *Wood Sci. Technol.*, 2006, 40(2), 159–171.
66. B. E. Warren, *X-Ray Diffraction*, Dover Publications, 1990.
67. X. Fan and C. Wang, *Chem. Soc. Rev.*, 2021, 18, 10001–10704.
68. S. S. Zhang, *J. Power Sources*, 2013, 231, 153–162.
69. T. Famprikis, P. Canepa, J. A. Dawson, M. Saiful Islam and C. Masquelier, *Nat. Mater.*, 2019, 18(12), 1278–1291.
70. H. K. Yoon, W. S. Chung and N. J. Jo, *Electrochim. Acta*, 2004, 50(2–3), 289–293.
71. J. Malathi, M. Kumaravadivel, G. M. Brahmanandhan, M. Hema, R. Baskaran and S. Selvasekarapandian, *J. Non-Cryst. Solids*, 2010, 356(43), 2277–2281.
72. M. S. Su'Ait, A. Ahmad, H. Hamzah and M. Y. A. Rahman, *J. Phys. D: Appl. Phys.*, 2009, 42(5), 055410.
73. Y. Wu, Y. Li, Y. Wang, Q. Liu, Q. Chen, M. Chen, *J. Energy Chem.*, 2022, 64, 62–84.
74. C. Nusrath Unnisa, S. Chitra, S. Selvasekarapandian, S. Monisha, G. Nirmala Devi, V. Moniha and M. Hema, *Ionics*, 2018, 24(7), 1979–1993.
75. K. P. Radha, S. Selvasekarapandian, S. Karthikeyan, M. Hema and C. Sanjeeviraja, *Ionics*, 2013, 19(10), 1437–1447.
76. M. A. Ashraf, M. J. Maah, A. K. Qureshi, M. Gharibreza and I. Yusoff, *Desalin. Water Treat.*, 2013, 51(16–18), 3650–3661.
77. S. Xiao, F. Wang, Y. Yang, Z. Chang and Y. Wu, *RSC Adv.*, 2014, 4(1), 76–81.
78. N. Vijaya, S. Selvasekarapandian, M. Sornalatha, K. S. Sujithra and S. Monisha, *Ionics*, 2017, 23(10), 2799–2808.

79. A. D. Azzahari, N. F. Abdul Mutalib, M. Rizwan, C. N. Abouloula, V. Selvanathan, F. Sonsudin and R. Yahya, *High Perform. Polym.*, 2018, 30(8), 993–1001.
80. F. G. Torres and G. E. De-la-Torre, *Carbohydr. Polym. Technol. Appl.*, 2021, 2, 100023.
81. M. F. Shukur, M. S. Azmi, S. M. M. Zawawi, N. A. Majid, H. A. Illias and M. F. Z. Kadir, *Phys. Scr.*, 2013, 157, 014049–014051.
82. M. F. Shukur, N. A. Majid, R. Ithnin and M. F. Z. Kadir, *Phys. Scr.*, 2013, 157, 014051–014055.
83. V. Moniha, M. Alagar, S. Selvasekarapandian, B. Sundaresan, R. Hemalatha and G. Boopathi, *J. Solid State Electrochem.*, 2018, 22(10), 3209–3223.
84. R. Chitra, P. Sathya, S. Selvasekarapandian, S. Monisha and V. Moniha and S. Meyvel, *Ionics*, 2019, 25(5), 2147–2157.
85. M. Muthukrishnan, C. Shanthi, S. Selvasekarapandian, R. Manjuladevi, P. Perumal and P. Christopher Selvin, *Ionics*, 2019, 25(1), 203–214.
86. S. Kiruthika, M. Malathi, S. Selvasekarapandian, K. Tamilarasan and T. Maheshwari, *Polymer Bull.*, 2020, 77(12), 6299–6317.
87. L. Hagman, P. Kierkegaard, P. Karvonen, A. Virtanen and J. Paasivirta, *Acta Chem. Scand.*, 1968, 22(6),1822–1832.
88. L. Cheng, E. J. Crumlin, W. Chen, R. Qiao, H. Hou, S. F. Lux, V. Zorba, R. Russo, R. Kostecki, Z. Liu, K. Persson, W. Yang, J. Cabana, T. Richardson, G. Chen, M. Doeff, *Phys. Chem. Chem. Phys.*, 2014, 16(34), 18294–18300.
89. D. T. Rogstad, M. A. Einarsrud and A. M. Svensson, *J. Electrochem. Soc.*, 2022, 169(11), 110531–1105337.
90. R. Miyazaki, *Front. Energy Res.*, 2020, 8, 171–176.
91. A. Manthiram, X. Yu and S. Wang, *Nat. Rev. Mater.*, 2017, 2(4), 16103–16119.
92. F. Zheng, M. Kotobuki, S. Song, M. O. Lai and L. Lu, *J. Power Sources*, 2018, 389, 198–213.
93. A. Kato, M. Yamamoto, F. Utsuno, H. Higuchi and M. Takahashi, *Commun. Mater.*, 2021, 2(1), 112–116.
94. X. Chi, Y. Zhang, F. Hao, S. Kmiec, H. Dong, R. Xu, K. Zhao, Q. Ai, T. Terlier, L. Wang, L. Zhao, L. Guo, J. Lou, H. L. Xin, S. W. Martin and Y. Yao, *Nat. Commun.*, 2022, 13(1), 2854–2863.
95. Z. Li, R. Yu, S. Weng, Q. Zhang, X. Wang and X. Guo, *Nat. Commun.*, 2023, 14(1), 482–492.
96. S. Choudhury, S. Stalin, D. Vu, A. Warren, Y. Deng, P. Biswal and L. A. Acher, *Nat. Commun.*, 2019, 10(1), 4398–4407.
97. J. Hu, C. Lai, K. Chen, Q. Wu, Y. Gu, C. Wu and C. Li, *Nat. Commun.*, 2022, 13(1), 7914–7920.
98. S. Shiraki, T. Shirasawa, T. Suzuki, H. Kawasoko, R. Shimizu and T. Hitosugi, *ACS Appl. Mater. Interfaces*, 2018, 10(48), 41732–41737.
99. Y. Guo, S. Wu, Y. B. He, F. Kang, L. Chen, H. Li and Q. H. Yang, *eScience*, 2022, 2(2), 138–163.
100. X. Lu, G. Xia, J. P. Lemmon and Z. Yang, *J. Power Sources*, 2010, 195(9), 2431–2442.
101. D. Karabelli, K. P. Birke and M. Weeber, *Batteries*, 2021, 7(1), 18–24.
102. A. Hayashi, T. Ohtomo, F. Mizuno, K. Tadanaga and M. Tatsumisago, *Electrochem. Commun.*, 2003, 5(8), 701–705.
103. Y. L. Ni'mah, M. Y. Cheng, J. H. Cheng, J. Rick and B. J. Hwang, *J. Power Sources*, 2015, 278, 375–381.

6 NASICON-type Li-ion Conducting Solid Electrolytes for All-Solid-State Li-ion Batteries

Sumit Khatua, Y. Bhaskara Rao, K. Ramakrushna Achary, Venkateswarlu Manne, and L. N. Patro

6.1 INTRODUCTION

All-solid-state Li-ion batteries (ASSLIBs) have been identified as a promising next-generation energy storage technology due to their potential for higher safety, more energy density, and good performance compared to traditional Li-ion batteries. Research studies on Li-ion batteries (LIBs) are continuously gaining significant scientific interest among the various metal-ion batteries such as Na^+, Mg^{2+}, Al^{3+}, and so on, because of their high energy densities, excellent cyclabilities, and low self-discharge rates [1, 2].

LIBs were commercialized by Sony in 1991. Later in 2019, the Nobel Prize in Chemistry was awarded to J.B. Goodenough, M.S. Wittingham, and A. Yoshino for their noteworthy contributions to the development of LIBs [3]. The demands on LIBs have grown rapidly since their commercialization because of their wide range of industrial and portable electronic device applications, such as mobiles, laptops, digital cameras, etc. In addition, they have a strong capability to store energy from different renewable energy sources (solar and wind power) towards the achievement of a fossil-fuel-free world [1, 4, 5]. Moreover, the electrification of vehicles would not be possible without the ever-growing performance of LIB cells together with their cost-effective mass production.

However, the present generation of LIBs in general faces several challenges, which include further advancement of their energy and power densities for large-scale applications, making them economically viable, recycling Li and other expensive elements such as Co, if any, from the cathode, and essentially improving their safety requirements [5–7]. Several safety issues such as leakage, flammability, and limited temperature range of operation could be resolved by designing an all-solid-state battery where the liquid electrolyte in the conventional battery will be replaced by a suitable solid electrolyte. The use of a solid electrolyte in a battery could also aid in a longer cycle life and higher energy density [7, 8].

DOI: 10.1201/9781003441182-7

Energy densities in the range of 200–250 W h kg^{-1} close to their physicochemical limits are achieved by using current LIBs. The formation of Li dendrites often limits the use of metallic Li as an anode with liquid electrolytes. The energy density of LIBs could be improved further by designing all-solid-state batteries that could allow the employment of high-capacity metallic Li as an efficient anode material [9, 10]. The solid electrolytes can eliminate the unwanted chemical reaction of the dissolved active materials that prevents the electrode crosstalk, improving the long-term stability of LIBs. The development of high-performance solid electrolytes remains a critical challenge for the practical implementation of ASSLIBs.

NASICON-type Li-ion conducting solid electrolytes (LICSEs) have attracted significant attention as a promising candidate for use in ASSLIBs because of their high Li-ion conductivity and excellent chemical stability. These materials have a unique crystal structure that enables fast ion transport while maintaining structural integrity, making them ideal for use in ASSLIBs. In recent years, significant progress has been made in the development and characterization of NASICON-type LICSEs for use in ASSLIBs. This review aims to provide an overview of recent research progress in this area and discuss the potential of NASICON-type LICSEs as a key component for the development of high-performance ASSLIBs.

Solid electrolytes are those materials that allow the macroscopic movement of ions through their structure, providing high ionic conductivity values ($>10^{-5}$ S cm^{-1}) at ambient temperature. They are electronic insulators and are often referred to as superionic conductors or fast ion conductors. They are categorized according to the type of mobile ions responsible for the conduction, namely, cation conductors (H^+, Li^+, Na^+, Cu^+, Ag^+, Mg^{2+}) and anion conductors (F^-, Cl^-, O^{2-}). Based on their microstructure, they are classified into various categories, such as crystalline or polycrystalline frameworks, glassy electrolytes, polymer electrolytes, composites, and gel electrolytes [11, 12].

Solid electrolytes attract various technological solid-state ionic device applications ranging from chemical sensors, smart windows, fuel cells, supercapacitors, and memristors to batteries [12, 13]. An ideal solid electrolyte should exhibit high room temperature (RT) ionic conductivity values to allow the fast diffusion of the mobile ions between the electrodes, lower electronic conductivity (ionic transport number close to one) to impede the short circuit, a wide electrochemical potential window to allow the electrolyte to be useful with various electrode materials, be thermally and chemically stable during the charge/discharge cycles, and have low interfacial resistance with the electrodes [7, 13, 14].

LICSEs are categorized into various groups such as LISICON (Ex: $Li_{10}GeP_2S_{12}$), argyrodite (Ex: Li_6PS_5Cl), NASICON (Ex: $LiTi_2(PO_4)_3$), garnet (Ex: $Li_7La_3Zr_2O_{12}$), and perovskite (Ex: $Li_{3x}La_{2/3-x}TiO_3$)-type materials, depending on their structure, and each group of materials has its own advantages and disadvantages [15, 16]. Oxide-based ceramic materials are always advantageous compared to sulphide-based ones, as the latter materials often suffer from instability with air and moisture, and, moreover, they are toxic. Garnet-based materials exhibit higher mechanical strength and good electrochemical stabilities; however, they are relatively unstable in air, and the raw materials are expensive [15, 17]. Perovskite-based materials often show good chemical stabilities; however, they exhibit relatively lower ionic conductivity values and involve harsh preparation conditions [15].

Li-ion conducting materials exhibiting NASICON (Na-super ion conductors, $Na_3Zr_2Si_2PO_{12}$) structure are widely studied because of their relatively good ionic conductivity values and thermal, mechanical, and electrochemical stabilities in the ambient atmosphere; moreover, the raw materials are relatively cheap [17–19]. In addition, the *dc* polarization study on these materials using the blocking electrodes confirmed that they are ionic conductors with ionic transport numbers close to one. NASICON-type materials in general have been known since their discovery in 1976 by J.B. Goodenough [20]. The general formula of NASICON-type solid electrolytes is $AMM'(PO_4)_3$; here, A is the alkali or alkaline-earth metal ion. The M and M' sites are di/tri/tetra/pentavalent cations to suitably balance the charge neutrality of the compound [21]. Among them, $LiM_2(PO_4)_3$; M: Zr, Ti, and Ge (LMP)-based materials have drawn substantial scientific interest because of their relatively higher Li-ionic conductivity values at ambient temperature and their potential applications in batteries and supercapacitors [17, 22]. A few reports are also known on the use of LMP-based systems as surface modification materials for different cathode materials, such as $LiCoO_2$, to improve the capacity and cycle performance of the cell [23, 24].

This chapter highlights the development of LMP-based materials and their applications in ASSLIBs. The first part of the chapter discusses the crystal structure of bare-LMP systems exhibiting rhombohedral structure and, further, the four polymorphs of $LiZr_2(PO_4)$ (LZP), namely, monoclinic, orthorhombic, triclinic, and rhombohedral phases of LZP. More often, NASICON-type solid electrolytes contain impurity phases, which are sometimes found to be unavoidable even after optimizing their many synthesis parameters [25]. Several synthesis methods, such as solid-state reaction, sol-gel, and synthesis conditions, such as sintering temperatures, sintering durations, sintering atmospheres, and so forth—known to achieve the phase pure material—are highlighted.

Major studies on NASICON-type solid electrolytes are known to improve their ionic conductivity values close to 10^{-4}–10^{-3} S cm^{-1}, similar to the liquid electrolytes. The ionic conductivity (σ) of a solid electrolyte is given by $\sigma = nq\mu$. Here, n is the density of the charge carriers with their mobility, μ and charge, q. There are two primary strategies known to improve the conductivity of a solid electrolyte: one by enhancing the number of mobile ions per unit cell and/or the second by increasing their mobility by enlarging the size of the bottleneck [26]. The influence of various dopants of different oxidation states and their levels on the sintering temperature and ionic conductivity of LMP systems is highlighted. In addition, the Li-ion conduction in LMP-based composite polymer electrolytes and thin film electrolytes is discussed. In the end, the use of LMP-based solid electrolytes on ASSLIBs with different types of well-known cathode materials such as $LiFePO_4$ (LFP) and Li-Ni-Mn-Co-O_2 (LNMC) is reviewed.

6.2 CRYSTAL STRUCTURE

The crystal structure of $LiM_2(PO_4)_3$ (M: Zr, Ti, and Ge) solid electrolytes exhibiting NASICON-type rhombohedral structure $\left(R\bar{3}c\right)$ is shown in Figure 6.1a. The structure consists of MO_6 octahedra units interconnected with the corner-sharing PO_4

tetrahedra units, forming a 3D network that provides channels for Li-ion conduction. There are two possible sites available for the Li-ions: the M1 site with six-fold coordination located directly between two MO_6 octahedra and the M2 site with eight-fold coordination lying between the two columns of MO_6 octahedra. Li-ions migrate along the M1-M2-M1 sites through the hopping process. The motion of Li-ions between the M1 and M2 sites is often restricted by the bottleneck made up of three O-atoms bound to the PO_4 tetrahedron and MO_6 octahedron pockets [27, 28].

The ionic size of the M cation in LMP solid electrolytes plays a crucial role in determining the effective width of the bottleneck, which is known to control the mobility of the Li-ions and hence the conductivity of the material [16]. Martínez-Juárez et al. established the relation between the size of the cations, M and M', and the size of the bottleneck responsible for the Li-ion conduction in $LiMM'(PO_4)_3$ (M, M′: Ge, Ti, Sn, and Hf) systems crystallizing in rhombohedral structure [29].

Among the LMP-systems, LZP exhibits four different polymorphs, namely triclinic (α'-phase), rhombohedral (α-phase), monoclinic (β'-phase), and orthorhombic (β-phase), depending on the type of the initial precursors and synthesis methodologies adopted (Figure 6.1b) [17, 28, 30, 31]. Higher preparation temperature ($\geq 1100^{\circ}$C) favors the formation of LZP in α'-phase (triclinic, $P\bar{1}$), whereas lower preparation temperature ($< 1100^{\circ}$C) results in the formation of LZP in β'-phase $\left(\text{monoclinic}, P2_1/n\right)$. The α'-phase of LZP transforms into α-phase (rhombohedral, $R\bar{3}c$) at around 40 °C, whereas the β'-phase of LZP transform into β-phase (orthorhombic, *Pbna*) at around 300 °C [30–32]. In addition, the structural phase transitions; $\alpha' \leftrightarrow \alpha$ and $\beta' \leftrightarrow \beta$ transitions are reversible. However, the direct transformations of LZP between the α and β phases are not known.

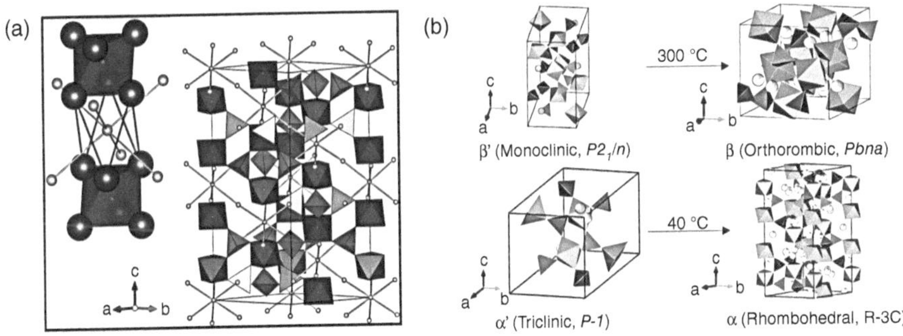

FIGURE 6.1 (a) Representations of a typical NASICON structure. Pathways for Li^+ motion is drawn between M1 and M2 sites. The structure in the upper-lefthand corner shows a close-up view of the conduction bottleneck region, with the constrictive window between O atoms; (b) schematic representation of $LiZr_2(PO_4)_3$ with the four polymorphs of the NASICON-type structure.

Source: (a) Reprinted with permission from ref [27]. Copyright 2014 American Chemical Society. (b) Reprinted with permission from ref [28]. Copyright 2017 American Chemical Society.

Among the four LZP polymorphs known, LZP crystallizing in rhombohedral structure shows the highest conductivity because of its larger unit cell volume. However, in monoclinic LZP, the Li-ions are fully ordered and usually located at the regular tetrahedral sites, with no neighboring vacant sites available for the hopping of the Li-ions [33, 34]. Various efforts, such as changes in the Zr-precursors, sintering temperature, aliovalent doping, and so forth, are attempted to stabilize LZP in rhombohedral structure at RT [35–38]. El-Shinawi et al. stabilized the LZP rhombohedral phase at RT by modifying the sol-gel method using EDTA/citrate as a complexing agent [35]. Balaya's group stabilized the LZP rhombohedral phase by doping with La^{3+}/Sr^{2+} at the Zr^{4+} site and reported RT Li-ion conductivity of 7.2×10^{-5} S cm^{-1} ($Li_{1.1}Zr_{1.9}La_{0.1}(PO_4)_3$) and 3.4×10^{-5} S cm^{-1} ($Li_{1.2}Zr_{1.9}Sr_{0.1}(PO_4)_3$) [36, 37]. Li et al. stabilized LZP exhibiting rhombohedral structure at RT using Zr acetate as the Zr-precursor and spark plasma sintering [38].

The ionic conductivity of a material signifies the transport behavior of the mobile ions and, furthermore, determines the functionality of the material in different solid-state ionic device applications [34–38]. It is usually measured by the impedance spectroscopy technique, which involves the measurement of the impedance in a wide range of frequencies (Hz-MHz) after applying a small potential to it. The Nyquist impedance plot is often drawn between the negative of the imaginary part of the impedance ($-Z''$) with its real part (Z'), which provides information about the bulk and/or grain boundary contributions to the total conductivity (Figure 6.2).

The capacitance value associated with the grain/bulk, grain boundary, and sample-electrode contributions is of the order of 10^{-12} F, 10^{-11}–10^{-8} F, and 10^{-7}–10^{-5} F, respectively [14]. The experimentally measured impedance data is often fitted with a

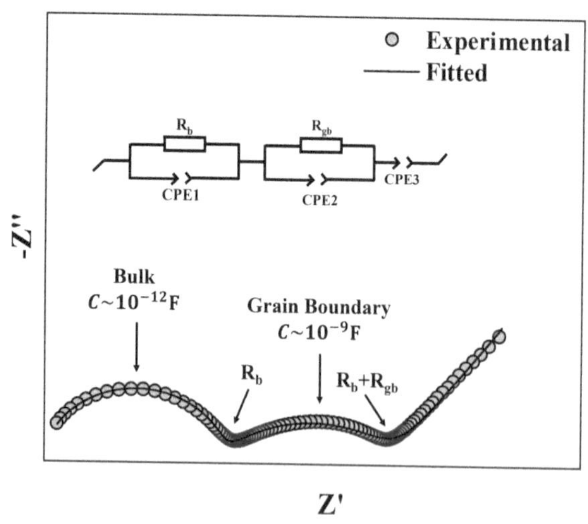

FIGURE 6.2 Schematic Nyquist impedance plot for a $LiM_2(PO_4)_3$ (M: Zr, Ti, and Ge) solid electrolyte showing both the bulk and grain boundary conductions. [R_b: bulk resistance, and R_{gb}: grain boundary resistance].

suitable equivalent circuit to find the bulk (R_b) and/or grain boundary (R_{gb}) resistances. The CPE in the equivalent circuit shown in Figure 6.2 is the constant phase element that explains the depression of the semicircles with their center below the real Z axis and non-ideal electrode geometry (at lower frequencies) [26]. The total conductivity (σ_t) is calculated by the relation, $\sigma_t = \frac{l}{(R_g + R_{gb})A}$; l and A are the thickness and area of the cross-section, respectively.

The plot of conductivity with the reciprocal of temperature ($\log \sigma_t$ vs. $1/T$) provides information about the activation energy required for the conduction of the mobile ions, whereas the frequency-dependent conductivity plot at a typical temperature is often fitted with the Almond-West equation to find the hopping rate of the mobile ions [26]. The temperature-dependent conductivity plot could also provide information about the characteristic phase transition, if any, from the observation of the change in the activation energy value. The latter parts of the chapter mainly discuss the preparation of LMP-based solid electrolytes and the role of various dopants on the ionic conductivity of LMP systems.

6.3 SYNTHESIS METHODOLOGY

The commercial applications of solid electrolytes mainly need a cost-effective synthesis procedure using inexpensive and easily available raw materials, lower sintering temperatures, and, importantly, a suitable provision for large-scale production. The kind of synthesis methodology often determines the phase formation, ionic conductivity, and densification of the material. The total ionic conductivity exhibited by a polycrystalline solid electrolyte is often limited by its grain boundary resistances. Various synthesis conditions are optimized to obtain a solid electrolyte with a dense microstructure to lower its grain boundaries.

The most common preparatory methods known for the synthesis of NASICON-type LICSEs include high-temperature solid-state reaction, aqueous solution-based, hydrothermal, and sol-gel [34–37, 39–42]. In general, the high-temperature solid-state reaction method is widely applied for the preparation of polycrystalline materials. The method involves the mixing of the starting materials in solid-state form using a pestle and mortar or ball mill, followed by high-temperature calcination. Furthermore, high-temperature sintering (700–1200 °C) is needed for the densification of the solid electrolyte in its pellet form. Although the preparatory method is simple, the need for high-temperature sintering could result in different secondary impurity phases, and additionally, the loss of Li during the high-temperature sintering could not be ruled out. Various reports are known on the synthesis of LMP-based systems by solid-state reaction [25, 39].

For the development of LTP- and LGP-based solid electrolytes, a low-cost aqueous solution-based method is known using water as the solution medium [40, 41]. The starting precursors and heat treatment temperature play a key role in determining the phase formation of the material. Ma et al. reported the effect of both sintering temperature and sintering duration on the synthesis of $Li_{1.3}Al_{0.3}Ti_{1.7}(PO_4)_3$ material using the aqueous solution-based method [40]. The hydrothermal method involves the creation of high-temperature and high-vapor-pressure reaction inside a closed vessel (autoclave), utilizing the aqueous solution as the reaction system. Kim et al. reported the influence of different preparation conditions, namely, reaction time, calcination, and sintering temperatures on the phase formation and conductivity of Al^{3+}-doped LTP by the hydrothermal method [42].

Among the mentioned preparatory methods, sol-gel is an efficient synthesis method for the preparation of NASICON-type solid electrolytes. In this method, colloidal solutions are made using the starting precursors in a solvent, and later a gel network is obtained after heating the solution. The gel is calcinated further to achieve phase-pure material. This method requires a comparatively lower sintering temperature and sintering duration than the solid-state reaction. Various reports are known on the preparation of LMP-based materials by the sol-gel route [28, 34–37, 43]. Both solid-state reaction and sol-gel are suitable preparatory methods known for industrial applications. Cassel et al. synthesized $Li_{1.2}Zr_{1.9}Ca_{0.1}(PO_4)_3$ by both solid-state reaction and sol-gel and reported that the sol-gel technique not only reduces the synthesis temperature from 1200 °C (solid-state reaction) to 1100 °C (sol-gel) but also improves the ionic conductivity from 1.2×10^{-7} S cm^{-1} to 7.17×10^{-7} S cm^{-1} [43]. Besides the synthesis methods, other parameters such as sintering temperature, sintering method, sintering duration, sintering aid, and sintering environment influence the structure and transport results of LMP-based materials (Figure 6.3).

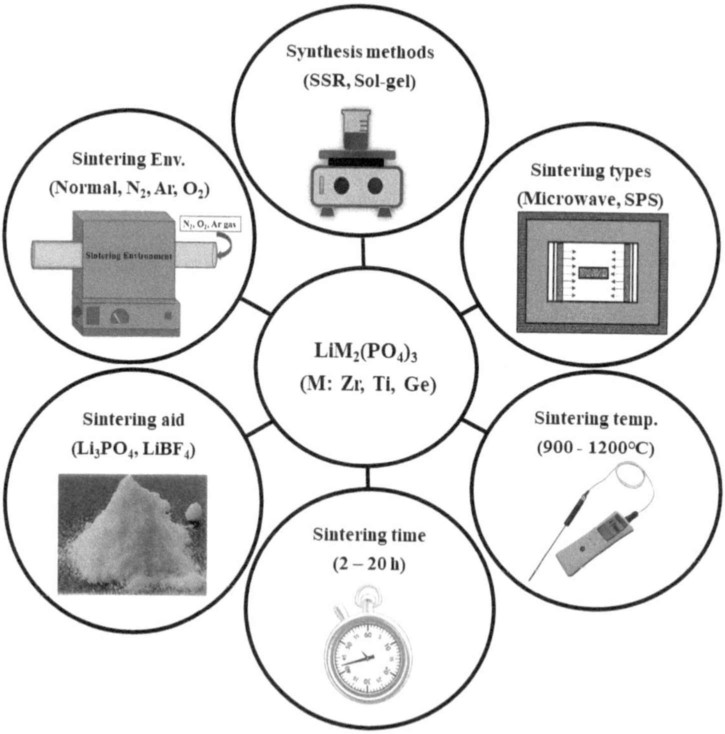

FIGURE 6.3 Various synthesis parameters known for the phase formation and conductivity of $LiM_2(PO_4)_3$ (M: Zr, Ti, and Ge)-based solid electrolytes.

The proper selection of the sintering temperature and sintering duration could result in the phase formation of the material with a dense microstructure. The uncontrolled use of these two parameters may result in the cracking of the pellet. The use of microwave-assisted or spark plasma sintering often reduces the sintering temperature and holding time in comparison to conventional/furnace sintering using the normal furnace. Among them, spark plasma sintering could increase densification, although the technique is expensive.

Chang et al. synthesized Al and V co-doped LTP ceramics sintered by both conventional sintering and spark plasma sintering and reported better densification of the material sintered in the latter method [44]. Lai et al. showed the synthesis of $Li_{1.2}Ca_{0.1}Zr_{1.9}(PO_4)_3$ material exhibiting RT ionic conductivity of 1.7×10^{-5} S cm^{-1} by microwave sintering, which reduced the sintering temperature from 1200 °C (conventional sintering) to 900 °C with only a 4 h holding time [45]. Kobayashi et al. reported an enhancement in conductivity of two orders (at 50 °C) in hydrothermally synthesized LTP using spark plasma sintering compared to conventional sintering [46]. The influence of sintering durations on the microstructure of $Li_{1+x}Al_xTi_{2-x}(PO_4)_3$ material at a fixed sintering temperature and their impact on the ionic conductivity are reported [47].

The ionic conductivity of NASICON-type solid electrolytes could be improved by using sintering aids such as Li_3PO_4 or $LiBF_4$, which are known to increase the contact area between the grains, reducing the grain boundary resistances [18, 48]. The selection of the precursors could also control the phase formation of LZP, which is known to have four different polymorphs. Li et al. reported the synthesis of LZP using Li_2CO_3, $(NH_4)_2HPO_4$, and four different types of Zr-precursors, namely, $Zr(AC)_4$, $ZrOCl_2$, $Zr(NO_3)_4$, and ZrO_2, and showed the successful formation of the LZP rhombohedral phase with $Zr(AC)_4$ as the Zr-source. However, the use of the other Zr-precursors results in the formation of the triclinic LZP phase [38]. A systematic study on the dependency of the annealing environment, such as normal, N_2, and O_2 atmospheres, on the ionic conductivity of Al-doped LZP system is reported, and the enhanced conductivity and Li-ion mobility are seen for the solid electrolyte prepared by using an O_2 environment [49].

Apart from the LMP-based polycrystalline electrolytes, various studies are also known on the synthesis and conductivity studies of LMP-based composite-polymer electrolytes, glass ceramics, and thin film electrolytes. Composite-polymer electrolytes are usually prepared by the solution casting technique [14, 50]. Glass ceramics are prepared by the controlled crystallization of the base glass at different temperatures [51, 52]. Reports are known on the development of LMP-based thin film electrolytes using different growth methods such as hydrothermal, aerosol deposition, and radio-frequency sputtering [53–55].

6.4 DOPING

Besides the synthesis methods and synthesis conditions, homogeneous doping is known to be an efficient approach to improve the conductivity of a solid electrolyte. Doping depends on the ionic radii and oxidation states of both the dopant ion and the site where it is doped. In LMP systems, the dopant elements are partially substituted, mainly at the M-site, which tunes the size of the bottleneck responsible for the Li-ion conduction. The substitution of the aliovalent ions often creates vacancies or interstitials, depending on both the oxidation states of the host-site and dopant ions. Various reports are known demonstrating an enhancement in the ionic conductivity of LMP systems after doping with trivalent ions such as Y^{3+}, Al^{3+}, La^{3+}, Cr^{3+}, etc., and bivalent ions such as Sr^{2+}, Ca^{2+}, etc., at the M^{4+} [Zr^{4+} (0.72 Å), Ti^{4+} (0.605 Å), and Ge^{4+} (0.53 Å)] site [28, 34, 36, 37, 56, 57].

Pareek et al. studied the influence of Al^{3+} substitution at the Zr^{4+} site of monoclinic LZP and reported an RT conductivity of 1.83×10^{-6} S cm^{-1} for $Li_{1.25}Zr_{1.75}Al_{0.25}(PO_4)_3$, which is about two orders higher than the LZP conductivity (2.27×10^{-8} S cm^{-1}) [34]. Ramar et al. reported the sol-gel synthesis of rhombohedral $Li_{1+x}Zr_{2-x}La_x(PO_4)_3$ (x = 0–0.2) materials with the best ionic conductivity for $x = 0.1$ compositions (7.2×10^{-5} S cm^{-1}) and later showed the chemical and structural stability of the material after exposing it to air, water, Li, acidic, and basic solutions [36]. A few reports are known on the substitution of bivalent ions such as Ca^{2+} or Sr^{2+} at the Zr-site of LZP prepared by different methods [37, 43, 57]. On LZP-based systems, a few reports are also known demonstrating the stabilization of the high Li-ion conducting rhombohedral phase at RT after aliovalent doping with $Y^{3+}/La^{3+}/Al^{3+}/Sr^{2+}$ at the Zr^{4+} site [28, 36, 37, 58]. On LTP solid electrolyte, various reports are mainly focused on Al^{3+} doping at the Ti^{4+} site synthesized by different methods [39, 59].

The successful synthesis of amorphous $Li_{1.3}Al_{0.3}Ti_{1.7}(PO_4)_3$ by mechanical milling at RT and, furthermore, the transformation of the amorphous material into its polycrystalline form exhibiting NASICON structure with RT ionic conductivity of 10^{-4} S cm^{-1} after sintering at 900 °C for 10 h is reported [59]. An earlier report by Hamdoune et al. showed the insertion of In^{3+} of different levels in $Li_{1+x}Ti_{2-x}In_xP_3O_{12}$, resulting in three different atomic arrangements preserving NASICON structure, namely, rhombohedral for $0 \leq x \leq 0.4$, orthorhombic for $0.4 < x < 1.0$, and monoclinic for $1.0 \leq x \leq 2.0$ with maximum ionic conductivity (2×10^{-2} S cm^{-1}) for $x = 0.35$ composition at 300 °C [60]. Kwatek and Nowińsk reported an increase in the ionic conductivity for LTP (1.51×10^{-8} S cm^{-1}) upon doping with Al^{3+}, $Li_{1.3}Al_{0.3}Ti_{1.7}(PO_4)_3$ (5.06×10^{-7} S cm^{-1}). Furthermore, they have also made LTP-based composite electrolytes by dispersing with an ionic liquid 1–butyl–3–methyl-imidazolium tetra-fluoroborate [BMIM][BF_4] of different weight percentages and reported an ionic conductivity of 2.62×10^{-4} S cm^{-1} and activation energy of 0.11 eV for the composite consisting of $Li_{1.3}Al_{0.3}Ti_{1.7}(PO_4)_3$ with 10 wt.% [BMIM][BF_4] [39].

Like LZP and LTP systems, the ionic conductivity of LGP is known to be further enhanced through doping and preparation processes. Kang et al. reported both the first principle density functional theory (DFT) calculations and experimental measurements on Al^{3+}-doped LGP exhibiting a four-order enhancement in conductivity value in comparison to the bare one. The enhanced result is further explained with respect to both the increase in Li-ion concentrations and Li-ion mobilities [61].

The double substitution approach is often found to be an efficient strategy for NASICON-based solid electrolytes, overcoming the limitations of traditional doping methods. Doping of Al^{3+} in place of the Ti^{4+} ion and V^{5+} in place of the P^{5+} ion in LTP shows an increase in ionic conductivity, not only by enhancing densification but also by changing the bottleneck size for Li-ion migration. By decreasing the grain boundary resistance through spark plasma sintering, an ionic conductivity of 2.6×10^{-4} S cm^{-1} has been achieved for the $Li_{1.3}Al_{0.3}Ti_{1.7}(PO_4)_{2.9}(VO_4)_{0.1}$ system [44].

Various reports are also known on LGP-based glass ceramics. Nikodimos investigated the dual doping of Al^{3+} and Sc^{3+} in the Ge^{4+} site of LGP by both computational and experimental methods. The doping not only helps to increase the crystal cell dimensions but also creates a new migration path. The Li-ion bulk conductivity calculated by the DFT and experiment is found to be 6.8×10^{-3} S cm^{-1} and 5.8×10^{-3} S cm^{-1}, respectively, for $Li_{1.5}Al_{0.33}Sc_{0.17}Ge_{1.5}(PO_4)_3$ at 25 °C [62]. The same group also reported the simultaneous doping of Al and Mg in LGP ($Li_{1.6}Al_{0.4}Mg_{0.1}Ge_{1.5}(PO_4)_3$) glass ceramics with RT bulk ionic conductivity of 7.4×10^{-3} S cm^{-1} [63]. Al-doped LGP glass ceramics $Li_{1.5}Al_{0.5}Ge_{1.5}(PO_4)_3$ exhibit an ionic conductivity of 5×10^{-4} S cm^{-1}, which is prepared via the melt quenching method, followed by sintering at 800 °C for 8 h [64].

Illbeigi et al. reported the simultaneous doping of Cr and Al in LGP, $Li_{1.5}Al_{0.4}Cr_{0.1}Ge_{1.5}(PO_4)_3$ glass ceramics prepared by the melt-quenching method followed by the crystallization at different temperatures and reported the highest Li-ion conductivity of 6.65×10^{-3} S cm^{-1} for the glass ceramic that was crystallized at 850 °C [65]. Table 6.1 summarizes the influence of various dopants and synthesis parameters on the ionic conductivity of LMP-based solid electrolytes. Among the various dopants known, Al^{3+} seems to be effective in LMP-based solid electrolytes by changing the lattice parameters that provide fast Li-ion conduction. Although the conductivity of LMP-based doped systems increases with an increase in dopant levels, the conductivity always passes through a maximum beyond which it decreases.

Apart from the framework polycrystalline form, solid electrolytes are further categorized into glassy electrolytes, polymer electrolytes, and composite electrolytes based on their phase, composition, and microstructure [12]. LMP-based solid electrolytes are often incorporated in various polymer matrices (composite-polymer electrolytes) not only to improve their mechanical and transport characteristics but also to reduce the interfacial resistance at the electrode-electrolyte interface in the device. Thus, a brief overview of polymer electrolytes in general and LMP-based composite polymer electrolytes is presented.

6.5 POLYMER ELECTROLYTES

Polymer electrolytes offer many advantages, such as flexibility, greater mechanical strengths, ease of preparation, and shape compactness, thus attracting promising applications in flexible wearable devices. In LIB's development, polymer-based electrolytes are classified mainly into three categories: dry solid-polymer electrolytes, gel-polymer electrolytes, and composite-polymer electrolytes [69].

In the case of a dry solid-polymer electrolyte, a Li-based salt is dispersed in a suitable polymer host of high molecular weight. PMMA (poly-methyl methacrylate), PEO (poly-ethylene oxide), PAN (poly-acrylonitrile), PVDF (poly-vinylidene fluoride), and so forth are often considered for the host of polymer electrolytes because of their high dielectric constant and strong Li-ion solubility. Among them, PEO is found to be the most commonly used polymer [70]. Although dry solid-polymer electrolytes have good electrochemical stability and compatibility, they still suffer from low ionic conductivity values [14, 50].

TABLE 6.1
Ionic Conductivity of LMP-Based Systems Prepared by Different Methods

S. No	Dopant	Compound	RT Conductivity (S cm^{-1})	Prep. Method	Calcination temperature	Sintering Temperature	Ref
			$LiZr_2(PO_4)_3$-based solid electrolytes				
1	—	$LiZr_2(PO_4)_3$ (monoclinic)	2.27×10^{-8}	Sol-gel	500 °C, 12 h	900 °C, 12 h (CS)	[34]
2	—	$LiZr_2(PO_4)_3$ (rhombohedral with additional triclinic reflections)	8.06×10^{-7}	Sol-gel	500 °C, 6 h 1000 °C, 6 h	1200 °C, 15 h (CS)	[37]
3	—	$LiZr_2(PO_4)_3$ (rhombohedral)	2.2×10^{-5} 3.8×10^{-5}	Solid-state reaction	900 °C, 10 h	1150 °C, 20 h (CS) 1000 °C, 10 min (SPS)	[38]
4	Al^{3+}	$Li_{1.25}Zr_{1.75}Al_{0.25}(PO_4)_3$ (monoclinic)	1.83×10^{-6}	Sol-gel	500 °C, 12 h	900 °C, 12 h (CS)	[34]
5	Al^{3+}	$Li_{1.275}Al_{0.275}Zr_{1.725}(PO_4)_3$ (rhombohedral)	3.1×10^{-6}	Solid-state reaction	900 °C, 6 h	1200 °C, 4 h (CS)	[58]
6	La^{3+}	$Li_{1.1}Zr_{1.9}La_{0.1}(PO_4)_3$ (rhombohedral)	7.2×10^{-5}	Sol-gel	500 °C, 6 h 1000 °C, 6 h	1280 °C, 18 h (CS)	[36]
7	Y^{3+}	$Li_{1.15}Y_{0.15}Zr_{1.85}(PO_4)_3$ (rhombohedral)	3.5×10^{-5}	Sol-gel	600 °C, 6 h	1200 °C, 2 mins (SPS)	[28]
8	Sr^{2+}	$Li_{1.2}Zr_{1.9}Sr_{0.1}(PO_4)_3$ (rhombohedral)	3.4×10^{-5}	Sol-gel	500 °C, 6 h 1000 °C, 6 h	1200 °C, 15 h (CS)	[37]
9	Ca^{2+}	$Li_{1.2}Zr_{1.9}Ca_{0.1}(PO_4)_3$ (rhombohedral)	4.9×10^{-5}	Solid-state reaction	500 °C, 1 h 800 °C, 6 h	1200 °C, 20 h (CS)	[57]
			$LiTi_2(PO_4)_3$-based solid electrolytes				
10	—	$LiTi_2(PO_4)_3$	1.51×10^{-8}	Solid-state reaction	900 °C, 10 h	—	[39]
11	—	$LiTi_2(PO_4)_3$	10^{-6} (50 °C)	Hydrothermal	—	1200 °C, 10 min (SPS)	[46]
12	Nb^{5+}	$Li_{1.5}Al_{0.5}Ti_{1.5}(PO_4)_3$ with 8wt.% of Nb_2O_5	3.54×10^{-4}	Sol-gel	600–800 °C, 3 h	880 °C, 5 h (CS)	[66]

13	Al^{3+}	$Li_{1.3}Al_{0.3}Ti_{1.7}(PO_4)_3$	1.21×10^{-3}	Solution-based	700 °C	800 °C, 12 h (CS)	[40]
14	Al^{3+}	$Li_{1.3}Al_{0.3}Ti_{1.7}(PO_4)_3$	$(3–4) \times 10^{-4}$	Sol-gel	700 °C, 1 h	1000 °C, 1 h (CS)	[67]
15	Al^{3+} and V^{5+}	$Li_{1.3}Al_{0.3}Ti_{1.7}(PO_4)_{2.9}(VO_4)_{0.1}$	2.6×10^{-4}	Solid-state reaction	900 °C for 2 h	1100 °C, 10 min (SPS)	[44]
16	Al^{3+} and Cr^{2+}	$Li_{1.4}Al_{0.3}Cr_{0.1}Ti_{1.6}(PO_4)_3$	1.06×10^{-3}	Sol-gel	500 °C, 4 h 800 °C, 4 h	1070 °C, 6 h (CS)	[56]
			$LiGe_2(PO_4)_3$-based solid electrolytes				
17		$LiGe_2(PO_4)_3$	7.2×10^{-7}	Solid-state reaction	400 °C, 2 h	900 °C, 10 h (CS)	[68]
18	Al^{3+}	$Li_{1.5}Al_{0.5}Ge_{1.5}(PO_4)_3$	3.02×10^{-4}	Solid-state reaction	—	800 °C, 2 h (CS)	[61]
19	Al^{3+} and Mg^{2+}	$Li_{1.6}Al_{0.4}Mg_{0.1}Ge_{1.5}(PO_4)_3$ Glass-ceramics	7.435×10^{-3}	Melting temperature: 1280 °C, 1 h Quenched temperature: 300 °C Crystallization temperature: 750 °C, 8 h			[63]
20	Al^{3+} and Cr^{2+}	$Li_{1.5}Al_{0.4}Cr_{0.1}Ge_{1.5}(PO_4)_3$ Glass-ceramics	6.65×10^{-3}	Melting temperature: 1450 °C, 1 h Quenched temperature: 300 °C Crystallization temperature: 850 °C, 8 h			[65]
21	Al^{3+} and Sc^{3+}	$Li_{1.5}Al_{0.33}Sc_{0.17}Ge_{1.5}(PO_4)_3$ Glass-ceramics	5.826×10^{-3}	Melting temperature: 1280 °C, 1 h Quenched temperature: 350 °C Crystallization temperature: 800 °C, 8 h			[62]

Note: CS: Conventional sintering, SPS: Spark plasma sintering.

Gel-polymer electrolytes are often made by incorporating large amounts of organic solvents into the polymer matrix. They offer good ionic conductivity values; however, they suffer from substantial corrosivity with metal electrodes [50]. The ionic conductivities of the dry solid-polymer electrolytes are often improved by dispersing with ceramic fillers such as NASICON/Al_2O_3/TiO_2 and are known as composite-polymer electrolytes. They exhibit an apparently higher Li-ion transference number (~0.5) compared to the dry solid-polymer and gel electrolytes [50, 70].

The use of composite-polymer electrolytes also adds stability at the electrode-electrolyte interfaces, reducing their interfacial resistances. Singh et al. reported the structural and transport behavior of $5LiCF_3SO_3$-95[$PEO_{0.3}LATP_{0.7}$] composite polymer electrolyte with a conductivity of 10^{-4} S cm^{-1} (45 °C) and Li^+ ionic transport number of 0.46. Furthermore, the group has also demonstrated the application of the composite-polymer electrolyte in an all-solid-state supercapacitor exhibiting a capacity of 2–4 F g^{-1}. Here, LATP stands for Al^{3-}doped LTP, $Li_{1.3}Al_{0.3}Ti_{1.7}(PO_4)_3$ composition [71]. Li and Wang reported the addition of 10 wt.% $Li_{1.5}Al_{0.5}Ti_{1.5}(PO_4)_3$, improving the ionic conductivity of PVDF-HFP polymer (poly-(vinylidene fluoride-cohexa-fluropropylene)) to 2.3×10^{-4} S cm^{-1} from 7.1×10^{-5} S cm^{-1} and also reported better cyclic performance and stability of the Li||$LiFePO_4$ battery using the composite-polymer electrolyte compared to the battery with the polymer electrolyte [72].

Das et al. synthesized a composite-polymer electrolyte using P(VDF-HFP) (poly (vinylidene fluoride-co-hexa-flurophosphate)), LiTFSI (lithium bis(tri-fluoro-methane-sulfonyimide)), EMITFSI (1-ethyl-3-methylimidazolium bis-(trifluoromethylsulfonyl-imide)), and $Li_{1.5}Al_{0.5}Ge_{1.5}P_{2.9}Si_{0.1}O_{12}$ glass ceramic, and further reported ionic conductivity of 4.49×10^{-3} S cm^{-1} with thermal stability up to 315 °C for the composite-polymer electrolyte [73].

Besides the LMP-based polymer electrolytes, a few articles are known on LTP-based thin film electrolytes grown by several methods such as hydrothermal, aerosol deposition, and radio-frequency sputtering [53–55]. Peng et al. reported Al-doped LTP thin films grown on the graphite substrate using the hydrothermal method by varying the Al sources, such as $Al_2(SO_4)_3$, Al metal, and Al $(OH)_3$, and their concentrations. The best conductivity (1.65×10^{-4} S cm^{-1}) is reported for the LTP thin film where 10 mM $Al_2(SO_4)_3$ (aluminium content) is used [53]. Inada et al. successfully deposited a $Li_{1.5}Al_{0.5}Ge_{1.5}(PO_4)_3$ thin film on a SUS316L stainless steel substrate by aerosol deposition, showing similar bulk conductivity (10^{-4} S cm^{-1}) with the $Li_{1.5}Al_{0.5}Ge_{1.5}(PO_4)_3$ powder [54]. In the following section, the application aspects of LMP-based inorganic, glass-ceramic, and composite-polymer electrolytes in all-solid-state batteries are highlighted.

6.6 APPLICATIONS OF $LIM_2(PO_4)_3$ (M: ZR, TI, AND GE)-BASED SOLID ELECTROLYTES IN BATTERIES

All-solid-state batteries are the next-generation batteries that could offer additional safety features, high energy densities, and a long cycle life. The development of an all-solid-state battery depends on the selection of the solid electrolyte. LMP-based solid electrolytes, particularly the doped solid solutions and composite polymer electrolytes, show promising ionic conductivity values, wide electrochemical stabilities, and ionic transport numbers close to unity, satisfying the necessary conditions for application in ASSLIBs.

The major challenges for ASSLIBs are towards reducing the interfacial impedance between the electrolyte and electrode interface that controls the cyclic performance of the battery. The large interfacial impedance occurs due to the unnecessary redox reactions emanating from the inter-diffusion, space charge layers, and electrochemical instabilities. In addition, poor mechanical contact between the electrode and electrolyte, volume expansion of the electrode, and surface micro-cracks could also give rise to the large interfacial impedance inhibiting the mobility of ions. This part of the chapter highlights the use of LMP-based solid electrolytes in ASSLIBs and various strategies known to suppress their electrode and electrolyte interfacial impedance values to achieve better electrochemical performance.

Xu et al. reported an all-solid-state $Li/Li_{1.15}Y_{0.15}Zr_{1.85}(PO_4)_3/LiMn_{1.5}Ni_{0.5}O_4$ battery delivering a capacity of 70 mA h g^{-1} at a current density of 30 μA cm^{-2} with a Coulombic efficiency of 96.6% after 100 cycles [28]. Zhou et al. demonstrated an all-solid-state $Li/Li_{1.2}Mg_{0.1}Zr_{1.9}(PO_4)_3/LiFePO_4$ battery delivering a discharge capacity of 155 and 140 mA h g^{-1} at current densities of 0.05 and 0.1 mA cm^{-2}, respectively, with 90% capacity retention after 50 charge/discharge cycles. This stable cyclic performance of the battery is explained with respect to the stable electrode-electrolyte interface originating from the increased wettability of the solid electrolyte with the Li-metal anode due to the formation of Li_3P interphases, which suppress the nucleation and growth of Li-dendrites [74]. A report is known on the use of $Li_{1.3}Al_{0.3}Ti_{1.7}(PO_4)_3$ as a surface modification material for the $LiCoO_2$ cathode material. The cell consisting of the surface-modified $LiCoO_2$ cathode material, Li-metal counter electrode, and $LiPF_6$ in EC (ethylene carbonate) and EMC (ethyl methyl carbonate) electrolytes exhibits a capacity of around 180 mA h g^{-1} with a high charge cut-off potential of 4.5 V (*vs.* Li/Li^+) [59].

Li et al. demonstrated an all-solid-state Li/LFP cell with a polymer catholyte exhibiting a discharge capacity of 140 mA h g^{-1} at a current density of 50 μA cm^{-2} at 80 °C [38]. Wu et al. demonstrated the use of LZP-based composite-polymer electrolyte in all-solid-state Li-metal batteries with LFP and $LiNi_{0.8}Mn_{0.1}Co_{0.1}O_2$ cathode materials exhibiting good cyclic performances at 40 °C [75]. Li and Wang made a composite solid electrolyte (CSE) consisting of 10 wt.% $Li_{1.5}Al_{0.5}Ti_{1.5}(PO_4)_3$ powder embedded in the PVDF-HFP (poly-(vinylidene fluoride-co-hexafluoropropylene)) polymer matrix and, furthermore, reported the electrochemical results of the LFP||CSE||Li cells exhibiting an initial specific capacity of 150 mA h g^{-1} and 130 mA h g^{-1} after 50 cycles at 0.2C rate. On the other side, LFP||polymer||Li cells showed an initial specific capacity of only 112 mA h g^{-1} and 54 mA h g^{-1} after 50 cycles [72].

Das et al. prepared a composite solid electrolyte (CSE) consisting of 20 wt.% of $Li_{1.5}Al_{0.5}Ge_{1.5}P_{2.9}Si_{0.1}O_{12}$ glass ceramic, P(VDF-HFP) (poly-(vinylidene fluoride-co-hexa-fluorophosphate)), LiTFSI, and EMITFSI, and later demonstrated the cell LFP/CSE/Li exhibiting a capacity of 151 mA h g^{-1} at 0.05C after the 50th cycle [73]. The discharge-specific capacities exhibited by various Li-batteries at different current densities using LMP-based solid electrolytes and well-known cathode materials such as LFP and LNMC are shown in Figure 6.4 [31, 41, 53, 64, 72–77].

Various approaches are often attempted by different groups to lower the interfacial resistances between the solid electrolyte and electrodes. The interfacial

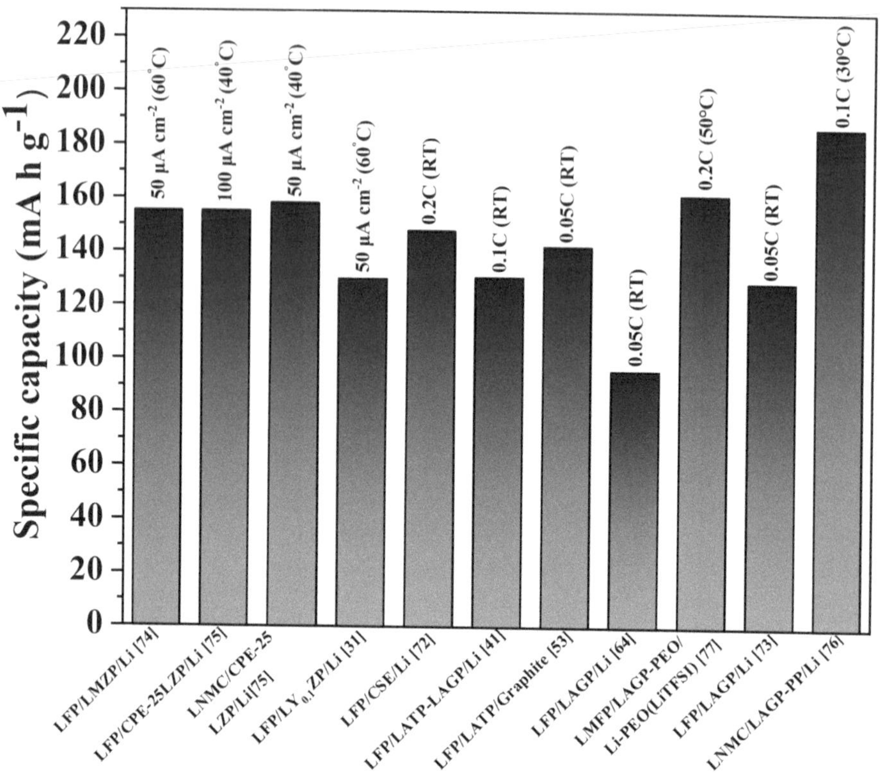

FIGURE 6.4 Discharge capacities reported by various Li-batteries using $LiM_2(PO_4)_3$ (M: Zr, Ti, and Ge)-based solid electrolytes [31, 41, 53, 64, 72–77].

resistance between the Li-metal and solid electrolyte could be reduced by introducing a Li-ion conducting polymer in between them. Li/polymer/$Li_{1.1}Y_{0.1}Zr_{1.9}(PO_4)_3$/LFP showed a discharge capacity of 130 mA h g^{-1} at a current density of 0.05 mA cm^{-2} and, furthermore, a capacity of 119 mA h g^{-1} even after 100 cycles at 60 °C. Here, the polymer is made by using both PEO and LiTFSI [31].

To reduce the interfacial resistance between $Li_{1.5}Al_{0.5}Ge_{1.5}(PO_4)_3$ (LAGP) solid electrolyte and Li metal and—furthermore, to avoid the corrosion of LAGP against Li metal—a polypropylene (PP) layer wetted with the liquid electrolyte is introduced at the electrode-electrolyte interface. The cell, LiNMC|LAGP-PP|Li, delivered an initial capacity of 187 mA h g^{-1} and maintained 98.5% of capacity retention even after 200 cycles with 100% Coulombic efficiency [76]. Wang et al. reported suppression of Li-dendrite formation after modifying the Li-anode with the PEO(LiTFSI) film and later assembled a cell of Li-PEO(LiTFSI)/$Li_{1.5}Al_{0.5}Ge_{1.5}(PO_4)_3$-PEO/$LiMn_{0.8}Fe_{0.2}PO_4$ delivering an initial discharge capacity of 160.8 mA h g^{-1} at 50 °C [77].

The mechanical contact between the electrode and electrolyte could be improved by combining organic and inorganic electrolytes together. A hybrid solid Li-metal battery is fabricated after adding about 5µl liquid electrolyte (1M $LiPF_6$ (EC/DEC-1:1 vol)) at the Li/$Li_{1.6}Al_{0.4}Mg_{0.1}Ge_{1.5}(PO_4)_3$ and $LiNi_{0.33}Mn_{0.33}Co_{0.33}O_2$/$Li_{1.6}Al_{0.4}Mg_{0.1}Ge_{1.5}(PO_4)_3$ interfaces. The cell delivered a reversible capacity of 145.4 mA h g^{-1} with an initial Coulombic efficiency of 86.7% at a current density of 0.2 mA cm^{-2}, and, moreover, the fabricated cell exhibited over 80.5% capacity retention even after 250 cycles [63].

The use of bi-layer solid electrolytes is another possible strategy known to enhance the stability of the solid electrolyte with Li-metal. Zhao et al. prepared a double-layer solid electrolyte combining the $Li_{1.3}Al_{0.3}Ti_{1.7}(PO_4)_3$ (LATP) and $Li_{1.3}Al_{0.3}Ge_{1.7}(PO_4)_3$ (LAGP) solid electrolytes and reported the electrochemical results of the cell LFP/LATP-LAGP/Li with initial discharge capacities of 130.6 mA h g^{-1}, corresponding to 76.8% of its theoretical capacity. However, after three charge-discharge cycles, the charge and discharge capacities increased to 145.3 and 141.4 mA h g^{-1}, respectively. The observation of initial low capacity is explained with respect to the higher electrolyte-electrode interfacial resistances of the cell [41].

6.7 CONCLUSIONS

ASSLIBs have broad prospects in different electronic devices, smart grids, and electric vehicles. The major component of an all-solid-state battery is the solid electrolyte, which is known to suffer from various challenges such as low ionic conductivity and high electrolyte-electrode interfacial resistance. LMP-based NASICON materials are observed to be suitable solid electrolytes for ASSLIB applications because of their good ionic conductivities, wide potential window, and stabilities in a broad range of temperatures. Rhombohedral LZP is found to exhibit higher ionic conductivity compared to its other polymorphs. It is also seen that high-temperature rhombohedral LZP exhibiting higher ionic conductivity could be stabilized at RT by selecting the appropriate precursor, sintering temperature, and/or dopant ions. Figure 6.5 shows the temperature-dependent conductivity plots for some $LiM_2(PO_4)_3$ (M: Zr, Ti, and Ge)-based polycrystalline, composite-polymer, glass-ceramic, and thin film electrolytes.

Various preparatory methods are discussed for controlling the phase formation and ionic conductivity results of LMP-based materials. Spark plasma sintering is found to improve the ionic conductivity of the solid electrolyte effectively in comparison to the conventional sintering method. Sintering aids are found to be effective as microstructure modifiers, reducing grain boundary resistances. An improvement in the ionic conductivity of LMP solid electrolytes is seen upon doping with the bivalent, trivalent, or pentavalent dopants at the M-site. Among them, a better improvement in conductivity is seen with Al^{3+} as the dopant ion. The double substitution of ions at the M-site shows better improvement in ionic conductivity values compared to the same with single ion doping. Doping improves the ionic conductivity of the LMP material by optimizing the bottleneck size for the movement of Li-ions and, in addition, increases the concentration of Li-ions due to the differences in the oxidation states of the host and dopant ions.

LTP- and LGP-based solid electrolytes exhibit higher ionic conductivities compared to LZP-based systems. However, the Ti^{4+} in LTP gets easily reduced to Ti^{3+} against Li-metal, and the presence of Ge in LGP is often expensive. Among the various LMP-based solid electrolytes known, $Li_{1.5}Al_{0.4}Cr_{0.1}Ge_{1.5}(PO_4)_3$ and $Li_{1.5}Al_{0.33}Sc_{0.17}Ge_{1.5}(PO_4)_3$ glass ceramics offer the maximum ionic conductivity values (Figure 6.5).

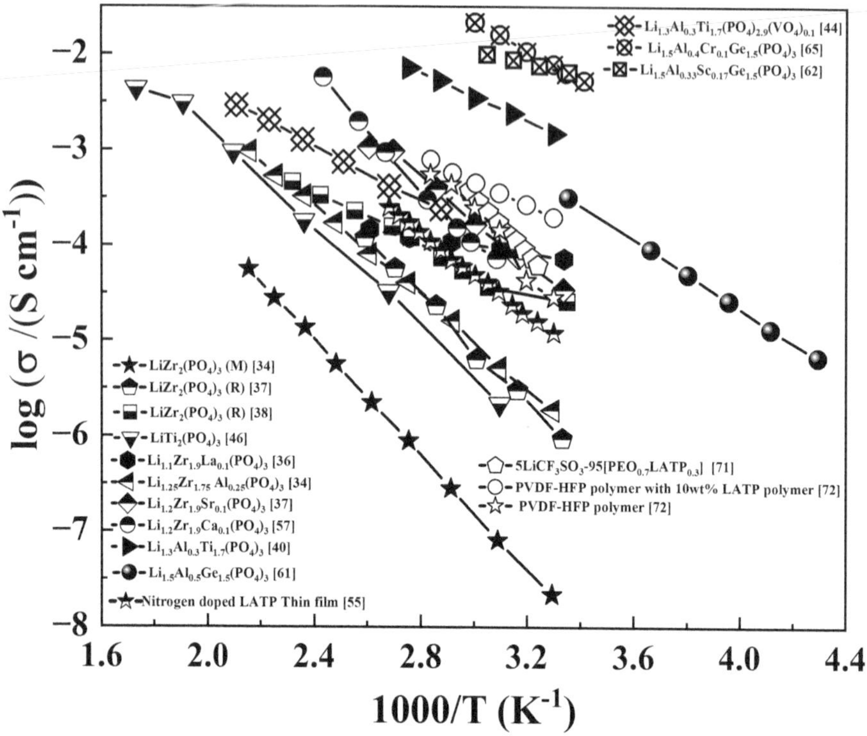

FIGURE 6.5 Temperature-dependent conductivity plots for $LiM_2(PO_4)_3$ (M: Zr, Ti, and Ge)-based polycrystalline, composite-polymer, glass-ceramic, and thin film electrolytes reported by the various groups [34, 36–38, 40, 44, 46, 55, 57, 61, 62, 65, 71, 72].

In order to facilitate large-scale production of solid electrolytes for commercial batteries, future studies should focus on developing cost-effective synthesis methodologies. Various reports are known demonstrating the use of LMP-based solid electrolytes, particularly composite polymer electrolytes, and LGP-based glass ceramics in all-solid-state Li-ion batteries employing high-performance cathode materials such as LFP and LNMC. It is seen that most of the cells are tested for 50–100 cycles to demonstrate their good Coulombic efficiency and cyclic performance. Future studies should involve testing the cells for a large number of cycles to show their capacity retention and Coulombic efficiency. The electrochemical performance of ASSLIBs often depend on the stability at the electrode-electrolyte interface, which manifests greater Coulombic efficiency and capacity retention. Systematic studies are thus vital for *in-situ* and/or postmortem analysis of the cells using various techniques such as XRD, HRTEM, Raman spectroscopy, and XPS to better understand the electrochemistry behind the capacity fading.

Various approaches have been highlighted in this chapter to improve the compatibility between the electrolyte and electrodes, resulting in improved cycle stability of LIBs using LMP-based solid electrolytes. Based on these results, it is clear that LMP-based materials have great potential for use as solid electrolytes in safe and advanced ASSLIBs.

REFERENCES

1. J.B. Goodenough and K.S. Park, *J. Am. Chem. Soc.*, 2013, 135, 1167–1176.
2. A. Arya and A.L. Sharma, *J. Solid State Electrochem.*, 2019, 23, 997–1059.
3. J.L. Brédas, J.M. Buriak, F. Caruso, K.S. Choi, B.A. Korgel, M.R. Palacín, K. Persson, E. Reichmanis, F. Schüth, R. Seshadri and M.D. Ward, *Chem. Mater.*, 2019, 31, 8577–8581.
4. B. Diouf and R. Pode, *Renew. Energy*, 2015, 76, 375–380.
5. A. Masias, J. Marcicki and W.A. Paxton, *ACS Energy Lett.*, 2021, 6, 621–630.
6. J.B. Goodenough and Y. Kim, *Chem. Mater.*, 2010, 22, 587–603.
7. Q. Zhao, S. Stalin, C.Z. Zhao and L.A. Archer, *Nat. Rev. Mater.*, 2020, 5, 229–252.
8. J. Janek and W.G. Zeier, *Nat. Energy*, 2016, 1, 16141.
9. X.B. Cheng, C.Z. Zhao, Y.X. Yao, H. Liu and Q. Zhang, *Chem*, 2019, 5, 74–96.
10. V. Miß, A. Ramanayagam and B. Roling, *ACS Appl. Mater. Interfaces*, 2022, 14, 38246–38254.
11. L.N. Patro and K. Hariharan, *Solid State Ion.*, 2013, 239, 41–49.
12. C.S. Sunandana, *Introduction to Solid State Ionics: Phenomenology and Applications*, CRC Press, 2015.
13. S.K. Heiskanen, J. Kim and B.L. Lucht, *Joule*, 2019, 3, 2322–2333.
14. Y.B. Rao, K.K. Bharathi and L.N. Patro, *Solid State Ion.*, 2021, 366–367, 115671.
15. F. Zheng, M. Kotobuki, S. Song, M.O. Lai and L. Lu, *J. Power Sources*, 2018, 389, 198–213.
16. J.C. Bachman, S. Muy, A. Grimaud, H.H. Chang, N. Pour, S.F. Lux, O. Paschos, F. Maglia, S. Lupart, P. Lamp, L. Giordano and Y. Shao-Horn, *Chem. Rev.*, 2016, 116, 140–162.
17. M. Hou, F. Liang, K. Chen, Y. Dai and D. Xue, *Nanotechnology*, 2020, 31, 132003.
18. L. Dai, J. Wang, Z. Shi, L. Yu and J. Shi, *Ceram. Inter.*, 2021, 47, 11662–11667.
19. J. Zhu, Y. Xiang, J. Zhao, H. Wang, Y. Li, B. Zheng, H. He, Z. Zhang, J. Huang and Y. Yang, *Energy Storage Mater.*, 2022, 44, 190–196.
20. J.B. Goodenough, H.Y.P. Hong and J.A. Kafalas, *Mater. Res. Bull.*, 1976, 11, 203–220.
21. N. Anantharamulu, K.K. Rao, G. Rambabu, B.V. Kumar, V. Radha and M. Vithal, *J. Mater. Sci.*, 2011, 46, 2821–2837.
22. S. Sharma, M.D. Singh and A. Dalvi, *J. Energy Storage*, 2022, 49, 104178.
23. A. Tron, T. Yoon, Y.D. Park, S.M. Oh and J. Mun, *J. Nanosci. Nanotechnol.*, 2017, 17, 4977–4982.
24. Z. Wang, H. Zhong and G. Song, *J. Alloys Compd.*, 2020, 849, 156467.
25. Z. Luo, C. Qin, W. Xu, H. Liang, W. Lei, X. Shen and A. Lu, *Ceram. Int.*, 2020, 46, 15613–15620.
26. L.N. Patro, *J. Solid State Electrochem.*, 2020, 24, 2219–2232.
27. B.E. Francisco, C.R. Stoldt and J.C. M'Peko, *Chem. Mater.*, 2014, 26, 4741–4749.
28. H. Xu, S. Wang, H. Wilson, F. Zhao and A. Manthiram, *Chem. Mater.*, 2017, 29, 7206–7212.
29. A. Martínez-Juárez, C. Pecharromán, J.E. Iglesias and J.M. Rojo, *J. Phys. Chem. B*, 1998, 102, 372–375.
30. M. Harada, H. Takeda, S. Suzuki, K. Nakano, N. Tanibata, M. Nakayama, M. Karasuyama and I. Takeuchi, *J. Mater. Chem. A*, 2020, 8, 15103–15109.
31. Q. Li, C. Xu, B. Huang and X. Yin, *Materials*, 2020, 13, 1719.
32. K. Kamali and T.R. Ravindran, *J. Phys. Chem. A*, 2016, 120, 1971–1977.
33. M. Catti, N. Morgante and R.M. Ibberson, *J. Solid State Chem.*, 2000, 152, 340–347.
34. T. Pareek, B. Singh, S. Dwivedi, A.K. Yadav, Anita, S. Sen, P. Kumar and S. Kumar, *Electrochim. Acta*, 2018, 263, 533–543.
35. H.E. Shinawi, C. Greaves and J. Janek, *RSC Adv.*, 2015, 5, 17054–17059.
36. V. Ramar, S. Kumar, S.R. Sivakkumar and P. Balaya, *Electrochim. Acta*, 2018, 271, 120–126.

37. S. Kumar and P. Balaya, *Solid State Ion.*, 2016, 296, 1–6.
38. Y. Li, W. Zhou, X. Chen, X. Lü, Z. Cui, S. Xin, L. Xue, Q. Jia and J.B. Goodenough, *Proc. Natl. Acad. Sci.*, 2016, 113, 13313–13317.
39. K. Kwatek and J.L. Nowiński, *Solid State Ion.*, 2017, 302, 54–60.
40. F. Ma, E. Zhao, S. Zhu, W. Yan, D. Sun, Y. Jin and C. Nan, *Solid State Ion.*, 2016, 295, 7–12.
41. E. Zhao, F. Ma, Y. Guo and Y. Jin, *RSC Adv.*, 2016, 6, 92579–92585.
42. K.M. Kim, D.O. Shin and Y.G. Lee, *Electrochim. Acta*, 2015, 176, 1364–1373.
43. A. Cassel, B. Fleutot, M. Courty, V. Viallet and M. Morcrette, *Solid State Ion.*, 2017, 309, 63–70.
44. C.M. Chang, Y. Lee, S.H. Hong and H.M. Park, *J. Am. Ceram. Soc.*, 2005, 88, 1803–1807.
45. Y. Lai, Z. Sun, L. Jiang, X. Hao, M. Jia, L. Wang and F. Liu, *Ceram. Int.*, 2019, 45, 11068–11072.
46. Y. Kobayashi, T. Takeuchi, M. Tabuchi, K. Ado and H. Kageyama, *J. Power Sources*, 1999, 81–82, 853–858.
47. T. Hupfer, E.C. Bucharsky, K.G. Schell, A. Senyshyn, M. Monchak, M.J. Hoffmann and H. Ehrenberg, *Solid State Ion.*, 2016, 288, 235–239.
48. X. Zhuo-Bing, C. Shang and G. Man-Man, *Trans. Nonferrous Met. Soc. China*, 2011, 21, 2454–2458.
49. I.N. Reddy, B. Akkinepally, C.V. Reddy, A. Sreedhar, T.J. Ko and J. Shim, *Ionics*, 2020, 26, 4287–4298.
50. R.C. Agrawal and G.P. Pandey, *J. Phys. D Appl. Phys.*, 2008, 41, 223001.
51. D. Safanama, N. Sharma, R.P. Rao, H.E.A. Brand and S. Adams, *J. Mater. Chem. A*, 2016, 4, 7718–7726.
52. Z. Liu, S. Venkatachalam, H. Kirchhain and L.V. Wüllen, *Solid State Ion.*, 2016, 295, 32–40.
53. C. Peng, Y. Kamiike, Y. Liang, K. Kuroda and M. Okido, *ACS Sustain. Chem. Eng.*, 2019, 7, 10751–10762.
54. R. Inada, K. Ishida, M. Tojo, T. Okada, T. Tojo and Y. Sakurai, *Ceram. Int.*, 2015, 41, 11136–11142.
55. G. Tan, F. Wu, L. Li, Y. Liu and R. Chen, *J. Phys. Chem. C*, 2012, 116, 3817–3826.
56. P. Zhang, H. Wang, Q. Si, M. Matsui, Y. Takeda, O. Yamamoto and N. Imanishi, *Solid State Ion.*, 2015, 272, 101–106.
57. H. Xie, Y. Li and J.B. Goodenough, *RSC Adv.*, 2011, 1, 1728–1731.
58. Y. Zhang, K. Chen, Y. Shen, Y. Lin and C.W. Nan, *Ceram. Int.*, 2017, 43, S598–S602.
59. H. Morimoto, H. Awano, J. Terashima, Y. Shindo, S. Nakanishi, N. Ito, K. Ishikawa and S.I. Tobishima, *J. Power Sources*, 2013, 240, 636–643.
60. S. Hamdoune, D.T. Qui and E.J.L. Schouler, *Solid State Ion.*, 1986, 18–19, 587–591.
61. J. Kang, H. Chung, C. Doh, B. Kang and B. Han, *J. Power Sources*, 2015, 293, 11–16.
62. Y. Nikodimos, M.C. Tsai, L.H. Abrha, H.H. Weldeyohannis, S.F. Chiu, H.K. Bezabh, K.N. Shitaw, F.W. Fenta, S.H. Wu, W.N. Su, C.C. Yang and B.J. Hwang, *J. Mater. Chem. A*, 2020, 8, 11302–11313.
63. Y. Nikodimos, L.H. Abrha, H.H. Weldeyohannes, K.N. Shitaw, N.T. Temesgen, B.W. Olbasa, C.J. Huang, S.K. Jiang, C.H. Wang, H.S. Sheu, S.H. Wu, W.N. Su, C.C. Yang and B.J. Hwang, *J. Mater. Chem. A*, 2020, 8, 26055–26065.
64. Y. Meesala, C.Y. Chen, A. Jena, Y.K. Liao, S.F. Hu, H. Chang and R.S. Liu, *J. Phys. Chem. C*, 2018, 122, 14383–14389.
65. M. Illbeigi, A. Fazlali, M. Kazazi and A.H. Mohammadi, *Solid State Ion.*, 2016, 289, 180–187.
66. Z. Cai, Y. Huang, W. Zhu and R. Xiao, *Solid State Ion.*, 2020, 354, 115399.
67. G.B. Kunshina, O.G. Gromov, E.P. Lokshin and V.T. Kalinnikov, *Russ. J. Inorg. Chem.*, 2014, 59, 424–430.

68. J. Feng, H. Xia, M.O. Lai and L. Lu, *J. Phys. Chem. C*, 2009, 113, 20514–20520.
69. A. Manthiram, X. Yu and S. Wang, *Nat. Rev. Mater.*, 2017, 2, 16103.
70. J.M. Tarascon and M. Armand, *Nature*, 2001, 414, 359–367.
71. M.D. Singh, A. Dalvi, D.M. Phase and Y. Kumar, *J. Mater. Sci.*, 2020, 55, 3951–3963.
72. Y. Li and H. Wang, *Ind. Eng. Chem. Res.*, 2021, 60, 1494–1500.
73. A. Das, M. Goswami, K. Illath, T.G. Ajithkumar, A. Arya and M. Krishnan, *J. Non-Cryst. Solids*, 2021, 558, 120654.
74. Q. Zhou, B. Xu, P.H. Chien, Y. Li, B. Huang, N. Wu, H. Xu, N.S. Grundish, Y.Y. Hu and J.B. Goodenough, *Small Methods*, 2020, 4, 2000764.
75. N. Wu, P.H. Chien, Y. Li, A. Dolocan, H. Xu, B. Xu, N.S. Grundish, H. Jin, Y.Y. Hu and J.B. Goodenough, *J. Am. Chem. Soc.*, 2020, 142, 2497–2505.
76. D. Bosubabu, J. Sivaraj, R. Sampathkumar and K. Ramesha, *ACS Appl. Energy Mater.*, 2019, 2, 4118–4125.
77. C. Wang, Y. Yang, X. Liu, H. Zhong, H. Xu, Z. Xu, H. Shao and F. Ding, *ACS Appl. Mater. Interfaces*, 2017, 9, 13694–13702.

Section II

Beyond Li-ion Batteries

7 Electrolytes and Additives for Sodium-ion Battery Systems

Ramesh K. Guduru, Neha Tiwari, and Robin Singh

7.1 INTRODUCTION

It is well recognized that the burning of fossil fuels contributes to global warming, and hence green or renewable sources, such as wind, solar, and tidal waves have received a lot of attention as they are sustainable and environmentally benign. However, these are time- and space-constrained due to various reasons and necessitate energy storage at the time and spot of harvesting [1]. Advances made in the field of electrical energy storage (EES) devices have made it possible to store energy/power using these auxiliary energy sources. Numerous inventions have been created along the way that eventually led to the development of modern batteries—the timeline of battery development is shown in Figure 7.1.

Currently, Li-ion batteries (LIBs) have captured the major market for batteries due to their rechargeability, cost, and availability. The cost of lithium resources has risen constantly over the years and is several times higher than the cost of sodium resources. Due to these circumstances, the need to investigate alternate battery chemistries has become very important. However, long-term sustainable and large-scale production of LIBs are threatened by the lack of raw materials such as lithium, cobalt, and nickel. Also, such materials come with high prices and potential toxicity [3]. To create durable, dependable, and affordable batteries that can fulfill rising universal energy demands, new battery chemistries are therefore necessary. In this pursuit, Na-ion batteries (SIBs) are the best candidates for the near future. The advantages for SIBs are that sodium has a low production cost [4]; sodium is the sixth most abundantly available metal in the Earth's rock (around 2.60%), and seawater provides a limitless source of sodium [5]. These points justify that SIBs can become practical replacements for LIBs. Figure 7.2 shows the availability of various elements in the Earth's rock.

Sodium is an alkali metal, just below lithium (1.02 Å for Na^+ Vs. 0.76 Å for Li^+) and is the second lightest element among the metals. Electrode materials such as graphite and layered oxides experience slow reaction kinetics as a result of the higher ionic radius of Na-ion compared to Li-ion, which often results in reduced rate capability and weak cyclic stability [5]. The standard electrode potential of Na/Na^+ (2.71 V vs. SHE) and Li/Li^+ (3.04 V vs. SHE) are comparable [7], and both SIBs and LIBs use comparable "rocking-chair" charging and discharging procedures.

DOI: 10.1201/9781003441182-9

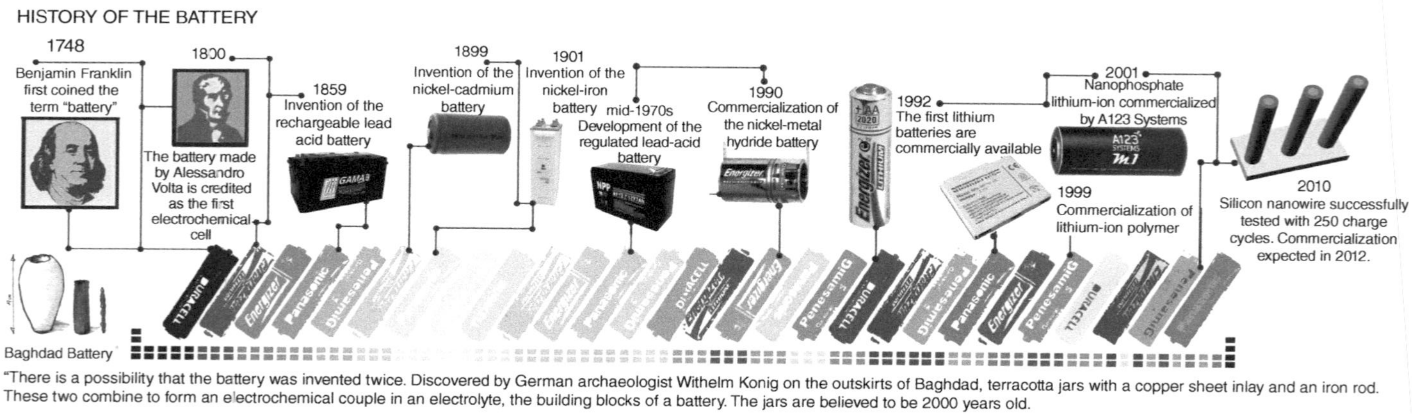

FIGURE 7.1 History and timeline of battery development.

Source: Reproduced with permission from Ref. [2].

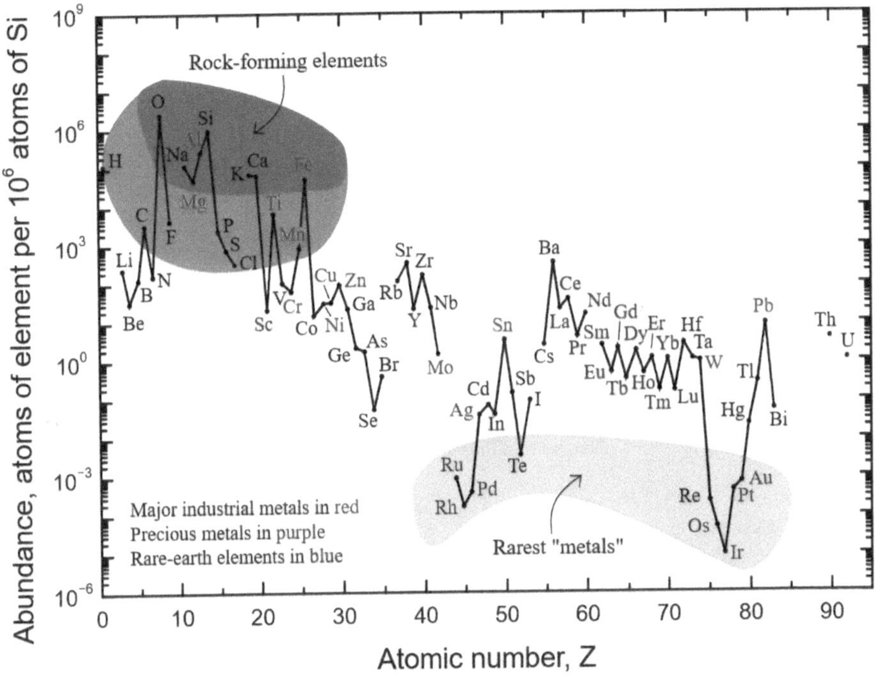

FIGURE 7.2 The number of elements in the crust of the Earth.

Source: Image from Haxel et al. (2002), courtesy of the U.S. Geological Survey [6].

SIBs have numerous advantages over Li-ion battery technologies. Although SIBs have relatively lower energy density than LIBs, they are inexpensive and safer. Table 7.1 compares Li and SIBs with respect to different attributes related to their performance.

SIBs use aluminum as a current collector for both electrodes, unlike LIBs, which use copper for the anode. The weak thermodynamic stability of the Na-ion intercalated graphite prevents the usage of graphite as an anode in SIBs. On top, the usage of ester-based solvents in SIBs makes graphite unsuitable for ion intercalation. The fabrication of a ternary graphite intercalation compound comes from the solvent-co-intercalation of Na-ions utilizing ether-based solvents. Hard carbon (HC), which has gained widespread acceptance as a carbonaceous anode for the fabrication of great-energy-density SIBs, was thus discovered to be an effective incorporation as host for Na-ions. The fabrication of high-efficiency SIBs needs optimization of electrolytes and integration of additives to enhance the rate performance and cyclic stability [8]. In this chapter, we will review the advancements in electrolytes and additives used for SIBs.

The electrolyte plays a vital role toward the performance and efficiency of SIBs. The ideal electrolyte for SIBs should have a broad electrochemical window with great ionic conductivity and good electrochemical stability [18]. Among various types of electrolytes, room-temperature ionic liquids (RTILs) and organic solvents are the most widely studied. RTILs have many advantages over conventional organic

TABLE 7.1
Comparison of Sodium and Lithium-ion Batteries

Attributes	Na-ion Battery	Lithium-ion Battery
Materials	Earth-abundant	Scarce
Safety	Low risk for aqueous batteries and a high risk for Na in carbon batteries	High risk
Volumetric energy density	250–375 Wh/L, based on prototypes [9]	200–683 Wh/L [10]
Gravimetric energy density (specific energy)	75–165 Wh/kg, based on prototypes and product announcements [9, 11]	120–260 Wh/kg [10]
Cost per kilowatt hour of capacity	$40–77 [12]	$137 (average in 2020) [13]
Cycles at 80% depth of discharge	Hundreds to thousands [14]	3,500 [15]
Cyclic stability	High (negligible self-discharge)	High (negligible self-discharge)
Direct current round-trip efficiency	Up to 92% [14]	85–95% [16]
Temperature range	–20 °C to 60 °C [14, 16]	Acceptable: –20 °C to 60 °CO ptimal: 15 °C to 35 °C [17]

solvents, including wide electrochemical windows, high thermal stability, and negligible vapor pressure. However, the high viscosity of RTILs hinders the diffusion of Na-ions, leading to poor rate performance. To overcome this issue, researchers have developed RTIL-based electrolytes with reduced viscosity by incorporating co-solvents or designing new ionic liquids with lower viscosity. In some cases, organic electrolytes—such as propylene carbonate (PC) and ethylene carbonate (EC)—are commonly used, due to their good compatibility with Na-ion battery electrodes and high ionic conductivity. However, these solvents are prone to decompose at high potentials, forming SEI layers that can hinder the movement of Na-ions. To address this issue, researchers have explored the use of additives to stabilize the SEI layer and improve cyclic stability [19].

One approach to enhance the performance of SIBs is the use of additives in the electrolyte. Additives are typically small molecules or compounds added to electrolyte to enhance its properties, such as conductivity, stability, and capacity retention. For instance, additives can form an additional layer on the surface of the electrode, preventing the irreversible degradation of the electrolyte and enhancing the cyclic stability of the battery. Additionally, additives can enhance the overall kinetics of SIBs, leading to improved rate capability and higher power output. The use of additives has been extensively investigated in LIBs, and their effectiveness has been well documented [20–23].

Similarly, recent studies have demonstrated that additives can also improve the efficiency of SIBs. This chapter gives an overview of recent research on the additives used for the electrolyte of SIBs. It also covers various types of additives, including solvents, salts, and other compounds, and their significance on the electrochemical efficiency of SIBs. The challenges and opportunities for suitable additives used for SIBs, highlighting the potential directions for future research [20], will also be discussed.

7.2 IMPORTANCE OF ADDITIVES IN SIBs

Additives are commonly used in the electrolytes of storage batteries to improve overall performance and reliability. Some of the key reasons are as follows:

1. Enhancing the stability of the electrolyte: The main challenge in the development of batteries is achieving stable cyclic performance. Additives can help stabilize the electrolyte by reducing the degradation of the electrolyte components and preventing unwanted side reactions that can lead to capacity loss.
2. Improving the solubility of sodium salts: Sodium salts are typically used as the active materials in the synthesis of cathode and anode of SIBs. However, these salts can be insoluble or have limited solubility in the electrolyte, which can limit the battery's performance. Additives can improve the solubility of these salts, making it easier for them to dissolve in the electrolyte.
3. Increasing the ionic conductivity of the electrolyte: Additives can increase ionic conductivity by decreasing the viscosity of the electrolyte and improving the mobility of the Na-ions.
4. Suppressing dendrite formation: SIBs are prone to the formation of small dendrites, which are needle-like structures formed at the anode, and can penetrate the separator, causing short circuits and leading to critical safety hazards. Additives can form an additional protective layer at the anode that suppresses dendrites or alter the chemical and physical properties of the electrolyte [24, 25].

Apart from the electrodes, the other critical element is an electrolyte, which plays a crucial role in both efficiency and safety for all battery systems. Table 7.2 shows the chemical and physical properties of different types of electrolytes utilized for SIBs [26]. The electrolyte must have high ionic conductivity, while still being electrically insulating. To prevent reduction or oxidation during battery usage, the electrolytes used in battery operation must also have good electrochemical stability. Given their operation in high voltage, lithium and sodium-ion batteries are particularly important for having good electrochemical stability. To ensure general safety, a good electrolyte should exhibit low volatility, good thermal stability, and low flammability. It ought to be made of non-toxic components, have great chemical stability, and be economical for large-scale applications.

7.3 AQUEOUS ELECTROLYTES

Aqueous electrolytes are extensively investigated for SIBs due to their low cost, great ionic conductivity, and low flammability compared to organic electrolytes. Aqueous electrolytes are typically composed of a sodium salt dissolved in water, although

TABLE 7.2
Physical and Chemical Properties of Electrolyte Use in Na-ion Batteries

Electrolyte	Composition	E.S.R. (V vs. Na^+/Na)	σ RT (mS cm^{-1})	E_a (kJ mol^{-1})	T_g (°C)	T_m (°C)	t_{Na+}	Ref.
Liquid (non-aqueous)	1 M $NaClO_4$:PC	0–5	6.4					[30]
	1 M $NaClO_4$-EC:PC	0–5.3	8					[30]
	1 M $NaClO_4$-EC:DME	0.5–4.5	12.55					[30]
	0.6 M $NaPF_6$-EC:DMC	1–4.5	6.8					[31]
Ionic liquid	[Bmim][Br_3]		8.93	33.73		8.2		[32]
	BMP-TFSI + 1 M $NaClO_4$		1					[33]
	$Na_{0.9}EMIm_{0.9}$TFSI		5.5		−85.7	−19		[34]
	BMP-TFSI + 1 M $NaBF_4$		1.9					[35]
	(0.3)Na[FSA] [AS(4.5)][FSA]		1.3					[36]
	NaFSA-C_1C_3pyrFSA (2:8)	−0.15–1.5	4.07					[37]
Glass ceramic	Cubic Na_3PS_4	0–5	0.2	27	260			[38]
	$94Na_3PS_4$-$6Na_4SiS_4$	0–5	0.74	27	220		0.96	[39]
Solid polymer	PVA-NaBr (70:30)		1.12×10^{-3}	42			0.95	[40]
	PVA-NaI (70:30)		1.02×10^{-2}	37				[42]
	PVA-NaF (80:20)		0.4	39			0.93	[42]
Gel polymer	PVDF-HFP:1 M $NaCF_3SO_3$ in EC:PC (1:1 vol%) + 3 wt.% SiO_2		4.1				0.43	[43]
	EMTF:PVdF-HFP (4:1 w/w) + 0.5 M NaTf	−2–2.4	5.74			95	0.23	[44]
	PMMA-EC-PC-1M $NaClO_4$	−2.5–2.5	3.4		<−80		0.23	[45]
	PVdF-HFP + 1 M $NaClO_4$	0–4.6	0.6	4.6	130		0.3	[46]

Notes: T_g: Glass transition temperature; T_m: Melting temperature; S.P.: Softening point; t_{Na+}: Transference number; ν: Viscosity; E_a: Activation energy; ESR : Electrochemical Stability Range; σ: Ionic conductivity.

the use of additives and co-solvents can also enhance their performance. The major challenge with aqueous electrolytes is the narrow electrochemical window. However, recent research has focused on developing new aqueous electrolytes using salts such as sodium bis(fluorosulfonyl)imide (NaFSI) and sodium trifluoromethane sulfonate (NaTFSI) for increased energy density and wide operating voltage of the batteries [27, 28]. There are many such studies conducted with different types of salts and electrolytes for enhanced performance of SIBs, which are summarized below.

- ***Salt-in-water Electrolytes*****:** These electrolytes consist of an inorganic salt dissolved in water; for example, sodium hydroxide (NaOH) or sodium sulfate (Na_2SO_4) are dissolved in water to create an aqueous electrolyte [29]. These could be both acidic and basic in nature.
- ***Salt-in-polymer Electrolytes*****:** These electrolytes have an inorganic salt dissolved in polyethylene oxide (PEO), a polymer matrix that helps enhance the stability and ionic conductivity of the electrolyte [41, 47].
- ***Salt-in-gel Electrolytes*****:** These electrolytes have an inorganic salt dissolved in a hydrogel matrix made of agarose or alginate and provide a stable and flexible matrix for the electrolyte.
- ***Hybrid Electrolytes*****:** These electrolytes are a mixture of aqueous and organic solvents. A composition of PC and water is used as a co-solvent for enhanced stability of the electrolyte at high voltages while maintaining the great ionic conductivity of the aqueous component.
- ***Deep Eutectic Solvents*****:** These electrolytes have a hydrogen bond acceptor and a quaternary ammonium salt, such as choline chloride or urea. Deep eutectic solvents have been investigated as a potential replacement for conventional organic or aqueous electrolytes because of their low flammability and high ionic conductivity [28].

The selection of an aqueous electrolyte depends on the specific application and performance requirements of the SIBs, and these electrolytes have different pros and cons. For example, aqueous electrolytes offer several advantages compared to organic electrolytes. These are low in cost and abundant and extend SIB's large-scale energy storage applications. In addition, aqueous electrolytes also provide the following advantages:

- High safety: Aqueous electrolytes are non-flammable and non-volatile and reduce the risk of fire and explosion compared to organic electrolytes.
- High ionic conductivity: Aqueous electrolytes can have high ionic conductivity, which is essential for efficient ion transport and better battery performance.
- Environmentally friendly: Aqueous electrolytes are generally considered to be more environmentally friendly than organic electrolytes, as they do not contain toxic or hazardous organic solvents [26, 48].
- Wide operating temperature range: Aqueous electrolytes can operate over a broad temperature range, from room temperature up to 100 °C, without compromising their performance.

- Improved safety in extreme conditions: Aqueous electrolytes can also improve safety in extreme conditions such as high temperatures, as water can act as a coolant and help dissipate heat [27].

The advantages of aqueous electrolytes make them a great candidate for safe, cost-effective, and high-performance SIBs [27, 49]. While the aqueous electrolytes offer several advantages, they also have some limitations that need to be considered:

- Narrow voltage window due to the electrochemical dissociation of water above 1.23 V, which causes lower energy density compared to the organic electrolytes.
- Aqueous electrolytes also cause corrosion of the electrodes and other battery components, leading to reduced battery performance and shorter life.
- Some sodium salts have low solubility in water, limiting their concentration in the electrolytes and thereby reducing ionic conductivity.
- Aqueous electrolytes can have a high-water activity leading to water uptake by the electrodes, resulting in poor battery performance.
- Aqueous electrolytes have a relatively high freezing point and hence cannot be used in cold environments.

These disadvantages highlight the need for further research to develop new aqueous electrolytes that can overcome these limitations and enable the fabrication of efficient SIBs. Recent advances in aqueous electrolytes include the development of novel electrolyte compositions and additives to improve their performance and overcome some of their limitations as summarized below:

- High-concentration electrolytes: Researchers have developed high-concentration aqueous electrolytes using novel sodium salts such as halides, sulfate, nitrate, carboxylate, and carbonates that have high solubility in water, which enables achieving high ionic conductivity and improved stability over a wide voltage range [50].
- pH-neutral electrolytes are also developed to reduce the corrosiveness of the electrolytes and improve the stability of the electrode. These electrolytes use weak acids and bases to maintain a neutral pH while providing high ionic conductivity [26].
- Hybrid electrolytes combine aqueous and organic solvents and exhibit better voltage stability and energy density compared to aqueous electrolytes in addition to high ionic conductivity with wider voltage windows [51].
- Additives improve the performance of aqueous electrolytes; for example, the addition of fluorinated compounds improves the solubility and stability of the electrolytes, while the addition of carbon-based materials improves electrode stability and performance [52].
- Redox mediators are also explored as a way to increase the voltage range of aqueous electrolytes. These mediators can reversibly shuttle between the anode and cathode, enabling the use of a wider voltage range [53].

Recent advances in aqueous electrolytes for SIBs include the development of novel electrolyte compositions and additives to improve their performance and overcome some of their limitations. Several of these are summarized below:

- High-concentration electrolytes: Researchers have developed high-concentration aqueous electrolytes using novel sodium salts that have high solubility in water, which enables achieving high ionic conductivity and improved stability over a wide voltage range.
- pH-neutral electrolytes: pH-neutral electrolytes have been developed to decrease the corrosiveness of the electrolyte and enhance the stability of the electrode. These electrolytes use weak acids and bases to maintain a neutral pH while providing high ionic conductivity.
- Hybrid electrolytes: Hybrid electrolytes combine aqueous and organic solvents and exhibit better voltage stability and energy density compared to aqueous electrolytes, in addition to high ionic conductivity with wider voltage windows.
- Additives: Additives improve the performance of aqueous electrolytes; for example, the addition of fluorinated compounds improves the solubility and stability of the electrolytes, while the addition of carbon-based materials improves the electrode stability and performance [48].
- Redox mediators: Redox mediators have been explored as a way to increase the voltage range of aqueous electrolytes. These mediators can reversibly shuttle between the anode and cathode, enabling the use of a wider voltage range.

These advances in aqueous electrolytes demonstrate the potential aqueous electrolytes for SIBs and pave the way for the fabrication of high-efficiency and cost-effective energy storage systems [54–56].

7.4 ORGANIC AQUEOUS ELECTROLYTES

Organic electrolytes are a potential class of electrolytes for SIBs due to their low viscosity, good compatibility, and high ionic conductivity with a broad range of electrode materials. However, organic electrolytes also suffer from several limitations, such as degradation and decomposition at high voltages, and deteriorate cyclic stability and safety issues.

Recent research has focused on developing novel organic electrolytes and modifying existing ones to overcome these limitations and enhance the efficiency of SIBs. One approach is to use novel sodium salts, such as sodium difluoro(oxalate)borate (NaDFOB), to enhance the stability and performance of the electrolyte. Another approach is to modify the organic solvent to improve its stability, such as using vinylene carbonate (VC) or FEC as co-solvents with conventional organic solvents. Researchers have also investigated the use of polymer electrolytes, which are composed of a polymer matrix and sodium salt. These materials have shown promising results in terms of high ionic conductivity and improved stability, although their practical implementation in SIBs is still limited.

In addition, the use of redox-active electrolytes was also explored, which can provide additional capacity beyond the specific capacity of the electrode materials; e.g., quinone-based electrolytes have shown great capacity and good stability in SIBs [28].

Organic electrolytes show a great opportunity for utilization in SIBs, but more research is needed to overcome their limitations and further improvement in their performance and safety. Future research directions could include the development of new sodium salts and solvents, the use of hybrid electrolytes combining organic and inorganic components, and the optimization of electrolyte composition and design for specific electrode materials and applications.

The common composition of organic electrolytes for SIBs is a mixture of sodium salts mixed with organic solvents. Cell performance is mostly determined by the characteristics of the dissolved sodium salt(s) and the solvent(s). Following are some of the solvent properties that must be considered:

- Low-boiling and high-melting temperatures
- Robust chemical and electrochemical stabilities
- Low viscosity to improve electrode wetting and increase ion mobility
- Large permittivity to solvate the salt(s) and dissociate

In SIBs, carbonate and ether-based organic solvents are the most commonly used, such as tri-ethylene glycol di-methyl ether (tri glyme), dimethoxyethane (DME), dimethyl carbonate (DMC), diethylene glycol dimethyl ether (diglyme), ethylmethyl carbonate (EMC), diethyl carbonate (DEC), EC, and PC [28], which are also nontoxic and provide chemical stability and secure the cyclic stability of cells. The sodium salts should show great solubility in the suitable solvents to permit electrolytes with an acceptable concentration of charge transfer.

The most vital sodium salts for SIBs are produced from the weakly coordinating anions (WCAs), i.e., highly delocalized negative charges, as used in Lithium-ion batteries electrolytes: ClO_4^-, PF_6^-, BF_4^-, TFSI, FSI, and OTf [57]. Salts $NaClO_4$ and $NaPF_6$ are by far the most commonly investigated for SIBs, although both raise serious safety issues. The perchlorate anion (ClO_4^-) is poisonous and extremely oxidative, and when combined with organic solvents, can produce explosive chemicals. Additionally, drying the salt takes a long time. In the presence of water, the PF_6^- anions undergo auto-decomposition, generating hazardous hydrofluoric acid. $NaClO_4$ is still the most often employed electrolyte salt due to its low cost.

NaFSI and NaTFSI are two additional and safer salt substitutes. These compounds produce non-hazardous electrolytes with great ionic conductivity and exhibit better thermal stability. Nevertheless, the cost of both salts is higher than $NaClO_4$ [58, 59]. Cyanide (-CN) is a strong donor ligand, the electron-withdrawing nitrile group, and serves as the basis for non-fluorinated salt such as sodium dicyanamide (NaDCA). However, the DCA anion's poor anodic stability restricts its use in batteries with moderate-voltage cathode materials.

There are several systematic studies on different electrolytes. The electrolytes, such as $NaClO_4$ in PC, NaFSI in PC, NaFSI in EC: DEC, $NaPF_6$ in EC: DEC, and $NaBF_4$ in PC, demonstrated significantly higher Coulombic efficiencies (~100%) when compared to NASICON electrodes ($Na_3V_2(PO4)_3/C$), as well as high-capacity retention and stability [28, 54]. Researchers have hypothesized that the decreased Coulombic efficiency seen for the other electrolytes could be caused by either the existence of side chemical reactions or electrolyte decomposition.

7.5 IONIC LIQUID ELECTROLYTES

Ionic liquids (ILs) are also investigated as a potential replacement for conventional organic liquid electrolytes. ILs offer several advantages over conventional electrolytes, including non-flammability, a broad range of electrochemical windows, and high thermal stability, which can allow the use of high-voltage cathode materials [28].

Several types of ionic liquid (IL) electrolytes have been explored for utilization in SIBs. Some of the most commonly studied IL electrolytes include:

- Imidazolium-based ILs have good thermal stability along with broad electrochemical windows, which makes them promising candidates for high-voltage cathode materials.
- Pyrrolidinium-based ILs show good thermal stability and form stable SEI layers, which improves the cyclic stability of SIBs.
- Ammonium-based ILs were explored as electrolytes in SIBs, due to their low cost and easy availability. These ILs also have good thermal stability and form stable SEI layers, which helps improve the cyclic stability of SIBs.
- Phosphonium-based ILs are less commonly studied than other types of ILs, but have shown promise as electrolytes in SIBs. These ILs have good thermal stability and a broad range of electrochemical windows, which can enable the use of high-voltage cathode materials.
- Choline-based ILs also have good thermal stability and low toxicity, which makes them attractive for use in environmentally friendly and sustainable SIBs.

In addition to these, many other types of ILs have also been explored for use in SIBs, including bicyclic and heterocyclic ILs, etc. Overall, the choice of IL electrolyte for SIBs will depend on a variety of factors, including cost, availability, and performance characteristics, such as thermal stability and ionic conductivity [60]. ILs offer several advantages, which include:

- ILs show excellent thermal stability, resisting decomposition and breakdown at high temperatures. This makes them attractive for use in high-temperature applications, to help enhance the safety and reliability of SIBs.
- ILs provide a broad electrochemical window and operate at high voltages, without breaking down or decomposing. This allows for the use of great voltage cathode materials in SIBs and thereby contributes to improving their energy density and performance.
- ILs have low volatility and do not evaporate or leak easily. This can help enhance the safety and stability of SIBs, especially in high-temperature applications.
- ILs have high ionic conductivity and transport ions quickly and efficiently. This helps enhance the power density of SIBs with low internal resistance.

- ILs have good solubility for a wide range of salts and other electrolytes and help enhance the compatibility of SIBs with different cathode and anode materials.
- ILs can be synthesized in a wide range of chemistries, which makes them possible to tailor their properties to specific applications. This allows for the development of IL electrolytes that are optimized for specific types of SIBs.

The unique characteristics of IL-based electrolytes make them attractive for usage in SIBs and other innovative energy storage systems. Nevertheless, the cost and availability of ILs, as well as their compatibility with different electrode materials, must be carefully considered when designing and developing SIBs. Despite many advantages, they also have many disadvantages, which limit the utility of ILs in SIBs. These are some of the main ones:

- High cost, due to the specialized equipment and expertise needed to make ILs.
- While ILs can be synthesized in the laboratory, their widespread use in commercial applications is limited due to their availability. This is because many ILs are not produced in large quantities and not readily available from commercial suppliers.
- ILs typically have a much higher viscosity than conventional electrolytes, which can limit their ionic conductivity and reduce the overall efficiency of SIBs.
- The use of IL electrolytes can result in the production of a thicker and more complex solid electrolyte interfacial layer on the electrode surface and increase the resistance, which in turn could reduce the efficiency of the battery.
- ILs are toxic to living organisms, which can limit their use in environmentally friendly and sustainable SIBs [28, 60].

Some ILs may suffer from low ionic conductivity at low temperatures, and their high viscosity may limit their application in high-power SIBs.

7.6 SOLID ELECTROLYTES

Solid electrolytes have emerged as a promising alternative to liquid electrolytes in SIBs. Solid electrolytes offer the advantages of safety and stability, as well as the potential for high energy density through the use of metallic sodium anodes [28]. The advantages of solid electrolytes are:

- High Stability: Solid electrolytes have high chemical and electrochemical stability, which means they are less likely to decompose or react with other components of the battery. This leads to longer battery lifetimes and better overall performance.
- Improved Safety: Solid electrolytes are generally safer than liquid electrolytes because they are non-flammable and less likely to leak or spill. This is particularly crucial for large-scale energy storage systems.

- High Ion Conductivity: Solid electrolytes can have high ionic conductivity, offering efficient transport of ions between the cathode and anode. This leads to higher battery efficiency and faster charging times.
- Wide Temperature Range: Solid electrolytes can operate in a broad range of temperatures, including extreme temperatures that can be problematic for liquid electrolytes. This makes them suitable for use in harsh environments and extreme weather conditions.
- Reduced Dendrite Formation: Solid electrolytes reduce the formation of dendrite and avoid shorting of the electrodes. This is because solid electrolytes are less likely to interact or promote dendrite formation.

While solid electrolytes offer several advantages over liquid electrolytes, they also have some disadvantages that can limit their applicability in SIBs. Some of the main disadvantages of solid electrolytes are listed below:

- High Cost: Solid electrolytes can be more expensive to produce compared to liquid electrolytes, which can limit their practical use, especially in large-scale applications.
- Limited Mechanical Stability: Some solid electrolytes can have limited mechanical stability, leading to cracking or breaking during use. This can decrease ion conductivity or short-circuit the battery electrodes.
- Limited Compatibility: Solid electrolytes have compatibility only with certain electrode materials or other components of the battery. This can limit the selection of electrode materials that can be used in a Na-ion battery with a solid electrolyte.
- The Difficulty of Manufacturing: Manufacturing solid electrolytes can be more complex and difficult compared to liquid electrolytes and result in lower yield and higher production costs.

Overall, solid electrolytes for SIBs have both pluses and minuses, and the choice of electrolyte material relies on the specific application and performance requirements. Further research and development are required to address the existing limitations and improve the function of solid electrolytes for use in SIBs [61].

Solid Na-ion battery electrolytes can be divided into several categories from the materials' perspective. They are oxide-based, sulfide-based, and halide-based materials. Some common examples of solid SIBs electrolytes are:

- Oxide-Based Solid Electrolytes: These materials are based on oxides such as $Na_3Zr_2Si_2PO_{12}$ (NZSP) or Na_3PO_4. They have high thermal and chemical stability and can exhibit high ionic conductivity at elevated temperatures. However, their low mechanical strength and poor low-temperature performance are a drawback.
- Sulfide-Based Solid Electrolytes: These materials are based on sulfides such as Na_3PS_4, Na_3SbS_4, or Na_3AsS_4. They have high ionic conductivity at room temperature, making them promising candidates for solid electrolytes in SIBs. However, their poor chemical stability and high reactivity with moisture and air limit their practical use.

- Halide-Based Solid Electrolytes: These materials are based on halides such as Na_3AlF_6 or β-$NaAlSiO_4$. They can exhibit great ionic conductivity and high chemical stability, making them promising candidates for use as solid Na-ion battery electrolytes. However, their mechanical properties can be a concern [1, 28].

Some emerging classes of solid electrolytes are based on other types of materials, such as polymer-based or composite-based solid electrolytes—which have shown promise in laboratory studies—but they still need to be optimized for practical use. Overall, solid electrolytes for SIBs are a vigorous area of research, and new materials are being developed.

7.7 HYBRID ELECTROLYTES

Hybrid electrolytes are a promising approach to enhancing the efficiency and safety of SIBs. Hybrid electrolytes combine the merits of both solid and liquid electrolytes, and a solid electrolyte is used as a separator between the cathode and anode, while a liquid electrolyte is used as a medium to transport ions between the electrodes [25].

The use of a hybrid electrolyte in SIBs has several advantages. Firstly, the solid electrolyte provides stability and prevents the formation of dendrites, which can cause short circuits and reduce battery performance. Secondly, the liquid electrolyte allows for high ion conductivity, which can improve battery performance. Finally, the use of a hybrid electrolyte can reduce the risk of thermal runaway, as the solid electrolyte is less flammable than a pure liquid electrolyte [28]. Some examples of hybrid electrolytes are discussed below:

- Polymer-Ceramic Hybrid Electrolyte: This type of hybrid electrolyte combines a solid polymer electrolyte with a ceramic electrolyte. The solid polymer electrolyte provides mechanical stability and prevents dendrite formation, while the ceramic electrolyte enhances ionic conductivity [62]. Various polymer-ceramic hybrid electrolytes for SIBs such as PEO-NASICON and PEO-LAGP have been studied [63].
- Liquid-Solid Hybrid Electrolyte: This type of hybrid electrolyte combines a liquid electrolyte with a solid electrolyte separator. The liquid electrolyte enhances ionic conductivity, while the solid electrolyte provides stability and prevents dendrite formation [64]. Examples for liquid-solid hybrid electrolytes for SIBs are NaTFSI-PC/P(VDF-HFP)-SiO_2 and NaTFSI-PC/PEO-LiTFSI-SiO_2 [65].
- Gel-Polymer Hybrid Electrolyte: This type of hybrid electrolyte combines a gel electrolyte with a polymer electrolyte. The gel electrolyte provides high ionic conductivity and prevents dendrite formation, while the polymer electrolyte provides stability. Examples are PEO/$LiClO_4$-PEG/$NaClO_4$ and PEO/$LiClO_4$-GO/$NaClO_4$ [66, 67].

Hybrid electrolytes show promise in enhancing the efficiency and safety of SIBs, and ongoing research is exploring different materials and structures to optimize their properties.

In addition to the above, below are a few more examples of recent advancements in hybrid electrolytes for SIBs:

- ***Self-Healing Hybrid Electrolytes*:** These electrolytes contain a material that can autonomously repair the damage of the electrolyte caused by mechanical stress or other factors [68]. A recent study reported the development of a self-healing hybrid electrolyte using a polymeric material that can rapidly self-repair after being damaged [67–68].
- ***Hybrid Electrolytes for High-Temperature Applications*:** SIBs are often limited by their operating temperature range, as high temperatures can cause degradation of the electrolyte and reduce battery performance. Recent research has produced high-temperature hybrid electrolytes that can operate at temperatures up to 120 °C by using a solid-state electrolyte combined with a liquid electrolyte [63], e.g., polyethylene oxide (PEO) polymer matrix and a lithium salt-doped ionic liquid [69].

These recent advances in hybrid electrolytes demonstrate their potential to enhance the stability and performance of these batteries, and ongoing research is exploring new materials and structures to optimize their properties. In conclusion, hybrid electrolytes offer a promising solution to the challenges associated with using SIBs.

7.8 ADDITIVES FOR ELECTROLYTES IN SIBS

For the enhanced electrochemical performance of rechargeable batteries, such as Li and SIBs, electrolytes need additives that not only improve the conductivity and stability of the electrolyte, but also help minimize undesired side effects during the electrode and electrolyte interactions. The additives are well renowned for being the most straightforward, efficient, and cost-efficient ways to improve real performance. In addition, the additives also stabilize the electrode-electrolyte interphase and thereby improve cell safety [12]. Table 7.3 shows the summary of electrolyte/electrode additives for SIBs [24].

7.8.1 Some of the Commonly Used Additives for Electrolytes in SIBs Include

- ***Salt Additives*:** These are salts added to the electrolyte to improve its conductivity and stability. Examples include sodium perchlorate ($NaClO_4$) and sodium hexafluorophosphate ($NaPF_6$) [79].
- ***Solvent Additives*:** These are organic compounds added to the electrolyte to improve its solubility and viscosity. Examples are EC and PC [80].
- ***Additives to Improve the Anode Stability*:** SIBs typically use graphite as the anode material, which can react with the electrolyte and cause degradation over time. Additives like vinylene carbonate (VC) and FEC can form a protective layer on the anode surface and improve its stability.

TABLE 7.3
Electrolyte/Electrode Additives for SIBs

Additive	Type	Cathode & Anode	Electrolyte	Capacity (mAh g^{-1})	Retention Rate (Capacity Increase)	Ref.
FEC	Electrolyte additive	$NaNi_{1/2}Mn_{1/2}O_2$/hard carbon	1 M $NaClO_4$ in PC	245	+35% (50 cycles)	[70]
FEC	Electrolyte additive	Hard carbon/Na	1 M $NaPF_6$ in PC	250	+45% (80 cycles)	[71]
FEC + TMSP	Electrolyte additive	$Sn4P_3$/Na	1 M $NaClO_4$ in EC/PC (1:1)	852	+70% (50 cycles)	[72]
FEC (5%)	Electrolyte additive	Black P/Na	1 M $NaPF_6$ in EC/DEC (1:1)	1587	+30% (25 cycles)	[73]
VC	Electrolyte additive	$Na_3V_2(PO_4)_3$@C/FeS@C	1 M $NaCF_3SO_3$-DGM	114.6		[74]
NaFSA	Safety additive	Hard carbon/Na	3.3 M NaFSA/TMP	230	+100% (1200 cycles)	[75]
NaTFSI	Safety additive	NVP/Na	1.2 M NaTFSI-TMP/BTFE/VC	100		[76]
NaN_3	Compensating additive	$Na_{2/3}Fe_{1/2}Mn_{1/2}O_2$/Na	1 M $NaClO_4$ in EC/PC (1:1)	170	+32 mAh g^{-1}	[77]
Na_2NiO_2	Compensating additive	$NaCrO_2$/Sb@C	1 M $NaClO_4$ in EC/DEC (1:1) with 10 vol.% FEC	90	+15 mAh g^{-1}	[78]

- ***Additives to Improve the Cathode Stability*:** Similar to the anode, the cathode material can also degrade over time due to reactions with the electrolyte. Additives like lithium bis (oxalate) borate (LiBOB) and lithium difluoro (oxalate) borate (LiDFOB) can improve the cathode stability and prolong the battery's life.
- ***Safety-Related Additives*:** Safety-related additives are compounds added to the electrolyte to improve its safety characteristics, such as stability, non-flammability, and non-toxicity. These additives can enhance the overall thermal stability of the battery, suppress the generation of toxic gases during thermal runaways, and reduce the risk of fire and explosion [48].

7.8.2 Salt Additives

One approach to the enhancement of the performance and durability of electrolytes for SIBs is to add salt additives to the electrolyte. These additives can increase the ionic conductivity of the electrolyte, enhance the stability of the SEI, prevent the corrosion of the anode material, and enhance the cyclic stability and capacity retention of the battery. Examples include Lithium bis (trifluoromethane sulfonyl) imide (LiTFSI), FEC, NaFSI, sodium difluoro (oxalate) borate (NaDFOB), and sodium bis (oxalate) borate (NaBOB), etc. Therefore, salt additives for electrolytes in SIBs are a promising area of research for improving the performance and durability of these batteries. By enhancing the properties of the electrolyte, these additives can help overcome the challenges associated with SIBs and make them a viable alternative to LIBs [79].

Some commonly used salt additives are discussed below.

7.8.2.1 Sodium Perchlorate ($NaClO_4$)

In a Na-ion battery, Na-ions are the charge carriers that flow between the anode and cathode through the electrolyte. The electrolyte must have good ionic conductivity to allow efficient transport of the Na-ions, and the addition of $NaClO_4$ can improve the electrolyte's conductivity. This white crystalline salt is often added to the electrolyte to improve its conductivity and stability. It has a high solubility in polar solvents like PC and EC, which are commonly used in Na-ion battery electrolytes. $NaClO_4$ can also help improve the battery's performance by reducing the evolvement of unwanted by-products during the charging and discharging cycles. This can help extend the battery's cycle life and improve its overall efficiency. However, it is important to note that $NaClO_4$ is a highly reactive substance that can be hazardous if mishandled or used improperly [53]. Proper safety measures must be taken when handling and using this substance, and it should only be used by trained professionals in appropriate laboratory settings.

7.8.2.2 Sodium Bis (Fluoro Sulfonyl) Imide (NaFSI)

NaFSI is a salt that has gained popularity in recent years due to its high anodic stability and excellent ionic conductivity. It is a highly stable and highly soluble salt that can improve the cycle life and rate capability of the battery by increasing the ionic conductivity of the electrolyte. NaFSI enhances ionic conductivity by creating a larger number of charge carriers and reducing the resistance of the electrolyte.

NaFSI stabilizes the battery's performance by reducing the formation of unwanted by-products during the charging and discharging cycles, which extend the battery's cycle life and improve its overall efficiency.

Compared to other salts like $NaClO_4$ and $NaPF_6$, NaFSI has lower toxicity and environmental impact, making it a more desirable electrolyte additive. It also has a wider electrochemical window, allowing it to be used in a wider range of battery chemistries and operating conditions. However, it is important to note that NaFSI can be highly reactive and pose a safety hazard if mishandled or used improperly. Proper safety measures must be taken when handling and using this substance, and it should only be used by trained professionals in appropriate laboratory settings [48, 81].

7.8.2.3 Sodium Hexafluorophosphate ($NaPF_6$)

The addition of $NaPF_6$ to the electrolyte can enhance ionic conductivity by creating a larger number of charge carriers and reducing the resistance of the electrolyte [82]. $NaPF_6$ also helps stabilize the battery's performance by reducing the formation of unwanted by-products during the charging and discharging cycles, which in turn helps extend the battery's cycle life and improve efficiency. However, $NaPF_6$ can also be highly reactive and pose a safety hazard if mishandled or used improperly. Proper safety measures must be taken when handling and using this substance, and it should only be used by trained professionals in appropriate laboratory settings. Additionally, the use of $NaPF_6$ as an electrolyte additive can have environmental implications, as the disposal of the salt can be challenging due to its toxicity. As such, alternative electrolyte additives with lower environmental impact are being explored [82, 83].

7.8.2.4 Sodium Trifluoromethane Sulfonate (NaOTf)

NaOTf is another salt that is used as an electrolyte additive, which improves the solubility and conductivity of the electrolyte while preventing the formation of unwanted passivation layers on the electrodes, thus improving the battery's performance. It is a highly soluble salt that plays an important role in facilitating the movement of Na^+ ions between the anode and cathode, resulting in the improved working of SIBs by increasing the ionic conductivity of the electrolyte [84]. In addition, NaOTf can enhance ionic conductivity by creating a larger number of charge carriers and reducing the resistance of the electrolyte. It also improves the battery's performance by reducing the formation of unwanted by-products during the charging and discharging cycles, which helps extend the battery's cycle life and improve the overall efficiency.

Compared to other salts like $NaClO_4$ and $NaPF_6$, NaOTf has lower toxicity and environmental impact, making it a more desirable electrolyte additive. It also has a wider electrochemical window, allowing it to be used in a wider range of battery chemistries and operating conditions. However, it is important to note that NaOTf can be highly reactive and pose a safety hazard if mishandled or used improperly. Proper safety measures must be taken when handling and using this substance, and it should only be used by trained professionals in appropriate laboratory settings [85]. Additionally, its higher cost compared to other salts may be a limiting factor for its widespread use.

7.8.3 Solvent Additives

Solvent additives are organic compounds that can be added to the electrolyte of SIBs to improve their solubility, viscosity, and stability.

Some commonly used solvent additives are discussed below:

7.8.3.1 Dimethyl Carbonate (DMC)

DMC is a polar aprotic solvent that can dissolve a wide range of salts, making it useful for improving the solubility of electrolyte salts and enhancing ionic conductivity. The addition of DMC to electrolyte can increase the concentration of the electrolyte salts, which can improve the battery's performance. DMC is a low-viscosity solvent that enhances the solubility of sodium salts and reduces the viscosity of the electrolyte. It also improves the stability of the SEI layer and the life cycle of SIBs. Additionally, DMC has lower toxicity compared to other commonly used solvent additives such as EC and PC, making it a safer alternative. DMC is also considered to be more environmentally friendly, as it is a renewable chemical and has a lower carbon footprint compared to other petroleum-derived solvents.

However, DMC has a lower boiling point compared to other solvents, which can make it more volatile and increase the risk of flammability. Proper safety measures must be taken when handling and using DMC in appropriate laboratory settings [24, 31]. Overall, DMC is a promising solvent additive for electrolytes in SIBs due to its ability to improve the battery's performance and its lower toxicity and environmental impact.

7.8.3.2 Diethyl Carbonate (DEC)

DEC is a commonly used polar aprotic solvent additive for electrolytes in SIBs. It can dissolve a wide range of electrolyte salts and improve the conductivity. DEC is a low-viscosity solvent often used in combination with EC and DMC to enhance the solubility, along with the conductivity of the electrolyte. It reduces the overall extent of reactivity of the electrolyte with the anode and cathode materials. It also enhances the solubility of the electrolyte salts, which can increase the concentration of the Na-ions in the electrolyte and enhance its ionic conductivity. DEC can also help stabilize the battery's performance by improving the passivation of the electrode surfaces and reducing the formation of unwanted by-products during the charging and discharging cycles. This can help to extend the battery's cycle life and improve its overall efficiency.

Compared to other commonly used solvent additives such as EC and PC, DEC has a higher boiling point, which can make it more stable at higher operating temperatures. However, it is less effective at preventing the formation of the SEI on the electrode surface, which can negatively impact the battery's performance. Furthermore, DEC is considered to be less toxic and more environmentally friendly compared to other commonly used solvent additives. However, it is still important to handle and use it properly in appropriate laboratory settings [31]. Overall, DEC is a promising solvent additive for electrolytes in SIBs due to its ability to improve the battery's performance, lower toxicity, and environmental impact. However, its effectiveness at preventing SEI formation should be considered in the overall design of the battery.

7.8.4 Additives for Improvement of Anode Stability

7.8.4.1 Sodium Oxalate

One of the major challenges with SIBs is the limited cycle life, which is due to the degradation of the electrode and electrolyte materials over time. Sodium oxalate is effective in reducing the degradation of the electrode materials and enhancing the stability of the electrolyte. This is due to the fact that it forms an additional protective layer at the electrode, which prevents further degradation and improves the performance of the SIBs. Sodium oxalate also improves power density and charging/discharging rates. This is because sodium oxalate can increase the number of free Na-ions in the electrolyte, which can enhance the movement of ions and electrons in the battery. Overall, sodium oxalate shows promising results as an additive for electrolytes in SIBs, and further research is needed to fully understand its potential benefits and limitations [86, 87].

7.8.4.2 Sodium Sulfite

Sodium sulfite can scavenge oxygen in the electrolyte, which reduces the formation of oxidative species that can attack the anode and improve its stability. The main challenge with SIBs is the stability of the electrolyte, which can degrade over time and reduce the actual performance of the battery. Sodium sulfite is effective in improving the stability of the electrolyte by reducing the oxidation of the electrolyte components, which can lead to a longer cycle life and improved performance. Overall, sodium sulfite shows potential as an additive for electrolytes in SIBs, and further research is needed to fully understand its benefits and limitations.

7.8.4.3 Alkyl Phosphonate – Bis(2,2,2-Trifluoroethyl) Methyl Phosphonate (TFMP)

TFMP has been studied as an additive due to its ability to improve the stability and electrochemical properties of the battery. It enhances the stability of the electrolyte by reducing the degradation of the electrolyte components and the electrode materials. This is because TFMP can act as a stabilizer and suppress the occurrence of unwanted side reactions in the electrolyte, which leads to longer cycle life and improved performance for the SIBs. Furthermore, TFMP improves the thermal stability of the electrolyte, which can prevent the battery from overheating or catching fire. This is particularly important for the safety of the battery, as it reduces the risk of thermal runaway and improves the overall reliability of the battery [88]. However, the use of TFMP and other fluorinated compounds may have potential environmental and health risks due to the release of toxic fluorinated compounds during their manufacture and use.

7.8.4.4 Sodium Nitrate

Sodium nitrate can form a passivated layer on the surface of the anode, which prevents the anode from interacting with the electrolyte and improves its stability. Additionally, sodium nitrate can also enhance the rate capability of the battery by reducing the resistance of the electrolyte. Furthermore, sodium nitrate can improve the stability of the electrolyte by reducing the oxidation of the electrolyte components and the electrode materials. This is because sodium nitrate acts as a scavenger of reactive oxygen species in the electrolyte, which prevents the degradation of the battery and increase its cycle life. Moreover, sodium nitrate has low toxicity and is inexpensive, which makes it a desirable additive for large-scale applications [89]. However, it is important to note that the use of sodium nitrate may lead to safety concerns due to its potential reactivity with other chemicals in the electrolyte. Therefore, further research is needed to fully understand the safety implications of using sodium nitrate as an additive in SIBs.

7.8.4.5 Sodium Perchlorate

$NaClO_4$ has been studied as an additive for electrolytes in SIBs due to its high solubility and good stability in the electrolyte. It can form a protective layer on the anode surface, which prevents the formation of harmful species that can attack the anode. $NaClO_4$ improves the performance of SIBs by boosting the concentration of free Na-ions in the electrolyte, which facilitates the movement of ions and electrons in the battery and improves its overall performance. Moreover, it can enhance the safety of the battery by reducing the flammability of the electrolyte. This is because it has a high decomposition temperature that prevents the electrolyte from catching fire in case of thermal runaway.

However, the use of $NaClO_4$ as an additive in SIBs may raise safety concerns, due to its potential reactivity with other chemicals in the electrolyte, which in turn can lead to the formation of explosive compounds. Therefore, further research is needed to fully understand the safety implications of using $NaClO_4$ as an additive in SIBs [53, 78]. Additionally, the use of $NaClO_4$ may raise environmental concerns due to its potential toxicity and persistence in the environment. Therefore, the use of $NaClO_4$ and other perchlorate compounds in batteries and other applications may need to be carefully evaluated in terms of their potential impact on the environment, safety, and human health.

It must be noted that the overall effectiveness of these additives depends upon several factors including the specific composition of the electrolyte and the anode material, so it is important to experimentally test and optimize the additive concentration and formulation for a given system.

7.8.5 Additives for Improvement of Cathode Stability

Several additives can improve the cathode stability for electrolytes in SIBs. A few samples are discussed below:

7.8.5.1 Vinylene Carbonate (VC)

VC is a cyclic carbonate molecule with a vinyl group attached to one of its carbon atoms. It forms an additional layer on the cathode surface that protects and helps to reduce the reactivity of the electrolyte with the cathode. When added to the electrolyte, VC improves the stability of the SEI layer, formed at the anode surface, which is crucial for the performance and life cycle of SIBs. Interaction of the electrolyte with the anode surface leads to the formation of the SEI layer, which acts as a potential hurdle to prevent additional interactions between the electrolyte and the anode.

However, the SEI layer can break down over time and cause degradation of the battery performance. In addition, VC can also improve the rate capability of SIBs by reducing the viscosity and increasing the ionic conductivity [90]. Overall, the use of VC as an additive in the electrolyte for SIBs can lead to improved performance, better cycle life, and enhanced safety.

7.8.5.2 Fluoroethylene Carbonate (FEC)

FEC is another common electrolyte additive used in SIBs. It is a cyclic carbonate molecule with a fluorine atom attached to one of its carbon atoms. The addition of FEC to the electrolyte enhances the stability of the SEI layer on the anode, similar to VC. However, FEC is particularly effective at enhancing the overall stability of the SEI layer in high-voltage cathode materials, such as those based on nickel or manganese.

The improvement in the stability of the SEI layer is observed due to the evolvement of a uniform and stable layer on the anode surface, as well as the reduction of side reactions that can occur at the cathode-electrolyte interface. This can lead to improved cyclic stability and higher capacity retention for the battery. FEC can also increase the oxidation potential of the electrolyte, which leads to improved safety of SIBs by reducing the risk of other hazardous reactions and thermal runaway [48, 72].

When used as an additive in the electrolyte for SIBs, FEC can lead to enhanced performance, better cycle life, and improved safety of the battery, particularly for high-voltage cathode materials, as shown in Figure 7.3 [70]. Several co-additives are utilized in addition to FEC to achieve enhanced electrochemical operations.

7.8.5.3 Lithium Bis (Fluoro Sulfonyl) Imide (LiFSI)

LiFSI is a lithium salt with an anion that contains fluoro sulfonyl groups. LiFSI can improve the performance of the battery in multiple ways. First, it can increase the

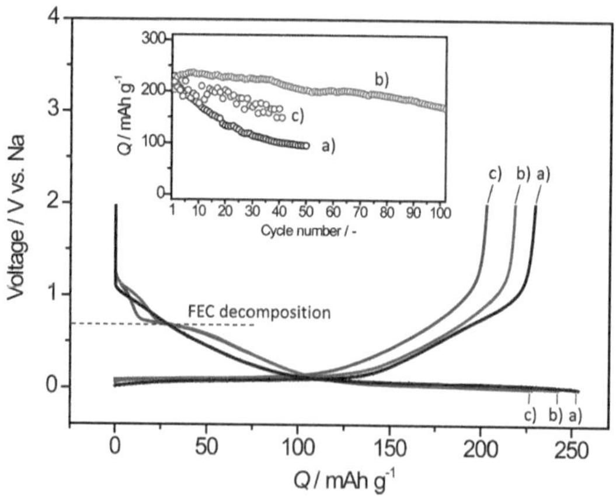

FIGURE 7.3 Charge/discharge curves (a) without FEC, (b) with 2% FEC, and (c) with 10% FEC of HC electrodes. This shows cyclic efficiency.

power and rate capability of the battery by enhancing the ionic conductivity of the electrolyte. LiFSI is particularly effective at enhancing the rate capability of the battery at low temperatures. Second, LiFSI can lead to improved cyclic stability and higher capacity retention for the battery by improving the stability of the SEI layer on the anode surface. Third, LiFSI can improve the thermal stability of the electrolyte, which increases the safety levels of the battery by reducing the risk of thermal runaway and other hazardous reactions. LiFSI has also been shown to be compatible with a wide range of cathode materials, which can increase the flexibility and versatility of Na-ion battery design. Overall, the use of LiFSI as an additive in the electrolyte for SIBs leads to improved performance, increased safety, and good compatibility with a range of cathode materials [91, 92].

7.8.5.4 Tris (Trimethylsilyl) Phosphite (TMSPi)

TMSPi is a colorless liquid that is soluble in organic solvents. It enhances the cyclic stability of the SIBs by lowering the decomposition of the electrolyte along with the evolvement of gas bubbles during the reaction. This can help prevent capacity loss and enhances the performance of the system. TMSPi demonstrated improved safety of the battery by reducing the risk of thermal runaway and other hazardous reactions. In addition, it is effective at reducing the flammability and reactivity of the electrolyte. TMSPi can improve the rate capability of the SIBs by increasing the conductivity of the electrolyte. This can help increase the power and rate capability of the battery.

The use of TMSPi as an additive in the electrolyte for SIBs leads to improved cyclic stability, increased safety, and improved rate capability. However, further research is required to fully understand the effects of TMSPi on the performance and safety of SIBs [93, 94]. Overall, the choice of additive depends on the specific cathode material and electrolyte used in the Na-ion battery, as well as the desired performance characteristics.

7.8.6 Safety-Related Additives

SIBs are promising alternatives to lithium-ion batteries, but the safety concerns of SIBs need to be improved for practical applications. One approach to enhancing the safety of these batteries is utilizing safety-related additives in the electrolyte. Safety-related additives are compounds added to the electrolyte to improve its safety characteristics, such as stability, non-flammability, and non-toxicity. These additives enhance the overall thermal stability of the battery, suppress the generation of toxic gases during thermal runaways, and reduce the risk of fire and explosion.

Some examples of safety-related additives that have been studied for use in SIBs include flame retardants (such as phosphorus-based compounds) and electrolyte stabilizers (such as VC and FEC). Flame retardants can suppress the thermal runaway of the battery by forming a preventive layer on the electrode surface, while electrolyte stabilizers can enhance the stability of the electrolyte and prevent the formation of undesirable by-products. Overall, the use of safety-related additives can improve the safety of SIBs, and ongoing research is focused on identifying and optimizing the most effective additives for this purpose [48].

7.8.6.1 Biphenyl (BP)

BP is a type of aromatic hydrocarbon that has been proposed as a safe additive for electrolytes in SIBs. While it is not commonly utilized in this way, it has been studied as a potential additive for improving the performance of these batteries. Investigations suggest the use of biphenyl as an additive in SIBs. The researchers found that, when added to the electrolyte, biphenyl resulted in enhanced cyclic stability and reduced capacity fading, which are common problems in SIBs. The study also suggested that biphenyl may help reduce the evolvement of passivation layers on the electrode surface, which can decrease the battery's performance.

Another published study investigated the use of biphenyl as an additive in Na-ion battery electrolytes, and the researchers found that biphenyl improved the electrochemical activities of the battery, particularly the rate capability and cyclic stability [95, 96]. Overall, while additional thorough research is required to investigate and conclude the potential benefits and limitations of using biphenyl as an additive in SIBs, these studies suggest it may be a promising avenue for improving the performance of these batteries.

7.8.6.2 Trimethyl Phosphate (TMP)

TMP is an organic flame-retardant compound that has been shown to improve thermal stability as a safe additive for electrolytes in SIBs. When added to electrolyte, TMP can increase the onset temperature of thermal decomposition and reduce the release of flammable gases. This helps prevent thermal runaway and enhances the safety of SIBs by increasing the ionic conductivity of the electrolyte, which can improve the efficiency of the battery and reduce its internal resistance. Using TMP as a safety additive in SIBs shows promise in improving the safety and performance of these batteries. Further research is needed to optimize the concentration of TMP and its effect on the long-term stability of SIBs [68, 97].

7.8.6.3 Tri(2,2,2-Trifluoromethyl) Phosphite

This is a phosphorus-containing organic compound with the molecular formula $C_6H_{11}F_9O_3P$ and is often used as an electrolyte additive in SIBs to improve their performance. One of the key benefits of tri(2,2,2-trifluoromethyl) phosphite **($(CF_3CH_2O)_3P$)** as an additive is its ability to increase the stability of the electrolyte. Tri(2,2,2-trifluoromethyl) phosphite helps stabilize the electrolyte by forming a passivated layer on the electrode surface, reducing unwanted side reactions, and improving the battery's overall performance. Tri(2,2,2-trifluoromethyl) phosphite also helps increase the battery's capacity and energy density, as it enables the Na-ions to move more efficiently through the electrolyte, reducing internal resistance and allowing more energy to be stored in the battery [98]. Therefore, tri(2,2,2-trifluoromethyl) phosphite as an additive in Na^+ ions can help improve their performance and make them more competitive with other types of batteries, such as lithium-ion batteries.

7.8.6.4 Methyl Nonafluorobutyl Ether (MFE)

MFE is a fluorinated organic compound having molecular formula C_5F_9O, which has been studied as a potential additive for electrolytes in SIBs. The main advantage of MFE as an additive is its ability to improve the stability and safety of the electrolyte.

MFE is highly stable and inert, which means it does not react easily with other components in the electrolyte or with the electrodes. This can help prevent unwanted side reactions and improve the overall safety of the battery. Additionally, MFE can also increase the performance of the Na-ion battery by increasing its capacity and energy density. This is because MFE can help improve the mobility of Na-ions in the electrolyte, reducing the internal resistance of the battery, and allowing more energy to be stored. Another benefit of MFE is its low vapor pressure, which means it is less likely to evaporate from the electrolyte and cause problems such as corrosion or gas formation. This can help increase the stability and reliability of the battery over time. Therefore, MFE shows promising potential for improving the stability, safety, and performance of SIBs [56, 99].

Additives for electrolytes in SIBs have shown the potential in improving the performance and stability of these batteries. The addition of certain electrolyte additives has been found to enhance the cyclic stability and capacity retention of SIBs. These additives can help mitigate issues such as capacity fading and the formation of films, which can degrade battery performance over time. However, more research is needed to fully understand the impact of different electrolyte additives on Na-ion battery performance, as well as to develop optimized electrolyte formulations for specific applications.

Additionally, there may be concerns regarding the environmental impact of certain additives, which may need to be addressed through further testing and development. Overall, while the use of additives in SIBs shows promise in improving performance and stability, continued research and development are needed to fully realize their potential.

7.9 FUTURE PROSPECTS

In terms of the future prospects of SIBs, research is ongoing to further improve the performance and stability of SIBs—and to develop new materials and new designs for the SIBs. Some of the key areas of research in SIBs include:

- ***Electrolytes*:** The electrolytes used in SIBs play a critical role in determining their performance and stability. Research is ongoing to develop new electrolytes that can provide better ion conductivity, stability, and safety.
- ***Cathode Materials*:** Research is ongoing to develop new cathode materials that can provide high capacity and long cycle life. Currently, some of the most promising cathode materials for SIBs include sodium vanadium phosphates, sodium manganese oxides, $NaFePO_4$, and sodium nickel cobalt oxides.
- ***Anode Materials*:** Research is also ongoing to develop new anode materials that can provide high capacity and stability. Currently, some of the most promising anode materials for SIBs include hard carbon, graphite, sodium titanate, and sodium metal.
- ***Scale-up and Commercialization*:** While SIBs show great potential, several challenges still need to be addressed before they can be commercialized on a large scale. One key challenge is to develop scalable and cost-effective production methods for SIBs.

The performance and safety of SIBs are greatly affected by electrolyte, and therefore, significant research efforts are being made to improve the electrolytes' properties. Here are some prospects for Na-ion battery electrolytes:

- ***High Ionic Conductivity*****:** The future electrolytes of SIBs should have high ionic conductivity to enhance battery performance. In this regard, solid electrolytes are being explored as an alternative to liquid electrolytes, which have low ionic conductivity.
- ***Stability*****:** The stability of the electrolyte is crucial for the long-term performance of SIBs. The future electrolytes should have high chemical and electrochemical stability to reduce the risk of decomposition and side reactions, which can lead to capacity fade and safety issues.
- ***Compatibility with Sodium Metal Anodes*****:** One of the significant challenges of SIBs is the dendrite growth of sodium metal anodes during charging, which can lead to short-circuiting and safety issues. The future electrolytes should be compatible with sodium metal anodes and prevent dendrite growth during charging.
- ***Environmental Friendliness*****:** The future electrolytes of SIBs should be environmentally friendly, nontoxic, and safe. Many conventional electrolytes use toxic and flammable materials, which can pose a risk to human health and the environment.

Overall, the prospects of SIBs and their electrolytes are promising, and ongoing research efforts are expected to further improve their performance, stability, and scalability.

REFERENCES

1. A. Chandra, A. Chandra, and R. S. Dhundhel, *Indian J. Pure Appl. Phys.*, 2020, 58, 113–119.
2. History of Batteries: A Timeline – News about Energy Storage, Batteries, Climate Change and the Environment at https://www.upsbatterycenter.com/blog/history-batteries-timeline/
3. Supply Chain and Raw Materials for Growing Battery Production at https://www.influitenergy.com/post/supply-chain-and-raw-materials-for-growing-battery-production
4. Sodium – Wikipedia at https://en.wikipedia.org/wiki/Sodium
5. L. Monconduit, and L. Croguennec, *Na-Ion Batteries*, 2021, xi–xviii. doi:10.1002/9781119818069
6. G. B. Haxel, J. B. Hedrick, G. J. Orris, P. H. Stauffer, and J. W. Hendley II, *Fact Sheet*, 2002. doi:10.3133/FS08702
7. Standard Reduction Potential – Chemistry Libre Texts at https://chem.libretexts.org/Bookshelves/Analytical_Chemistry/Supplemental_Modules_(Analytical_Chemistry)/Electrochemistry/Redox_Chemistry/Standard_Reduction_Potential
8. Z. Tian, Y. Zhang, J. Zhu, Q. Li, T. Liu, and M. Antonietti, *Adv. Energy Mater.*, 2021, 11, 2102489.
9. K. M. Abraham, *ACS Energy Lett.*, 2020, 5, 3544–3547.
10. Y. Ding, Z. P. Cano, A. Yu, J. Lu, and Z. Chen, *Electrochem. Energy Rev.*, 2019, 2, 1–28.

11. CATL Unveils Its Latest Breakthrough Technology by Releasing Its First Generation of Sodium-ion Batteries at https://www.catl.com/en/news/665.html
12. J. F. Peters, A. P. Cruz, and M. Weil, *Batteries*, 2019, 5, 10.
13. Battery Pack Prices Cited Below $100/kWh for the First Time in 2020, While Market Average Sits at $137/kWh | Bloomberg NEF at https://about.bnef.com/blog/battery-pack-prices-cited-below-100-kwh-for-the-first-time-in-2020-while-market-average-sits-at-137-kwh/
14. Strong Performance – Faradion at https://faradion.co.uk/technology-benefits/strong-performance/
15. K. Mongird, V. Fotedar, V. Viswanathan, V. Koritarov, P. Balducci, B. Hadjerioua, and J. Alam, *Energy storage technology and cost characterization report*. No. PNNL-28866. Pacific Northwest National Lab (PNNL), Richland, WA (United States), 2019.
16. ITP Battery Test Centre Reports – Australian Renewable Energy Agency (ARENA) at https://arena.gov.au/knowledge-bank/itp-battery-test-centre-reports/
17. S. Ma, M. Jiang, P. Tao, C. Song, J. Wu, J. Wang, T. Deng, and W. Shang, *Prog. Nat. Sci.: Mater. Int.*, 2018, 28, 653–666.
18. D. Monti, E. Jónsson, A. Boschin, M. R. Palacín, A. Ponrouch, and P. Johansson, *Phys. Chem. Chem. Phys.*, 2020, 22, 22768–22777.
19. H. Che, S. Chen, Y. Xie, H. Wang, K. Amine, X. Z. Liao, and Z. F. Ma, *Energy Environ. Sci.*, 2017, 10, 1075–1101.
20. A. Tornheim, C. Peebles, J. A. Gilbert, R. Sahore, J. C. Garcia, J. Bareño, H. Iddir, C. Liao, and D. P. Abraham, *J. Power Sources*, 2017, 365, 201–209.
21. S. H. Arshad, E. K. Murphy, J. M. Callahan, P. Janssen, J. Kasnatscheew, B. Streipert, Y. Ji, S. Li, and G. Zhong, *J. Electrochem. Soc.*, 2020, 167, 120535.
22. B. Tong, Z. Song, H. Wan, W. Feng, M. Armand, J. Liu, H. Zhang, Z. Zhou, and C. Heng Zhang, *InfoMat*, 2021, 3, 1364–1392.
23. J. J. Walton, T. Hiasa, H. Kumita, K. Takeshi, and G. Sandford, *ACS Appl. Mater. Interfaces*, 2020, 12, 15893–15902.
24. Z. X. Huang, X. L. Zhang, X. X. Zhao, Y. Y. Zhao, V. Aravindan, Y. H. Liu, H. Geng, and X. L. Wu, *Inorg. Chem. Front.*, 2022, 10, 37–48.
25. B. Mosallanejad, S. S. Malek, M. Ershadi, A. A. Daryakenari, Q. Cao, F. Boorboor Ajdari, and S. Ramakrishna, *J. Electroanal. Chem.*, 2021, 895, 115505.
26. K. Vignarooban, R. Kushagra, A. Elango, P. Badami, B. E. Mellander, X. Xu, T. G. Tucker, C. Nam, and A. M. Kannan, *Int. J. Hydrogen Energy*, 2016, 41, 2829–2846.
27. J. Yi, and Y. Xia, *MRS Energy Sustain.*, 2022, 9, 106–128.
28. F. Makhlooghiazad, C. Pozo-Gonzalo, P. Johansson, and M. Forsyth, *Na-ion Batteries*, 2021, 205–242.
29. J. Zhang, C. Cui, P. F. Wang, Q. Li, L. Chen, F. Han, T. Jin, S. Liu, H. Choudhary, S. R. Raghavan, N. Eidson, A. Von Cresce, L. Ma, J. Uddin, D. Addison, C. Yang, and C. Wang, *Energy Environ. Sci.*, 2020, 13, 2878–2887.
30. A. Ponrouch, E. Marchante, M. Courty, J. M. Tarascon, and M. R. Palacín, *Energy Environ. Sci.*, 2012, 5, 8572–8583.
31. A. Bhide, J. Hofmann, A. Katharina Dürr, J. Janek, and P. Adelhelm, *Phys. Chem. Chem. Phys.*, 2014, 16, 1987–1998.
32. A. Bagno, C. Butts, C. Chiappe, F. D'Amico, J. C. D. Lord, D. Pieraccini, and F. Rastrelli, *Org. Biomol. Chem.*, 2005, 3, 1624–1630.
33. C. H. Wang, Y. W. Yeh, N. Wongittharom, Y. C. Wang, C. J. Tseng, S. W. Lee, W. S. Chang, and J. K. Chang, *J. Power Sources*, 2015, 274, 1016–1023.
34. D. Monti, E. Jónsson, M. R. Palacín, and P. Johansson, *J. Power Sources*, 2014, 245, 630–636.

35. N. Wongittharom, C. H. Wang, Y. C. Wang, C. H. Yang, and J. K. Chang, *ACS Appl. Mater. Interfaces*, 2014, 6, 17564–17570.
36. K. Matsumoto, R. Taniki, T. Nohira, and R. Hagiwara, *J. Electrochem. Soc.*, 2015, 162, A1409–A1414.
37. C. Ding, T. Nohira, K. Kuroda, R. Hagiwara, A. Fukunaga, S. Sakai, K. Nitta, and S. Inazawa, *J. Power Sources*, 2013, 238, 296–300.
38. A. Hayashi, K. Noi, A. Sakuda, and M. Tatsumisago, *Nat. Commun.*, 2012, 3, 1–5.
39. N. Tanibata, K. Noi, A. Hayashi, and M. Tatsumisago, *RSC Adv.*, 2014, 4, 17120–17123.
40. P. B. Bhargav, V. M. Mohan, A. K. Sharma, and V. V. R. N. Rao, *Ionics (Kiel)*, 2007, 13, 441–446.
41. P. B. Bhargav, V. M. Mohan, A. K. Sharma, and V. V. R. N. Rao, *Int. J. Polym. Mater. Polym. Biomater.*, 2007, 56, 579–591.
42. P. B. Bhargav, V. M. Mohan, A. K. Sharma, and V. V. R. N. Rao, *Curr. Appl. Phys.*, 2009, 9, 165–171.
43. D. Kumar, M. Suleman, and S. A. Hashmi, *Solid State Ion.*, 2011, 202, 45–53.
44. D. Kumar, and S. A. Hashmi, *Solid State Ion.*, 2010, 181, 416–423.
45. D. Kumar, and S. A. Hashmi, *J. Power Sources*, 2010, 195, 5101–5108.
46. Y. Q. Yang, Z. Chang, M. X. Li, X. W. Wang, and Y. P. Wu, *Solid State Ion.*, 2015, 269, 1–7.
47. L. Qiao, X. Judez, T. Rojo, M. Armand, and H. Zhang, *J. Electrochem. Soc.*, 2020, 167, 070534.
48. R. Rajagopalan, and L. Zhang, *Adv. Mater. Sodium Ion Storage*, 2019, 131–142.
49. Are Aqueous Sodium Batteries Safer-Tycorun Batteries at https://www.tycorun.com/blogs/news/are-aqueous-sodium-batteries-safer
50. M. Liu, H. Ao, Y. Jin, Z. Hou, X. Zhang, Y. Zhu, and Y. Qian, *Mater. Today Energy*, 2020, 17, 100432.
51. H. Zhang, B. Qin, J. Han, and S. Passerini, *ACS Energy Lett.*, 2018, 3, 1769–1770.
52. H. Guo, Z. Shao, Y. Zhang, X. Cui, L. Mao, S. Cheng, M. Ma, W. Lan, Q. Su, and E. Xie, *J. Colloid Interface Sci.*, 2022, 608, 1481–1488.
53. G. G. Eshetu, G. A. Elia, M. Armand, M. Forsyth, S. Komaba, T. Rojo, and S. Passerini, *Adv. Energy Mater.*, 2020, 10, 2000093.
54. D. Lakshmi, N. Palaniyandy, S. Bohm, and B. B. Mamba, *Curr. Opin. Electrochem.*, 2023, 38, 101217.
55. J. Wu, H. Liu, H. Bu, X. Zhang, H. Zhang, W. Wang, L. Yang, C. Zeng, and S. Zhu, *J. Alloys Compd.*, 2023, 934, 167872.
56. Y. Fang, L. Xiao, Z. Chen, X. Ai, Y. Cao, and H. Yang, *Electrochem. Energy Rev.*, 2018, 1, 294–323.
57. Y. Xu, M. Zhou, and Y. Lei, *Mater. Today*, 2018, 21, 60–78.
58. Y. Yu, H. Che, X. Yang, Y. Deng, L. Li, and Z. F. Ma, *Electrochem. Commun.*, 2020, 110, 106635.
59. Z. Lin, Q. Xia, W. Wang, W. Li, and S. Chou, *InfoMat*, 2019, 1, 376–389.
60. Y. Zheng, D. Wang, S. Kaushik, S. Zhang, T. Wada, J. Hwang, K. Matsumoto, and R. Hagiwara, *EnergyChem*, 2022, 4, 100075.
61. Y. Wang, S. Song, C. Xu, N. Hu, J. Molenda, and L. Lu, *Nano Mater. Sci.*, 2019, 1, 91–100.
62. F. Gebert, J. Knott, R. Gorkin, S. L. Chou, and S. X. Dou, *Energy Storage Mater.*, 2021, 36, 10–30.
63. J. Feng, L. Wang, Y. Chen, P. Wang, H. Zhang, and X. He, *Nano Converg.*, 2021, 8, 1–12.
64. S. Song, M. Kotobuki, F. Zheng, C. Xu, S. V. Savilov, N. Hu, L. Lu, Y. Wang, and W. D. Z. Li, *J. Mater. Chem. A*, 2017, 5, 6424–6431.

65. H. M. Woolley, and N. M. Vargas-Barbosa, *J. Mater. Chem. A*, 2023, 11, 1083–1097.
66. A. Gabryelczyk, H. Smogór, and A. Swiderska-Mocek, *Electrochim. Acta*, 2023, 439, 141645.
67. K. Zhu, Z. Sun, Z. Li, P. Liu, X. Chen, and L. Jiao, *Energy Storage Mater.*, 2022, 53, 523–531.
68. Z. Zeng, X. Jiang, R. Li, D. Yuan, X. Ai, H. Yang, Y. Cao, Z. Q. Zeng, X. Y. Jiang, D. D. Yuan, X. P. Ai, H. X. Yang, Y. L. Cao, and R. Li, *Adv. Sci.*, 2016, 3, 1600066.
69. K. B. Hueso, M. Armand, and T. Rojo, *Energy Environ. Sci.*, 2013, 6, 734–749.
70. S. Komaba, T. Ishikawa, N. Yabuuchi, W. Murata, A. Ito, and Y. Ohsawa, *ACS Appl. Mater. Interfaces*, 2011, 3, 4165–4168.
71. M. Dahbi, T. Nakano, N. Yabuuchi, S. Fujimura, K. Chihara, K. Kubota, J. Y. Son, Y. T. Cui, H. Oji, and S. Komaba, *ChemElectroChem*, 2016, 3, 1856–1867.
72. J. Y. Jang, Y. Lee, Y. Kim, J. Lee, S. M. Lee, K. T. Lee, and N. S. Choi, *J. Mater. Chem. A*, 2015, 3, 8332–8338.
73. M. Dahbi, N. Yabuuchi, M. Fukunishi, K. Kubota, K. Chihara, K. Tokiwa, X. F. Yu, H. Ushiyama, K. Yamashita, J. Y. Son, Y. T. Cui, H. Oji, and S. Komaba, *Chem. Mater.*, 2016, 28, 1625–1635.
74. J. Shi, L. Ding, Y. Wan, L. Mi, L. Chen, D. Yang, Y. Hu, and W. Chen, *J. Energy Chem.*, 2021, 57, 650–655.
75. J. Wang, Y. Yamada, K. Sodeyama, E. Watanabe, K. Takada, Y. Tateyama, and A. Yamada, *Nat. Energy*, 2017, 3, 22–29.
76. Z. Yang, J. He, W. H. Lai, J. Peng, X. H. Liu, X. X. He, X. F. Guo, L. Li, Y. Qiao, J. M. Ma, M. Wu, and S. L. Chou, *Angew. Chem. Int. Ed.*, 60, 27086–27094.
77. G. Singh, B. Acebedo, M. C. Cabanas, D. Shanmukaraj, M. Armand, and T. Rojo, *Electrochem. Commun.*, 2013, 37, 61–63.
78. K. Park, B. C. Yu, and J. B. Goodenough, *Chem. Mater.*, 2015, 27, 6682–6688.
79. J. H. Jo, J. U. Choi, Y. J. Park, J. K. Ko, H. Yashiro, and S. T. Myung, *Energy Storage Mater.*, 2020, 32, 281–289.
80. G. G. Eshetu, M. Martinez-Ibañez, E. Sánchez-Diez, I. Gracia, C. Li, L. M. Rodriguez-Martinez, T. Rojo, H. Zhang, and M. Armand, *Chem. Asian J.*, 2018, 13, 2770–2780.
81. X. Qi, Q. Ma, L. Liu, Y. S. Hu, H. Li, Z. Zhou, X. Huang, and L. Chen, *ChemElectroChem*, 2016, 3, 1741–1745.
82. P. Barnes, K. Smith, R. Parrish, C. Jones, P. Skinner, E. Storch, Q. White, C. Deng, D. Karsann, M. L. Lau, J. J. Dumais, E. J. Dufek, and H. Xiong, *J. Power Sources*, 2020, 447, 227363.
83. D. M. C. Ould, S. Menkin, C. A. O'Keefe, F. Coowar, J. Barker, C. P. Grey, and D. S. Wright, *Angew. Chem. Int. Ed.*, 2021, 60, 24882–24887.
84. C. Chandra, W. Devina, A. D. M. Sarofil, and J. Kim, *Chem. Eng. J.*, 2022, 438, 135411.
85. K. Nakamoto, R. Sakamoto, Y. Sawada, M. Ito, and S. Okada, *Small Methods*, 2019, 3.
86. D. Shanmukaraj, K. Kretschmer, T. Sahu, W. Bao, T. Rojo, G. Wang, and M. Armand, *ChemSusChem*, 2018, 11, 3286–3291.
87. Y. J. Guo, Y. Bin Niu, Z. Wei, S. Y. Zhang, Q. Meng, H. Li, Y. X. Yin, and Y. G. Guo, *ACS Appl. Mater. Interfaces*, 2021, 13, 2772–2778.
88. W. Zhang, F. Zeng, H. Huang, Y. Yu, M. Xu, L. Xing, and W. Li, *Nano Res.*,2022, 1–9.
89. J. J. Fan, P. Dai, C. G. Shi, Y. Wen, C. X. Luo, J. Yang, C. Song, L. Huang, and S. G. Sun, *Adv. Funct. Mater.*, 2021, 31, 2010500.
90. B. Shen, R. Zhan, C. Dai, Y. Li, L. Hu, Y. Niu, J. Jiang, Q. Wang, and M. Xu, *J. Colloid Interface Sci.*, 2019, 553, 524–529.
91. S. Uchida, and M. Ishikawa, *J. Power Sources*, 2017, 359, 480–486.
92. D. Reber, R. Figi, R. S. Kühnel, and C. Battaglia, *Electrochim. Acta*, 2019, 321, 134644.

93. C. Yan, Y. Xu, J. Xia, C. Gong, and K. Chen, *J. Energy Chem.*, 2016, 25, 659–666.
94. Y. K. Han, J. Yoo, and T. Yim, *Electrochim. Acta*, 2016, 215, 455–465.
95. W. C. Sun, S. S. Wang, and S. Dong, *J. Mater. Sci.*, 2021, 56, 13763–13771.
96. Y. Kim, J. Jung, H. Yu, G. T. Kim, D. Jeong, D. Bresser, S. J. Kang, Y. Kim, and S. Passerini, *Adv. Funct. Mater.*, 2020, 30, 2001249.
97. X. Liu, X. Jiang, Z. Zeng, X. Ai, H. Yang, F. Zhong, Y. Xia, and Y. Cao, *ACS Appl. Mater. Interfaces*, 2018, 10, 38141–38150.
98. Y. Gu, S. Fang, L. Yang, and S. I. Hirano, *ACS Appl. Energy Mater.*, 2021, 4, 4919–4927.
99. Y. Li, F. Wu, Y. Li, M. Liu, X. Feng, Y. Bai, and C. Wu, *Chem. Soc. Rev.*, 2022, 51, 4484–4536.

8 Electrolytes and Additives for Zinc-ion Systems

Guozhao Fang, Zhexuan Liu, and Shuquan Liang

8.1 INTRODUCTION

Zinc-ion batteries (ZIBs) have shown promising development potential for large-scale electrochemical energy storage owing to their high economic efficiency. In the past ten years, there has been a surge in research on ZIBs, as shown in Figure 8.1a, which has been explored and discussed from the perspectives of electrodes, electrolytes, and reaction mechanism. Like other battery systems, reliable and durable operation depends on the properties of bulk electrolyte with electrode-electrolyte interface, and this demand has stimulated the development of suitable electrolytes. Accordingly, the number of articles on the electrolyte design of zinc-ion batteries is also increasing. Over the years, different types of electrolytes—including alkaline electrolyte, aqueous electrolyte, ionic liquid electrolyte, gel electrolyte, organic electrolyte, solid electrolyte, etc.—have been developed. These electrolytes and their features are shown in Figure 8.1b.

As early as the 1960s, researchers had already experimented with alkaline electrolytes in ZIBs. Alkaline electrolytes have gradually become the most widely used electrolytes in early ZIBs for their cheap price and relatively high ionic conductivity. However, Zn anode in alkaline electrolytes causes irreversible side reactions, leading to severe capacity reduction and unsatisfactory CE [1]. These difficulties have hindered the commercial prospects of ZIBs with alkaline electrolytes.

Yamamoto et al. first achieved rechargeable aqueous Zn/MnO_2 batteries in 1986, using a mild $ZnSO_4$ electrolyte rather than an alkaline electrolyte. Compared to alkaline electrolytes, mild acid electrolytes offer superior energy density and better compatibility. However, these electrolytes have been unable to solve the problems of dissolving cathode electrode material and intermediate by-products, as well as the narrow electrochemically stable potential window (ESPW) and poor cycling stability [2].

Ionic liquid electrolytes benefit from having favorable thermal and chemical stability and an outstanding ESPW, but the high viscosity of ionic liquids contributes to weak electrical conductivity, trace impurities, and inferior versatility, limiting their application [3]. Most ionic liquid electrolytes, on the other hand, are expensive and unsuitable for commercial employment. Organic electrolytes have a wider range of ESPW than aqueous electrolytes and are compatible with Zn metal anodes

DOI: 10.1201/9781003441182-10

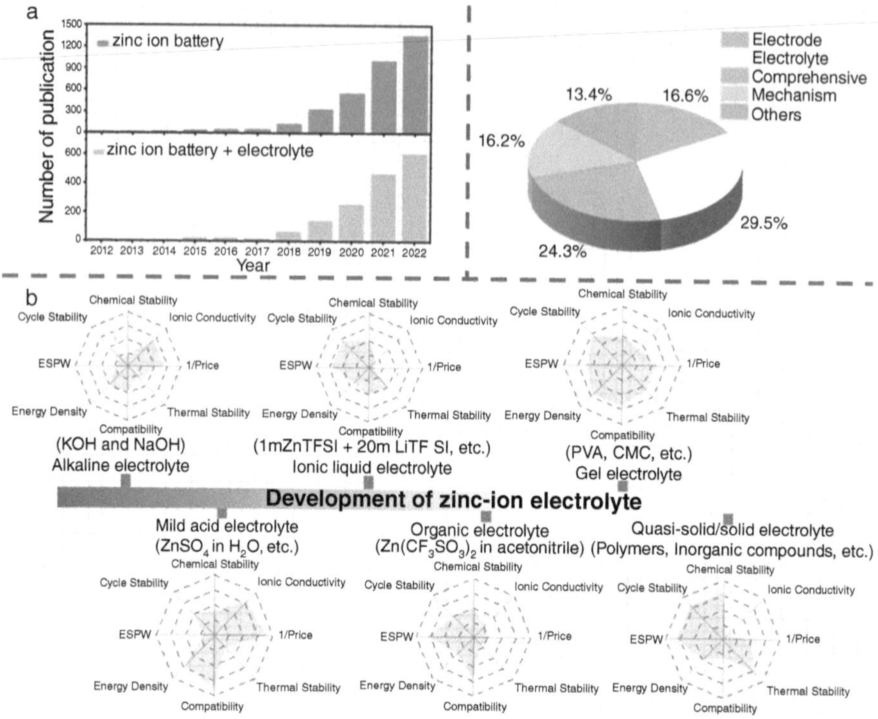

FIGURE 8.1 (a) Changes in the number of publications based on the keywords "zinc-ion battery" and "zinc-ion battery + electrolyte," and a brief statistic about the published reviews until January 18, 2023; (b) different types of zinc-ion electrolytes and their features.

Source: Data from the Web of Science.

simultaneously. However, this goes against the original intention of designing ZIBs, especially with safety and environmental friendliness.

Compared to liquid electrolytes, gel electrolytes provide excellent mechanical strength and flexibility, offering the potential to prepare flexible energy storage devices. Inevitably, they remain greatly limited due to their chemical and thermal instability and poor interfacial physical contact. Quasi-solid and solid electrolytes may be promising in solving such issue, given their high ESPW and chemical stability [4]. However, in solid electrolyte, the poor water content and poor compatibility between the electrolyte and the electrode—as well as the resulting low zinc ion conductivity and high interfacial impedance—are an important challenge for solid electrolyte. In addition, large-scale preparation of gel electrolyte and solid electrolyte is still a great challenge.

Although these electrolytes have been mentioned, weakly acidic aqueous electrolyte and its derivatives have been the focus of research over the years due to their safety, low cost, and easy preparation [5]. Nevertheless, the difficulties faced by researchers are huge. During the charge/discharge cycle, most of the cathode material or intermediate by-products produced can be dissolved in the weakly acidic electrolyte [6]. As a consequence, the cathode electrode material is degraded by disordered deformation of the fragile structure, resulting in a severe decline in battery capacity. Secondly, the electrolyte exhibits poor chemical and electrochemical stability, tends to undergo side reactions such as hydrogen and oxygen precipitation, and performs poorly at specific temperatures. The decomposition of the electrolyte can lead to potential problems with deformation and expansion of the battery. Moreover, a number of other problems caused by electrolyte-electrode interactions are more deadly and may reduce Coulombic efficiency or even lead to battery failure.

In summary, these complications may ultimately contribute to a reduction in energy performance and shorten the life of the rechargeable AZIBs. In addition to academic research on the electrolyte itself, many practical issues are associated with the use of electrolytes in practical production applications, for instance, production costs and efficiency, safety assessments, electrolyte consumption and compensation, adaptability to extreme temperatures, gases generated by electrolyte decomposition, and other practical issues associated with electrolytes [7]. These will impact on the actual production and usage of the aqueous ZIBs. Considering that these problems encountered in practical production are different from those encountered in academic research, a comprehensive and insightful understanding of this aspect is urgently needed.

In this chapter, we provide a system summary of the evolution of aqueous zinc-ion electrolytes, including basic characteristics, issues, optimization strategies, and particularly the interfacial compatibility between electrolytes and electrode materials, etc. Furthermore, it provides an outlook on fundamental understanding and practical considerations of aqueous zinc-ion electrolytes. In the future, the design of zinc-ion electrolyte is expected to be fully developed with practical features. At the same time, for the commercialization of ZIBs, an electrolyte with more affordable and environmentally friendly, as well as excellent, electrochemical properties needs to be found.

8.2 BASIC FEATURES OF ZINC-ION ELECTROLYTES

As a significantly important component of a battery, the electrolyte is mainly responsible for ion diffusion and transport properties, and chemical and electrochemical stability (Figure 8.2). This not only affects some basic electrochemical features, but it also even changes the reaction mechanism in some cases.

8.2.1 Salts, Solvents, and Additives

Various zinc salts have been reported, including $ZnSO_4$, $ZnCl_2$, $Zn(BF_4)_2$, $Zn(CH_3COO)_2$, $Zn(ClO_4)_2$, $Zn(CF_3SO_3)_2$, etc. Due to the abundant water molecules and ionized hydrogen ions in the aqueous environment, the electrode reaction is usually accompanied by fluctuation of the local pH value. Hence, the anions of SO_4^{2-}, $CF_3SO_3^-$, etc., which could promote the formation of hydroxysulfate to balance the reaction environment, have been widely applied. The corrosion and oxidation of NO_3^-, Cl^- and CH_3COO^- anions determine their incompatibility in conventional aqueous electrolyte. However, such anions contribute to the preparation of special novel electrolytes by optimization of the ion solvation structure and hydrogen bond

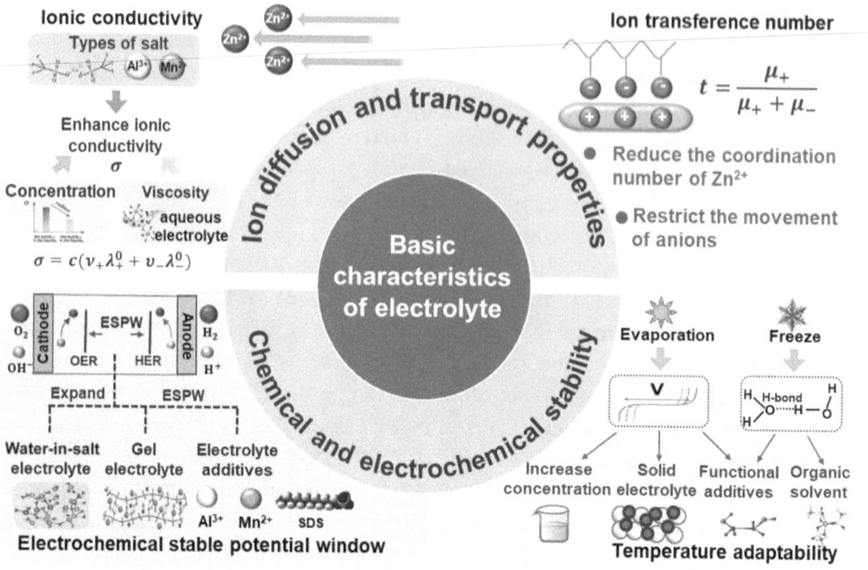

FIGURE 8.2 Basic features of zinc-ion electrolytes.

analogue interactions between groups. As well known, the construction of hydrogen bond networks can regulate the electrochemical behavior of water molecules in aqueous electrolytes, which has unexpected benefits in compensating the intrinsic inferiority of anions and improving stability.

Aqueous electrolyte in zinc-ion systems is the leading research for low cost and high safety. While it also suffers from the narrow electrochemical stability window (~1.23 V), side reactions will inevitably lead to the decay of battery performance [8]. In order to suppress the water-induced side reactions, many studies have been conducted on pure organic or novel hybrid solvents. Various organic solvents that can be miscible with water have been tried, including alcohols, ethers, esters, nitriles, etc. Even some organic solvents that are difficult to be miscible with water can also form a homogeneous inorganic-organic hybrid electrolyte under the mediation of some special zinc salts. This has been proved to be effective in broadening the electrochemical stable window, inhibiting water decomposition and antifreezing. In some cases that are extremely sensitive to water molecules, the solutes can also play the role of solvent to form ionic liquids through hydrogen bond analogue interactions between different organic groups. However, this kind of electrolyte still has some problems, such as low ionic conductivity and high viscosity, which still need to be improved.

The optimization of zinc salt and solvent cannot guarantee the operation of battery reaction completely, which requires the development of various functional additives simultaneously. A small amount of electrolyte additives assists in promoting the electrochemical performances of electrode and electrolyte [9], mainly including organic additives, ionic additives, metal additives and inorganic additives.

Based on properties such as ion valence, radius, and electrophilicity, the ionic additives play different roles in changing the reaction mechanism, inhibiting the Zn anode dendrite growth, optimizing cathode electrochemical performance, improving the ionic conductivity of electrolyte, etc. [10]. The small molecule additives mainly modify the zinc anode surface morphology and the interface reaction kinetics. Abundant groups can also participate in the ion solvation structure, which also changes the cathode interfacial electrochemical process in some cases. Similar to other small organic molecules, the polymer additives mainly change the overpotential of zinc anode deposition and inhibit corrosion and dendritic growth [11]. Especially, the unique regulation of hydrogen bonding networks in aqueous electrolytes by polymers can sometimes greatly alter the state of electrolyte bulk and interface, thereby stimulating satisfactory functions.

Organic acids-type additives such as tartaric acid and succinic acid are mainly increase hydrogen evolution overpotential [12]. Compared with organic additives and ionic additives, metal additives and other inorganic additives have limited solubility. Benefitting from the zincophilic property, metal electrolyte additives primarily refer to lead and tin [13], which has the potential for alloying to change the zinc deposition behavior and affect the texture growth. Inorganic additives can be divided into oxide and inorganic acid [12], and this type of additive is relatively niche, with limited applicability. Depending on the battery system, it could suppress the cathode dissolution, improve the reaction depth, regulate zinc anode crystal orientation or morphology, and prohibit the zinc anode side reactions.

8.2.2 Ion Diffusion and Transport Properties

Generally, there are three types of mass transfer in electrolytes: diffusion mass transfer, p-transfer mass transfer, and electromigration mass transfer, among which diffusion mass transfer plays a leading role in the electrochemical reaction at the electrode-electrolyte interface. Ionic diffusion and transport characteristics are the most basic properties of electrolytes.

During the charging and discharging process, the electric field and concentration between electrodes drive the diffusion and transfer in electrolyte. For instance, in $ZnSO_4$ aqueous electrolyte, the positive Zn^{2+} migrates to the anode under the electric field directed towards it, whereas the SO_4^{2-} with negative charge migrates to the cathode. But in reality, only the increase in migration numbers of working ions is meaningful, such as zinc ions in ZIBs.

In general, liquid electrolytes do not restrict the diffusion and transport of any ions or molecules [14]. However, due to solvation, the mobility of Zn^{2+} ions will form a solvent shell on the surface, while the anions almost do not solvate, which leads to accumulation on the cathode surface. In this case, the gradually increasing ion concentration polarization between electrodes will significantly reduce energy density and cycling stability of ZIBs [15]. Therefore, it is promising to increase the ionic migration number by limiting anion migration and reducing the coordination degree of working ions [16]. For instance, the interaction between cations and Lewis base sites in polymer polar groups correspondingly reduce the Zn^{2+} migration number in the gel electrolyte and solid electrolyte [14]. To solve this problem, the stimulation of the interaction between additives and groups and the occupation of coordination sites can theoretically release more free Zn^{2+}, thus promoting the Zn^{2+} ion migration number [17].

The electrolyte conductivity represents the ion transport capacity in electrolyte, and generally the conductivity of liquid electrolytes ranks first due to their low viscosity and solvation energy. Taking a strong electrolyte system as an example, the conductivity of the strong electrolyte solution with low concentration complies with Kohlrausch's Law, which could be described as Equation (8.1):

$$\sigma = c\left(v_+\lambda_+^0 + \upsilon_-\lambda_-^0\right) \tag{8.1}$$

where c is the concentration of the electrolyte, υ_+ and v_- are the number of moles of cation and anion dissociated by dissolving one mole of electrolyte respectively, and λ_+^0 and λ_-^0 are the limiting molar conductivity of cations and anions.

It can be recognized from Equation (8.1) that electrolyte ion conductivity is related to the type of solvent and the type and quantity of solute ions. High concentrations of electrolytes provide high ionic conductivity when the zinc salt type is fixed [18]. When the type of zinc salt changes, the conductivity of electrolytes will also be affected, such as the difference between $CF_3SO_3^-$-based and SO_4^{2-}-based aqueous electrolyte. Similarly, solvent viscosity and solvation effect also have obvious influence on ionic conductivity. In liquid electrolytes, the viscosity of organic electrolytes is significantly higher than that of aqueous electrolytes, and the ionic conductivity is much lower than that of aqueous electrolytes.

In general, any additive that can improve ion richness and salt solubility is beneficial for improving ion conductivity [19]. The strong solvation effect separates cations from anions and magnifies the distance between ions, thus obtaining high conductivity. There are inevitable interactions between ions and functional groups in gel electrolytes and solid electrolytes, and their low mobility results in relatively low ionic conductivity of the two electrolytes. Therefore, the strong interaction between appropriate additives and the functional groups of polymers can effectively replace the restricted working ions, thereby improving ion conductivity [17]. It is worth noting that not all ionic mixing can improve ionic conductivity, and the interaction criteria between electrolytes need to be considered.

8.2.3 Chemical and Electrochemical Stability

In order to achieve high energy density storage requirements, electrolyte must ensure high chemical and electrochemical stability. The electrochemical stability window (ESW) determines the operating voltage range, limiting the selection of cathode and anode materials. Generally, the ESW is influenced by two aspects: the lowest unoccupied molecular orbital (LUMO) and highest occupied molecular orbital (HOMO) of electrolyte, which corresponds to the chemical potential μ_C of the cathode and μ_A of the anode, respectively [20]. According to front-line orbital theory, the cathode material obtains electrons and is reduced by electrolyte when the electrochemical potential μ_C of the cathode material is lower than the HOMO, and this will lead to the cathode polarization. On the contrary, the electrolyte captures the electrons in the anode and is reduced when the electrochemical potential μ_A of the anode material is higher than the LUMO. At the same time, the reduced product forms a layer of "solid-electrolyte interface layer" (SEI film) on the surface of the negative electrode material.

The electrochemical stability window is mainly affected by the solvent and zinc salt in the electrolyte. From the solvent point of view, the electrochemical stability window of aqueous electrolyte theory is 1.23 V, which is obviously narrower than that of non-aqueous electrolyte. However, other measures are needed to broaden the electrochemical stabilization window due to the optimal overall performance of water electrolytes. From the perspective of zinc salts, some have higher solubility, which allows more water molecules in the electrolyte to participate in the solvation structure, thereby inhibiting decomposition and gas production, such as the 2.3 V high voltage in concentrated $ZnSO_4$, $Zn(CF_3SO_3)_2$ or $Zn(TFSI)_2$ electrolyte.

At present, many attempts have been made to widen the window of electrochemical stability of aqueous electrolytes. It has been reported that building an interface layer at the interface between electrolyte and electrode can significantly widen the electrolyte stabilization window. The main methods include non-*in-situ* physical or chemical deposition, functional additives, and preparation of high concentration electrolyte.

In addition, the window for electrochemical stabilization can be widened by reducing the water content of electrolytes, for example, by preparing "water in salt" electrolytes, gel electrolytes with less water content, or quasi-solid electrolytes. These have also become a research hotspot in recent years. Finally, since the potential of oxygen evolution reactions (OER) and hydrogen evolution reactions (HER) varies with the pH value of the electrolyte, the electrochemical stability window can also be kept within the appropriate range by adjusting the pH value of electrolyte [21]. Similarly, electrolyte additives can broaden the electrochemical window by inhibiting OER and HER.

8.3 ELECTROLYTE STRATEGIES IN AQUEOUS ZINC-ION BATTERIES

Although significant progress has been achieved in the field of ZIBs in recent years, there are still several major challenges that limit their practical application. These challenges mainly include: (1) The serious side reactions on zinc anode. The formation of zinc dendrites will puncture the diaphragm and eventually short-circuit the battery, the corrosion will reduce the Coulombic efficiency, and the hydrogen evolution reaction will cause battery expansion and electrolyte leakage; (2) Instability and dissolution of cathode materials. Due to the continuous ion (de)intercalation, the collapse and transformation of crystal lattice make battery performance decline rapidly; (3) Low chemical and electrochemical stability of electrolyte. These are mainly reflected in the decomposition of electrolyte and narrow electrochemical stability window. As the medium of ion transport and reaction environment, the electrolyte determines not only its intrinsic properties, but also the electrochemical storage behaviors of anode and cathode, as shown in Figure 8.3. Obviously, the use of electrolyte additives and the development of new electrolytes are the most feasible, economic, and effective ways to solve the above problems faced by aqueous ZIBs.

8.3.1 Electrolyte Additives

Developing suitable electrolyte additives have proved to be an effective strategy to solve problems, such as electrolyte decomposition, dendrite growth of zinc anode, and structure collapse of cathode, etc. As shown in Figure 8.3, the electrolyte additives can be roughly divided into ionic additives, organic additives, inorganic additives, and metal additives, as mentioned above.

For ionic electrolyte additives, it changes the type and quantity of working ions, which play a crucial role in improving ion conductivity (Figure 8.3a). The ionic additives can also affect the anode deposition morphology and suppress the cathode dissolution side reaction. For example, the additive ions are preferentially adsorbed on initial tips to change the local electric field and construct an electrostatic shield on the Zn anode surface during the charging process, which promotes the Zn^{2+} migration from the tip to adjacent deposition sites until a smooth deposition layer is formed [22]. For cathode reaction, the valence change of transition metal elements during the reaction process tends to make it dissolve in electrolyte irreversibly, such as the Jahn-Teller effect of Mn^{3+} in manganese-based cathode materials. Therefore, adding an appropriate amount of Mn^{2+} to the electrolyte is beneficial for maintaining the balance between MnO_2 cathode and Mn^{2+} in the electrolyte [23]. In addition, it has been reported that ionic additives give new functions to the electrolyte; for example, aqueous electrolyte containing Al^{3+} increases the working voltage of Zn-MnO_2 battery (Figure 8.3a) [24].

For organic electrolyte additives, there are also popular subjects in additive research because of their variety [25]. Their main functions can be summarized as modulating the electrochemical stability window of the electrolyte, optimizing the surface deposition morphology, and promoting the zinc nucleation and improving cathode/electrolyte interface compatibility, etc. According to the energy band theory, expanding the ESW of aqueous ZIBs—which is the band gap E_g between LUMO and HOMO energy levels—can be achieved in two methods, including reducing HOMO and increasing LUMO, corresponding to increasing OER potential and decreasing HER potential, respectively [20].

A typical strategy is to construct a hydrophobic layer at the electrode-electrolyte interface, providing a high energy barrier for water electrolysis and expanding the ESW (Figure 8.3b) [26]. For anode, the suppression of dendrites is the main purpose of regulating the zinc deposition process. The organic additives can inhibit dendrite formation not only by electrostatic shielding as described earlier, but also by increasing nucleation overpotential. The higher thc nucleation overpotential means the higher the nucleation number and nucleation speed, and the more the dendrite growth can be inhibited. The additive can increase the overpotential by adsorbing to the active site, resulting in a reduction of zinc deposition at the active site, thus making the zinc deposition denser and more uniform [27]. For cathode, the additives can improve the interfacial compatibility by constructing *in-situ* film according to surface adsorption or chemical passivation, which will be emphasized in the next section.

For inorganic and metal electrolyte additives, mainly including metal, oxide, and inorganic acid, there is relatively little research on them due to their relatively small solubility in aqueous solution. Metal electrolyte additives primarily refer to lead and

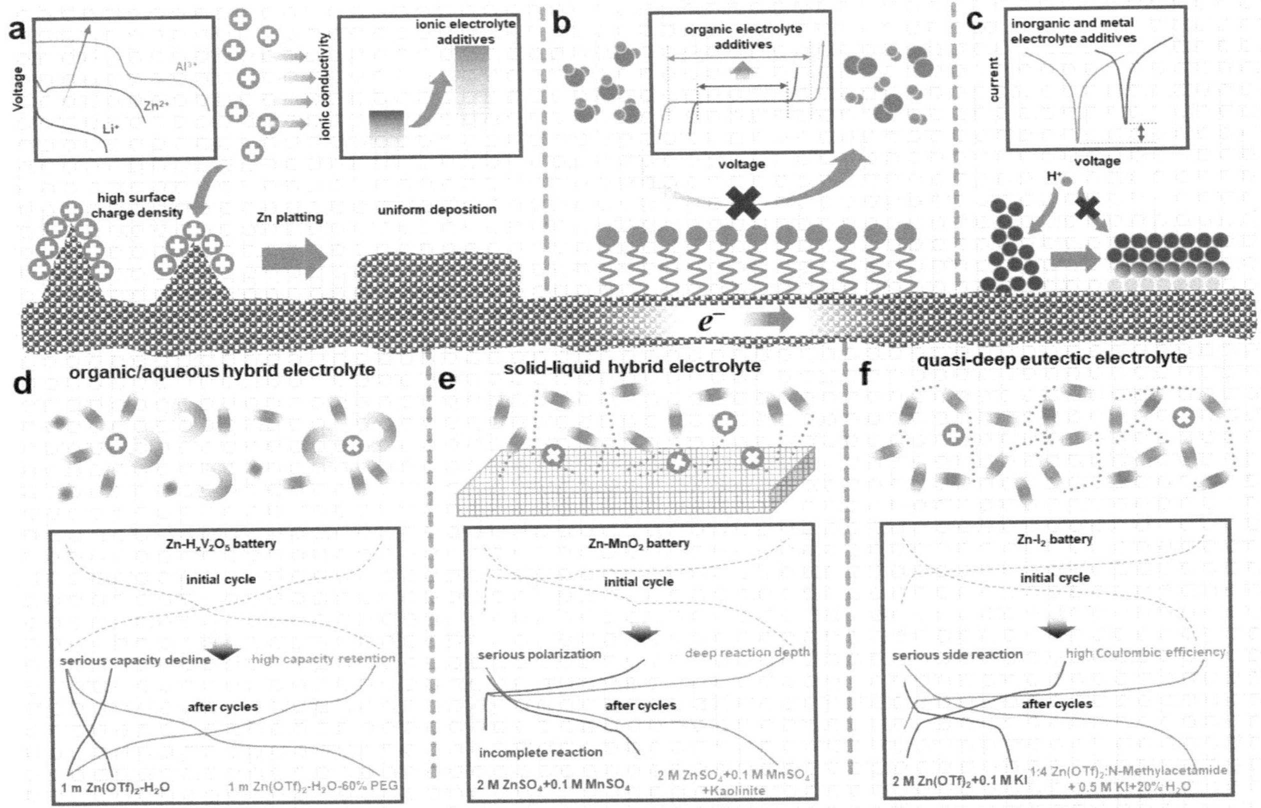

FIGURE 8.3 Schematic illustration of function of electrolyte additives and the structure of new electrolytes. (a) Enhanced operating voltage and ionic conductivity by ionic additives; (b) suppressed electrolyte decomposition by organic additives; (c) inhibited dendrite and corrosion by inorganic and metal additives; (d) inhibited dissolution of vanadium-based cathode by organic solvent; (e) assurance of the integrity of manganese-based cathode reaction; (f) prohibited irreversible iodine redox reaction.

tin, with the principal function of changing the zinc deposition texture growth, and it can also regulate the surface local galvanic cells to inhibit the corrosion side reaction [25]. For instance, adding trace Pb into the electrolyte can significantly change the zinc deposition potential, promoting the finer and uniform deposition morphology (Figure 8.3c) [13]. Therefore, inorganic and metal electrolyte additives are two very convenient and effective ways to regulate zinc deposition behavior.

8.3.2 New Electrolyte

As mentioned above, the battery has different properties when different additives are used. Therefore, theoretically designing new electrolytes with corresponding properties based on the demand for battery functions is the future trend of electrolyte development. For instance, in order to solve the dissolution of vanadium-based cathode materials and the sluggish ion intercalation kinetics in pure organic electrolyte, the organic/aqueous hybrid electrolyte was proposed with suitable ionic conductivity, moderate de-solvation barrier, and superior inhibiting ability of vanadium dissolution.

A typical case is the strong interaction between the rich hydroxyl groups in polyethylene glycol (PEG) and other species [28]. It can replace the water molecules in the zinc ion solvate sheath to inhibit the side reaction during the zinc plating/stripping process, and, more importantly, it could also reduce the content of free water to inhibit the vanadium dissolution (Figure 8.3d). The organic/aqueous hybrid electrolyte approach can shed a new light to the capacity decay mechanism of ion intercalation-type cathodes, which is helpful to the construction of long-life zinc-ion systems.

The metal anode in the aqueous electrolyte system exhibits high reactivity with the electrolyte to produce by-products and cause the instability of the electrode-electrolyte interface [29]. Hence, mixing liquid electrolyte with insoluble inert inorganic polymer fillers to prepare solid-liquid hybrid electrolyte has also attracted wide attention. The electrolyte sufficiently transfers the zinc ions due to the construction of the unobstructed ions transportation network composed of intergranular liquid phase migration and interlayer diffusion. The distinctive *in-situ* formation of dual interphases—especially with heterojunction charge aggregation—cleverly settles the issues of interfacial compatibility, which achieves a stable Zn^{2+} plating/stripping [30].

Especially for the deposition/dissolution reaction process of manganese-based cathode, its multi-step electrochemical behaviors depend largely on the state of the interfacial ions. The different transport properties of different ions caused by the solid filler in solid-liquid hybrid electrolyte can play a "quasi-decoupling" effect [31]. This creates the suitable interface reaction environments for the cathode and anode, respectively, thus improving the reaction reversibility and stability of each reaction step of the whole battery system, as shown in Figure 8.3e.

From another point of view, the binding of molecules in the electrolyte is not limited to the adsorption effect of inert fillers, and the interaction between rich groups can also build a bond network in the electrolyte. A simple mixture of Lewis acids, Lewis bases, and water in specific molar ratios emerged as the quasi-deep eutectic electrolyte. The abundant organic molecules and groups can not only coordinate with ions, but they can also function as a hydrogen bond acceptor to join with adjacent water molecules [32].

The cations in quasi-deep eutectic electrolyte possessed a unique double-shell solvated structure with a tighter anhydrous inner layer. In the case that the amount of free H_2O is effectively restricted, the water-induced side reaction—such as gas production and cathode dissolution—could also be suppressed even in such electrolyte with higher water content. Based on these advantages, the quasi-deep eutectic electrolyte could stimulate ideal electrochemical performances in some special zinc-ion systems. For instance, in the case that the amount of free water molecules is effectively restricted, I^- has a looser solvated environment and the formation of I_3^- as an intermediate product during I_2 reduction is inhibited. Hence, except for the stable uniform zinc plating/stripping, the quasi-deep eutectic electrolyte can inhibit the irreversible and unnecessary redox reactions and corresponding voltage platforms initiated by water molecules at high operating voltage in zinc-iodine batteries (Figure 8.3f) [33].

In conclusion, new electrolytes can be designed by adjusting electrolyte state, composition, additive content, and other factors. However, whether changing the state of the electrolyte or using the electrolyte additives—while inhibiting the growth of anode dendrites and slowing the dissolution of cathode materials—it is also necessary to select a new type of electrolyte with long service life and strong ion transport capacity so that the battery can meet different use requirements and also have better performance.

8.4 COMPATIBILITY BETWEEN ELECTROLYTES AND ELECTRODES

Due to the high electrochemical activity of aqueous electrolytes, the reaction process at the interface is very complex. Based on the properties of electrolyte as discussed above, this section summarizes and elaborates on the different requirements of different electrode materials, emphasizing the compatibility consideration from various aspects of electrolytes with anode and cathode interfaces.

8.4.1 Compatibility between Electrolyte and Anode

Although zinc metal can remain relatively stable in weak acid electrolyte, the change of ion concentration at the interface during cycles will also cause the zinc plating/stripping-beyond reactions related to water molecules. Hence, the high compatibility for anode mainly focuses on the construction of interfacial water-poor environment.

8.4.1.1 Interface with Low Water Content

An obvious strategy emerged by increasing the concentration of zinc salt, in which more coordination sites of Zn^{2+} are occupied by anions instead of water molecules [18]. Such contact ion pair (CIP) structure could reduce the water content in the interface electric double layer (EDL) and reduce the charge transfer of water molecules during the interfacial process. Without enough water molecules to support the formation of solvent separation ion pair (SSIP), such unique high concentration electrolyte could promote stable zinc plating/stripping generally.

Similarly, the quasi or deep eutectic electrolytes are all dedicated to creating a water-deficient or water-limited Zn^{2+} solvation structure [34]. The water molecules will participate in the internal interaction (hydrogen bond and coordination), isolating the direct contact between active water molecules and zinc anode from the perspective of both thermodynamic and electrochemistry. The further decrease of water content and increase of organic components in electrolyte make pure organic electrolyte feasible.

There is no doubt that pure organic electrolytes, including ionic liquid and liquefied gas electrolytes, could provide better thermodynamic stability and wider electrochemical stable potential window, and basically cut off the occurrence of side reactions. However, organic electrolytes expose another fatal problem, i.e., sacrificing the high electrochemical activity that depends on water molecules of aqueous zinc ion systems [35]. This generally involves the cathode interface reaction, which will be discussed later.

8.4.1.2 Functional Interfacial EDL Structure

A large number of small organic additives has been reported to improve the stability of zinc anodes, which can change the zinc ion solvation structure and reduce active water molecules. Benefitting from the functional groups in small organic molecules, such as hydroxyl, carboxyl, ether, etc., the water molecules can be excluded from EDL to optimize the deposition process [36]. On the one hand, the interfacial two-dimensional ion diffusion could induce continuous growth of dendrites; on the other hand, the organic molecules adsorbed at the interface can provide a suitable nucleation barrier and reduce the nucleated grain radius. With the synergetic effect of suppressing dendrite and refining grain, the uniform and flat surface morphology of zinc anode could be finally achieved.

In addition to the adsorption and bonding of functional groups, the large organic molecules would result in the formation of molecular crowding electrolytes [37]. The significant spatial repulsion cannot be ignored between close macromolecules due to their impenetrability, which makes the adjacent space "inaccessible." In general, the volume of macromolecules is generally at least two orders of magnitude higher than that of water molecules. It can be considered that the significant excluded-volume effect of macromolecules at the interface will limit the free diffusion of water molecules and related side reactions [38].

Especially for the chain polymer, its abundant hydrophilic groups could rearrange the "cation–H_2O–anion–polymer–anode" bonding network, reconstructing the space charge region and promoting reaction stability. Such a cross-linked network can also endow the electrolytes with obvious viscoelasticity at low concentration, which could dampen electroconvective flow induced by strong polarization at the interface, especially under high current densities [39]. It has been proved to reduce the convective fluid rolls associated with hydrodynamic instability, thus inhibiting the loose dendrites caused by preferential zinc deposition and gas production at localized regions on anode.

8.4.1.3 Electrolyte-Induced Regulation of Zinc Anode Surface Structure

As competing cations, some cations would be preferentially adsorbed on the zinc anode surface, which induces an electrostatic shielding effect and prevents the agglomerative zinc deposition. From the perspective of interface EDL, the adsorbed high-valence competitive cations could compress the EDL and reduce the Debye length. This weakens the EDL repulsive force between the zinc deposition, and thus changes the deposition orientation [40]. For those metals that can form alloys with zinc, trace additives in the electrolyte or electroplating solution can promote the formation of thin alloy layers during cycles, such as nickel, lead, indium, etc. Similar to the empirical Sabatier principle of chemical heterogeneous catalysis, the chemical interaction between the deposition phase and the substrate phase needs to be controlled within a suitable range to achieve the plating/stripping redox process with the highest reversibility and stability [41].

Inspired by the spontaneous formation of solid electrolyte interface (SEI) in lithium- and sodium-ion batteries, the *in-situ* formed film on the zinc metal surface could also fit well with anode and electrolyte. For example, the abundant zincophilic-O groups not only make it self-adsorb to the zinc surface to provide active sites for uniform deposition, but also participate in the formation of robust solid-electrolyte interface film. It can be expected that the formation of an *in-situ* SEI can play a significant role in suppressing dendrites. In addition, the spontaneous reactivity between additives and zinc metal can also construct the interfacial film [42]. It should be emphasized that some anions—especially Cl^-, ClO_4^-, and OTf^-, which tend to form special ion solvation structure or rich hydrogen bonds—could induce co-decomposition with other components in electrolyte to form various organic-inorganic interfacial *in-situ* film at the initial cycles, promoting zinc anode stability [43, 44].

In fact, the superiority of the solid-solid interface between zinc metal and interfacial film may mainly be attributed to the isolated active water molecules. This is highly consistent with various interface coating strategies. Hence, the gel electrolytes, solid-liquid hybrid electrolytes and solid electrolytes with low water content, also exhibit attractive potential in this regard. Due to the abundant functional groups, hydrogen bonds, and cross-linked polymer molecular chains in hydrogels, they usually exhibit excellent mechanical properties. The bulk adaptability and high modulus can well fit the surface morphology of zinc anode, especially for those *in-situ* gelatin strategies that chemically "weld" zinc anode and electrolyte, solving the problem of incompatible solid-solid electrolyte/electrode interface [45]. It also can mechanically inhibit the growth of dendrites due to the pressure given by its inherent elastoplasticity [46].

Meanwhile, the coordination environment between zinc ions and the functional groups on the polymer chains constructs abundant ion migration channels, thus promoting uniform deposition. Similarly, the free water molecules in the solid-liquid hybrid electrolyte with mineral clay are also absorbed by the abundant interface and lattice interlayers. Cations and water molecules can migrate along mineral grains and lattice interlayers [47, 48], in which the zinc ion solvation structure is significantly affected by the lattice channel size. Hence, the special structure of the mineral can not only ensure the ionic conductivity and the electrochemical activity of the interface, but it can also significantly inhibit the side reaction.

8.4.2 Compatibility between Electrolyte and Cathode

8.4.2.1 Regulation of Cathode Dissolution

The electrolyte design for cathode is mainly to solve the intrinsic inferiority of cathode reaction, and one of the most critical problems is the dissolution of cathode materials. Specifically, research on manganese-based cathode materials in recent years has found that the manganese deposition/dissolution reaction may dominate the whole interfacial process ($Mn^{2+} + H_2O \leftrightarrow MnO_2 + H^+$), which exhibits high requirement for water content, manganese ions, and pH value [49]. However, the deposition/dissolution reaction is accompanied by severe fluctuation of pH value and obvious change in manganese ion concentration, according to the reaction equation.

This is unfavorable to the electrolyte environment and the interfacial reaction reversibility thermodynamically [50], thus it is necessary to introduce various buffers, ion complexing agents, or other additives to stabilize the interface state. In this case, although the mitigation of the interfacial ion concentration change can stimulate a satisfactory reaction depth, the low conductivity of the deposited manganese oxides also leads to polarization increase and capacity decline. Hence, the further introduction of redox mediator, that may participate in the joint interfacial redox reaction, plays a catalytic role and also improves the energy density [51].

On the contrary, the reaction mechanism of vanadium-based phosphate, Prussian blue analogues, and organic cathode materials mainly depends on ion (de)intercalation. The main objective of electrolyte design is to inhibit the dissolution of these cathode materials. Therefore, it is often required that the electrolyte is in a waterless state. As mentioned in the previous section on electrolyte design for zinc anode, various high concentration electrolytes, gel electrolytes, hybrid electrolytes, and organic electrolytes exhibit immediate effect in this regard. This is why most of the waterless electrolyte designs for zinc anode choose non-manganese-based cathode as the counter electrode.

In addition, due to the (de)intercalation competition between zinc ions or hydrogen ions—especially in the aqueous zinc-ion system—different electrolytes even determine the energy storage behavior of cathode materials directly. For instance, the (de)intercalation of hydrated zinc ions could be adjusted to that of hydrogen ions in highly concentrated electrolyte, which significantly inhibits the internal mechanical stress of the cathode structure [52]. However, the insertion of hydrogen ions may also induce the dissolution of the cathode, especially the formation of soluble $VO_2(OH)_2^-$, in vanadium-based materials [53]. Hence, it is also proposed to adjust the (de)intercalation of hydrated zinc ions to that of zinc ions modify the electrochemical activity of water molecules in organic-inorganic hybrid electrolyte [54].

It can be considered that the intercalation-type cathode material does not have the same rigid requirements for sufficient water molecules as the manganese-based materials. More attention should be paid to the stability of the electrode structure and the reversibility of the reaction affected by intercalated charge carriers. This may also involve the co-intercalation of various charge carriers, which will be further demonstrated in the following content.

Zinc-iodine, zinc-bromine, and zinc-sulfur battery systems have attracted wide attention recently due to the high specific capacity of the multi-electron transfer reaction mechanism [55, 56]. Similar to the solid-liquid conversion reaction of aqueous zinc-manganese batteries, these battery systems also suffer from irreversible interfacial reactions caused by the dissolution of intermediate products. However, the absence of special requirements for active water molecules makes it available to apply high concentration electrolytes, eutectic electrolytes, gel electrolytes, organic electrolytes, etc. It should be noted that the special coordination chemistry of the halogen molecule allows it to form a linear (or nearly linear) trihalide structure with the halogen ion and are thermodynamic stable [57]. Hence, the combination with highly electronegative ions or other complex organic molecules can effectively control the dissolution, adsorption, and other properties of the product, thus improving the energy density and Coulombic efficiency [58].

8.4.2.2 Stimulation of Extra Redox Reaction

In fact, a more ideal battery system may not be limited to a single reaction. The smart coupling of multiple reaction mechanisms in a specific electrolyte environment may achieve something greater than the whole. For instance, for the inert manganese oxide deposition with low conductivity in the zinc-manganese battery, the iodine ion in electrolyte can act as a mediator. According to the equation ($MnO_2 + I^- + H^+ \leftrightarrow Mn^{2+} + I_3^- + H_2O$), the iodide mediator chemically promotes the reaction reversibility of manganese deposition/dissolution and contributes part of the capacity [59]. Bromine medium can also be incorporated into the solvation sheath and modulate the coordination environment of zinc and manganese ions, promoting the desolvation and cation transport kinetics [60]. More importantly, the capacity provided by Br_3^-/Br^- redox pair also increases the energy density of such hybrid battery system.

Unfortunately, the narrow ESW greatly limits the availability of cathode materials. The reversible redox reaction of oxygen can be realized in the zinc-ion system, but it also requires highly concentrated electrolyte [61]. The introduction of extra redox mediator is expected to be grafted on the intrinsic reaction of cathode materials, which means the redox process can be activated at a proper potential [51].

8.4.2.3 Self-Adaptive Optimization of Cathode Materials

Considering the difference of cathode electrochemical behavior in different electrolytes, it is usually difficult to match the appropriate electrode-electrolyte pair artificially. In this case, the self-adaptive optimization of electrolyte to electrode is proposed [62]. A typical case is the abundant defect structures produced in the original electrochemically inert cathode material under the effect of water molecules in aqueous electrolyte, including interlayer doping, vacancy, amorphous phase, etc. Especially for the compounds with low valence, the water molecules in the electrolyte will facilitate the transition to the electrochemically active phase after the first charge process [63, 64]. The electrolyte-cathode compatibility and interfacial kinetics are responsible for these different electrochemical activation behaviors. Alternatively, magnesium ions, aluminum ions, ammonium ions, etc., in the electrolyte will also be irreversibly intercalated into the cathode during the cycles to achieve lattice reconstruction. The bonding between intercalated ions and anions in the lattice can significantly enhance the robustness of cathode materials [65]. Therefore, understanding the structural self-adaptability of cathode materials may be from a cathode-electrolyte system.

For other cathode materials, the competitive reaction between different ions may be the key to solve the electrochemical stability. The different electrochemical energy storage behavior is not only related to the metal cations added in the electrolyte, but it also depends on the species of anions. Typically, during the (de)intercalation of lithium or sodium ions in phosphate, partial ion intercalation sites may be occupied by zinc ions [66]. Such intercalation of zinc ions may be beneficial for constructing novel lattice structure with low strain, thus stimulating high electrochemical performances [67]. However, the detailed reaction mechanism of such hybrid ion batteries and the relationship with the electrolyte composition are still unclear, which needs further exploration.

Finally, it can be concluded that most electrolyte strategies can achieve high stability of zinc anode. This is because the interfacial reaction is mainly zinc plating/stripping and does not involve various ions or complex interface processes. What needs to be emphasized is to reduce the contact between free water molecules and zinc metal, whether by reducing water content, ion solvation regulation, interface film, or others. However, the electrolyte design for cathode is more challenging, which requires further consideration of the interface and bulk reaction mechanism of cathode materials. Up to now, the energy storage mechanism of some cathode materials is not completely clear, which may even change with different electrolyte systems.

8.5 SUMMARY AND PERSPECTIVES

Although zinc-ion electrolyte has attracted much attention, it also has many problems, including basic scientific research and advanced practical application. Especially after taking into account the adaptability of the electrolyte to the cathode and the anode, respectively, the different requirements for the reaction environment may be the main factor impeding further development. Here, the key issues that need more attention in the future are summarized as follows (Figure 8.4), including the development of multi-component electrolyte, assessment of cost and safety, optimization of extreme temperature adaptability, and consideration of other practical problems.

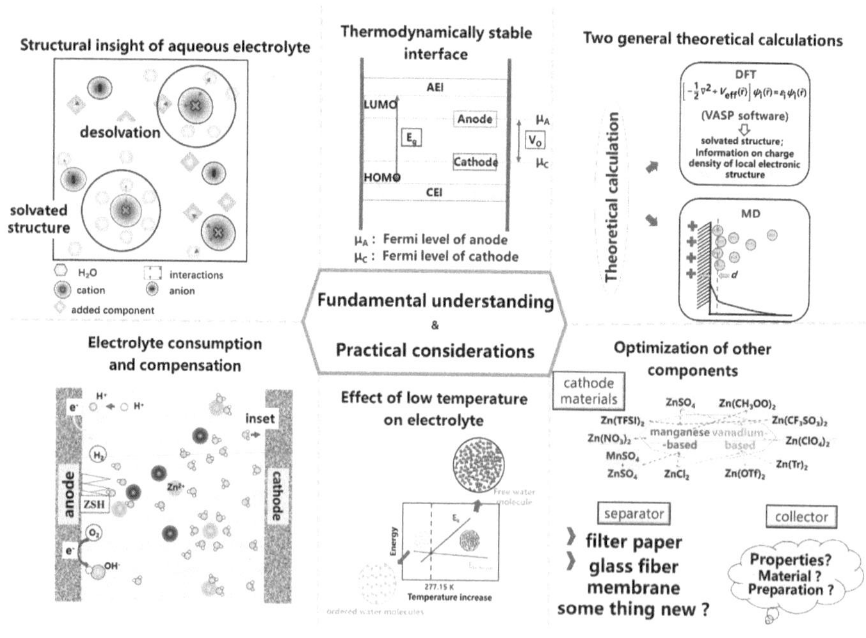

FIGURE 8.4 Summary and perspectives for electrolyte.

8.5.1 Fundamental Understanding of Zinc-Ion Electrolytes

8.5.1.1 Structural Insight of Aqueous Electrolyte

The rational design of electrolytes requires a deep understanding of the basic structure of electrolytes, as it plays an important role in the structure-property relationship between electrolyte composition and electrochemical performance. The ion solvation structure of the bulk electrolyte and the electric double layer structure of the electrode-electrolyte interface are usually the dominant factors determining electrochemical behavior. In addition to the static properties, revealing the *in-situ* evolution process of these structures during electrochemical reactions also contributes to better electrolyte design. For example, the precipitation of hydroxyzinc sulfate generated at the electrode interface in conventional liquid electrolytes can maintain relative stability of the interface environment, but it may also result in significant battery internal resistance. As an important characteristic of aqueous electrolytes, local pH values often trigger complex electrochemical behaviors. Therefore, the state of electrolyte, such as amount of electrolyte and ion concentration, is variable. It is important to investigate this *in-situ* state by using *in-situ* spectroscopy technologies and *in-situ* detection means.

8.5.1.2 Relationship between Electrolyte and Electrochemical Performance

Although many reports have made satisfactory progress in improving various aspects of electrochemical performance, there is still a lack of reasonable demonstration of the precise relationship between electrolyte structure and electrochemical performance. Considering that the reaction between electrolyte and electrode usually leads to the generation of interfacial substances, the different electrodes/electrolytes can lead to differences in electrolyte/electrode interface (EEI). Therefore, it is meaningful and valuable to study the influence of electrolyte structure and its parameters on electrochemical performance from the perspective of EEI structure. For example, the solvation structure evolution dynamics and the diffusion and transport of ions in EEI are theoretically closely related to the electrochemical stability and rate performance, respectively. In summary, it not only needs to determine the functional additives or novel electrolyte component to ensure an effective EEI, but it is also necessary to be aware of how the EEI achieve good electrochemical performance of a battery.

8.5.1.3 Synergy between Experimental and Theoretical Analysis

Due to the complexity of the electrolyte system, a close synergy between experimental and theoretical analysis deserves more attention. For instance, more information of local electronic structure and charge density of solvation structure can be achieved with the assist of the First principles calculation. The molecular dynamics also has been widely used to evaluate ion migration and diffusion in bulk electrolyte and interface. However, the matching degree between experiment conclusions and calculation results is limited by the complex electrochemical behaviors.

In this regard, machine learning would hopefully be applied to predict the relationship between electrolyte characteristics and electrochemical performances based on the experimental or theoretical data base. The learning results can also be fed back into the better electrolyte design. However, it should be noted that at present, machine learning for liquid electrolyte still has great challenges because of the complexity and uncertainty of the model.

8.5.2 Practical Considerations for Zinc-Ion Electrolytes

8.5.2.1 Electrolyte Consumption and Compensation

Excessive electrolyte will reduce the energy density of the battery system, but less electrolyte may not be sufficient to support the consumption in practical applications. The derivative reaction at electrode-electrolyte interface, such as HER/OER, water molecule intercalation, new phase formation, etc., all result in the obvious electrolyte consumption. Hence, the dosage of electrolyte also needs to be determined accurately for different systems or technology.

The consumption and successive replenishment of electrolyte may require a relatively open reaction environment, which also means that the impact of external conditions should be minimized. To address these issues, the quasi-solid hybrid electrolytes developed in recent years seems to be promising to promote the interfacial reaction reversibility, thus retaining water content and insulating from external influences on the premise of ensuring the electrochemical performances. However, the properties of solid fillers and the affinity between electrolyte and electrode interface need to be further explored.

8.5.2.2 Adaptability to Extreme Temperature

The depressing truth is that one of the most difficult challenges for zinc-ion batteries in practice is extreme temperatures. If the ambient temperature is too high, the enhancement of hydrogen evolution reaction will aggravate the capacity attenuation and reduce its service life. The addition of flame-retardant organic electrolyte additives and the use of gel electrolyte can improve the high temperature electrochemical performance and fire resistance of batteries. The electrolyte solidifies at very low temperatures, which slows down the reaction kinetics and reduces the charge transfer ability. Although high concentration electrolytes, organic additives, cosolvents, and gel electrolytes can modify the hydrogen-bond network to improve the electrochemical performance of low-temperature batteries, it also suffers from unsatisfactory compatibility with electrodes, especially due to the limited water molecules. Moreover, some solutions only achieve excellent performance at low temperatures and cannot operate in a wide temperature range. This work still needs further exploration.

8.5.2.3 Optimization of System Collocation

The realization of practical zinc-ion batteries is not only for the selection of electrolytes and their compatibility with electrodes, but also for the optimization of the whole system. For different cathode materials, the selection of current collector and separator should also be noted instead of the simple compatibility with electrolyte. Specifically, the metal current collector suffers from the oxidation under the effect of aqueous environment and electrons, and the thickness of conventional glass fiber separator does not support the construction of high energy density devices.

8.5.2.4 Other Practical Concerns

The relative oxidation resistance of the zinc-ion system determines its advantages of assembly in atmosphere, but even a small amount of dissolved oxygen will accelerate the battery decline. From this perspective, the competition between the cost of optimizing electrolyte with antioxidant ability and the levelized cost of energy determines its necessity in practical application. Considering the inherent low cost and high safety advantages of the aqueous zinc-ion system, it may be unreasonable to apply expensive electrolyte blindly to ensure stable electrochemical performances. Uneconomical zinc salts $Zn(TFSI)_2$, flammable organic components, and immature ionic liquid electrolytes do not take advantage in this respect. In addition, the low saturated vapor pressure of the aqueous solution may lead to partial evaporation in the vacuum environment inside the pouch cells, resulting in the lack of tight fitting between internal components. The assembly process and application environment of battery devices also need to take this into account.

In all, the electrolyte of the zinc-ion system plays an important role in the whole battery. The compatibility between the electrolyte and other components dominates the electrochemical behavior and performance. Hence, in order to design practical zinc-ion batteries with high specific energy and long-life stability for actual large-scale energy storage, it is urgent to deeply understand the basic scientific mechanism and practical application technology simultaneously.

REFERENCES

1. Q. Zhang, J. Luan, Y. Tang, X. Ji and H. Wang, *Angew. Chem. Int. Ed.*, 2020, 59, 13180–13191.
2. S. Hosseini, S. J. Han, A. Arponwichanop, T. Yonezawa and S. Kheawhom, *Sci. Rep.*, 2018, 8, 11273.
3. A. Naveed, H. Yang, Y. Shao, J. Yang, N. Yanna, J. Liu, S. Shi, L. Zhang, A. Ye, B. He and J. Wang, *Adv. Mater.*, 2019, 31, 1900668.
4. L. Ma, S. Chen, N. Li, Z. Liu, Z. Tang, J. A. Zapien, S. Chen, J. Fan and C. Zhi, *Adv. Mater.*, 2020, 32, 1908121.
5. Y. Wang, J. Yi and Y. Xia, *Adv. Energy Mater.*, 2012, 2, 830.
6. P. Yu, Y. Zeng, H. Zhang, M. Yu, Y. Tong and X. Lu, *Small*, 2019, 15, 1804760.
7. D. Chao, W. Zhou, F. Xie, C. Ye, H. Li, M. Jaroniec and S.-Z. Qiao, *Sci. Adv.*, 2020, 6, https://doi.org/10.1126/sciadv.aba4098
8. H. Kim, J. Hong, K.-Y. Park, H. Kim, S.-W. Kim and K. Kang, *Chem. Rev.*, 2014, 114, 11788–11827.
9. F. Ding, W. Xu, G. L. Graff, J. Zhang, M. L. Sushko, X. Chen, Y. Shao, M. H. Engelhard, Z. Nie, J. Xiao, X. Liu, P. V. Sushko, J. Liu and J.-G. Zhang, *J. Am. Chem. Soc.*, 2013, 135, 4450–4456.
10. Y. Song, J. Hu, J. Tang, W. Gu, L. He and X. Ji, *ACS Appl. Mater. Inter.*, 2016, 8, 32031–32040.
11. B. Tang, L. Shan, S. Liang and J. Zhou, *Energy Environ. Sci.*, 2019, 12, 3288–3304.
12. C. W. Lee, K. Sathiyanarayanan, S. W. Eom, H. S. Kim and M. S. Yun, *J. Power Sources*, 2006, 159, 1474–1477.
13. T. Otani, Y. Fukunaka and T. Homma, *Electrochim. Acta*, 2017, 242, 364–372.
14. K. Wu, J. Huang, J. Yi, X. Liu, Y. Liu, Y. Wang, J. Zhang and Y. Xia, *Adv. Energy Mater.*, 2020, 10, 1903977.

15. K. M. Diederichsen, E. J. McShane and B. D. McCloskey, *ACS Energy Lett.*, 2017, 2, 2563–2575.
16. L. Suo, Y.-S. Hu, H. Li, M. Armand and L. Chen, *Nat. Commun.*, 2013, 4, 1481.
17. K. Leng, G. Li, J. Guo, X. Zhang, A. Wang, X. Liu and J. Luo, *Adv. Funct. Mater.*, 2020, 30, 2001317.
18. B. W. Olbasa, F. W. Fenta, S.-F. Chiu, M.-C. Tsai, C.-J. Huang, B. A. Jote, T. T. Beyene, Y.-F. Liao, C.-H. Wang, W.-N. Su, H. Dai and B. J. Hwang, *ACS Appl. Energy Mater.*, 2020, 3, 4499–4508.
19. F. Wan, L. Zhang, X. Dai, X. Wang, Z. Niu and J. Chen, *Nat. Commun.*, 2018, 9, 1656.
20. J. B. Goodenough and Y. Kim, *Chem. Mater.*, 2010, 22, 587–603.
21. D. Chao, C. Ye, F. Xie, W. Zhou, Q. Zhang, Q. Gu, K. Davey, L. Gu and S. Z. Qiao, *Adv. Mater.*, 2020, 32, https://doi.org/10.1002/adma.202001894
22. F. Wan, L. L. Zhang, X. Dai, X. Y. Wang, Z. Q. Niu and J. Chen, *Nat. Commun.*, 2018, 9, 1656.
23. M. Liao, J. W. Wang, L. Ye, H. Sun, Y. Z. Wen, C. Wang, X. M. Sun, B. J. Wang and H. S. Peng, *Angew. Chem. Int. Ed.*, 2020, 59, 2273–2278.
24. S. Guo, J. Li, B. Zhang, W. Chen, G. Fang, M. Long and S. Liang, *Sci. Bull.*, 2022, 67, 626–635.
25. S. Guo, L. P. Qin, T. S. Zhang, M. Zhou, J. Zhou, G. Z. Fang and S. Q. Liang, *Energy Storage Mater.*, 2021, 34, 545–562.
26. Z. G. Hou, X. Q. Zhang, X. N. Li, Y. C. Zhu, J. W. Liang and Y. T. Qian, *J. Mater. Chem. A*, 2017, 5, 730–738.
27. J. Zhu, Y. Zhou and C. Gao, *J. Power Sources*, 1998, 72, 231–235.
28. Y. Chen, S. Guo, L. Qin, Q. Wan, Y. Pan, M. Zhou, M. Long, G. Fang and S. Liang, *Batteries & Supercaps*, 2022, 5, https://doi.org/10.1002/batt.202200412
29. Y. Mei, Y. Liu, W. Xu, M. Zhang, Y. Dong and J. Qiu, *Chem. Eng. J.*, 2023, 452, 139574.
30. S. Guo, L. Qin, C. Hu, L. Li, Z. Luo, G. Fang and S. Liang, *Adv. Energy Mater.*, 2022, 12, 2200730.
31. Y. Pan, Z. Liu, S. Liu, L. Qin, Y. Yang, M. Zhou, Y. Sun, X. Cao, S. Liang and G. Fang, *Adv. Energy Mater.*, 2023, 13, 2203766.
32. M. Yang, J. Zhu, S. Bi, R. Wang and Z. Niu, *Adv. Mater.*, 2022, 34, https://doi.org/10.1002/adma.202201744
33. Y. Yang, S. Liang, B. Lu and J. Zhou, *Energy Environ. Sci.*, 2022, 15, 1192–1200.
34. M. Han, J. Huang, X. Xie, T. C. Li, J. Huang, S. Liang, J. Zhou and H. J. Fan, *Adv. Funct. Mater.*, 2022, 2110957.
35. S.-D. Han, S. Kim, D. Li, V. Petkov, H. D. Yoo, P. J. Phillips, H. Wang, J. J. Kim, K. L. More, B. Key, R. F. Klie, J. Cabana, V. R. Stamenkovic, T. T. Fister, N. M. Markovic, A. K. Burrell, S. Tepavcevic and J. T. Vaughey, *Chem. Mater.*, 2017, 29, 4874–4884.
36. K. Zhao, G. Fan, J. Liu, F. Liu, J. Li, X. Zhou, Y. Ni, M. Yu, Y. M. Zhang, H. Su, Q. Liu and F. Cheng, *J. Am. Chem. Soc.*, 2022, 144, 11129–11137.
37. J. Xie, Z. Liang and Y. C. Lu, *Nat. Mater.*, 2020, 19, 1006–1011.
38. J. Wei, P. Zhang, T. Shen, Y. Liu, T. Dai, Z. Tie and Z. Jin, *ACS Energy Lett.*, 2022, 762–771.
39. D. Zhang, A. J. Warren, G. Li, Z. Cheng, X. Han, Q. Zhao, X. Liu, Y. Deng and L. A. Archer, *Macromolecules*, 2020, 53, 2694–2701.
40. R. Zhao, H. Wang, H. Du, Y. Yang, Z. Gao, L. Qie and Y. Huang, *Nat. Commun.*, 2022, 13, 3252.
41. J. Zheng, Y. Deng, W. Li, J. Yin, P. J. West, T. Tang, X. Tong, D. C. Bock, S. Jin, Q. Zhao, R. Garcia-Mendez, K. J. Takeuchi, E. S. Takeuchi, A. C. Marschilok and L. A. Archer, *Sci. Adv.*, 2022, 8, eabq6321.
42. J. Xu, W. Lv, W. Yang, Y. Jin, Q. Jin, B. Sun, Z. Zhang, T. Wang, L. Zheng, X. Shi, B. Sun and G. Wang, *ACS Nano*, 2022, 16, 11392–11404.

43. L. Cao, D. Li, E. Hu, J. Xu, T. Deng, L. Ma, Y. Wang, X. Q. Yang and C. Wang, *J. Am. Chem. Soc.*, 2020, 142, 21404–21409.
44. H. Qiu, X. Du, J. Zhao, Y. Wang, J. Ju, Z. Chen, Z. Hu, D. Yan, X. Zhou and G. Cui, *Nat. Commun.*, 2019, 10, 5374.
45. Y. Qin, H. Li, C. Han, F. Mo and X. Wang, *Adv. Mater.*, 2022, 34, https://doi.org/10.1002/adma.202207118
46. Y. P. Lin, B. Xia, S. Hu, Y. Zhong, Y. E. Huang, Z. Z. Zhang, N. Wu, Y. W. Wu, X. H. Wu, X. Y. Huang, Z. Xiao and K. Z. Du, *Energy Environ. Mater.*, 2020, 3, 535–540.
47. S. Guo, L. Qin, C. Hu, L. Li, Z. Luo, G. Fang and S. Liang, *Adv. Energy Mater.*, 2022, 12, 2200730.
48. X. Xie, H. Fu, Y. Fang, B. Lu, J. Zhou and S. Liang, *Adv. Energy Mater.*, 2021, 12, 2102393.
49. X. Guo, J. Zhou, C. L. Bai, X. K. Li, G. Z. Fang and S. Q. Liang, *Mater. Today Energy*, 2020, 16, 100396.
50. Z. Liu, L. Li, L. Qin, S. Guo, G. Fang, Z. Luo and S. Liang, *Adv. Mater.*, 2022, 34, e2204681.
51. T. Zhang, Q. Chen, X. Li, J. Liu, W. Zhou, B. Wang, Z. Zhao, W. Li, D. Chao and D. Zhao, *CCS Chem.*, 2022, 1–29.
52. X. Tang, P. Wang, M. Bai, Z. Wang, H. Wang, M. Zhang and Y. Ma, *Adv. Sci.*, 2021, 8, https://doi.org/10.1002/advs.202102053
53. D. S. Liu, Y. Zhang, S. Liu, L. Wei, S. You, D. Chen, M. Ye, Y. Yang, X. Rui, Y. Qin and C. C. Li, *Adv. Funct. Mater.*, 2022, 32, 2111714.
54. F. Wang, L. E. Blanc, Q. Li, A. Faraone, X. Ji, H. H. Chen Mayer, R. L. Paul, J. A. Dura, E. Hu, K. Xu, L. F. Nazar and C. Wang, *Adv. Energy Mater.*, 2021, 11, 2102016.
55. X. Zheng, Z. Liu, J. Sun, R. Luo, K. Xu, M. Si, J. Kang, Y. Yuan, S. Liu, T. Ahmad, T. Jiang, N. Chen, M. Wang, Y. Xu, M. Chuai, Z. Zhu, Q. Peng, Y. Meng, K. Zhang, W. Wang and W. Chen, *Nat. Commun.*, 2023, 14, 76.
56. H. Shuai, R. Liu, W. Li, X. Yang, H. Lu, Y. Gao, J. Xu and K. Huang, *Adv. Energy Mater.*, 2022, 13, 2202992.
57. G.-M. Weng, Z. Li, G. Cong, Y. Zhou and Y.-C. Lu, *Energy Environ. Sci.*, 2017, 10, 735–741.
58. X. Li, M. Li, Z. Huang, G. Liang, Z. Chen, Q. Yang, Q. Huang and C. Zhi, *Energy Environ. Sci.*, 2021, 14, 407–413.
59. J. Lei, Y. Yao, Z. Wang and Y.-C. Lu, *Energy Environ. Sci.*, 2021, 14, 4418–4426.
60. J. Wu, Y. Li, J. Huang, X. Chi, J. Yang and Y. Liu, *J. Mater. Chem. A*, 2021, 9, 21888–21896.
61. F. Wan, Y. Zhang, L. Zhang, D. Liu, C. Wang, L. Song, Z. Niu and J. Chen, *Angew. Chem. Int. Ed.*, 2019, 58, 7062–7067.
62. J. Huang, X. Xie, K. Liu, S. Liang and G. Fang, *Energy Environ. Mater.*, 2023, 6, e12309.
63. G. Fang, S. Liang, Z. Chen, P. Cui, X. Zheng, A. Pan, B. Lu, X. Lu and J. Zhou, *Adv. Funct. Mater.*, 2019, 29, 1905267.
64. X. Xie, G. Fang, W. Xu, J. Li, M. Long, S. Liang, G. Cao and A. Pan, *Small*, 2021, 17, https://doi.org/10.1002/smll.202101944
65. M. Li, Y. Zhang, J. Hu, X. Wang, J. Zhu, C. Niu, C. Han and L. Mai, *Nano Energy*, 2022, 100, 107539.
66. L. Ou, H. Ou, M. Qin, Z. Liu, G. Fang, X. Cao and S. Liang, *ChemSusChem*, 2022, 15, https://doi.org/10.1002/cssc.202201184
67. D. Zhao, S. Chen, Y. Lai, M. Ding, Y. Cao and Z. Chen, *Nano Energy*, 2022, 100, 107520.

9 Electrolytes and Additives for Zinc-Air Systems

Shiva Rezaei Motlagh, Mohan Gopalakrishnan, Mohammad Etesami, Ramin Khezri, Ahmad Azmin Mohamad, Jitti Kasemchainan, and Soorathep Kheawhom

9.1 INTRODUCTION

With the increasing industrialization of modern civilization, there is a growing demand for energy [1]. This has led to the emergence of issues linked to the use of fossil fuels and the consequent emission of excessive carbon dioxide into the atmosphere [2]. Of late, the progress in developing rechargeable batteries as energy storage systems (ESSs) has accelerated, largely because of the irregularity of wind and solar energy. These batteries offer an effective solution for both portable devices and grid-scale energy storage, serving a wide range of applications. To successfully apply these findings on a larger scale, it is crucial to consider several factors, including resource availability, costs, energy density, safety considerations, and product lifespan. [3, 4].

Zinc (Zn) presents itself as a readily available and economically viable material for utilization in secondary batteries, owing to its abundant presence as the most prevalent element in the Earth's lithosphere. Its remarkable reactivity and the simplicity of its deposition on an electrode from an aqueous electrolyte add to its attractiveness as a feasible alternative. The distinctive features of Zn-air batteries (ZABs) have generated significant interest in the energy storage field, as they may hold promise as a valuable contender for cutting-edge battery technologies [5–7]. Among commercially available metal-air batteries, ZABs stand out as highly accessible and well-established technology.

When it comes to next-generation ESSs, rechargeable ZABs (RZABs) are considered particularly attractive compared to other electrically rechargeable metal-air technologies [8]. Back in 1878, Maiche obtained a patent for a primary ZAB that employed a neutral NH_4Cl aqueous solution as its electrolyte. Heise and Schumacher were pioneers in producing commercial-neutral ZABs that found their usage in various fields. When CO_2 reacts with NH_4Cl, a neutral NH_4Cl solution does not produce carbonates. Nearly a century later in 1973, a quasi-neutral electrolyte in a 5 M NH_4Cl aqueous solution was developed that had enough buffering capacity for primary ZABs [9].

 DOI: 10.1201/9781003441182-11

Starting from 1997, there has been significant interest in RZABs that employ alkaline electrolytes. A recent surge of interest in bi-functional air cathodes has similarly been seen, as they can simultaneously enhance both oxygen reduction reaction (ORR) during discharge and oxygen evolution reaction (OER) during charging. This development aims to improve the overall efficiency and durability of the battery [12, 13]. To address Zn-electrode issues in RZABs, sophisticated methodologies used have concentrated on selecting a suitable electrolyte, which can lead to significant improvements in rechargeability.

The function of electrolytes in batteries is essential as they help regulate the fundamental electrochemical reactions and improve ionic conductivity. Electrolytes are analogous to the "blood" of batteries. The rechargeability and redox reactions of RZABs are reliant on the proper functioning of electrolytes. To enhance ZABs' efficiency, it is vital to employ an electrolyte that possesses high ionic conductivity, good stability in both electrochemical and thermal environments, low viscosity, a broad operating potential range, is environmentally friendly, non-flammable, and cost-effective [14]. ZABs exhibit versatility when it comes to electrolyte compatibility, enabling their operation with a wide range of electrolyte options. These options include aqueous electrolytes with different pH levels (such as alkaline, near-neutral, or acidic), as well as non-aqueous electrolytes composed of organic solvents, ionic liquids, gel polymers, deep eutectic solvents, solid polymer electrolytes, and quasi-solid-state electrolytes. Figure 9.1 provides an illustration of the various electrolyte types utilized in ZABs.

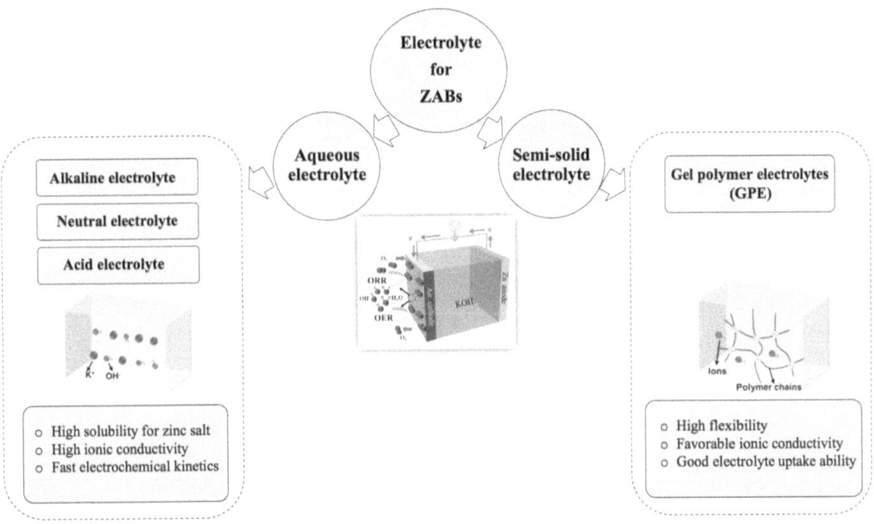

FIGURE 9.1 Schematics of electrolyte classification used in this study for ZABs.

Source: Reproduced with permission from [12]. Copyright 2022 Elsevier (Open Access).

9.2 AQUEOUS ELECTROLYTES

Aqueous electrolytes used in alkaline, acid, and neutral media are predominantly KOH, H_2SO_4, and Na_2SO_4. Alkaline electrolytes are used in ZABs more than acid or neutral electrolytes. The aqueous nature of ZABs makes them safe to operate and handle [15]. As seen in Table 9.1, depending upon the pH of the electrolyte used, the reaction of Zn anodes (oxidation/reduction) and air electrodes, i.e., ORR and OER, differ.

9.2.1 Aqueous Alkaline Electrolytes

KOH, NaOH, and LiOH are commonly used alkaline electrolytes for ZABs [10]. Utilizing KOH electrolytes in a ZAB offers a distinct advantage, owing to their outstanding electrochemical kinetics and mass transfer properties at a temperature of 25 °C. This advantage stems from their efficient ionic conductivity (550 mS/cm) and low viscosity (2.20 mPa.s), both of which contribute to the superior performance of the battery [11]. However, the performance of the battery can be affected by leakage, water loss, and carbonation of the alkaline electrolyte [16]. Figure 9.2 displays the typical issues that can arise on Zn anodes when used in an aqueous electrolyte.

When employing a concentrated KOH electrolyte in a ZAB, several challenges related to the anode can potentially affect the performance of the battery. These

TABLE 9.1
The Reaction of Zn-Air in Aqueous Electrolytes at Varied pH Levels

Aqueous Electrolyte	Zn Electrode Reaction	E^0,V	Air Electrode Reaction	E^0,V
Alkaline	$Zn + 2OH^- \rightleftharpoons ZnO + H_2O + 2e^-$	1.250	$O_2 + 2H_2O + 4e^- \rightleftharpoons 4OH^-$	0.401
Neutral	$Zn \rightleftharpoons Zn^{2+} + 2e^-$	−0.762	$2H^+ + \frac{1}{2}O_2 + 2e^- \rightleftharpoons 2H_2O$	1.229
Acid	$Zn \rightleftharpoons Zn^{2+} + 2e^-$	−0.762	$O_2 + 4H^+ + 2e^- \rightleftharpoons 2H_2O$	1.229

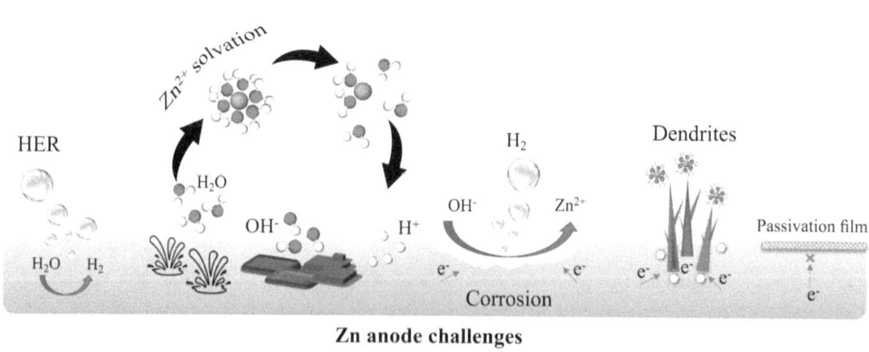

FIGURE 9.2 Concerns associated with the Zn anode in an aqueous alkaline electrolyte.

challenges include Zn dissolution, electrode passivation, shape-change, dendritic growth, and the Hydrogen Evolution Reaction (HER) [17]. The pH of the electrolyte is a crucial factor influencing Zn dissolution.

Figure 9.3a illustrates the reaction mechanisms and potential issues that can arise with Zn anodes in either an acidic or alkaline environment. The Pourbaix diagram showcases the different reaction by-products that result from a pH variation. According to Eq. (9.1) and as depicted in the diagram, at pH levels of 9 or above, a portion of $Zn(OH)_2$ and Zn undergoes interaction with OH^- ions, leading to the formation of zincate ($Zn(OH)_4^{2-}$) [18]. During the charging process, the accumulation of Zn metal on the surface of the anode leads to the formation of dendrites. These dendrites have the potential to grow and penetrate the separator, leading to a decline in performance and the risk of a short circuit.

$$Zn + 4OH^- \rightleftharpoons Zn(OH)_4^{2-} + 2e^- \tag{9.1}$$

$$Zn(OH)_4^{2-} \rightleftharpoons ZnO + 2OH^- + H_2O \tag{9.2}$$

$$Zn(OH)_2 \rightleftharpoons ZnO + H_2O \tag{9.3}$$

The importance of controlling the formation of ZnO on the Zn electrode has been highlighted [19]. The mechanism of zincate generation, which is influenced by the orientation of the Zn electrode, is depicted in Figure 9.3b and c. When a Zn electrode comes into contact with a concentrated alkaline electrolyte, the zincate's solubility increases, and once it surpasses the electrolyte's solubility limit, it converts into a solid (ZnO). During operation, the surface of an electrode may develop a layer of ZnO, which is known as passivation. This phenomenon can lead to a decrease in the discharge capacity of Zn anodes [21].

Zn passivation can be reduced by using highly concentrated alkaline electrolytes. Nonetheless, the utilization of such electrolytes induces morphological alterations and dendritic development. During the process of discharging in ZABs having alkaline electrolytes, $Zn(OH)_4^{2-}$ is seen. During charging, Zn returns to the electrode surface. Due to charging and discharging, changes in the shape and surface harshness of the electrode are observed, inducing dendritic growth and unequal thicknesses. Dendritic shapes, which have pointed and needle-like appearances, can penetrate the separators, causing an internal short circuit [22].

The surface of a Zn electrode experiences the HER process because of its lower potential than that of hydrogen. During charging, hydrogen gas utilizes a small number of electrons, and its production at the Zn surface consumes the electrolyte, producing swelling in the battery. The HER process can cause a decrease in the battery's life cycle [23]. Besides, the interaction between carbon dioxide and the hydroxyl group of the alkaline electrolytes results in the creation of insoluble carbonate salts at the electrode's surface. This carbonate/bicarbonate is primarily accountable for diminishing the performance of the cathode, decreasing ionic conductivity, restricting air access, and causing a shortage of oxygen at the cathode [22]. In Table 9.2, the

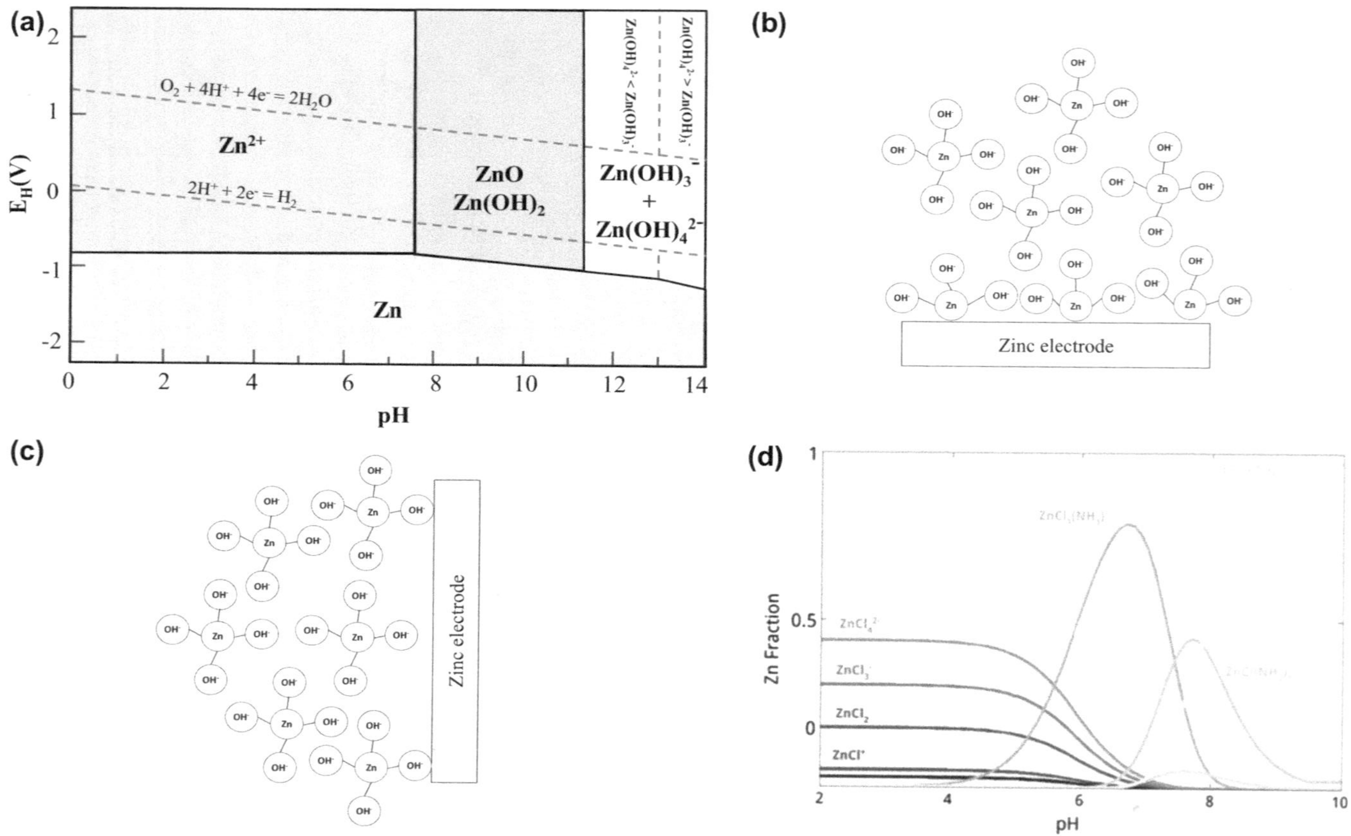

FIGURE 9.3 (a) Pourbaix diagram of electrolytes with various pH values; Formation of zincate ions in Zn electrodes having, (b) horizontal, (c) vertical orientations; (d) Zn^{2+} ion fractionation vs. pH in a neutral electrolyte. Reproduced with permission from [20].

Source: (a)–(c) Reproduced with permission from [26]. Copyright 2016 Wiley-VCH. (d) Copyright 2017 Wiley-VCH.

TABLE 9.2
Major Aqueous Electrolyte Issues and Potential Solutions for RZABs.

RZABs Issues	Description	Causes	Potential Failures	Proposed Solutions Based on Electrolyte Modification	Ref.
Formation of carbonate	Carbonate precipitation in the battery electrolyte	OH^- ions from the concentrated alkaline electrolytes interact with atmospheric CO_2	• Depletion of OH • pH fluctuations of the electrolytes	• Utilize the adsorption techniques of physisorption and chemisorption	[18]
Formation of dendrites	Irregular Zn deposition occurs through charging at active sites	Ion concentration gradient between electrode and electrolyte	• Short circuit • Capacity depletion	• Metal ion additives with smaller reduction potentials • Ionic additives • Non-ionic additives consist of organic additives with strongly polarized ether molecules	[24]

(Continued)

TABLE 9.2 (Continued)

RZABs Issues	Description	Causes	Potential Failures	Proposed Solutions Based on Electrolyte Modification	Ref.
Passivation	Oxide layer forms on the surface of an electrode	ZnO precipitates beyond the solubility limit of $Zn(OH)_4^{2-}$ ions in the electrolyte; then, during discharge ZnO is deposited onto the electrode surface	• Reduced electrode reactivity • Decreased efficiency • Impaired ion transport and diffusion • Increased internal resistance • Limitations in battery performance and capacity	• The solubility of $Zn(OH)_4^{2-}$ ions in the electrolyte can be reduced by incorporating an additive into the electrolyte solution	[25]
Self-discharge	Corrosion of Zn leads to the occurrence of partial discharge and early degradation of the battery	Zn corrosion as exposed to alkaline electrolyte and the formation of Zn hydroxide or Zn oxide compounds evaporation and HER	• Reduction of charge retention • Reducing the efficiency	• Metal additives with a high HER over potential	[24]
Shape changes of Zn anode	Change of electrode's surface structure during cycling	Accumulation and redistribution of Zn material during charge and discharge cycles	• Permanent loss of capacity • Densification of the battery	• The solubility of discharged Zn types is decreased by the presence of an electrolyte additive	[26]

Flooding of the air electrode	Electrolyte penetrates into the gas diffusion layer of the air cathode	• Air cathode binder degradation • Reduction in hydrophobicity of the air cathode	• Electrolyte loss • Reduced activity of air cathode • Reduced battery life	• Modified cathode-electrolyte interface	[27]
Dehydration of electrolyte	Evaporation of the electrolyte's water content	• Increased battery temperature • Battery is exposed to the open air	• Change in electrolyte concentration	• Using gel polymer electrolyte	[28]

main challenges of RZABs with aqueous alkaline electrolytes are presented. Using electrode and electrolyte additives, the issues of alkaline electrolytes can be solved.

9.2.1.1 Electrolyte Additives for Aqueous Alkaline Electrolytes

Integration of additives in alkaline electrolytes can enhance the performance of RZABs. Thus, related Zn dissolution, passivation, dendritic growth, and shape-change issues can be diminished. A suitable additive can boost the cyclability of alkaline RZABs by facilitating the generation of compact and homogeneous Zn electrodeposition. Additives have been found to help reduce the accumulation of material on the anode by improving the dissolution of Zn deposition throughout each charge and discharge cycle. However, the presence of electrolyte additives can cause early ZnO precipitation by reducing the solubility of $Zn(OH)_4{}^{2-}$ through their ability to adsorb onto sites where active hydrogen evolution occurs [29, 30].

Typically, additives employed in batteries can be classified into three main categories [31, 32]:

1. Organic additives, including alcohols, formaldehyde, perfluorocarboxylic acids, polyamines, surfactants, malonic acid, tetraalkylammonium salt, among others.
2. Inorganic additives such as Tl_2O_3, HgO, Ga_2O_3, Bi_2O_3, PbO, CdO, $Ca(OH)_2$, $In(OH)_3$, In_2O_3, $In(OH)_3$, SiO_2, and more.
3. Pure metals like Hg, Cd, Bi, Sn, In, Mg, Al, Ni, and Pb, which are utilized as electrode modifiers to enhance the structural properties.

However, high-concentration alkaline electrolytes used in air electrodes can cause carbon materials to corrode and reduce their performance. This phenomenon is represented in Eq. (9.4) [33]:

$$C + 6OH^- \rightarrow CO_3^{2-} + 3H_2O + 2e^- \tag{9.4}$$

The utilization of a highly concentrated alkaline electrolyte can reduce the passivation of Zn, yet it simultaneously leads to a deformation in the shape of the Zn anode. However, incorporating electrolyte additives containing silicates has shown promise in improving the surface characteristics of electrodes and reducing the occurrence of passivation. This is due to the ability of silicate compounds to adhere to ZnO particles, thereby inhibiting the precipitation of Zn [34]. The addition of sodium dodecyl benzene sulfonate (SDBS) to an alkaline electrolyte solution offers the potential to reduce surface passivation of a Zn electrode and enhance discharge efficiency [35]. Moreover, slowing down the discharge process can increase the solubility of the protective layer that develops on the surface of the anode [36].

In a Zn anode, shape-change can be avoided using particular compounds, which can decrease the dissolution of Zn in an electrolyte. Such compounds form insoluble mixtures that decrease the pH of an electrolyte, thus preventing shape-change in the Zn electrode. The compounds arsenate, borate, carbonate, fluoride, and phosphate have also been used to decrease Zn solubility. Additionally, the organic/inorganic electrolyte CH_3OH-KOH can help reduce Zn passivation [37, 38]. The introduction of surfactants into an alkaline electrolyte can potentially regulate the electrode's passivation, which has an impact on the mass of Zn and the formation of dendrites [39, 40]. Moreover, incorporating specific polymers such as polyvinyl alcohol (PVA), poly(vinylidene fluoride), and carboxymethyl cellulose into the electrolyte can hinder the deformation of a Zn electrode.

In order to prevent the formation of carbonates in alkaline electrolytes, different substances like piperazine (PZ), 2-(2-aminoethylamino) ethanol (AEEA), and monoethanolamide (MEA) have been utilized [10, 41, 42]. These amines are highly reactive with CO_2, preventing the formation of carbonate/bicarbonate in electrolytes. Ionic liquids or deep eutectic solvents can be added to electrolytes to limit or halt the movement of CO_2 within the solution. A few authors claim that the addition of K_2CO_3 in alkaline electrolytes decreases the CO_2 effect [11]. Moreover, it is seen that the addition of K_2CO_3 in electrolytes leads to a decrease in kinetics for the formation of carbonates [12]. The addition of K_2CO_3 can also cause a loss of battery performance.

HER can be mitigated when additives are used. The compounds-based antimony, bismuth, cadmium, gallium, indium, and lead can be used for decreasing HER in electrolytes [43]. The addition of organic surfactants and acids (H_3PO_4, $C_4H_6O_6$, $C_4H_6O_4$, and $C_6H_8O_7$) to alkaline solutions can elevate the overpotential of HER [44]. The rise in hydrogen overpotential is a result of the adsorption of acids on the surface of a Zn electrode, impeding its active sites. The presence of fluor-surfactants as additives in alkaline electrolytes has been observed to elevate the HER overpotential. In alkaline electrolytes, polyethylene glycol (PEG) has demonstrated the ability to hinder dendrite formation. Additionally, ethanol has been employed as an environmentally friendly additive in alkaline electrolytes to enhance the electrochemical performance of RZABs. It is noteworthy that alcohol can undergo conversion into alkoxide ions, which may compete with hydroxide ions for binding to Zn ions in alkaline solutions. [25]. The inclusion of alcohol in experiments has demonstrated a significant reduction in the formation of ZnO.

A prior investigation has indicated that incorporating dimethyl sulfoxide (DMSO) into an alkaline electrolyte can enhance the electrochemical properties of a Zn electrode. By increasing the surface area of the Zn electrodes, it becomes feasible to create additional deposition space, thereby mitigating dendritic growth [45]. Moreover, adding $K_2S_2O_8$ or KPS to the electrolyte is found to improve the electrochemical characteristics and overall performance of RZABs [30]. Furthermore, the introduction of sodium dodecyl sulfate (0.2 mM) and Pluronic F-127 to a concentrated KOH electrolyte has been observed to yield a substantial increase in the capacity of a Zn anode, with improvements of 30% and 24%, respectively [46].

The creation of a passive film that is loose and porous can facilitate ion exchange through diffusion. The incorporation of SDS and P127 transformed the layer from being compact into being loose. In order to improve the charge and discharge performance of ZABs, researchers have explored the use of various additives in a KOH (6M) electrolyte, some of which include thiourea (CH_4N_2S), sodium dodecyl benzene sulfonate (SDBS), and benzotriazole (BTA) [47]. The incorporation of BTA resulted in an enhanced life span and performance of the ZABs. Table 9.3 provides a variety of electrolyte additives used in alkaline ZABs.

TABLE 9.3
Alkaline Electrolyte Additives are Used in ZABs

Electrolyte and Additive Materials	Electrodes	Current Density, mA/cm^2	Specific Capacity, mAh/g	Ref.
KOH (7M); KPS (450 ppm)	Anode: Zn granules Cathode: MnO_2/Ni foam	15	710	[30]
KOH (6M); *$Zn(AC)_2$ (0.2M)	Anode: Zn foil Cathode: Co_3O_4-x/CP	230	800	[48]
KOH (7M); DMSO (20% vol.%)	Anode: Zn granules Cathode: MnO_2/Ni foam	150	550	[49]
KOH; water-suspended graphene (0.1 g)	Anode: Zn strip Cathode: Co-Sn-CNP/SS	26.43	212.6	[50]
KOH (8M); C_2H_5OH (0–50% v/v)	Anode: Zn granules Cathode: MnO_2/Ni foam	30	470	[25]
KOH (6M); $ZnCl_2$ (0.2M)	Anode: Zn plate Cathode: Pt–SCFP/Carbon cloth	214	781	[50]

* $Zn(AC)_2$: Zn acetate; Ni: Nickel.

9.2.2 Near Neutral Aqueous Electrolytes

An electrolyte for ZABs must meet multiple requirements. Such an electrolyte must be ionically conductive, must solubilize oxygen well, have modest viscosity, and have low vapor pressure. For near-neutral ZABs, materials for electrolytes must ensure ZnO can precipitate and have a stable pH. Electrolytes that have been buffered to a pH between 4 and 12 are considered to be in the near-neutral range. It is noted that Zn^{2+} complexing agents can increase the solubility of Zn ion, which is typically relatively low in the pH region: near neutral. To prevent both ZnO precipitation and undesired Zn electrode deformation, it is essential to meticulously select the formation of Zn-complexes. This process also aids in increasing Zn ion solubility in the electrolyte.

Extensive investigations have been carried out on carboxylic or amino carboxylic acids and their salts, which are weak organic acids, to reduce HER and enhance the quality of Zn deposition [51]. These acids are typically polyprotic and can form bonds with positive counter-ions: conjugate bases are negatively charged. Usually, Zn-organic salts have solubilities high enough to prevent ZnO precipitation from being at risk. The extensive variety of organic weak acids offers electrolytes more design options [52]. Due to the variability of intermediate products, organic acids and their salts exhibit complex redox properties and lower ionic conductivity.

Coin-cells often suffer from poor power performance when using electrolytes composed of aqueous solutions of inorganic salts, such as NH_4Cl and $ZnCl_2$. These electrolytes, which are typically neutral or close to neutral, exhibit low ionic conductivity and are prone to corrosion [20]. $ZnCl_2$ is responsible for supplying Zn^{2+} for the plating/stripping of the Zn electrode, while NH_4Cl acts as a buffer to prevent Zn electrode corrosion and maintain a neutral pH level within the electrolytes. The use of neutral electrolytes can overcome dendritic development and carbonate precipitation. Moreover, when Zn solubility is increased along with CO_2 absorption, the cycle life of ZABs is enhanced.

Different clusters—such as $ZnCl^+$, $ZnCl_2$, $ZnCl_3^-$, and $ZnCl_4^{2-}$ that are linked to Cl in aqueous $ZnCl_2$ solutions—are observed at varying pH values in the near-neutral electrolyte, i.e., $ZnCl_2$ (0.51 M) with NH_4Cl (2.34 M), as shown in Figure 9.3d [53]. In Eq. (9.5), it is demonstrated that at lower pH levels, the higher concentration of H^+ ions hinders the buffering capacity of NH_4^+. As a result, the mixture of electrolytes in ZABs is seen to contain Zn^{2+} clusters with a composition resembling that of a $ZnCl_2$ solution. As observed, the pH of the mixed electrolyte is kept close to neutral by the NH_4^+ ions:

$$NH_4^+ \rightleftharpoons NH_3 + H^+ \tag{9.5}$$

The process of the ZAB charge/discharge is seen to cause significant pH changes. During discharge, pH rises due to the consumption of H^+, as in Eq. (9.6), while during charging, the generation of H^+ causes a decrease in pH, as in Eq. (9.7):

$$2Zn + O_2 + 12NH_4OH + 4H^+ \rightarrow 2\left[Zn\left(NH_3\right)_6\right]^{2+} + 14H_2O \tag{9.6}$$

$$2\left[Zn\left(NH_3\right)_6\right]^{2+} + 14H_2O \rightarrow 2Zn + O_2 + 12NH_4OH + 4H^+ \tag{9.7}$$

To predict how the structure of Zn^{2+} clusters is transformed via changes in pH—and NH_3 concentration together with Cl^- concentration—thermodynamic calculations based on the stability of hydrates were carried out [20]. It was found that small variations in concentration and pH levels of electrolytes can cause significant changes in ionic conductivity and discharge products. To avoid significant pH variations during charge/discharge, researchers suggested to begin with a slightly alkaline initial pH of around 8 for the combined electrolyte. Reducing the overall concentration of Cl^- can also increase pH stability. Nonetheless, such a technique may impede the evaporation of charging precipitates, thus decreasing the reversibility of the battery.

The following neutral electrolytes—NH_4Cl, KCl, KNO_3, and Na_2SO_4—have all been investigated as pH buffers in aqueous $ZnCl_2$ solutions. However, the addition of different anions raises several issues. For example, adding NO_3^- can result in the formation of the oxidizing salt ($Zn(NO_3)_2$), which can quickly oxidize the Zn surface [54]. ZABs utilizing an aqueous solution of $Zn(NO_3)_2$ as an electrolyte display excellent electrochemical capabilities, although the rapid oxidation of the Zn electrodes requires frequent replacement.

Comprehending the underlying principles of ORR and OER is essential in the development of proficient oxygen electrocatalysts that can operate efficiently in electrolytes with neutral pH [55]. Redox metal centers located on electrocatalyst surfaces are the sites where ORR/OER reactions occur [56]. Therefore, to achieve an optimal M–O bond strength during reactions, it is desirable to employ metals or metal ions with diverse oxidation states that fall within a moderate range of strength. This principle applies to all electrocatalysts used in ORR and OER, as well as electrolytes with a pH level that is neutral or close to neutral. In both alkaline and acidic electrolytes, the initial reactants involved in ORR are H^+ and OH^-. However, neutral electrolytes exhibit lower ORR activity due to the relatively low concentration of H^+ and OH^- ions.

To compare the properties of different aqueous organic salt solutions, continuum modeling in conjunction with thermodynamic descriptors was examined. A previous study discovered that amino and carboxylic acids are efficient electrolytes for ZABs [57]. These acids include organic ions capable of binding to Zn^{2+} and regulating pH levels. When anhydrous citric acid and 2-aminoacetic acid were used, the ZABs were able to cycle up to 500 h and for another 400 h after the electrolyte was replaced (Figure 9.4b). A pH level within the range of 8.25 to 10.25 was maintained by the electrolyte, acting as a buffer, as illustrated in Figure 9.4c. In Figure 9.4d, three distinct stages are seen during the discharge of the ZAB: nucleation, conversion, and passivation. As a discharge product, ZnO can be thermodynamically induced by the electrolyte, which regulates the interfacial structure of Zn electrodes and increases the lifespan of ZABs.

The addition of MnO_x catalyst on carbon paper in the presence of neutral chloride-based electrolytes has the potential to enhance the cycling stability of RZABs for up to 90 days, which was enabled by decreased carbonation and corrosion of the carbon

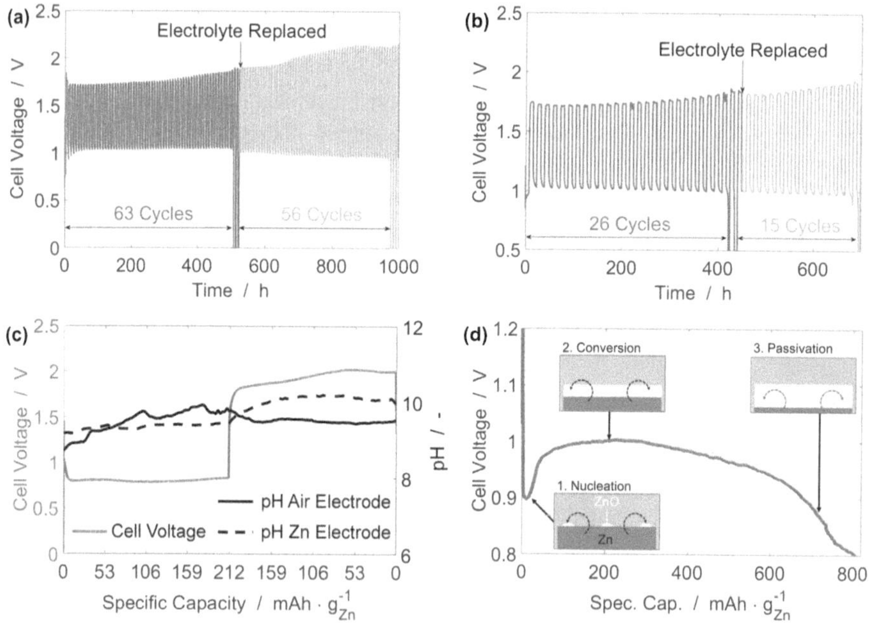

FIGURE 9.4 Performance of ZABs in an aqueous electrolyte comprising anhydrous citric acid and 2-aminoacetic acid when a current density of 0.5 mA/cm^2 is applied; cycle life after (a) 4 h; (b) 8 h; at a current density of 2 mA/cm^2 after 25 h; (c) cell voltage, pH of Zn electrode and pH of air electrodes, and at a current density of 0.5 mA/cm^2; (d) discharging cell voltage 1.2–0.8 V.

Source: Reproduced by permission [58]. Copyright 2020 Wiley-VCH (Open Access).

in the air cathode [58]. A near-neutral electrolyte for ZABs has been employed and exhibited an extended cycle life of 70 h for the battery [59]. Furthermore, the use of RZABs with a GPE based on PVA and a near-neutral electrolyte solution has been previously explored [60]. The RZABs utilized a bifunctional air cathode composed of $Ru_{0.6}Sn_{0.4}O_2$. Moreover, RZABs incorporating a GPE consisting of $ZnCl_2$ and NH_4Cl demonstrated remarkable performance. Even when the RZABs bent at a 90° angle, they maintained a high current density capability of 10 mA/cm^2 during both charging and discharging processes.

9.2.3 Acidic Electrolytes

Metal-based batteries such as RZABs that use acidic electrolytes such as H_3PO_4, HNO_3, H_2SO_4, and HCl have a short lifespan due to various challenges, including the air electrode's limited catalytic activity. To overcome these issues, appropriate carriers and catalysts need to be employed [18]. Dissimilarities in the electrochemical behavior of acidic and alkaline electrolytes are well-known. Dendritic formation takes place at a greater current density in aqueous acidic electrolytes in contrast to alkaline electrolyte solutions. Moreover, the passivation process of the Zn surface is confined to the existence of certain substances that form a thin film [61]. In non-complex acidic and neutral solutions, the electrode reaction (Zn/Zn^{2+}) involves two successive charge transfer phases, which are illustrated in Eqs. (9.8) and (9.9). Subsequently, Zn_{ad}^{+} adsorbs to metal surfaces and functions as an intermediary material:

$$Zn \rightleftharpoons Zn_{ad}^{+} + e^{-} \tag{9.8}$$

$$Zn_{ad}^{+} \rightleftharpoons Zn^{2+} + e^{-} \tag{9.9}$$

In Eq. (9.8), the anodic dissolution or cathodic deposition occurred quickly, but the reaction is slower in Eq. (9.9) [48]. In Eq. (10), the hydrogen evolution process can be expressed as:

$$2H_3O^{+} + 2e^{-} \rightarrow H_2 + 2H_2O \tag{9.10}$$

An acidic electrolyte comprised of H_2SO_4, H_3PO_4, and boric acid (H_3BO_3) for an RZAB was investigated [62]. Due to the specific electrolyte utilized, no formation of dendrites was reported, and the batteries did not exhibit any issues with carbon dioxide adsorption. Alkaline and natural electrolytes provide alternative options for use in ZABs, offering advantages over acidic electrolytes, such as the requirement for proper catalysts and support materials, as well as their relatively higher lifespan.

9.3 QUASI-SOLID-STATE ELECTROLYTES

Quasi-solid-state electrolytes (QSSEs) exist in an intermediate state between liquid and solid electrolytes, offering the benefits of both while circumventing their limitations. QSSEs enjoy the benefits of greater mechanical stiffness and a safer environment due to their non-flammability than liquid electrolytes. Furthermore, these electrolytes exhibit superior ionic conductivity and enhanced interfacial properties when compared to solid electrolytes. As a result, they have garnered significant attention in the field of ZABs. Owing to their noteworthy advantages, this class of electrolytes has demonstrated remarkable performance in ZABs. Researchers have also employed a range of polymers—including polyacrylic acid (PAA), PVA, and gelatin, as well as modified polymers and copolymers—in conjunction with near-neutral or alkaline electrolytes to produce QSSEs for ZABs.

The implementation of quasi-solid-state electrolytes in ZABs offers a substantial reduction in the leakage of liquid electrolytes. The ionic conductivity of an electrolyte plays a vital role in influencing the electrochemical behavior of battery systems, and it is influenced by factors such as the chemical structure of the polymer and the inclusion of additives. Polymer-based electrolytes, such as those based on PVA, PAN, and PAM, have found widespread use in various applications due to their favorable properties [63].

A flexible quasi-solid-state electrolyte was created by employing sodium polyacrylate (PANA) as the polymer [64]. In Figure 9.5 the outstanding performance of the electrolyte in a ZAB is shown, exhibiting impressive ionic conductivity of 81.2 mS/cm and remarkable water content retention of 95% over a span of 240 h [65]. Following 1000 cycles, the battery maintained 93.62% of its initial capacity. The hydrogel electrolyte proved to be suitable for use over a broad range of temperatures. The performance of ZABs can significantly be improved using an organo-hydrogel electrolyte based on polyamide (PAM). ZABs utilizing PAM electrolyte offer the advantage of maintaining exceptional performance even under demanding conditions, such as high current densities of 100 mA/cm^2 and prolonged cycling periods surpassing 50 h at 25 °C [66]. Impressively, the ZAB exhibits remarkable stability even at -60 °C, retaining over 90% of its capacity after more than 300 h of cycling at a current density of 0.5 mA/cm^2.

9.3.1 Gel Polymer Electrolytes (GPEs)

In addition to porous polymer separators, solid polymer electrolytes, and anion exchange membranes, another category of membranes known as GPEs has been utilized in metal-based batteries like ZABs [67]. The advantage of utilizing GPEs is that due to their three-dimensional cross-linked polymeric structure, numerous spaces are exposed for the absorption of electrolytes and ions to take place in addition to their incredible mechanical properties and flexibility. However, GPEs—due to their restricted water retention—provide lower conductivity of hydroxide ions. To overcome this drawback, hydroxide conductivity must be improved.

A GPE based on PVA was synthesized by incorporating PAA and PAM with KOH and later utilized in a ZAB [68]. PVA, PAA, and PAM all exhibited unique ionic conductivities in KOH at a temperature of 20 °C, with values of 161 mS/cm, 204 mS/cm, and 21 mS/cm, respectively. In another study, the utilization of a gel electrolyte membrane consisting of porous PVA and silica in a flexible ZAB showed excellent properties such as high ionic conductivity and exceptional retention capacity [70]. The gel electrolyte membrane containing 5 wt.% of silica demonstrated reliable mechanical and thermal stability during ZAB cycling as well as an optimal ionic

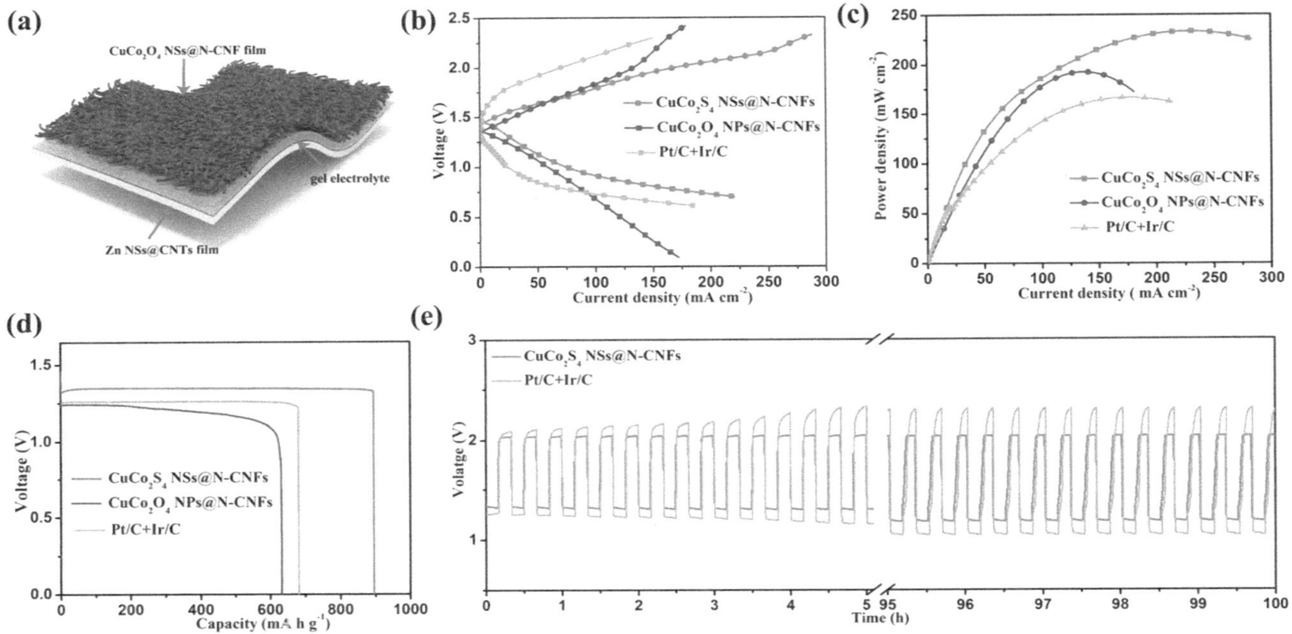

FIGURE 9.5 (a) Schematic representation of flexible quasi-solid-state ZABs; (b) charge-discharge polarization curves; (c) power density curve; (d) voltage vs. capacity curves; and (e) charge/discharge cycling curves via using various catalysts.

Source: Reproduced with permission from [65]. Copyright 2019 Wiley-VCH.

conductivity of 57.3 mS/cm. Surface hydroxyl groups of silica particles were found to enhance water retention in the PVA-silica electrolyte.

A PVA-based polymer with strong water retention capacity established an impressive performance in a chlorine-based solution with a pH close to neutral [59]. Initially, the electrolyte displayed a significant ionic conductivity of 76.1 mS/cm. However, after a span of three days, the ionic conductivity decreased to 41.9 mS/cm. In their study, Wu et al. [70] prepared a membrane for a ZAB by synthesizing a mixture of PVA and PAA. Through experimental investigations, the optimal composition ratio of 10:7.5 was determined, resulting in the PVA-PAA electrolyte exhibiting a maximum ionic conductivity of 301 mS/cm. According to an experiment carried out by Jia et al. [71], a PVA-based GPE electrolyte containing PEO and graphene oxide was investigated. Thus, it was found that a ZAB with this electrolyte recorded a peak power density of 57 mW cm^{-2} and an open circuit voltage (OCV) of +1.49 V. By incorporating graphene oxide (GO) into the polymeric electrolyte of PVA+GO+PEO, which contains oxygen functional groups, both water retention and ionic conductivity of the electrolyte were improved.

In Figure 9.6a and b, the use of gelatin and alkaline KOH as the electrolyte in a newly developed cable-type battery has been examined [72]. The study found that the all-solid-state ZAB proved to be highly flexible, signifying exceptional performance and having an ionic conductivity of 3.1 mS/cm. Propenoic acid (PA) and methylenebisacrylamide (MBAm) were used to create a gel-based electrolyte in an alkaline environment (KOH) for use as a membrane in a ZAB [73]. By introducing the Al_2O_3 nanolayer to the cathode, the battery's catalytic activity was enhanced, as evident from the rise in OCV compared to the one that did not have alumina in the

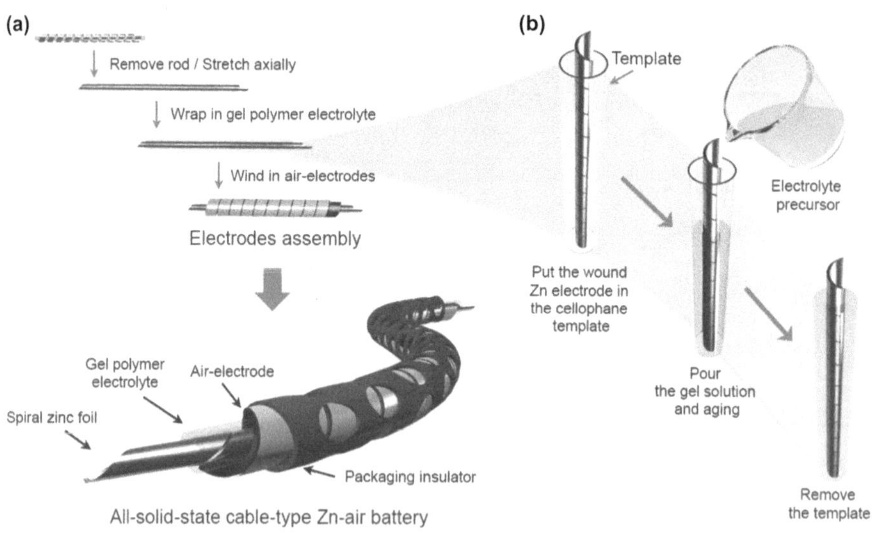

FIGURE 9.6 (a) Schema of the all-solid-state ZAB with cable-type configuration; and (b) coating process of the gelatin-based electrolyte.

Source: Reproduced with permission [72]. Copyright 2015 Wiley-VCH.

TABLE 9.4
Some of the Additives used in Quasi-Solid-State Electrolytes

Electrolyte and Additive Material(s)	Application	Remark	Ref.
PANA soaked in KOH; $Zn(Ac)_2$	QSSEs ZAB	Conductivity: 81.2 mS/cm; Water retention of 95% after 240 h	[63]
KOH; PVA gel	Flexible ZAB	Specific capacity: 896 mAh/g; Capacity retention: 93.62%. OCV = 1.46 V	[64]
PAM soaked in KOH; $Zn(Ac)_2$	QSSEs ZAB	Conductivity of 0.40, 0.087, 2.6, and 475.6 mS/cm at −40, −60, −20, and −60 °C, respectively; High cycling stability	[65]
PVA + silica	ZAB	Conductivity of 57.3 mS/cm; Cycling stability: 48 h	[69]
PVA in chlorine solution	Flexible ZAB	Conductivity of 76.1 mS/cm	[58]
PVA + PAA	Zn-air; Al-air cells	Conductivity of 142–301 mS/cm; Power density: 50 mW/cm^2	[70]
PVA + PEO + GO	All-solid-state ZAB	Power density: 57 mW/cm^2; OCV=1.49 V	[71]
Gelatin + KOH	Flexible all-solid-state cable type ZAB	Conductivity of 3.1 mS/cm	[72]
PA + MBAm + KOH	ZABs with Co-NC@ Al_2O_3 cathode	Power density: 72.4 mW/cm^2	[73]
PAM + MBAm	Flexible ZAB	Power density: 105 mW/cm^2	[74]
PAMPS-K penetrating MC in alkaline	Flexible ZAB	Conductivity of 105 mS/cm; Specific capacity: 7.647×10^2 mAh / g; Power density: 850.2 mWh/g	[75]

cathode (whose OCV was only 1.33 V). When operated at a current density of 94.2 mA/cm^2, the ZAB employing Co-NC@Al_2O_3 exhibited the highest power density of 72.4 mW/cm^2. Furthermore, a PAM-based polymer electrolyte that contained an MBAm crosslinker was able to achieve a satisfactory conductivity level of 2.156×10^2 mS/cm [74]. Owing to the use of a flexible solid electrolyte that contained PAM, the ZAB was seen to obtain the highest ever recorded power density of 105 mW/cm^2. In a different study, the combination of polyelectrolytes for achieving high ionic conductivity was investigated. The study found the optimal combination of methyl cellulose and cross-linked poly(2-acrylamido-2-methylpropanesulfonic acid) potassium salt [75]. This particular combination yielded an impressive ionic conductivity of 105 mS/cm at room temperature. Table 9.4 presents information related to quasi-solid-state and gel polymer electrolytes.

9.4 CONCLUSION AND OUTLOOK

Due to their numerous advantages, Zn-air batteries (ZABs) are widely regarded as promising candidates for future energy storage solutions, with the potential to surpass lithium-ion batteries in various aspects. Due to their reversible nature, eco-friendliness, low cost, and abundant availability of active materials, these batteries are particularly appropriate for ESSs. Such advantages make them a potential game-changer in the field of energy storage. Although ZABs hold significant promise, their commercial feasibility has been hindered by their short cycle life and low energy efficiency. Furthermore, the insufficient power density of bifunctional air electrodes is caused by the excessive overpotential of ZABs. To accomplish the industrialization of ZABs, affordable innovative materials are considered vital. The efficiency of circulation, power density, and performance capacity of ZABs are significantly affected by the electrolytes, making them a crucial component.

So far, alkaline electrolytes have been preferred mainly because of their superior interfacial properties and ionic conductivity. However, the external environment's CO_2 content and relative humidity can still affect the performance of alkaline electrolytes. To enhance the efficiency of alkaline electrolytes, it is essential to study the various types and quantities of additives used. Neutral aqueous electrolytes, which do not involve the formation of dendrites or carbonates, have been identified as viable substitutes.

Maintaining pH stability is a major challenge when utilizing neutral aqueous electrolytes, such as those containing inorganic or organic salt solutions during charge and discharge cycling. If ORR/OER rates are slow, the significance of this issue increases. To address this issue, Cl^- containing inorganic solutions such as $ZnCl_2$ + NH_4Cl are commonly used in research laboratory tests, as they offer a relatively stable pH environment and ensure reasonable safety and cycling stability for RZABs. Large-scale RZABs face challenges with side reactions, namely, carbon evolution reactions.

One possible solution is to use neutral electrolytes made of different inorganic salts. Nevertheless, it is necessary to thoroughly investigate their efficiency in RZABs. Despite the fact that utilizing neutral aqueous electrolytes containing organic salts in RZABs looks to be a feasible strategy, further research is necessary to comprehend the oxidation of these salts during redox reactions. Although their ionic conductivity must be enhanced, the development of quasi-solid electrolytes allows for the resolution of interface performance difficulties and the construction of portable and adaptable ZABs.

This chapter summarizes the categorization, difficulties, and future possibilities of electrolyte and electrolyte additives in ZABs. To address the requirement for a new generation of energy storage, electrolyte research should receive more consideration. It is hoped that the material presented in this book will serve as a source of inspiration for researchers exploring electrochemical energy conversion and storage.

ACKNOWLEDGEMENT

The Program Management Unit for Human Resources & Institutional Development, Research, and Innovation (B16F640166) is acknowledged. S.R.M. thanks Chulalongkorn Academic Advancement for its 2nd Century Project for Postdoctoral Fellowship.

REFERENCES

1. R. Georgious, R. Refaat, J. Garcia and A.A. Daoud, *Electronics*, 2021, 10(17), 2134.
2. P. Tan, B. Chen, H. Xu, H. Zhang, W. Cai, M. Ni, M. Liu and Z. Shao, *Energy Environ. Sci.*, 2017, 10(10), 2056–2080.
3. Z. Zhu, T. Jiang, M. Ali, Y. Meng, Y. Jin, Y. Cui and W. Chen, *Chem. Rev.*, 2022, 22, 16610–16751.
4. M. Etesami, R. Khezri, A. Abbasi, M.T. Nguyen, T. Yonezawa, S. Kheawhom and A. Somwangthanaroj, *J. Alloys Compd.*, 2022, 922, 166287.
5. S.R. Motlagh, R. Khezri, A.A. Mohamad, R. Pornprasertsuk, P. Kidkhunthod, M.T. Nguyen, T. Yonezawa and S. Kheawhom, *Mater. Sci. Technol.*, 2023, 6, 178–191.
6. R. Khezri, S. Rezaei Motlagh, M. Etesami, A.A. Mohamad, F. Mahlendorf, A. Somwangthanaroj and S. Kheawhom, *Chem. Eng. J.*, 2022, 449, 137796.
7. R. Khezri, K. Jirasattayaporn, A. Abbasi, T. Maiyalagan, A.A. Mohamad and S. Kheawhom, *Energies*, 2020, 13(6), 1429.
8. R. Khezri, A. Parnianifard, S.R. Motlagh, M. Etesami, W. Lao-Atiman, A. Abbasi, A. Arpornwichanop, A.A. Mohamad, S. Olaru and S. Kheawhom, *J. Ind. Eng. Chem.*, 115, 570–582.
9. J. Jindra, J. Mrha and M. Musilová, *J. Appl. Electrochem.*, 1973, 3, 297–301.
10. D. Schröder, N.N. Sinai Borker, M. König and U. Krewer, *J. Appl. Electrochem.*, 2015, 45, 427–437.
11. N. Radenahmad, R. Khezri, A.A. Mohamad, M.T. Nguyen, T. Yonezawa, A. Somwangthanaroj and S. Kheawhom, *J. Alloys Compd.*, 2021, 883, 160935.
12. Q. Liu, R. Liu, C. He, C. Xia, W. Guo, Z.L. Xu and B.Y. Xia, *EScience*, 2022, 5, 453–466.
13. J. Fu, Z.P. Cano, M.G. Park, A. Yu, M. Fowler and Z. Chen, *Adv. Mater.*, 2017, 7, 1604685.
14. J.T. Cullinane and G.T. Rochelle, *Ind. Eng. Chem. Res.*, 2006, 8, 2531–2545.
15. T.C. Adler, F.R. McLarnon and E.J. Cairns, *Ind. Eng. Chem. Res.*, 1998, 8, 3237–3241.
16. R.Y. Wang, D.W. Kirk and G.X. Zhang, *J. Electrochem. Soc.*, 2006, 153, C357.
17. X. Chen, Z. Zhou, H.E. Karahan, Q. Shao, L. Wei and Y. Chen, *Small*, 2018, 44, 1801929.
18. A.R. Mainar, E. Iruin, L.C. Colmenares, A. Kvasha, I. Meatza, M. Bengoechea, O. Leonet, I. Boyano, Z. Zhang and J.A. Blazquez, *J. Energy Storage*, 2018, 15, 304–328.
19. A.R. Mainar, L.C. Colmenares, J.A. Blázquez and I. Urdampilleta, *Int. J. Energy Res.*, 2018, 42(3), 903–918.
20. S. Clark, A. Latz and B. Horstmann, *ChemSusChem*, 2017, 10(23), 4735–4747.
21. M. Bockelmann, M. Becker, L. Reining, U. Kunz and T. Turek, *J. Electrochem. Soc.*, 2019, 166(6), A1132.
22. J. Yi, P. Liang, X. Liu, K. Wu, Y. Liu, Y. Wang, Y. Xia and J. Zhang, *Energy Environ. Sci.*, 2018, 11(11), 3075–3095.
23. K.E.K. Sun, T.K.A. Hoang, T.N.L. Doan, Y. Yu, X. Zhu, Y. Tian and P. Chen, *ACS Appl. Mater. Interfaces*, 2017, 11, 9681–9687.
24. M.A. Rahman, X. Wang and C. Wen, *J. Electrochem. Soc.*, 2013, 160, A1759.
25. S. Hosseini, S.J. Han, A. Arponwichanop, T. Yonezawa and S. Kheawhom, *Sci. Rep.*, 2018, 8, 11273.
26. A.R. Mainar, O. Leonet, M. Bengoechea, I. Boyano, I. Meatza, A. Kvasha, A. Guerfi and J.A. Blázquez, *Int. J. Energy Res.*, 2016, 40(8), 1032–1049.
27. J. Pan, Y.Y. Xu, H. Yang, Z. Dong, H. Liu, B.Y. Xia, *Adv. Sci.*, 2018, 5, 1700691.
28. A.K. Worku, D.W. Ayele and N.G. Habtu, *Mater. Today Adv.*, 2021, 7, 100116.

29. A.R. Mainar, L.C. Colmenares, H.J. Grande and J.A. Blázquez, *Batteries*, 2018, 4(3), 46.
30. R. Khezri, S. Hosseini, A. Lahiri, S.R. Motlagh, M.T. Nguyen, T. Yonezawa and S. Kheawhom, *Int. J. Mol. Sci.*, 2020, 21(19), 7303.
31. Z. Zhao, X. Fan, J. Ding, W. Hu, C. Zhong and J. Lu, *ACS Energy Lett.*, 2019, 4(9), 2259–2270.
32. S. Hosseini, S. Masoudi Soltani, Y.Y. Li, *Chem. Eng. J.*, 2021, 408, 127241.
33. C. Wang, J. Li, Z. Zhou, Y. Pan, Z. Yu, Z. Pei, S. Zhao, L. Wei and Y. Chen, *Energy Chem.*, 2021, 4, 100055.
34. F.R. McLarnon and E.J. Cairns, *J. Electrochem. Soc.*, 1991, 138, 645.
35. Y. Li and H. Dai, *Chem. Soc. Rev.*, 2014, 43, 5257–5275.
36. M. Cai and S.M. Park, *J. Electrochem. Soc.*, 1996, 143, 2125.
37. J.T. Nichols, F.R. McLarnon and E.J. Cairns, *Chem. Eng. Commun.*, 1985, 1, 355–379.
38. H.L. Lewis, T. Danko, A. Himy and W. Johnson, *J. Power Sources*, 1999, 1, 61–65.
39. M. Liang, H. Zhou, Q. Huang, S. Hu and W. Li, *J. Appl. Electrochem.*, 2011, 41, 991–997.
40. T. Cohen-Hyams, Y. Ziengerman and Y. Ein-Eli, *J. Power Sources*, 2006, 157(1), 584–591.
41. A. Aroonwilas and V. Veawab, *Ind. Eng. Chem. Res.*, 2004, 9, 2228–2237.
42. H.H. Cheng and C.S. Tan, *J. Power Sources*, 2006, 162(2), 1431–1436.
43. L.M. Hsien, C.J. Huang, P.H. Cheng, J.H. Cheng and C.C. Wang, *J. Mater. Chem. A*, 2020, 8, 20637–20649.
44. C.W. Lee, K. Sathiyanarayanan, S.W. Eom, H.S. Kim and M.S. Yun, *J. Power Sources*, 2006, 159(2), 1474–1477.
45. A. Naveed, H. Yang, Y. Shao, J. Yang, N. Yanna, J. Liu, S. Shi, L. Zhang, A. Ye, B. He and J. Wang, *Adv. Mater.*, 2019, 31, 1900668.
46. S. Hosseini, W. Lao-Atiman, S.J. Han, A. Arpornwichanop, T. Yonezawa and S. Kheawhom, *Sci. Rep.*, 2018, 8, 14909.
47. Y. Xiao, J. Shi, F. Zhao, Z. Zhang and W. He, *J. Electrochem. Soc.*, 2018, 165, A47.
48. F.A. Getie, D.W. Ayele, N.G. Habtu, F.A. Yihun and T.A. Yemata, *SN Appl. Sci.*, 2022, 4, 270.
49. S. Hosseini, A. Abbasi, L.O. Uginet, N. Haustraete, S. Praserthdam, T. Yonezawa and S. Kheawhom, *Sci. Rep.*, 2019, 9, 14958.
50. P. Chen, K. Zhang, D. Tang, W. Liu, F. Meng, Q. Huang and J. Liu, *Front. Chem.*, 2020, 8, 372.
51. Y. Li, M. Gong, Y. Liang, J. Feng, J.E. Kim, H. Wang, G. Hong, B. Zhang and H. Dai, *Nat. Commun.*, 2013, 4, 1805.
52. F. Wang, O. Borodin, T. Gao, X. Fan, W. Sun, F. Han, A. Faraone, J.A. Dura, K. Xu and C. Wang, *Nat Mater.*, 2018, 17, 543–549.
53. D.J. Harris, J.P. Brodholt, J.H. Harding and D.M. Sherman, *Mol. Phys.*, 2001, 10, 825–833.
54. C. Wang, Z. Pei, Q. Meng, C. Zhang, X. Sui, Z. Yuan, S. Wang and Y. Chen, *Angew. Chem. Int. Ed.*, 2021, 60, 990.
55. N.H. Abdullah, K. Shameli, P.M. Nia, M. Etesami, E.C. Abdullah and L.C. Abdullah, *Arab. J. Chem.*, 2020, 1, 1297–1308.
56. M. Etesami, M. Oroujzadeh and S. Mehdipour-Ataei, *Mol. Catal.*, 2020, 485, 110834.
57. A. Sumboja, X. Ge, G. Zheng, F.T. Goh, T.A. Hor, Y. Zong and Z. Liu, *J. Power Sources*, 2016, 332, 330–336.
58. S. Clark, A.R. Mainar, E. Iruin, , L. Colmenares, *Adv. Energy Mater.*, (2020), 10(10), 1903470.

59. Y. Li, X. Fan, X. Liu, S. Qu, J. Liu, J. Ding, X. Han, Y. Deng, W. Hu and C. Zhong, *J. Mater. Chem. A*, 2019, 7, 25449–25457.
60. C.-Y. Lai, Y.T. Lu, W.Y. Jao, H.Y. Chen and C.C. Hu, *Electrochem. Commun.*, 2022, 136, 107240.
61. L. Lei, Y. Sun, X. Wang, Y. Jiang and J. Li, *Front. Mater.*, 2020, 96, 2296–8016.
62. K. F. Blurton & A. F. Sammells, Washington, DC: U.S. Patent and Trademark Office, 1980, U.S. Patent No. 4,220,690.
63. M.M. Amaral, R. Venâncio, A.C. Peterlevitz and H. Zanin, *J. Energy Chem.*, 2022, 67, 697–717.
64. Z. Pei, Y. Huang, Z. Tang, L. Ma, Z. Liu, Q. Xue, Z. Wang, H. Li, Y. Chen and C. Zhi, *Energy Storage Mater.*, 2019, 20, 234–242.
65. Z. Pan, H. Chen, J. Yang, Y. Ma, Q. Zhang, Z. Kou, X. Ding, Y. Pang, L. Zhang, Q. Gu, C. Yan and J. Wang, *Adv. Sci.*, 2019, 6, 1900628.
66. Q. Wang, Q. Feng, Y. Lei, S. Tang, L. Xu, Y. Xiong, G. Fang, Y. Wang, P. Yang, J. Liu, W. Liu and X. Xiong, *Nat. Commun.*, 2020, 13, 3689.
67. A. Abbasi, Y. Xu, R. Khezri, M. Etesami, C. Lin, S. Kheawhom and Y. Lu, *Mater. Today Sustain.*, 2022, 18, 100126.
68. T.N.T. Tran, H.J. Chung and D.G. Ivey, *Electrochim. Acta.*, 2019, 327, 135021.
69. X. Fan, J. Liu, Z. Song, X. Han, Y. Deng, C. Zhong and W. Hu, *Nano Energy*, 2019, 56, 454–462.
70. G.M. Wu, S.J. Lin and C.C. Yang, *J. Membr. Sci.*, 2006, 1, 802–808.
71. H.L. Jia, J. Zhao, L. Gu, Z.J. Peng, Z.L. Bao, X.L. Sun and M.Y. Guan, *Sustain. Energy Fuels*, 2020, 4, 6165–6173.
72. J. Park, M. Park, G. Nam, J. S. Lee and J. Cho, *Adv. Mater.*, 2015, 27, 1396–1401.
73. L. Zhu, D. Zheng, Z. Wang, X. Zheng, P. Fang, J. Zhu, M. Yu, Y. Tong and X. Lu, *Adv. Mater.*, 2018, 30, 1805268.
74. H. Miao, B. Chen, S. Li, X. Wu, Q. Wang, C. Zhang, Z. Sun and H. Li, *J. Power Sources*, 2020, 450, 227653.
75. N. Sun, F. Lu, Y. Yu, L. Su, X. Gao and L. Zheng, *ACS Appl. Mater. Interfaces*, 2020, 10, 11778–11788.

10 Electrolytes for Rechargeable Aluminum Batteries

Challenges and Opportunities

T. Sri Devi Kumari and P. W. Jaschin

10.1 INTRODUCTION

Battery technology is metal-intensive. Passive elements such as current collectors and casings are usually made of aluminum, copper, nickel, or stainless steel. Active elements are also based on metals (e.g., zinc, lead, lithium, cadmium, iron as anode-active materials) and their compounds (e.g., MnO_2, PbO_2, $LiCoO_2$, $LiFePO_4$, NiOOH as cathode-active materials). Considering the profligacy with which humankind has exploited metal resources that Mother Earth has bestowed on it, it is only to be expected that one day natural resources will run out. The situation is exacerbated by our throwaway culture and poor record of recycling spent batteries. Thus, a shift from a cowboy economy to a spaceman economy is needed—that is, a move from a linear economy to a circular economy [1]. Furthermore, uncertainty in the availability and bottlenecks in the supply chains of critical metals [2–6] make it prudent to find ways to switch allegiance to earth-abundant metals or even to organics. Recent trends in battery research focusing on earth-abundant elements such as sodium, magnesium, aluminum, oxygen, and sulfur are thus welcome moves towards sustainability.

10.2 POST-LITHIUM BATTERIES

Looking beyond lithium-based batteries that are today's workhorses covering a panoply of applications, we have viable choices in those based on low-equivalent weight metals such as aluminum (8.993 g/eqvt.), magnesium (12.153 g/eqvt.), calcium (20.039 g/eqvt.), sodium (23.989 g/eqvt.), and potassium (39.098 g/eqvt.). While chemistries based on these metals may not match lithium-ion in performance, they could be cheaper, safer, and environmentally friendly. Figure 10.1 compares the abundance, standard electrode potentials, and specific capacities of these metals. Aluminum as an anode clearly stands out. It should be pointed out that the bulk of

DOI: 10.1201/9781003441182-12

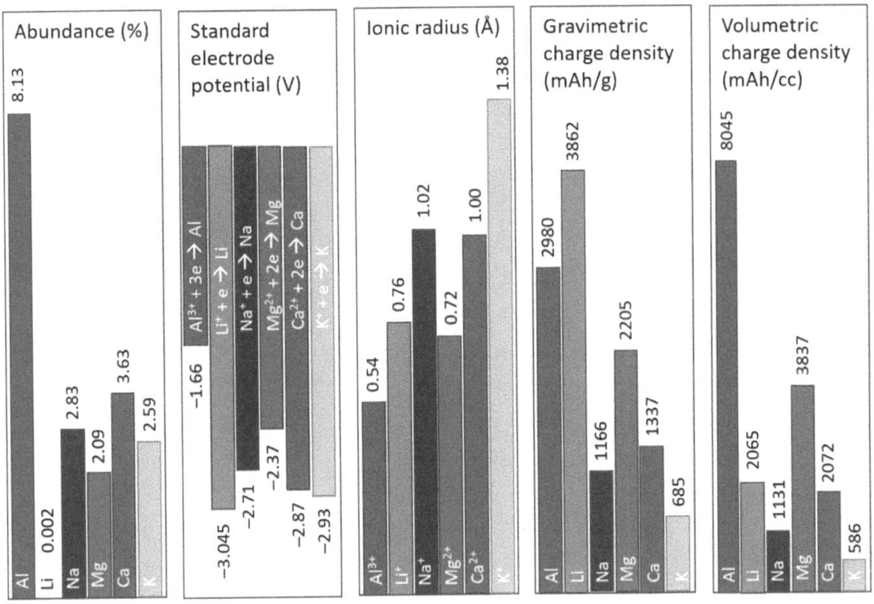

FIGURE 10.1 A conspectus detailing the abundance, electrode potentials, ionic radii, and charge densities of Al, Li, Na. Mg, Ca, and K.

the research into post-lithium batteries concentrates on intercalation chemistries with an eye on battery architecture and fabrication compatible with the well-established lithium-ion technology. Putting this scenario in context, Canepa et al. [7] reviewed progress and challenges made in the intercalation chemistry of multivalent cations as applied to batteries.

Dendrite-free magnesium can be plated, but scientists are still in search of suitable electrolytes and cathode materials for magnesium intercalation [8, 9]. Batteries based on calcium, the third most abundant metal in the Earth's crust, are in their infancy. Calcium has a melting point of 842 °C (Li: 180.5 °C), which is a safety incentive. It allows for faster electron transfer kinetics than magnesium due to its lower polarizing character, but the performance of the calcium anode is surface-film controlled [10, 11]. Recent work by Ponroauch et al. [12], which reports reversible calcium stripping and plating through a stable SEI on the metal at moderate temperatures (>75 °C), provides some hope for calcium-based batteries. However, identifying suitable cathodes for calcium-ion batteries remains knotty [13, 14], with most studies being devoted to V_2O_5 [15] and Prussian blues [16, 17].

As for univalent cations, an immediate proposition could be sodium-based batteries. Sodium-ion batteries, based on low-cost (one-tenth that of lithium) and abundant sodium, offer high energy densities and enhanced thermal stability [18]. However, sodium-ion battery technology is at the fledgling stage and yet to noticeably hit the market. Accounting for as much as 2.09 wt.% in the Earth's crust, potassium is a choice anode. Similar to lithium, potassium forms graphite intercalation compounds with capacities of up to 279 mAh/g [19]. Solvated potassium ions are smaller than solvated lithium and sodium ions due to weaker Lewis acidity, which allows faster ion diffusion and lower ion de-solvation energy. Laboratory-scale potassium-ion batteries containing graphite anodes and Prussian blue cathodes deliver more than 120 Wh/kg, values comparable with those of the first-generation lithium-ion batteries of the early 1990s. However, potassium-based batteries are in their early developmental stages with several issues relating to cathode-active materials and electrolytes remaining to be sorted out [20, 21].

10.2.1 Aluminum as an Anode

Aluminum as an anode material is bestowed with several attractive properties. It is the third most abundant element and the most abundant metal in the Earth's crust. It is inexpensive and environmentally benign, and electro-winning and processing of the metal are well-established. Aluminum forms a trivalent cation (Al^{3+}), releasing three electrons per atom, and with an atomic weight of 26.9815 g/mole, it boasts a gravimetric charge density of 2980 mAh/g (Li: 3862 mAh/g). Aluminum has a density of 2.7 g/cc (Li: 0.54 g/cc). Thus, its volumetric charge density is 8045 mAh/cc (Li: 2065 mAh/cc), the highest among known anodes. However, its standard electrode potential is −1.662 V (Li: −3.045 V).

Although this would mean that cells sporting aluminum would exhibit lower voltages than those with lithium, the aluminum anode would be much less taxing on the electrolyte than lithium. Another advantage aluminum has over lithium is that it can be repeatedly stripped and deposited without forming dendrites that could lead to short circuits [22]—unlike lithium [23]. Aluminum is processable under ambient conditions; thus, there are clear safety advantages in moving from lithium to aluminum.

There is also a strong economic case for recycling as recycling of aluminum from spent batteries requires only 5% of the energy needed for primary production [24], and disassembly of aluminum batteries is straightforward given that the electrolytes are non-flammable and the anode a plain metal foil. All said, it remains a challenge to realize the remarkable properties of aluminum as an anode. The challenge stems from its small ionic radius and high charge density, which, from an electrolyte point of view, favors the formation of large complexes such as chloroaluminate ions rather than 'free' Al^{3+} ions. The penalty of having to employ such complex ions is a reduction in the capacity of rechargeable aluminum batteries [25]. There are also problems associated with the corrosion of the anode [26] and current collectors [27], in addition to cathode stability in known electrolytes [28] and parasitic reactions in the system [29, 30].

10.3 ELECTROLYTES FOR ALUMINUM BATTERIES

The exploitation of aluminum as an anode material is hindered primarily by a lack of suitable electrolytes, which explains why aluminum batteries are largely confined to laboratory prototypes. Some recent papers review progress made in electrolytes towards harnessing the advantages of this next-generation anode in rechargeable batteries [31–43]. The earliest known use of aluminum as an anode was in a Buff cell, an aqueous primary cell constructed by Tomassi et al. in 1857 in which it was coupled with carbon [44]. There are, however, reports that an impure form of the metal was used in a Voltaic cell constructed by Hulot et al. in 1855 [45–48]. Subsequently, aluminum featured as an anode in Leclanche-type cells and ended up in dry cell configurations [49–52]. However, aluminum dry cells could not measure up to a more developed zinc dry cell technology. The mechanically rechargeable Al–air battery was conceptualized in 1962 [53–55]. The electrolytes thus far had been aqueous ones. In 1988 a rechargeable Al–Cl_2 battery was developed with an inorganic molten salt electrolyte [56]. Except for the Al–air battery, the others had to be waylaid primarily because of issues related to corrosion and passivation. However, aluminum as an anode material for primary, mechanically rechargeable, and electrically rechargeable chemistries is far too appealing. Thus began fresh investigations into aqueous and non-aqueous electrolytes for aluminum batteries [31, 32, 38, 47, 54–59]. Today, apart from the aqueous Al–air system, we have non-aqueous systems (e.g., Al–S and Al-ion) based primarily on organic ionic liquids.

10.3.1 Aqueous Electrolytes

Water-based electrolytes are highly conductive (~10^{-2}–10^{-1} S/cm). They are rather inexpensive, non-flammable, and generally environmentally benign. Until recently, the application of aqueous electrolytes in aluminum batteries has largely been limited to primary or mechanically rechargeable batteries with working voltages less than 1.5 V, or those with aluminum alloy anodes [60, 61]. Because the standard electrode potential of aluminum (–1.662 V) is more negative than the electrochemical stability window of water (Figure 10.2), it is impossible to plate the metal from aqueous solutions. In these solutions, hydrogen evolution takes preference, which has more positive *p*H-dependent standard potentials (–0.83 to 0.00 V between *p*H 0 and 14), resulting in the decomposition of the electrolyte. In fact, this is the basis of the hydrothermal oxidation of aluminum in water (300 °C, 15 MPa) for the production of hydrogen.

The Pourbaix diagram presented in Figure 10.2 [62] shows that aluminum is stable in aqueous media between *p*H values of 4.0 and 8.6. Below *p*H 4.0, it dissolves to produce Al^{3+} ions, and above *p*H 8.6, it produces AlO_2^- ions. Thus, highly acidic and highly basic electrolytes are not compatible with aluminum. However, between *p*H 4.0 and 8.6 the metal gets covered with a stable passive layer of Al_2O_3 [62], which can introduce a contact resistance as high as 35 kΩ [63]. The passive layer can lead to large reaction over potentials, severely undermining the Coulombic efficiency of the cell. The Al_2O_3 layer is a uniform, extremely thin (~2–10 nm) conformal film of amorphous alumina [64]. When aluminum is used as a cathode current collector in lithium-ion batteries, electronic conduction between the metal and the cathode coating occurs by tunneling [65]. However, when the metal is used as an anode, the passive layer acts as a barrier for Al^{3+} ion solvation and its diffusion, which poses a challenge in developing aluminum in batteries [66]. Today, the most prominent face of aqueous aluminum chemistry is the Al–air battery, which requires a bifunctional electrode to qualify as a truly electrically rechargeable battery.

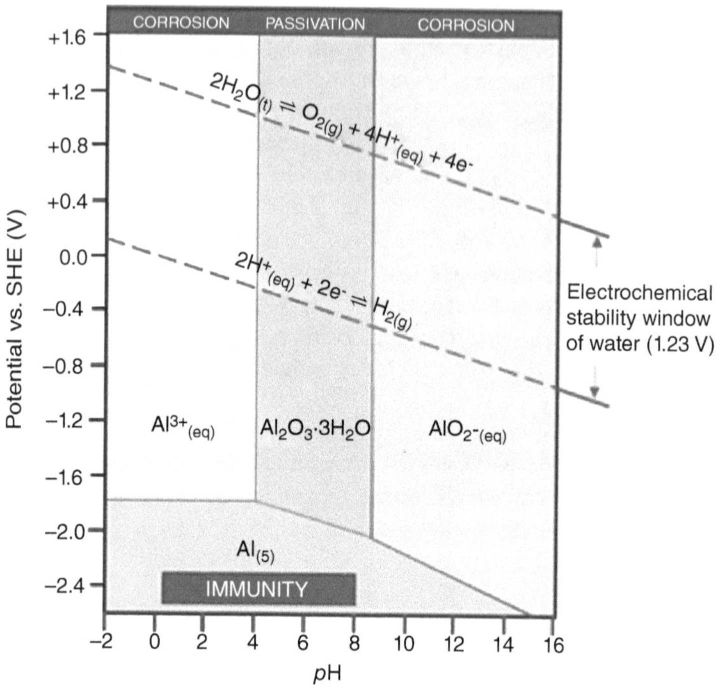

FIGURE 10.2 Pourbaix diagram of aluminum in water at 25 °C.

Source: Adapted from Ref. [62].

10.3.1.1 Water-in-Salt Electrolytes

One strategy for mitigating the effect of the passive layer on the aluminum anode is to employ highly concentrated electrolytes called WiSEs. Another is to pre-treat the metal so as to develop an SEI through which ion transport is facilitated. WiSEs are offshoots of aqueous electrolytes. Aqueous electrolytes are natural choices given the many desirable properties of water as a solvent: its high polarity and dielectric constant (allowing high concentrations of salts), low density and viscosity, wide liquidus range (extendable by addition of solutes), abundant availability, extremely high safety, and non-toxicity. However, its applicability is limited by its narrow thermodynamic voltage window of 1.23 V.

WiSEs contain extremely high concentrations of salts in water at or close to their solubility limits. They have structures similar to those of ionic liquids [67], with no 'free' water molecules, and they are being used up to form loose but intimate ion clusters. Such ion clusters are believed to suppress hydrogen and oxygen evolution reactions [68]. The aggregates of water molecules serve as conductive channels for cation transport [69], allowing for ionic conductivities comparable with those of conventional, less concentrated salt-in-water electrolytes [69]. Furthermore, such solutions show wider electrochemical stability windows [32, 70–72]. For example, a WiSE of 5M AlOTF was shown to be stable up to 2.65 V [68]. In another study, a WiSE of $AlCl_3{\cdot}6H_2O$ (the amount of the solute slightly above its solubility limit of

3.382 mol/kg) was shown to be stable up to 4 V [73]. Pan et al. [73], who demonstrated an aluminum-ion cell with this electrolyte, reported a discharge voltage of 1.44 V and a capacity of 165 mAh/g at 500 mA/g (3C) and a cycle life of 1000 cycles. However, the Coulombic efficiency was ~95%, which was an indication of side reactions that could be detrimental to practical cells. Nevertheless, WiSEs present obvious advantages such as higher safety, lower cost, and sustainability, as compared to organic ionic liquid electrolytes [60, 61, 74, 75].

WiSEs are a way to extend the electrochemical stability window of water (1.23 V) to more than 2 V. Molecular dynamics simulations have shown that their higher stability is associated with a water-depleted zone formed at the electrode surface [76]. The first reported use of a WiSE was that of a 21 M aqueous solution of LiTFSI (electrochemical stability window: 2.3 V) in an aqueous lithium-ion cell [77]. WiSEs have not been investigated much, particularly for aluminum-based batteries. Reports so far, however, look encouraging [68, 72, 73, 78]. Hu et al. [78] reported the use of a mixture of 1 M AlOTF, 17 M AlTFSI, and 0.02 M HCl as a WiSE in an Al–S cell. This electrolyte was reported to have an electrochemical stability window of 3.1 V. The authors stated that HCl was added to make the electrolyte slightly acidic with the aim of removing the passivation layer on the aluminum. The cell delivered a capacity of 420 mAh/g (of sulfur) and cycled at a Coulombic efficiency of 97% over 30 cycles [78].

10.3.1.2 SEI on Aluminum Anode

The Al_2O_3 film on aluminum is a true passive layer [79], being both electronically and ionically insulating. It is an impediment in the development of aluminum batteries, both with aqueous and ionic liquid electrolytes [75, 80–82]. It is thus desirable that the electrolyte dissolves the passive layer and permits access to a fresh anode surface [38, 75, 80–84] or forms an SEI of appropriate composition.

In their work on the behavior of the oxide barrier during the operation of an aluminum anode in an $AlCl_3$-based WiSE, Rastabi et al. [85] found that those portions on the aluminum surface that had iron-containing impurity phases tended to corrode preferentially. They concluded that locations with Al_3Fe in the metal were favorite spots for passive oxide barrier fissures to allow access to the electrolyte during the operation of aluminum batteries. An examination of the Al–Fe phase diagram would show that these spots correspond to Al_3Fe [86, 87].

Iron is a common impurity in aluminum (solubility: <0.05 wt.%) and is present as Al_3Fe [86–89], even in very pure aluminum (≥99.9%) [88]. Therefore, the suggestion that spots containing intermetallic impurity phases may be beneficial is of much commercial value for practical cells, considering that the production cost of high-purity aluminum (99.99%) is significantly higher than that of commercial-grade metal (99.5%) [90]. It also emerges that metallurgical modification of the aluminum surface could be a way to circumvent issues with anode passivation in aluminum batteries. However, detrimental effects, if any, from dissolved iron species on cell performance are not known.

SEIs have been confirmed to be beneficial for active ion conduction while simultaneously electronically insulating anode surfaces [91]. Wu et al. [81] adopted a surface pre-treatment process to generate an artificial SEI on an aluminum surface. The interphase reconstruction involved dipping aluminum in an ionic liquid-based electrolyte and subsequently washing it with water/ethanol. The corrosive electrolyte (in this case, BMImCl–$AlCl_3$) removed the native Al_2O_3 on the metal surface and replaced it with one containing Al_2O_3 and $Al(OH)_3$, which the authors claim to facilitate excellent reversibility for the plating/stripping of aluminum (high Coulombic efficiency: 98.7%; and low overpotential: 15.2 mV) [81]. Thus, surface reformation with suitable interphase for ion diffusion can transform materials previously considered inactive into highly active materials for battery application [83].

Zhao et al. [80] used an ionic liquid-enriched SEI in an aqueous aluminum-ion battery. An SEI thus formed with EMImCl–$AlCl_3$ was shown to predominantly consist of $AlCl_3$ and imidazolium ions. The SEI-modified anode showed an order of magnitude lower value for polarization during plating/stripping as compared to an anode without pre-treatment. In another experiment, the authors used an anode with a reconstructed SEI in an Al–MnO_2 cell containing 2 M AlOTF in water as an electrolyte and recorded an average discharge voltage of 1.3 V and a capacity of 380 mAh/g [80].

10.3.2 Non-Aqueous Liquid Electrolytes

10.3.2.1 Organic Solutions

Despite the availability of a large volume of literature on plating aluminum from liquid organic baths, their translation into electrolytes for aluminum-based batteries remains a distant goal. Among the first obstacles in using molecular organic solvents is the poor solubility of aluminum salts in them. The Al^{3+} ion (radius: 0.535 Å [92]) has a surface charge density six times that of Li^+ ions (0.76 Å [93]). The high charge density produces strong Coulombic interactions between Al^{3+} ions and common anions, rendering the dissociation of aluminum salts in low-dielectric constant organic solvents difficult or impossible. This explains why some of the earliest attempts at electroplating aluminum from baths with aprotic solvents such as ACN, DMF, PC, and pyridine failed [94].

Some etheric solvents dissolve aluminum salts by forming Al^{3+}-ether complexes through strong ion–dipole interactions. However, there is the disadvantage of high polarization during plating due to the high activation energy required for de-solvation [93]. Commonly reported plating baths are based on solutions of $AlCl_3$ in sulfones [95–100], glymes [101–104], EC [105], and γ-BL [106]; $AlCl_3$ + LiH in diethyl ether [107]; $AlCl_3$ + $LiAlH_4$ in THF [108–111], THF–benzene [112], and THF–toluene [113]; $AlBr_3$ in aromatic hydrocarbons [94, 114, 115] and *N, N*-dimethylaniline [116]; and $AlBr_3$ + KBr in ethylbenzene [117]. An important requirement is to obtain dendrite-free deposits of the metal, which may be realized in several bath compositions [99, 118–120]. Currently, however, satisfactory deposits can be obtained only at temperatures above room temperature [121]. This limits their application in batteries, where the use of highly volatile and flammable solvents can present undesirable safety problems. Zhao and VanderNoot [122] have reviewed aluminum electrodeposition from organic solvents.

The first report on aluminum deposition from an organic solvent-based bath was by Couch and Brenner [107]. However, their electrolyte—a solution of $LiA1H_4$ + $A1Cl_3$ in diethylether—is unsuitable for technological exploitation, as it is both unstable and flammable. Replacement of diethylether with the less-volatile THF resulted in baths with plating efficiencies of up to 100% [108, 109, 112]. Later, plating baths with organometallic aluminum compounds and those with aromatic solvents began to be explored. Baths containing aromatic hydrocarbons and etheric solvents [123], in particular, are not only combustible but also yield dendritic deposits. One report suggests that solutions of $AlBr_3$ in alkylbenzenes are beset with problems of unwanted chemical and electrochemical processes that occur during solution preparation and electrolysis, rendering them useless [115].

Electrolytes containing $AlCl_3$ dissolved in $DMSO_2$ are attractive as they are more stable than those based on conventional molecular organic solvents. $AlCl_3$ undergoes a solvolysis reaction in $DMSO_2$ to form $AlCl_4^-$ and $Al[DMSO_2]_3^{3+}$ species, the latter being amenable to electro-reduction to metallic aluminum [95, 124–127]. $DMSO_2$ as a solvent barely coordinates, is stable at moderately elevated temperatures, and dissolves a number of salts [128, 129]. An $AlCl_3$ (2 mol)–LiCl (1 mol)–$DMSO_2$ (1 kg) electrolyte supports aluminum plating and stripping, while no deposition was observed in an $AlCl_3$ (1 mol)–LiCl (1 mol)–$DMSO_2$ (1 kg) electrolyte [95]. Legrand et al. [97] showed that $AlCl_3$ (1 mol)–$DMSO_2$ (1 kg) (conductivity: 1.6×10^{-2} S/cm at 400 K) could reversibly plate and strip aluminum. Nakayama et al. [99] investigated solutions of $AlCl_3$ in sulfone-based electrolytes containing organic solvents as diluents. Their ^{27}Al NMR results showed that the concentration of the electroactive species, $Al_2Cl_7^-$, increased with increasing temperature, which explained their observation of higher plating efficiencies as the temperature was raised. Their optimized composition of $AlCl_3$ in ethyl *n*-propyl sulfone–toluene enabled plating/stripping at room temperature. It also showed negligible corrosion of a stainless steel current collector [99].

Liquid organic electrolytes have now largely been overlooked due to their volatility, flammability, and elevated working temperatures. It must, however, be noted that at the high concentrations of solutes in these electrolytes, the volatile properties of the solvent could be subdued. As for flammability, the use of flame-retardant additives could come in handy [130]. It appears that liquid organic electrolytes based on AlOTF hold much potential, particularly with respect to their safety and much reduced corrosivity, allowing a wide choice of current collectors and casing materials. Reed et al. [93], who investigated AlOTF in diethylene glycol dimethyl ether as an electrolyte (conductivity: 2.5×10^{-2} S/cm; wide electrochemical stability window), however, warned that plating/stripping efficiency in this electrolyte is not satisfactory. There is also a likelihood of passivation of the aluminum surface due to electrolyte decomposition [93].

10.3.2.2 Ionic Liquids

Because most practical systems have been demonstrated in these electrolytes so far, this section will command maximum attention. RTILs are solvent-free and are organic salts generally composed of bulky asymmetric organic cations and weakly coordinating organic/inorganic anions that exist in the liquid state at temperatures below 100 °C [93, 131–134]. In solutions, ions are generated by solvent–solute interaction, while in ionic liquids ions are formed by self-ionization. RTILs have wide liquid ranges and low vapor pressures and are non-volatile and non-flammable. Moreover, they are thermally stable over large temperature ranges. RTILs employed in aluminum-based batteries are generally Lewis acidic eutectic melts of $AlCl_3$ in an organic salt, R^+X^-, where R^+ is a cation such as an imidazolium, pyridinium, pyrrolidinium or sulfonium, and X^- is a halide ion, usually chloride. These RTILs have low-stability ion–solvent complexes, allowing facile Al^{3+}-ion solvation–de-solvation [135]. These melts also have wide electrochemical stability windows (up to 4.0 V) [136] and high ionic conductivities similar to aqueous electrolytes (1–2.5 × 10^{-3} S/cm) [80].

A vast majority of published literature reports aluminum batteries based on $AlCl_3$-based eutectics containing alkyl imidazolium cations, primarily the state-of-the-art cation $[EMIm]^+$ or $[BMIm]^+$ cation [137–140]. These electrolytes support the deposition of dense, crystalline aluminum at moderate current densities with high Coulombic efficiencies [140]. A major disadvantage of these RTILs is that they are highly hygroscopic [141, 142]. Exposure to moisture can lead to exothermic reactions, releasing HCl, among other products [143, 144]. They must, therefore, be handled in glove boxes or dry rooms [145, 146].

Cyclic voltammograms of aluminum in a Lewis acidic chloroaluminate ionic liquid of $AlCl_3$ and EMImCl (2:1 molar ratio) showed anodic and cathodic peaks centering around 0 V vs. Al^{3+}/Al, establishing that aluminum deposition and dissolution occurred in the electrolyte [147]. The highest value of Coulombic efficiency was about 70% at 25 °C, which decreased with an increase in temperature. Ingale et al. [148] demonstrated the stripping and plating of an aluminum anode with 100% depth of discharge in an EMImCl–$AlCl_3$ electrolyte. Improvement in cycle life under minimal compositional change was achieved by using organic solvents such as acetonitrile and benzene as additives. Solvent addition helped improve the mobility of electroactive species in the electrolyte. However, it would increase the risk of batteries going unsafe due to the flammability and high vapor pressure of the solvents.

10.3.2.2.1 Speciation in Chloroaluminate Electrolyte

Aluminum does not exist as Al^{3+} in chloroaluminate melts. The physical and electrochemical properties of the melts are governed by their Lewis acidity, which is determined by the molar ratio (r) of $[AlCl_3]$/[organic salt]. Speciation in these melts is defined by the following equations:

$$AlCl_3 + Cl^- \leftrightarrow AlCl_4^-$$

$$2AlCl_4^- \leftrightarrow Al_2Cl_7^- + Cl^-$$

$$3Al_2Cl_7^- \leftrightarrow 2Al_3Cl_{10}^- + Cl^-$$

Under Lewis basic conditions ($r < 1$), $AlCl_4^-$ and uncoordinated Cl^- ions (which serve as electron-pair donors) are present. Equimolar mixtures ($r = 1$) contain mostly neutral $AlCl_4^-$ ions. Under Lewis acidic conditions ($1 < r < 2$), melts contain primarily $AlCl_4^-$ and $Al_2Cl_7^-$ ions. Note that $Al_2Cl_7^-$ is a Lewis acid, as it can accept electron pairs. In very acidic melts ($r > 1.5$) oligomeric $Al_3Cl_{10}^-$ ions are also generated [143, 149]. $AlCl_3$ is not soluble at molar ratios above $r = 2$.

It must be noted that the tetrachloroaluminate anion ($AlCl_4^-$) is not an active species here; this anion gets reduced at a lower potential than the organic cation in the ionic liquid. In other words, electrolyte decomposition occurs before $AlCl_4^-$ can be reduced to Al. Thus, neutral and basic chloroaluminate melts, in which Cl^- and $AlCl_4^-$ are the predominant anionic species, are not amenable to aluminum electrodeposition. This limits the useful range of chloroaluminate melts to Lewis acidic compositions ($1 < r < 2$), in which $Al_2Cl_7^-$ lends itself to reduction to Al according to the equation:

$$4Al_2Cl_7 + 3e^- \leftrightarrow Al + 7AlCl_4^-$$

It, therefore, goes without saying that if the organic ionic liquid is chosen such that the reduction potential of its cation is more negative than that of the $AlCl_4^-$ anion, aluminum deposition can be realized from basic or neutral melts. The r value also determines the stability window of the melts [26]. For example, the anodic stability limit of BMImCl–$AlCl_3$ is about 2.6 V vs. Al^{3+}/Al for r values of 1.0 to 2.0, while it is 1.75 V when $r = 0.8$. The cathodic stability limit of the melt is −0.2 V vs. Al^{3+}/Al for r values of 1.5 and 2.0, and −2.0 V for r values of 0.8 to 1.1. The cathodic limit in the highly acidic electrolytes ($r = 1.5$ and 2.0) is associated with the deposition of aluminum. However, in the case of melts with $r = 0.8$, 1.0, and 1.1, the organic cation is reduced at the cathodic limit. The anodic stability limit is associated with chlorine evolution at the cathode according to one of the following reactions dictated by the predominant ionic species in the electrolyte [150]:

$$(\text{Basic})\, 2Cl^- \rightarrow Cl_2 + 2e^-$$

$$(\text{Neutral})\ 4AlCl_4^- \rightarrow Cl_2 + 2Al_2Cl_7^- + 2e^-$$

$$(\text{Acidic})\ 6Al_2Cl_7^- \rightarrow Cl_2 + 4Al_3Cl_{10} + 2e^-$$

10.3.2.2.2 *Lewis Acidity and Anode Capacity*

It can be deduced from the previous discussion that the capacity of the anode increases as the Lewis acidity of the chloroaluminate melt is increased, for at high acidities there is a higher concentration of hepta dichloroaluminate anions ($Al_2Cl_7^-$) that can be reduced to metallic aluminum during cell charging. Indeed, this was found to be empirically true, although at low current densities of about 20 mA/g [150]; at higher rates of discharge, the low diffusivity of the large$Al_2Cl_7^-$ ions leads to localized depletion of the anions at the electrode surface and hence a lowering of acidity at the electrode-electrolyte interface. It is clear that the charging process (aluminum plating) would end when no more $Al_2Cl_7^-$ ions are in the electrolyte; that is when the melt turns neutral ($r = 1$). In practice, a low-to-moderate Lewis acidity region ($r =$ 1.1–1.3) has been found to be optimal [151–161]. The mildly acidic melts also help remove the oxide film on the anode [26, 83]. An added advantage of this compositional region is its high levels of ionic conductivity [26].

10.3.2.2.3 Deposit Morphology

The morphology of deposits from Lewis acidic chloroaluminates depends on electrolyte composition and operating conditions such as current density [162, 163]. In addition, the coordination chemistry of the electrolyte and the nucleation and deposit growth also has a crucial role [164]. Moreover, repetitive plating and stripping can result in roughening of the aluminum surface without dendritic growth [163]. Roughening can lead to shedding and loss of active material and possible short circuits between the anode and the cathode. The commonly employed EMImCl–$AlCl_3$ electrolyte produces dense and adherent deposits at low current densities (typically < 40 mA/cm^2), while at higher current densities the deposits are generally less adherent and even dendritic [165, 166]. A recent report claims that the use of porous aluminum can suppress dendrite growth in the same electrolyte [120]. The higher surface area of the porous aluminum reduces the current density, allowing the formation of smooth deposits. The authors have shown cycling stability of more than 18,000 charge/discharge cycles [167].

10.3.2.2.4 Candidate RTILs for Aluminum Batteries

Although the non-aqueous chemistry of aluminum batteries has focused on imidazolium-based ionic liquids, the high cost and particularly the hygroscopicity of the electrolytes have pushed scientists to look for alternatives. In terms of cost, mixtures of $AlCl_3$ and Et_3NHCl [168] are an immediate choice, considering that Et_3NHCl is one order cheaper than EMImCl. Being organic compounds, the properties of ionic liquids can be tuned by introducing substituent groups in their structures [169–171]. Zhu et al. [170] investigated MPPyrCl–$AlCl_3$ as an electrolyte in an Al/graphite cell. The performance of the electrolyte was inferior to that of the conventional EMImCl–$AlCl_3$, which the authors attributed to the former's higher viscosity (and lower conductivity). The authors concluded that a weaker organic cation–chloroaluminate anion interaction expected with the larger $[MPPyr]^+$ cation (142 Å^3) ($[EMIm]^+$ cation: 118 Å^3) could be a determinant in the reversibility of the plating/stripping reactions, as well as the morphology of the deposits [26]. Another approach to alter the properties of ionic liquid-based electrolytes is to employ them as mixtures with inert organic solvents such as DCM. Although this can reduce the viscosity of the electrolyte and thus help improve its conductivity [172], it can also adversely affect its safety.

10.3.2.2.5 Chloride-Free and Less Corrosive Electrolytes

A pressing problem with RTIL-based electrolytes is their high corrosivity [173–176]. It is believed that the $Al_2Cl_7^-$ anions are mainly responsible for the high corrosivity [164]. Corrosion of the SUS430 stainless steel current collector by EMImCl–$AlCl_3$ was reported by Reed et al. [27], although it was less severe with carbon steel and titanium [173]. Among alternatives suggested to mitigate corrosivity are ionic liquids containing water-stable anions such as BMPyr–TFSI and EMIm–TFSI [177–179]. Another driver for a chloride-free electrolyte derives from the fact that ions with low electron donor ability (a low Gutman donor number) can lower the solubility of polysulfides generated during the discharge of the sulfur electrode in Al–S cells. Thus, electrolytes free from ions with high donor numbers, such as chloride, can help decrease polysulfide dissolution [180, 181].

Immediate alternatives to chloride are PF_6^-, TFSI, and OTF. In fact, $Al(PF_6)_3$ in DMSO [182] and AlTFSI in ACN [119, 183] have been attempted, while AlOTF has been studied with THF–PC [184], diglyme [93, 120], *N*-methyl acetamide+urea [185], and THF [186]. AlOTF has also been studied with aqueous systems [187, 188] and ionic liquids [80]. However, to date, clear directions on AlOTF-based electrolytes have not emerged. Other water-stable anions have also been proposed, although no definite conclusions have been reached on their applicability: HdMSO–AlOTF, HdMSO–AlTFSI, and HMIm–AlTFSI [180]. Other electrolytes such as those based on BF_4^-, NO_3^-, and CH_3COO^- anions made transient appearances [189], but chloroaluminates held sway, chiefly due to their ability to efficiently plate aluminum. It must be mentioned in passing that the plating process is made possible by the ability of acidic chloroaluminates to dissolve the adherent and thermodynamically stable Al_2O_3 native layer on the metal [26, 83, 190]. However, modification of the aluminum surface by forming suitable SEIs can enable the plating/stripping process in chlorine-free electrolytes [74, 82, 191].

An important contributor to electrode polarization and hence reduced energy efficiency is the fact that the electro-active $Al_2Cl_7^-$ anions travel against the electric field (away from the anode) during charging (plating reaction: $4Al_2Cl_7^- + 3e^- \rightarrow Al + 7AlCl_4^-$). Li et al. [164] came up with an ionic liquid electrolyte with aluminum-containing cations as the electro-active species. They complexed $AlCl_3$ with 4-ethylpyridine, a neutral ligand, and showed by Raman spectral studies that not only was this electrolyte free from the highly corrosive $Al_2Cl_7^-$ anions, but that it also teemed with $AlCl_4^-$ anions and Al^{3+}-containing cations, $[AlCl_2(4\text{-ethylpyridine})_n]^+$ [176, 192], generated as per the following scheme [193]:

$$2AlCl_3 + 2\left[4-\text{ethylpyridine}\right] \rightarrow AlCl_2\left[4-\text{ethylpyridine}\right]_2^+ + AlCl_4^-$$

$$2AlCl_3 \rightarrow AlCl_2^+ + AlCl_4^-$$

$$AlCl_2^+ + AlCl_2\left[4-\text{ethylpyridine}\right]_2^+ \rightarrow 2[AlCl_2\left[4-\text{ethylpyridine}\right]^+$$

A 1:3 molar mixture of 4-ethylpyridine–$AlCl_3$ exhibited an ionic conductivity of 8.9×10^{-4} S/cm. An Al–graphite cell assembled with this electrolyte gave a capacity of 95 mAh/g at a current drain of 25 mA/g. Furthermore, it exhibited considerably lower corrosivity toward aluminum, copper, and nickel, as compared to EMImCl–$AlCl_3$. 4-Ethylpyridine–$AlCl_3$ is also moisture-insensitive [164].

In a similar study, but with no proof of application, Yao et al. [194] demonstrated good conductivity (2.86×10^{-5} S/cm) and thermal stability (up to 150 °C) with $Al(NO_3)_3$ dissolved in PTHF ($M_n = 250$) and encapsulated in a matrix of a cycloaliphaticepoxy (3, 4-epoxycyclohexylmethyl-3, 4-epoxycyclohexanecarboxylate).

Unlike PEO, which shows strong electrostatic interactions with multivalent cations and thus renders dissociation and cation mobility difficult, PTHF has ether oxygens at greater spacing, allowing freer ion movement. Being a liquid at room temperature, PTHF has greater chain flexibility, which means ionic conductivity by segmental motion is far more facile. Electrolytes based on PTHF themselves have moderate conductivities [195–197], which can be preserved in their encapsulated form in a mechanically robust resin [198]. The use of PTHF has direct relevance to aluminum plating from organic baths [108–113], where the use of THF as a solvent is discouraged due to its high volatility, flammability, low flash/boiling points, and susceptibility to generate explosive peroxides.

10.3.2.2.6 Disadvantages of Ionic Liquid Electrolytes

As noted earlier, corrosivity and moisture sensitivity are major limiters in using ionic liquid-based electrolytes. The highly acidic electrolytes preclude the use of several common current collectors, separators, and polymeric binders. This has triggered the search for RTILs free from chloride and attempts to reduce reactivity by immobilization in suitable matrices. The electrochemical stability windows of chloroaluminate melt (generally 2.4 V) are also sought to be widened. Chloroaluminate melts are also prone to oxidation ($AlCl_4^-$ to Cl_2) and to reactions with cathode-active materials [199]. Although the passive layer of Al_2O_3 can restrict the growth of dendritic aluminum, it is vulnerable to disintegration [22] and can lead to dendrite formation, especially at high current densities [27]. In this context it must be recalled that a non-corrosive electrolyte of BMIm–AlOTF could not remove the passive layer [83], limiting the plating/stripping of aluminum. The surface, however, was rendered active by pre-treatment in EMImCl–$AlCl_3$. The high cost of RTILs and the toxicity of certain classes of RTILs must also be addressed in their commercial exploitation in batteries. Also, as for the specific capacity of aluminum in chloroaluminate electrolytes, it is instructive to note that the discharge reaction is

$$\mathrm{Al} + 7\mathrm{AlCl_4}^- \rightarrow 4\mathrm{Al_2Cl_7}^- + 3\mathrm{e}^-,$$

which shows that an additional seven Al atoms and 28 Cl atoms from the electrolyte are required to release three electrons from one Al atom. Thus, the theoretical specific capacity of aluminum in chloroaluminate melts is only 67 mAh/g. This is meager when juxtaposed with 2980 mAh/g projected based on the commonly assumed reaction

$$\mathrm{Al} \rightarrow \mathrm{Al}^{3+} + 3\mathrm{e}^-.$$

Similarly, the participation of the chloroaluminate electrolyte in the discharge of the sulfur cathode in Al–S batteries should not be discounted. The sulfur reduction reaction,

$$8\mathrm{Al_2Cl_7}^- + 6\mathrm{e}^- + 3\mathrm{S} \rightarrow \mathrm{Al_2S_3} + 14\mathrm{AlCl_4}^-$$

which is a multi-step reaction: [151, 199]

$$4Al_2Cl_7^- \rightarrow 4AlCl_3 + 4AlCl_4^- \rightarrow 7AlCl_4^- + Al^{3+};$$

$$2Al^{3+} + 3S + 6e^- \rightarrow Al_2S_3,$$

involves the participation of 16 Al atoms and 56 Cl atoms from the electrolyte for transferring six electrons to three S atoms. Thus, the theoretical capacity of sulfur in chloroaluminate melts should be 64 mAh/g, and not 1672 mAh/g, as governed by

$$S + 2e^- \rightarrow S^{2-}.$$

10.3.2.3 Deep Eutectics

Deep eutectics (sometimes called ionic liquid analogues) are low-cost substitutes for traditional ionic liquids. They are mixtures of Lewis or Brønsted acids and bases containing a variety of anions and/or cations [141, 200–203]. When a strong Lewis acidic metal halide is mixed with a Lewis basic ligand, heterolytic cleavage of Al_2Cl_6 (dimers of $AlCl_3$) occurs, leading to various ions including $Al_xCl_{3x+1}^-$ anions and $[AlCl_2{\cdot}(ligand)_n]^+$ cations [204–207]. The resulting electrolyte supports reversible plating/stripping of aluminum [208, 209]. Commonly employed deep eutectics are those formed by mixing $AlCl_3$ (a Lewis acidic salt) with a Lewis basic ligand such as urea or acetamide:

$$2AlCl_3 + nL \rightarrow \left[AlCl_2{\cdot}nL\right]^+ + \left[AlCl_4\right]^-,$$

where L is a ligand. Deep eutectics generally have lower ionicity and higher viscosity (and, therefore, lower conductivity) than ionic liquids. Several reports have appeared on aluminum cells using such electrolytes [203, 210–218]. A standout property of deep eutectics as applied to Al–S batteries is the lower solubility of sulfur in these electrolytes than in EMImCl–$AlCl_3$.

According to Wang et al. [214], the aluminum plating reactions in a urea-based deep eutectic are

$$2\left[AlCl_2{\cdot}n(\text{urea})\right]^+ + 3e^- \rightarrow Al + AlCl_4^- + 2n\,\text{urea}\,(\text{at high urea concentrations})$$

and

$$4Al_2Cl_7^- + 3e^- \rightarrow Al + 7AlCl_4^-\,(\text{at low urea concentrations}).$$

As battery electrolytes, deep eutectics have several key properties similar to ionic liquids (wide liquid range, low vapor pressures, non-flammability), but, being viscous, they have lower conductivities [215]. Replacing urea with its derivatives (e.g., *N*-methyl urea, *N*-ethyl urea) in these mixtures can result in drastically lowered viscosities. For example, a 1.4 $AlCl_3$/urea mixture has a viscosity of 133 cP at 25 °C, while $AlCl_3$–*N*-methyl urea and $AlCl_3$-*N*-ethyl urea mixtures have viscosities of 67 cP and 45 cP, respectively. However, most studies with deep eutectics deal with those containing urea [211, 212, 214, 215, 217, 218], acetamide [216, 218], or both [208, 209, 217]. Al–S batteries with such electrolytes have been reported to deliver consistently high initial capacities (500+ mAh/g, and even >2000 mAh/g) and sustain extended cycling (60–100 cycles) [203, 219]. This is an improvement over Al–S cells that employ EMImCl–$AlCl_3$ (~20 cycles) [151] whose inferior performance is attributed to the electrolyte reacting with the sulfur-active material.

Several authors have reported very high capacities for sulfur, often more than the theoretical value (1672 mAh/g), in deep eutectics based on urea and acetamide: 2359 mAh/g with urea [218]; and 2100 [203] and 2129 mAh/g [218] with acetamide. Some authors suggested that the high capacities were related to side reactions involving carbon in the cathode [203]. However, Lampkin et al. [218] showed that the contribution to capacity by carbon was insignificant. According to them the high capacities, often exceeding 1672 mAh/g, are ascribable to side reactions, including electrolyte degradation. Furthermore, Lampkin et al. [218] showed that no electrolyte degradation occurred in the absence of sulfur and suggested that polysulfide species formed by the reduction of sulfur triggered the degradation reactions. Also, the discharge voltage and capacity of Al–S cells with urea/acetamide decreased markedly upon cycling (as compared to one with an EMImCl–$AlCl_3$ electrolyte). The resulting changes in electrolyte composition and rise in electrolyte viscosity can seriously affect cell performance.

That Al–S cells with amide-based deep eutectics display discharge voltage plateaus of 1.65–1.80 V (e.g., [203, 219]) makes claims of sulfur alone as the cathode-active material in such cells dubious. For, the cell discharge reaction is:

$$2Al^{3+} + 3S + 6e^- \rightarrow Al_2S_3.$$

The ΔG^0 of the formation of Al_2S_3 at 298.15 K is −153.0 kcal/mol [220], which means that the E^0 of the reaction is 1.12 V. Even going by a later phase equilibrium data of Sharma and Chang [221], the E^0 value of the reaction would only be 1.20 V. Thus, the higher discharge voltages would only mean that the cell reactions in these deep eutectic-based systems are different.

10.3.2.4 Inorganic Molten Salts

Anhydrous $AlCl_3$ exists as an ionic crystal [222]. Upon melting, two $AlCl_3$ molecules can dimerize to yield an Al_2Cl_6 molecule. In Al_2Cl_6 a chlorine atom from each of the $AlCl_3$ molecules forms a dative covalent bond (with one of its two lone electron pairs) with the aluminum center. Al_2Cl_6 is a molecular liquid with poor ionic conductivity but is a strong Lewis acid. It can thus react with other halogen-containing organic and inorganic compounds to form chloroaluminate ionic liquids in which the anions have a general formula $Al_xCl_{3x+1}^-$ [223].

Although inorganic chloroaluminate melts have been exploited for the electrodeposition of aluminum for more than half a century [224–227], the industry has been slow to embrace it for rechargeable battery applications [228–231]. One impediment to their adoption is the high melting points of the melts (e.g., LiCl–NaCl–KCl: 347 °C [232]). However, the propensity of $AlCl_3$ to form large $[Al_xCl_{3x+1}]^-$ species with alkali chlorides leads to low-melting point eutectics (for example, m.p. of NaCl–KCl–$AlCl_3$: 93 °C [225]). Such low-melting eutectics lend themselves to exploitation in moderately elevated temperature batteries, which can be kept operational with water-based heating systems. Moreover, aluminum does not react spontaneously with alkali-chloroaluminate melts, which can translate to high anode cycling efficiencies, close to 100%.

Molten mixtures of NaCl and $AlCl_3$ are particularly appealing for intermediate-temperature aluminum batteries, given their low cost and high safety [211, 212, 233–235]. However, an impediment in developing such batteries is the strong corrosivity of the electrolyte. In an attempt to circumvent this disadvantage, Wang et al. [236] replaced the metallic aluminum with a liquid anode of gallium. An additional advantage of using gallium is that the aluminum deposited during charging alloys with gallium, and the alloy, being in the liquid state, preclude the formation of dendritic aluminum.

Song et al. [233] and Chen et al. [237], who used electrolytes of NaCl–$AlCl_3$ and $AlCl_3$–NaCl–KCl melts, respectively, in Al–graphite cells at 120 °C, found no evidence for dendritic growth at the anode. While the former cell delivered capacities of up to 136 mAh/g at 500 mA/g, the latter gave 128 mAh/g at 100 mA/g. In another study, Wang et al. [235] reported an aluminum–graphite paper cell with an $AlCl_3$–LiCl–KCl electrolyte (99 °C) delivering a capacity of 107 mAh/g at 200 mA/g with high Coulombic efficiencies. In a recent study, Sadoway et al. [231] used a eutectic mixture of NaCl–KCl–$AlCl_3$ in Al–S and Al–Se cells. They demonstrated that this electrolyte could support aluminum plating/stripping at 180 °C with low overpotentials and Coulombic efficiencies of 99.0–99.8%.

10.3.3 Solid Electrolytes

Just as solution kinetics dominate studies of reaction kinetics, liquid electrolytes dominate battery research from where the knowledge could be extended to studies on solid-state batteries. Among the standout features offered by liquid electrolytes are the simplicity of cell construction, high ionic conductivity, and excellent electrode-electrolyte contact. However, for constructions where leak proofness, orientation freedom, flexibility, safety, and weight and volume advantages are of capital importance (for example to power heart pacemakers), solid or quasi-solid electrolytes hold sway. Discussions on solid-state batteries are often overly prejudiced, bordering on extreme optimism or pessimism. There are major problems to be tackled, even with single-ion solid conductors where challenges remain in areas such as poor electrode-electrolyte interfacial kinetics and optimization of mechanical pressure needed for stable cell operation. Nevertheless, solid electrolytes do present intrinsic advantages that cannot be overlooked such as high thermal stability, extremely wide electrochemical stability windows, and the singularity of the ionically conducting species and sustaining high current densities without concentration polarization.

10.3.3.1 Ceramic Solid-State Electrolytes

Because of their high elastic moduli, ceramics are difficult to process and are thus useful only for rigid battery designs. Ionic mobility in ceramic inorganic conductors has a strong dependence on the valency and size of the mobile ion. In other words, designing solid conductors for mono- and divalent cations is easier considering the low activation energies associated with their conduction process. However, the mobility of trivalent cations is limited by high Coulombic interactions between the cation and the surrounding atomic framework or covalently bonded sublattice in the crystal [238]. The resulting high migration barriers rather favor fixing the cations in their crystallographic sites. Thus, it is no wonder that there are not many reports of migration of trivalent ions such as Al^{3+} or those of rare earth in the solid state [239–252].

Building on their work on the trivalent ion conductivity of Sc^{3+} [253], Y^{3+} [254], and Er^{3+} [254], Adachi et al. [240, 241] demonstrated Al^{3+} conduction in aluminum tungstate, $Al_2(WO_4)_3$, with the $Sc_2(WO_4)_3$-type structure. Typical conductivity values were about 2×10^{-5} S/cm at 800 °C [240]. Later, in 2002, they showed a one-order increase in Al^{3+} ion conductivity (along the *c* axis of the crystalline phase) in an $Al_2(WO_4)_3$–$Sc_2(WO_4)_3$ solid solution and attributed the change to competing effects of lattice expansion and a decrease in the number of mobile Al^{3+} ions [255]. Imanaka et al. [256, 257] further showed that in a NASICON-type $(Al_{0.2}Zr_{0.8})_{20/19}Nb(PO_4)_3$ [general formula: $(Al_xZr_{1-x})_{4/(4-x)}Nb(PO_4)_3$], the predominant migrating species was Al^{3+} (transference number >0.998) and that it had a conductivity of 4.5×10^{-4} S/cm at 600 °C (two orders more than that of $Al_2(WO_4)_3$ at 600 °C and comparable with those of anion-conducting solids). However, temperatures of 600 °C and above at which these conductivity values are realized are unacceptable for battery systems.

While experimentalists in general vouch for the veracity of the claims on Al^{3+} ion conductivity, some researchers, especially theoreticians, seek to differ [258–265]. That there are only a handful of experimental studies—and that unequivocal evidence for trivalent ion conductivity seems challenging—adds fuel to the debate. One point of contention is the contribution, if any, from cations of lower net charge, proton conduction (in hygroscopic compounds), or from O^{2-} ions [265]. Nestler et al. [265] even contend that there are no solid electrolytes for Al^{3+} ions identified as yet. Further experimental and theoretical studies are required to establish the species responsible for ionic conduction.

10.3.3.2 Polymer Electrolytes/Ionomers

The application of electrolytes based on polymers in lithium batteries and other electrochemical systems began with the seminal work of Fenton et al. [266] in 1973. Humongous research efforts since then have fructified into commercial lithium-polymer cells. However, the direct transfer of the knowledge thus gained into aluminum-based batteries is mired in both the complexity of the species responsible for aluminum deposition and in the high Lewis acidity of the electrolyte. Chloroaluminate electrolytes tend to react with a host of polymers. For example, PE, PP, PVA, polyimide, PEO, PMMA, and even PTFE and PVdF [31] are not compatible with chloroaluminate moieties [139, 267–269]. In fact, this is one reason why organic ionic liquid-containing aluminum batteries use glass fiber mats as separators, and the use of PVdF or polyolefins as the binder is shunned.

An ionic liquid-based polymer composite electrolyte with PEO, SiO_2, $AlCl_3$, and EMIm-TFSI appeared promising with a conductivity of 9.6×10^{-4} S/cm and an electrochemical stability window of 3 V vs. Al^{3+}/Al, but could not support plating/stripping [168]. Apparently, oxygen-laden groups (here, ether group) can alter the coordination of $Al_2Cl_7^-$, making them unavailable for electrochemical processes. A similar value of conductivity (2.5×10^{-4} S/cm) was measured with a gel polymer electrolyte of PAN and EMImCl–$AlCl_3$ with high dielectric constant EC as a plasticizer [270]. Free-standing films of this electrolyte (200–300 μm) prepared by solvent casting were shown to mitigate chloroaluminate corrosivity against a stainless steel 316 current collector (corrosion current an order less than those with conventional ionic liquids). An Al/pyrolytic graphite cell assembled with this electrolyte delivered a capacity of 63 mAh/g at 25 mA/g, a value similar to those obtained with conventional electrolytes [167, 271, 272].

The electroactive $Al_2Cl_7^-$ ion is known to react with molecules bearing lone electron pairs to produce $AlCl_4^-$ ions or neutral species [267]. This property precludes the use of certain solvents in the solution casting and *in-situ* polymerization processes involved in forming polymer electrolytes. In certain cases, solvent-free processing may be feasible, for example, in the preparation of ionogels of PEO with EMImCl–$AlCl_3$ [273, 274]. In this case, however, the electrolyte is made by mixing EMImCl–$AlCl_3$ with PEO heated to about 80 °C. Sun et al. [267] found that solvents such as acetone, ACN, and THF that have lone pair electrons are incompatible with EMImCl–$AlCl_3$ melts. The lone pairs act as Lewis bases, and their interactions with chloroaluminate species dampen the ability of the electrolyte to plate and strip aluminum.

However, solvents such as toluene and DCM seem to favor the plating/stripping processes, possibly due to enhancement in ionic mobility and electrolyte conductivity. Several research groups have, therefore, reported the use of DCM for polymerizing acrylamide monomers in the presence of EMImCl–$AlCl_3$ [267] and Et_3NHCl–$AlCl_3$ [275, 276] melts. Both the electrolytes showed improved resistance to moisture. The one with Et_3NHCl–$AlCl_3$ also showed appreciable low-temperature performance [276]. An Al/graphite cell with the Et_3NHCl–$AlCl_3$ polymer electrolyte gave an initial capacity of 120 mAh/g and sustained a hundred cycles at a current density of 60 mA/g.

As noted in Section 10.3.2.2.1, electrodeposition of aluminum from basic or chloroaluminate ionic liquids containing $AlCl_4^-$ is not possible, as the reduction of the ionic liquid precedes the reaction

$$AlCl_4^- + 3e^- \rightarrow Al + 4Cl^-.$$

The decomposition of Lewis acidic EMImCl–$AlCl_3$ ionic liquids occurs at potentials more positive than −2 V vs. Al^{3+}/Al—that is, before the potential for the reduction of the tetrachloroaluminate ion is reached. Hence, in such liquids, aluminum plates form the heavier $Al_2Cl_7^-$ anions. However, if the electrolyte is stable enough to permit the reduction of $AlCl_4^-$, the possibility opens a route for a higher voltage aluminum cell (higher by at least 1 V [268]) and exploitation of a lighter and less corrosive electrolyte. As noted earlier, the electrochemical stability window of ionic liquids is limited by those of its constituent ions, i.e., its anodic stability is limited by the oxidation of the chloroaluminate anion to chlorine, while the cathodic stability is limited by the reduction of the imidazolium cation [277]. Gelification of the Lewis neutral EMImCl–$AlCl_3$ with a polymer can widen the potential stability window beyond −2 V vs. Al^{3+}/Al, allowing the deposition of aluminum from $AlCl_4^-$ anions. Such gel forms of ionic liquids immobilized in a solid inorganic or polymeric host, or a combination of both, are called ionogels (ionic liquid + gel).

Indeed, ionogels tap the merits of solid and liquid electrolytes. These quasi-solid electrolytes are known to have good structural integrity [278], thermal stability [279], moisture sensitivity [267, 276], and extended electrochemical stability windows [273, 275]. Ionogels eliminate the use of separators, which can result in cells with lower internal resistance. Batteries with ionogels would be lighter and smaller—and leakproof—as ionogels are semi-solid. With Lewis neutral ionic liquids, aluminum batteries that use ionogels can opt for choices other than molybdenum, tantalum, and tungsten as current collectors. Schoetz et al. [273], who examined the feasibility of using an EMImCl–$AlCl_3$-PEO ionogel as an electrolyte in aluminum batteries, showed aluminum could be deposited from a Lewis neutral ionogel without dendrite growth within an extended potential stability window of 5 V, although at a low Coulombic efficiency of $\leqslant$60%. It must be noted that in electrolytes with PEO, PAN, PMMA, and PVdF, the concentration of the electrochemically active $Al_2Cl_7^-$ ions is reduced due to coordination [242]. Sun et al. [267] showed that an ionogel of EMImCl–$AlCl_3$ in PAN prepared by polymerizing acrylamide in a solution of EMImCl–$AlCl_3$ in DCM exhibited enhanced moisture stability.

10.4 SUMMARY AND OUTLOOK

Aluminum as an anode material is too attractive to be sidelined. However, beset with a host of issues especially relating to electrolytes, aluminum-based rechargeable batteries have not graduated beyond the laboratory bench. Tapping the merits of this next-generation anode requires efforts towards the development of electrolytes with wide electrochemical stability windows that can efficiently support reversible plating/stripping of aluminum without dendrite formation. It would be a good idea to wean away from expensive and corrosive chloroaluminate-based ionic liquids so as to employ a wide choice of materials for binders, current collectors, separators, and casings.

Particular focus should be laid on Lewis neutral ionic liquid electrolytes that permit aluminum plating/stripping from tetrachloroaluminate species. New ionic liquids, water-in-salt electrolytes, and chloride-free electrolytes would emerge with further research, aiding in the launch of aluminum batteries in commerce. The development of ionomers and polymer electrolytes should be given priority so as to design safe, leakproof, lightweight batteries. Breaking new ground in electrolyte chemistry is crucial.

ABBREVIATIONS

ACN acetonitrile
γ-BL gamma-butyrolactone
BMIm 1-butyl-3-methylimidazolium
BMPyr 1-butyl-1-methylpyrrolidinium

DCM	dichloromethane
DMF	dimethyl formamide
DMSO	dimethyl sulfoxide
$DMSO_2$	dimethylsulfone
EC	ethylene carbonate
EMIm	1-ethyl-3-methylimidazolium
Et_3NHCl	trimethylamine hydrochloride
HdMSO	hexa-dimethyl sulfoxide
HMIm	hexa-methylimidazolium
MPPyr	1-methyl-1-propylpyrrolidinium
OTF	trifluoromethanesulfonate
PC	propylene carbonate
PE	polyethylene
PEO	polyethylene oxide
PMMA	polymethyl methacrylate
PP	polypropylene
PTFE	polytetrafluoroethylene
PTHF	polytetrahydrofuran
PVA	polyvinyl alcohol
PVdF	polyvinylidene difluoride
RTIL	room temperature ionic liquid
SEI	solid electrolyte interphase
TFSI	bis(trifluoromethane sulfonyl)imide
WISE	water-in-salt electrolyte

REFERENCES

1. K.E. Boulding, The Economics of the Coming Spaceship Earth. In: H. Jarrett (ed.), *Environmental Quality in a Growing Economy*, Johns Hopkins University Press, Baltimore, MD (1966) pp. 3–14.
2. E.A. Olivetti, G. Ceder, G.G. Gaustad and X. Fu, *Joule*, 2017, 1, 229–243.
3. P. Greim, A.A. Solomon and C. Breyer, *Nat. Commun.*, 2020, 11, 1–11.
4. K. Turcheniuk, D. Bondarev, G.G. Amatucci and G. Yushin, *Mater. Today*, 2021, 42, 57–72.
5. G.M. Mudd, *Sustainability*, 2021, 13, 10855.
6. A. Zeng, W. Chen, K.D. Rasmussen, X. Zhu, M. Lundhaug, D.B. Muller, J. Tan, J.K. Keiding, L. Liu, T. Dai and A. Wang, *Nat. Commun.*, 2022, 13, 1–11.
7. P. Canepa, G.S. Gautam, D.C. Hannah, R. Malik, M. Liu, K.G. Gallagher, K.A. Persson and G. Ceder, *Chem. Rev.*, 2017, 117, 4287–4341.
8. C.B. Bucur, T. Gregory, A. G. Oliver and J. Muldoon, *J. Phys. Chem. Lett.*, 2015, 6, 3578–3591.
9. P.W. Jaschin, Y. Gao, Y. Li and S.-H. Bo, *J. Mater. Chem. A*, 2020, 8, 2875–2897.
10. D. Wang, X. Gao, Y. Chen, L. Jin, C. Kuss and P.G. Bruce, *Nat. Mater.*, 2018, 17, 16–20.
11. D. Aurbach, R. Skaletsky and Y. Gofer, *J. Electrochem. Soc.*, 1991, 138, 3536–3545.
12. A. Ponrouch, C. Frontera, F. Barde and M.R. Palacin, *Nat. Mater.*, 2016, 15, 169–173.
13. A. Ponrouch and M.R. Palacin, *Curr. Opin. Electrochem.*, 2018, 9, 1–7.
14. A.L. Lipson, B. Pan, S.H. Lapidus, C. Liao, J.T. Vaughey and B.J. Ingram, *Chem. Mater.*, 2015, 27, 8442–8447.

15. M. Hayashi, H. Arai, H. Ohtsuka and Y. Sakurai, *Electrochem. Solid State Lett.*, 2004, 7, A119–A121.
16. T. Shiga, H. Kondo, Y. Kato and M. Inoue, *J. Phys. Chem. C*, 2015, 119, 27946–27953.
17. T. Tojo, Y. Sugiura, R. Inada and Y. Sakurai, *Electrochim. Acta*, 2016, 207, 22–27.
18. J.Y. Hwang. S.T. Myung and Y.K. Sun, *Chem. Soc. Rev.*, 2017, 46, 3529–3614.
19. Z. Jian, W. Luo and X. Ji, *J. Am. Chem. Soc.*, 2015, 137, 11566–11569.
20. D. Su, A. McDonagh, S.Z. Qiao and G. Wang, *Adv. Mater.*, 2017, 29, 1604007.
21. L. Jiang, Y. Lu, C. Zhao, L Liu, J. Zhang, Q. Zhang, X. Shen, J. Zhao, X. Yu, H. Li, X. Huang, L. Chen and Y.-S. Hu, *Nat. Energy*, 2019, 4, 495–503.
22. H. Chen, H. Xu, B. Zheng, S. Wang, T. Huang, G. Guo, W. Gao and C. Gao, *ACS Appl. Mater. Interfaces*, 2017, 9, 22628–22634.
23. D. Lin, Y. Liu and Y. Cui, *Nat. Nanotechnol.*, 2017, 12, 194–206.
24. Committee TGAR. *Global Aluminum Recycling: A Cornerstone of Sustainable Development.* International Aluminum Institute, London (2009).
25. E. Faegh, B. Ng, D. Hayman and W.E. Mustain, *Nat. Energy*, 2021, 6, 21–29.
26. H. Wang, S. Gu, Y. Bai, S. Chen, N. Zhu, C. Wu and F. Wu, *J. Mater. Chem. A*, 2015, 3, 22677–22686.
27. L.D. Reed and E. Menke, *J. Electrochem. Soc.*, 2013, 160, A915–A917.
28. H. Tian, S. Zhang, Z. Meng, W. He and W.-Q. Han, *ACS Energy Lett.*, 2017, 2, 1170–1176.
29. P.K. Lai and M. Skyllas-Kazacos, *J. Electroanal. Chem.*, 1988, 248, 431–440.
30. R.T. Carlin, W. Crawford and M. Bersch, *J. Electrochem. Soc.*, 1992, 139, 2720–2726.
31. Y. Zhang, S. Liu, Y. Ji, J. Ma and H. Yu, *Adv. Mater.*, 2018, 30, 1706310.
32. H. Yang, H. Li, J. Li, Z. Sun, K. He, H.-M. Cheng and F. Li, *Angew. Chem. Int. Ed.*, 2019, 58, 11978–11996.
33. J. Shi, J. Zhang and J. Guo, *ACS Energy Lett.*, 2019, 4, 2124–2129.
34. B. Craig, T. Schoetz, A. Cruden and C.P. de Leon, *Renew. Sustain. Energy Rev.*, 2020, 133, 110100.
35. R. Buckingham, T. Asset and P. Atanassov, *J. Power Sources*, 2021, 498, 229762.
36. O.M. Leung, T. Schoetz, T. Prodromakis and C.P. de Leon, *J. Electrochem. Soc.*, 2021, 168, 056509.
37. A. Miguel, P. Jankowski, J.L. Pablos, T. Corrales, A. Lopez-Cudero, A. Bhowmik, D. Carrasco-Busturia, G. Ellis, N. García, J.M. García-Lastra and P. Tiemblo, *Polymer*, 2021, 224, 123707.
38. G.A. Elia, K.V. Kravchyk, M.V. Kovalenko, J. Chacon, A. Holland and R.G.A. Wills, *J. Power Sources*, 2021, 481, 228870.
39. J. Tu, W.-L. Song, H. Lei, Z. Yu, L.-L. Chen, M. Wang and S. Jiao, *Chem. Rev.*, 2021, 121, 4903–4961
40. S.M. Ali, N.A. Arif, M.M. Hashmi, M.B. Khan and Z.H. Khan, Recent developments in electrolyte materials for rechargeable batteries. In: Z.H. Khan (ed.), *Nanomaterials for Innovative Energy Systems and Devices. Materials Horizons: From Nature to Nanomaterials*, Springer, Singapore (2022).
41. L. Shen, X. Du, M. Ma, S. Wang, S. Huang and L. Xiong, *Adv. Sustain. Syst.*, 2022, 6, 2100418.
42. Q. Zhou, Y. Wu, J. Gautam, D. Wang, X. Jiang, Z. Ma, H. Zhang, L. Ni and G. Diao, *Coord. Chem. Rev.*, 2023, 474, 214856.
43. K.V. Kravchykand and M.V. Kovalenko, *Commun. Chem.*, 2020, 3, 120.
44. D. Tommasi, *Traite des Piles Electriques*, Georges Carre, Paris (1889) p. 185.
45. M. Hulot and M. Dumas, *C.R. Acad. Sci.*, 40 (1855) 1148.

46. P. Benjamin, *The Voltaic Cell: Its Construction and Its Capacity*, John Wiley, New York (1893).
47. G.A. Elia, K. Marquardt, K. Hoeppner, S. Fantini, R. Lin, E. Knipping, W. Peters, J.F. Drillet, S. Passerini and R. Hahn, *Adv. Mater.*, 2016, 28, 7564–7579.
48. L. Fu, N. Li, Y. Liu, W. Wang, Y. Zhu and Y. Wu, *Chin. J. Chem.*, 2017, 35, 13–20.
49. D.E. Sargent, Voltaic cell. US Pat. 2,554,447 (1951).
50. S. Ruben, Primary cell. US Pat. 2,638,489 (1953).
51. S. Ruben, Primary cell anode. US Pat. 2,796,456 (1957).
52. N.C. Cahoon and M.P. Korver, *J. Electrochem. Soc.*, 1959, 106, 469–471.
53. S. Zaromb, *J. Electrochem. Soc.*, 1962, 109, 1125–1130.
54. S. Zaromb and R.A. Foust, *J. Electrochem. Soc.*, 1962, 109, 1191–1192.
55. L. Bockstie, D. Trevethan and S. Zaromb, *J. Electrochem. Soc.*, 1963, 110, 267–271.
56. P. Giford and J. Palmisano, *J. Electrochem. Soc.*, 1988, 135, 650–654.
57. T. Nestler, E. Roedern, N.F. Nikolai, J. Hanzig, G.A. Elia and M. de Vivanco, *Phys. Sci. Rev.*, 2018, 4, 20170115.
58. Y. Hu, D. Sun, B. Luo and L. Wang, *Energy Technol.*, 2019, 7, 86–106.
59. S.M.A. Nayem, A. Ahmad, S.S. Shah, A.S. Alzahrani, A.J.S. Ahammad and M.A. Aziz, *Chem. Record*, 2022, 22, https://doi.org/10.1002/tcr.202200181
60. A. Holland, R.D. McKerracher, A. Cruden and R.G.A. Wills, *J. Appl. Electrochem.*, 2018, 48, 243–250.
61. P. Wang, Z. Chen, Z. Ji, Y. Feng, J. Wang, J. Liu, M. Hu, H. Wang, W. Gan and Y. Huang, *Chem. Eng. J.*, 2019, 373, 580–586.
62. E. Deltombe and M. Pourbaix, *Corrosion*, 1958, 14, 16–20.
63. H.-J. Oh, K.-W. Jang and C.-S. Chi, *Bull. Korean Chem. Soc.*, 1999, 20, 1340–1344.
64. C. Vargel, *Corrosion of Aluminum*, 1st ed., Elsevier Science, Oxford (2004).
65. S.-T. Myung, Y. Hitoshi and Y.-K. Sun, *J. Mater. Chem.*, 2011, 21, 9891–9911.
66. D. Yuan, J. Zhao, W. Manalastas Jr., S. Kumar and M. Srinivasan, *Nano Mater. Sci.*, 2020, 2, 248–263.
67. V.L. Martins and R.M. Torresi, *Curr. Opin. Electrochem.*, 2020, 21, 62–68.
68. A. Zhou, L. Jiang, J. Yue, Y. Tong, Q. Zhang, Z. Lin, B. Liu, C. Wu, L. Suo, Y.S. Hu, H. Li and L. Chen, *ACS Appl. Mater. Interfaces*, 2019, 11, 41356–41362.
69. O. Borodin, L. Suo, M. Gobet, X. Ren, F. Wang, A. Faraone, J. Peng, M. Olguin, M. Schroeder, M.S. Ding, E. Gobrogge, A. von Wald Cresce, S. Munoz, J.A. Dura, S. Greenbaum, C. Wang and K. Xu, *ACS Nano*, 2017, 11, 10462–10471.
70. K. Zhang, K.O. Kirlikovali, J.M. Suh, J.W. Choi, H.W. Jang, R.S. Varma and M. Shokouhimehr, *ACS Appl. Energy Mater.*, 2020, 3, 6019–6035.
71. R. Borah, F.R. Hughson, J. Johnston and T. Nann, *Mater. Today Adv.*, 2020, 6, 100046.
72. Y. Wang, X. Meng, J. Sun, Y. Liu and L. Hou, *Front. Chem.*, 2020, 8, art. 595.
73. W. Pan, Y. Wang, Y. Zhang, H.Y.H. Kwok, M. Wu, X. Zhao and D.Y.C. Leung, *J. Mater. Chem. A*, 2019, 7, 17420–17425.
74. D. Chao, W. Zhou, F. Xie, C. Ye, H. Li, M. Jaroniec and S.Z. Qiao, *Sci. Adv.*, 2020, 6, eaba4098.
75. T. Dong, K.L. Ng, Y. Wang, O. Voznyy and G. Azimi, *Adv. Energy Mater.*, 2021, 11, 2100077.
76. J. Vatamanu and O. Borodin, *J. Phys. Chem. Lett.*, 2017, 8, 4362–4367.
77. L. Suo O. Borodin, T. Gao, M. Olguin, J. Ho, X. Fan, C. Luo, C. Wang and K. Xu, *Science*, 2015, 350, 938–943.
78. Z. Hu, Y. Guo, H. Jin, H. Ji and L.J. Wan, *Chem. Commun.*, 2020, 56, 2023–2026.
79. Q. Li and N.J. Bjerrum, *J. Power Sources*, 2002, 110, 1–10.

80. Q. Zhao, M.J. Zachman, W.I. Al Sadat, J. Zheng, L.F. Kourkoutis and L.A. Archer, *Sci. Adv.*, 2018, 4, https://doi.org/10.1126/sciadv.aau8131
81. F. Wu, N. Zhu, Y. Bai, Y. Gao and C. Wu, *Green Energy Environ.*, 2018, 3, 71–77.
82. S. Kumar, V. Verma, H. Arora, W. Manalastas Jr. and M. Srinivasan, *ACS Appl. Energy Mater.*, 2020, 3, 8627–8635.
83. H. Wang, S. Gu, Y. Bai, S. Chen, F. Wu and C. Wu, *ACS Appl. Mater. Interfaces*, 2016, 8, 27444–27448.
84. T. Leisegang, F. Meutzner, M. Zschornak, W. Münchgesang, R. Schmid, T. Nestler and D.C. Meyer, *Front. Chem.*, 2019, 7, 268.
85. S.A. Rastabi, G. Razaz, M. Hummelgård, T. Carlberg, N. Blomquist, J. Ortegren and H. Olin, *J. Power Sources*, 2022, 523, 231066.
86. J.R. Davis (ed.), *Aluminum and Aluminum Alloys*, ASM International, Materials Park, Ohio (1993) p. 547.
87. G. Razaz and T. Carlberg, *Metall. Mater. Trans. A*, 2019, 50, 1873–1887.
88. O. Seri, *Mater. Sci. Forum*, 2006, 519, 729–734.
89. C. Vargel, *Corrosion of Aluminum*, 2nd ed., Elsevier, New York (2020) pp. 127–155.
90. Y.J. Cho, I.J. Park, H.J. Lee and J.G. Kim, *J. Power Sources*, 2015, 277, 370–378.
91. J. Zheng, P. Yan, D. Mei, M.H. Engelhard, S.S. Cartmell, B.J. Polzin, C. Wang, J.G. Zhang and W. Xu, *Adv. Energy Mater.*, 2016, 6, 1502151.
92. R.D. Shannon, *Acta Cryst. A*, 1976, 32, 751–767.
93. L.D. Reed, A. Arteaga and E.J. Menke, *J. Phys. Chem. B*, 2015, 119, 12677–12681.
94. E. Peled and E. Gileadi, *J. Electrochem. Soc.*, 1976, 123, 15–19.
95. L. Legrand, A. Tranchant and R. Messina, *Electrochim. Acta*, 1994, 39, 1427–1431.
96. L. Legrand, A. Tranchant and R. Messina, *J. Electrochem. Soc.*, 1994, 141, 378–381.
97. L. Legrand, M. Heintz, A. Tranchant and R. Messina, *Electrochim. Acta*, 1995, 40, 1711–1716.
98. L. Legrand, A. Tranchant and R. Messina, *Electrochim. Acta*, 1996, 41, 2715–2720.
99. Y. Nakayama, Y. Senda, H. Kawasaki, N. Koshitani, S. Hosoi, Y. Kudo, H. Morioka and M. Nagamine, *Phys. Chem. Chem. Phys.*, 2015, 17, 5758–5766.
100. M. Miyake, H. Fujii and T. Hirato, *Surf. Coat. Technol.*, 2015, 277, 160–164.
101. A. Kitada, K. Nakamura, K. Fukami and K. Murase, *Electrochemistry*, 2014, 82, 946–948.
102. A. Kitada, K. Nakamura, K. Fukami and K. Murase, *Electrochim. Acta*, 2016, 211, 561–567.
103. A. Kitada, Y. Kato, K. Fukami and K. Murase, *J. Surf. Finish. Soc. Jpn.*, 2018, 69, 310–311.
104. Z. Zhang, A. Kitada, S. Gao, K. Fukami, N. Tsuji, Z. Yao and K. Murase, *ACS Appl. Mater. Interfaces*, 2020, 12, 43289–43298.
105. B. Zhang, Z. Shi, L. Shen, A. Liu, J. Xu and X. Hu, *J. Electrochem. Soc.*, 2018, 165, D321–D327.
106. X. Wen, Y. Liu, D. Xu, Y. Zhao, R.K. Lake and J. Guo, *J. Phys. Chem. Lett.*, 2020, 11, 1589–1593.
107. D.E. Couch and A. Brenner, *J. Electrochem. Soc.*, 1952, 99, 234–244.
108. M. Yoshio and N. Ishibashi, *J. Appl. Electrochem.*, 1973, 3, 321–325.
109. M.W.M. Graef, *J. Electrochem. Soc.*, 1985, 132, 1038–1046.
110. M.C. Lefebvre and B.E. Conway, *J. Electroanal. Chem.*, 2000, 480, 34–45.
111. M.C. Lefebvre and B.E. Conway, *J. Electroanal. Chem.*, 2000, 480, 46–58.
112. N. Ishibashi and M. Yoshio, *Electrochim. Acta*, 1972, 17, 1343–1352.
113. W.A. Badawy, B.A. Sabrah and N.H.Y. Hilal, *J. Appl. Electrochem.*, 1987, 17, 357–369.
114. G.A. Capuano and W.G. Davenport, *J. Electrochem. Soc.*, 1971, 118, 1688–1694.

115. S.P. Shavkunov and T.L. Strugova, *Russ. J. Electrochem.*, 2003, 39, 642–649.
116. T. Hisano, T. Terazawa, I. Takeuchi, S. Inohara and H. Ikeda, *Bull. Chem. Soc. Jpn.*, 1971, 44, 599–603.
117. N. Yitzhack, P. Tereschuk, N. Sezin, D. Starosvetsky, A. Natan and Y. Ein-Eli, *J. Solid State Electrochem.*, 2020, 24, 2833–2846.
118. M. Chiku, H. Takeda, S. Matsumura, E. Higuchi and H. Inoue, *ACS Appl. Mater. Interfaces*, 2015, 7, 24385–24389.
119. M. Chiku, S. Matsumura, H. Takeda, E. Higuchi and H. Inoue, *J. Electrochem. Soc.*, 2017, 164, A1841–A1844.
120. L.D. Reed, S.N. Ortiz, M. Xiong and E.J. Menke, *Chem. Commun.*, 2015, 51, 14397–14400.
121. E. Zhang, W. Cao, B. Wang, X. Yu, L. Wang, Z. Xu and B. Lu, *Energy Storage Mater.*, 2018, 11, 91–99.
122. Y. Zhao and T.J. VanderNoot, *Electrochim. Acta*, 1997, 42, 3–13.
123. M. Galova, *Surf. Technol.*, 1980, 11, 357–359.
124. L. Legrand, A. Tranchant, R. Messina, F. Romain and A. Lautie, *Inorg. Chem.*, 1996, 35, 1310–1312.
125. T. Hirato, J. Fransaer and J.-P. Celis, *J. Electrochem. Soc.*, 2001, 148, C280–C283.
126. J. Fransaer, E. Leunis, T. Hirato and J-P. Celis, *J. Appl. Electrochem.*, 2002, 32, 123–128.
127. S. Shiomi, M. Miyake and T. Hirato, *J. Electrochem. Soc.*, 2012, 159, D225–D229.
128. T.R.J. Griffiths, *Chem. Commun.*, 1967, 1222–1223.
129. J. Hennion and J. Nicole, *Bull. Soc. Chim. Fr.*, 1978, 426–434.
130. P.G. Balakrishnan, R. Ramesh and T. Prem Kumar, *J. Power Sources*, 2006, 155, 401–414.
131. M. Galiński, A. Lewandowski and I. Stępniak, *Electrochim. Acta*, 2006, 51, 5567–5580.
132. M. Armand, F. Endres, D.R. MacFarlane, H. Ohno and B. Scrosati, *Nat. Mater.*, 2009, 8, 621–629.
133. Z. Lei, B. Chen, Y.-M. Koo and D.R. MacFarlane, *Chem. Rev.*, 2017, 117, 6633–6635.
134. J. Pešić, M. Watson, S. Papović and M. Vraneš, *Recent Pat. Nanotechnol.*, 2021, 15, 225–244.
135. G. Kamath, B. Narayanan, S.K.R.S. Sankaranarayanan, *Phys. Chem. Chem. Phys.*, 2014, 16, 20387–20391.
136. T. Schoetz, C. Ponce de Leon, M. Ueda and A. Bund, *J. Electrochem. Soc.*, 2017, 164, A3499–A3502.
137. M.C. Lin, M. Gong, B. Lu, Y. Wu, D.Y. Wang, M. Guan, M. Angell, C. Chen, J. Yang, B.-J. Hwang and H. Dai, *Nature*, 2015, 520, 325–328.
138. H. Sun, W. Wang, Z. Yu, Y. Yuan, S. Wang and S. Jiao, *Chem. Commun.*, 2015, 51, 11892–11895.
139. H. Wang, Y. Bai, S. Chen, X. Luo, C. Wu, F. Wu, J. Lu and K. Amine, *ACS Appl. Mater. Interfaces*, 2015, 7, 80–84.
140. T. Tsuda, G.R. Stafford and C.L. Hussey, *J. Electrochem. Soc.*, 2017, 164, H5007–H5017.
141. D.R. Egan, C.P. de Leon, R.J.K. Wood, R.L. Jones, K.R. Stokes and F.C. Walsh, *J. Power Sources*, 2013, 236, 293–310.
142. S.Z. El Abedin, P. Giridhar, P. Schwab and F. Endres, *Electrochem. Commun.*, 2010, 12, 1084–1086.
143. T. Welton, *Chem. Rev.*, 1999, 99, 2071–2084.
144. P. Wasserscheid and T. Welton, *Ionic Liquids in Synthesis*, Wiley-VCH Verlag GmbH & Co. KGaA, Germany (2008).
145. N. V. Plechkova and K.R. Seddon, *Chem. Soc. Rev.*, 2008, 37, 123–150.
146. D.S. Silvester and R.G. Compton, *Int. J. Res. Phys. Chem. Chem. Phys.*, 2006, 220, 1247–1274.

147. N. Kishimoto, T. Fukutsuka, Y. Uchimoto, T. Abe, Z. Ogumi, M. Inaba and A. Tasaka, *214th ECS Meeting*, Honolulu, Hawaii (2008) abstr. 571.
148. N.D. Ingale, J.W. Gallaway, D. Steingart, S. Banerjee and A. Couzis, *220th ECS Meeting*, Boston, Massachusetts (2011) abstr. 2281.
149. S. Takahashi, L.A. Curtiss, D. Gosztola, N. Koura and M.-L. Saboungi, *Inorg. Chem.*, 1995, 34, 2990–2993.
150. K.V. Kravchyk, C. Seno and M.V. Kovalenko, *ACS Energy Lett.*, 2020, 5, 545–549.
151. G. Cohn, L. Ma and L.A. Archer, *J. Power Sources*, 2015, 283, 416–422.
152. T. Gao, X. Li, X. Wang, J. Hu, X. Fan, L. Suo, A.J. Pearse, S.B. Lee, G.W. Rubloff, K.J. Gaskell, M. Noked and C. Wang, *Angew. Chem. Int. Ed.*, 2016, 55, 9898–9901.
153. W. Wang, Z. Cao, G.A. Elia, Y. Wu, W. Wahyudi, E. Abou-Hamad, A.H. Emwas, L. Cavallo, L.J. Li and J. Ming, *ACS Energy Lett.*, 2018, 3, 2899–2907.
154. Y. Guo, H. Jin, Z Qi, Z Hu, H. Ji and L.-J. Wan, *Adv. Funct. Mater.*, 2019, 29, 1807676.
155. K. Zhang, T.H. Lee, J.W. Cha, R.S. Varma, J.-W. Choi, H.W. Jang and M. Shokouhimehr, *Sci. Rep.*, 2019, 9, 13573.
156. Y. Guo, Z. Hu, J. Wang, Z. Peng, J. Zhu, H. Ji, and L.-J. Wan, *Angew. Chem. Int. Ed.*, 2020, 132, 23163–23167.
157. X. Zheng, R. Tang, Y. Zhang, L. Ma, X. Wang, Y. Dong, G. Kong and L. Wei, *Sustain. Energy Fuel.*, 2020, 4, 1630–1641.
158. J. Smajic, S. Wee, F.R.F. Simoes. M.N. Hedhili, N. Wehbe, E. Abou-Hamad and P.M.F.J. Costa, *ACS Appl. Energy Mater.*, 2020, 3, 6805–6814.
159. Y. Zhang, L. Ma, R. Tang, X. Zheng, X. Wang, Y. Dong, G. Kong, F. Zhao and L. Wei, *Int. J. Hydrog. Energy*, 2021, 46, 4936–4946.
160. X. Xiao, J. Tu, Z. Huang and S. Jiao, *Phys. Chem. Chem. Phys.*, 2021, 23, 10326–10334.
161. M. Klimpel, M.V. Kovalenko and K.V. Kravchyk, *Commun. Chem.*, 2022, 5, 77.
162. Y. Liang, H. Dong, D. Aurbach and Y. Yao, *Nat. Energy*, 2020, 5, 646–656.
163. G.A. Elia, I. Hasa, G. Greco, T. Diemant, K. Marquardt, K. Hoeppner, R.J. Behm, A. Hoell, S. Passerini and R. Hahn, *J. Mater. Chem. A*, 2017, 5, 9682–9690.
164. C. Li, J. Patra, J. Li, P.C. Rath, M.H. Lin and J.K. Chang, *Adv. Funct. Mater.*, 2020, 30, 1909565.
165. K.K. Maniam and S. Paul, *Coatings*, 2021, 11, 80.
166. T. Jiang, M.J. ChollierBrym, G. Dubé, A. Lasia and G.M. Brisard, *Surf. Coat. Tech.*, 2006, 201, 1–9.
167. Y. Long, H. Li, M. Ye, Z. Chen, Z. Wang, Y. Tao, Z. Weng, S.-Z. Qiao and Q.-H. Yang, *Energy Storage Mater.*, 2021, 34, 194–202.
168. K.L. Ng, T. Dong, J. Anawati and G. Azimi, *Adv. Sustain. Sys.*, 2020, 4, 2000074.
169. Q. Wang, Q. Zhang, X. Lu and S. Zhang, *Ionics*, 2017, 23, 2449–2455.
170. G. Zhu, M. Angell, C.-J. Pan, M.-C. Lin, H. Chen, C.-J. Huang, J. Lin, A. J. Achazi, P. Kaghazchi, B.-J. Hwang and H. Dai, *RSC Adv.*, 2019, 9, 11322–11330.
171. C. Yang, S. Wang, X. Zhang, Q. Zhang, W. Ma, S. Yu and G. Sun, *J. Phys. Chem. C*, 2019, 123, 11522–11528.
172. S. Xia, X.-M. Zhang, K. Huang, Y.-L. Chen and Y.-T. Wu, *J. Electroanal. Chem.*, 2015, 757, 167–175.
173. C.-H. Tseng, J.-K. Chang, J.-R. Chen, W.T. Tsai, M.-J. Deng and I-W. Sun, *Electrochem. Commun.*, 2010, 12, 1091–1094.
174. P.C. Lin, I.W. Sun, J.K. Chang, C.J. Su and J.C. Lin, *Corros. Sci.*, 2011, 53, 4318–4323.
175. Y. Oh, G. Lee and Y. Tak, *ChemElectroChem*, 2018, 5, 3348–3352.
176. H. Xu, T. Bai, H. Chen, F. Guo, J. Xi, T. Huang, S. Cai, X. Chu, J. Ling, W. Gao, Z. Xu and C. Gao, *Energy Storage Mater.*, 2019, 17, 38–45.
177. N.M. Rocher, E.I. Izgorodina, T. Ruther, M. Forsyth, D.R. MacFarlane, T. Rodopoulos, M.D. Horne and A.M. Bond, *Chem. Eur. J.*, 2009, 15, 3435–3447.

178. P. Eiden, Q. Liu, S. Z. El Abedin, F. Endres, I. Krossing, S. Zein El Abedin, F. Endres and I. Krossing, *Chem. Eur. J.*, 2009, 15, 3426–3434.
179. S.Z. El Abedin, E.M. Moustafa, R. Hempelmann, H. Natter and F. Endres, *ChemPhysChem*, 2006, 7, 1535–1543.
180. T. Mandai and P. Johansson, *J. Phys. Chem. C*, 2016, 120, 21285–21292.
181. Z. Slim and E.J. Menke, *J. Phys. Chem. C*, 2022, 126, 2365–2373.
182. X. Wen, J. Zhang, H. Luo, J. Shi, C. Tsay, H. Jiang, Y.-H. Lin, M.A. Schroeder, K. Xu and J. Guo, *J. Phys. Chem. Lett.*, 2021, 12, 5903–5908.
183. J. Krummacher, L.H. Hess and A. Balducci, *J. Electrochem. Soc.*, 2019, 166, A1763–A1768.
184. N. Jayaprakash, S.K. Das and L.A. Archer, *Chem. Commun.*, 2011, 47, 12610–12612.
185. T. Mandai and P. Johansson, *J. Mater. Chem. A*, 2015, 3, 12230–12239.
186. Z. Slim and E.J. Menke, *J. Phys. Chem. B*, 2020, 124, 5002–5008.
187. C. Wu, S. Gu, Q. Zhang, Y. Bai, M. Li, Y. Yuan, H. Wang, X. Liu, Y. Yuan, N. Zhu and F. Wu, *Nat. Commun.*, 2019, 10, 1–10.
188. C. Yan, C. Lv, L. Wang, W. Cui, L. Zhang, K.N. Dinh, H. Tan, C. Wu, T. Wu and Y. Ren, *J. Am. Chem. Soc.*, 2020, 142, 15295–15304.
189. J.S. Wilkes and M.J. Zaworotko, *Chem. Commun.*, 1992, 965–967.
190. H.J.T. Ellingham, *J. Soc. Chem. Ind.*, 1944, 63, 125–133.
191. C. Wu, S. Gu, Q. Zhang, Y. Bai, M. Li, Y.Y. Yuan, H. Wang, X. Liu, Y.Y. Yuan, N. Zhu, F. Wu, H. Li, L. Gu and J. Lu, *Nat. Commun.*, 2019, 10, 73.
192. M.C. Huang, C.H. Yang, C.C. Chiang, S.C. Chiu, Y.F. Chen, C.Y. Lin, L.Y. Wang, Y.L. Li, C.C. Yang and W.S. Chang, *Energies*, 2018, 11, 2760.
193. Y. Fang, K. Yoshii, X. Jiang, X. G. Sun, T. Tsuda, N. Mehio and S. Dai, *Electrochim. Acta*, 2015, 160, 82–88.
194. T. Yao, F.S. Genier, S. Biria and I.D. Hosein, *Res. Phys.*, 2018, 10, 529–531.
195. M. Alamgir, R.T. Moulton and K.M. Abraham, *Electrochim. Acta*, 1991, 36, 773–782.
196. C.A. Furtado, G. Silva, M.A. Pimenta and J.C. Machado, *Electrochim. Acta*, 1998, 43, 1477–1480.
197. O. Akbulut, I. Taniguchi, S. Kumar, Y. Shao-Horn and A. M. Mayes, *Electrochim. Acta*, 2007, 52, 1983–1989.
198. A. Hartwig and M. Sebald, *Eur. Polym. J.*, 2003, 39, 1975–1981.
199. T. Gao, X. Li, X. Wang, J. Hu, F. Han, X. Fan, L. Suo, A.J. Pearse, S.B. Lee, G.W. Rubloff, K.J. Gaskell, M. Noked and C. Wang, *Angew. Chem. Int. Ed.*, 2016, 128, 10052–10055.
200. E. L. Smith, A. P. Abbott and K. S. Ryder, *Chem. Rev.*, 2014, 114, 11060–11082.
201. H.M.A. Abood, A.P. Abbott, A.D. Ballantyne and K.S. Ryder, *Chem. Commun.*, 2011, 47, 3523–3525.
202. C. Russ and B. Konig, *Green Chem.*, 2012, 14, 2969–2982.
203. W. Chu, X. Zhang, J. Wang, S. Zhao, S. Liu and H. Yu, *Energy Storage Mater.*, 2019, 22, 418–423.
204. F. Coleman, G. Srinivasan and M. Swadzba-Kwasny, *Angew. Chem. Int. Ed.*, 2013, 52, 12582–12586.
205. P.C. Hu, R. Zhang, X.H. Meng, H.Y. Liu, C.M. Xu and Z.C. Liu, *Inorg. Chem.*, 2016, 55, 2374–2380.
206. C.Y. Liu, W.T. Chen, Z.M. Wu, B.L. Gao, X.W. Hu, Z.N. Shi and Z.W. Wang, *J. Mol. Liq.*, 2017, 247, 57–63.
207. P.C. Hu, W. Jiang, L.J. Zhong and S.F. Zhou, *RSC Adv.*, 2018, 8, 13248–13252.
208. A.P. Abbott, R.C. Harris, Y.T. Hsieh, K.S. Ryder and I.W. Sun, *Phys. Chem. Chem. Phys.*, 2014, 16, 14675–14681.
209. M. Li, B.L. Gao, C.Y. Liu, W.T. Chen, Z.W. Wang, Z.N. Shi and X.W. Hu, *J. Solid State Electrochem.*, 2017, 21, 469–476.

210. L.Y. Zhang, C.K. Zhang, Y. Ding, K. Ramirez-Meyers and G.H. Yu, *Joule*, 2017, 1, 623–633.
211. H.D. Jiao, C. Wang, J.G. Tu, D.H. Tian and S.Q. Jiao, *Chem. Commun.*, 2017, 53, 2331–2334.
212. M. Angell, C. J. Pan, Y. Rong, C. Yuan, M. C. Lin, B. J. Hwang and H. Dai, *Proc. Natl Acad. Sci. USA*, 2017, 114, 834–839.
213. C.K. Zhang, Y. Ding, L.Y. Zhang, X.L. Wang, Y. Zhao, X.H. Zhang and G.H. Yu, *Angew. Chem. Int. Ed.*, 2017, 56, 7454–7459.
214. C. Wang, J. Li, H. Jiao, J. Tu and S. Jiao, *RSC Adv.*, 2017, 7, 32288–32293.
215. M. Angell, G. Zhu, M. C. Lin, Y. Rong and H. Dai, *Adv. Funct. Mater.*, 2019, 30, 1901928.
216. N. Canever, N. Bertrand and T. Nann, *Chem. Commun.*, 2018, 54, 11725–11728.
217. M. Li, B.L. Gao, W.T. Chen, C.Y. Liu, Z.W. Wang, Z.N. Shi and X.W. Hu, *Electrochim. Acta*, 2015, 185, 148–155.
218. J. Lampkin, H. Li, L. Furness, R. Raccichini and N. Garcia-Araez, *ChemSusChem*, 2020, 13, 3514–3523.
219. Y. Bian, Y. Li, Z. Yu, H. Chen, K. Du, C. Qiu, G. Zhang, Z. Lv and M.-C. Lin, *ChemElectroChem*, 2018, 5, 3607–3611.
220. H.C. Ko, *Low-Temperature Heat Capacities and Enthalpy of Formation of Aluminumsulfide (Al_2S_3)*, Department of the Interior, Bureau of Mines, USA (1976).
221. R.C. Sharma and Y.A. Chang, *Bull. Alloy Phase Diagr.*, 1987, 8, 128–131.
222. K. Sasvari, *Acta Phys. Acad. Sci. Hungar.*, 1958, 9, 195–202.
223. C.R. Boston, Advances in molten salt chemistry. In: J. Braunstein, G. Mamantov, G.P. Smith (eds.), *Advances in Molten Salt Chemistry*, Springer, New York (1971) pp. 129–163.
224. F.H. Hurley and T.P. Wier, *J. Electrochem. Soc.*, 1951, 98, 207–212.
225. R. Midorikawa, *Denki Kagaku*, 23 (1955) 127–129.
226. G.R. Stafford, *J. Electrochem. Soc.*, 1989, 136, 635–639.
227. R.C. Howie and D.W. Macmillan, *J. Appl. Electrochem.*, 1972, 2, 217–222.
228. Q. Li, H.A. Hjuler, R.W. Berg and N.J. Bjerrum, *J. Electrochem. Soc.*, 1990, 137, 2794–2798.
229. N. Koura, *J. Electrochem. Soc.*, 1980, 127, 1529–1531.
230. M. Berrettoni, R. Tossici, S. Zamponi, R. Marassi and G. Mamantov, *J. Electrochem. Soc.*, 1993, 140, 969–973.
231. Q. Pang, J. Meng, S. Gupta, X. Hong, C.Y. Kwok, J. Zhao, Y. Jin, L. Xu, O. Karahan, Z. Wang, S. Toll, L. Mai, L.F. Nazar, M. Balasubramanian, B. Narayanan and D.R. Sadoway, *Nature*, 2022, 608, 704–711.
232. G. Lu, T. Lai, M. He and X. Liu, *Chin. Sci. Bull.*, 2019, 65, 641–648.
233. Y. Song, D. Jiao, J. Tu, J. Wang, Y. Liu, H. Jiao, X. Mao, Z. Guo and D.J. Fray, *J. Mater. Chem. A*, 2017, 5, 1282–1291.
234. J. Tu, S. Wang, S. Li, C. Wang, D. Sun and S. Jwiao, *J. Electrochem. Soc.*, 2017, 164, A3292–A3302.
235. J. Wang, X. Zhang, W. Chu, S. Liu and H. Yu, *Chem. Commun.*, 2019, 55, 2138–2141.
236. J. Wang, H. Jiao, W.-L. Song, M. Wang, J. Tu, Z. Tangand H. Zhu, *ACS Appl. Mater. Interfaces*, 2020, 12, 15063–15070.
237. C.Y. Chen, T. Tsuda, S. Kuwabata and C.L. Hussey, *Chem. Commun.*, 2018, 54, 4164–4167.
238. V.V. Kulish, D. Koch and S. Manzhos, *Phys. Chem. Chem. Phys.*, 2017, 19, 6076–6081.
239. S. Sattar, B. Ghosal, M. Underwood, H. Mertwoy, M. Saltzberg, W. Frydrych, G. Rohrer and G. Farrington, *J. Solid State Chem.*, 1986, 65, 231–240.

240. Y. Kobayashi, T. Egawa, S. Tamura, N. Imanaka and G. Adachi, *Chem. Mater.*, 1997, 9, 1649–1654.
241. N. Imanaka, M. Hieaiwa, S. Tamura, G. Adachi, H. Dabkowska, A. Dabkowski and J.E. Greedan, *Chem. Mater.*, 1998, 10, 2542–2545.
242. S. Tamura, T. Egawa, Y. Okazaki, Y. Kobayashi, N. Imanaka and G. Adachi, *Chem. Mater.*, 1998, 10, 1958–1962.
243. N. Imanaka, Y. Kobayashi, K. Fujiwara, T. Asano, Y. Okazaki and G. Adachi, *Chem. Mater.*, 1998, 10, 2006–2012.
244. J. Kohler, N. Imanaka and G. Adachi, *J. Mater. Chem.*, 1999, 9, 1357–1362.
245. H. Dabkowska, A. Dabkowski, J.E. Greedan, N. Imanaka, M. Hiraiwa, S. Tamura and G. Adachi, *J. Cryst. Growth*, 1999, 197, 879–882.
246. Y. Miyajima, T. Miyoshi, J. Tamaki, M. Matsuoka, Y. Yamamoto, C. Masquelier, M. Tabuchi, Y. Saito and H. Kageyama, *Solid State Ion.*, 1999, 124, 201–211.
247. N. Sakai, K. Toda and M. Sato, *Electrochem.*, 2000, 68, 504–506.
248. N. Imanaka, Y. Kobayashi, S. Tamura and G. Adachi, *Solid State Ion.*, 2000, 136–137, 319–324.
249. S. Tamura, N. Imanaka and G. Adachi, *J. Mater. Sci. Lett.*, 2001, 20, 2123–2125.
250. S. Tamura, N. Imanaka and G. Adachi, *Solid State Ion.*, 2002, 154–155, 767–771.
251. Y. Hasegawa, S. Tamura, N. Imanaka and G. Adachi, *J. Alloys Compd.*, 2004, 379, 262–265.
252. N. Imanaka, S. Tamura and N. Nunotani, Multivalence cation conductors. In: *Encyclopedia of Applied Electrochemistry*, G. Kreysa, K. Ota and R.F. Savinell (eds.), Springer, New York (2014) pp. 1334–1339.
253. N. Imanaka, Y. Kobayashi and G. Adachi, *Chem. Lett.*, 1995, 24, 433–434.
254. N. Imanaka and G. Adachi, *J. Alloys Compds.*, 1997, 250, 492–500.
255. N. Imanaka, M. Hiraiwa, S. Tamura, G. Adachi, H. Dabkowska and A. Dabkowski, *J. Mater. Sci.*, 2002, 37, 3483–3487.
256. N. Imanaka, Y. Hasegawa, M. Yamaguchi, M. Itaya, S. Tamura and G. Adachi, *Chem. Mater.*, 2002, 14, 4481–4483.
257. N. Imanaka and S. Tamura, *Bull. Chem. Soc. Jpn.*, 2011, 84, 353–362.
258. D.J. Driscoll, M.S. Islam and P.R. Slater, *Solid State Ion.*, 2005, 176, 539–546.
259. S. Adams, *Solid State Ion.*, 2006, 177, 1625–1630.
260. Y. Zhou, S. Adams, R.P. Rao, D.D. Edwards, A. Neiman and N. Pestereva, *Chem. Mater.*, 2008, 20, 6335–6345.
261. Y. Zhou, R.P. Rao and S. Adams, *Monatshefte für Chemie – Chemical Monthly*, 2009, 140, 1017–1023.
262. Y. Zhou, R.P. Rao and S. Adams, *Solid State Ion.*, 2011, 192, 34–37.
263. Y. Zhou, A. Neiman and S. Adams, *Phys. Status Solid.*, 2011, 248, 130–135.
264. A.Y. Neiman, N. Pestereva, Y. Zhou, D. Nechaev, E. Koteneva, K. Vanec, B. Higgins, N. Volkova and I. Korchuganova, *Rus. J. Electrochem.*, 2013, 49, 895–907.
265. T. Nestler, S. Fedotov, T. Leisegang and D.C. Meyer, *Crit. Rev. Solid State Mater. Sci.*, 2019, 44, 298–323.
266. D.E. Fenton, J.M. Parker and P.V. Wright, *Polymer*, 1973, 14, 589–591.
267. X. Sun, Y. Fang, X. Jiang, K. Yoshii, T. Tsuda and S. Dai, *Chem. Commun.*, 2016, 52, 292–295.
268. S. Song, M. Kotobuki, F. Zheng, Q. Li, C. Xu, Y. Wang, W.D.Z. Li, N. Hu and L. Lu, *Solid State Ion.*, 2017, 300, 165–168.
269. G.A. Elia, J.-B. Ducros, D. Sotta, V. Delhorbe, A. Brun, K. Marquardt and R. Hahn, *ACS Appl. Mater. Interfaces*, 2017, 9, 38381–38389.
270. G.A. Elia, C.I. Acevedo, R. Kazemi, S. Fantini, R. Lin and R. Hahn, *Energy Technol.*, 2021, 9, 2100208.

271. G. Greco, D. Tatchev, A. Hoell, M. Krumrey, S. Raoux, R. Hahn and G.A. Elia, *J. Mater. Chem. A*, 2018, 6, 22673–22680.
272. G.A. Elia, N.A. Kyeremateng, K. Marquardt and R. Hahn, *Batteries Supercaps*, 2018, 2, 83–90.
273. T. Schoetz, O.M. Leung, I. Efimov, C. Zaleski, A.M. Ortega, N.G. García, P.T. Magro and C.P. de Leon, *J. Electrochem. Soc.*, 2020, 167, 040516.
274. S.K. Chaurasia, R.K. Singh and S. Chandra, *Solid State Ion.*, 2011, 183, 32–39.
275. Z. Yu, S. Jiao, J. Tu, W.-L. Song, H. Lei, H. Jiao, H. Chen and D. Fang, *J. Mater. Chem. A*, 2019, 7, 20348–20356.
276. Z. Yu, S. Jiao, S. Li, X. Chen, W.-L. Song, T. Teng, J. Tu, H.-S. Chen, G. Zhang and D.-N. Fang, *Adv. Funct. Mater.*, 2019, 29, 1806799.
277. T. Schoetz, C.P. de Leon, A. Bund and M. Ueda, *Electrochim. Acta*, 2018, 263, 176–183.
278. J. Le Bideau, L. Viau and A. Vioux, *Chem. Soc. Rev.*, 2011, 40, 907–925.
279. S.A.M. Noor, P.M. Bayley, M. Forsyth and D.R. MacFarlane, *Electrochim. Acta*, 2013, 91, 219–226.

11 2D MXene-Based Electrolytes for All-Solid-State Batteries

Jeevan Kumar Reddy Modigunta, G. Murali, Young Ho Park, Hariharan Udhayakumar, A. M. Shanmugharaj, Seung Jun Lee, and Insik In

11.1 INTRODUCTION

The demand for energy storage devices is increasing day by day with enhanced applications in interdisciplinary fields such as portable electronics, electric vehicles, space applications, and storage grids for emergency purposes [1–5]. Current available commercial secondary batteries have major applications, but due to safety and durability issues, there is a constant need for new alternatives [6, 7]. In recent years the development of energy storage devices for electric vehicles (EVs) as a replacement for fossil fuels has seen a drastic rise. The utility of energy storage devices in EVs, especially secondary storage systems like lithium-ion batteries (LIBs) and fuel cells, has made them eco-friendly.

Current LIBs mostly involve the utilization of iron- or cobalt- or/and nickel-based electrodes and organic solvent-based electrolyte, along with polymer separators. The use of liquid electrolytes in LIBs is more favorable for achieving high performance in EVs applications; however, they are more prone to side reactions within the battery with rigorous usage at different conditions like higher operating temperatures and different current rates. The side reactions in LIBs further lead to dendrite growth and damage to the separator and electrode surfaces, ultimately resulting in cell or battery failure [8, 9]. In recent times, there have been frequent reports of battery failure and explosion due to factors like overcharging, high-temperature operations, and so on. As a solution to the above issues, alternative electrolytes like polymer- or ceramic-based or a hybrid of them have been developed—and researchers are still looking for the best electrolyte combinations for commercially available electrodes.

The replacement of liquid electrolytes with solid-state electrolytes has opened the path for all-solid-state batteries (ASSBs), which addresses most of the issues generated by liquid electrolytes. ASSBs have been researched extensively to attain commercial production and applications [10]. The use of all solid components in ASSBs provides better safety and higher energy density during applications.

There have been many types of solid type electrolytes such as ceramics, oxides, sulfides, halides, phosphates, nitrides, and polymer-based solid electrolytes synthesized by either wet or dry synthesis methods [10–16]. In the dry process, mostly the

DOI: 10.1201/9781003441182-13

raw material powders are mixed and ball milled, followed by heating at elevated temperatures till the desirable structure is obtained. This is the most economical and scalable process in comparison to the wet chemical process, which involves the mixing of starting materials in appropriate solvents and finally heating under restricted conditions [17, 18]. The possible advantages and challenges in ASSBs due to components such as cathode, anode, solid electrolyte, current collector, and binders are schematically depicted in Figure 11.1a.

The ideal conditions for conventional ASSB with a dry cathode, anode, and electrolyte are schematically depicted in Figure 11.1b. The ideal current collector thickness should be ~10 μm, the anode (Li metal foil) thickness should be between 40 to 45 μm, the electrolyte should be in the range of 1 to 20 μm, and the cathode thickness should be between 45 to 200 μm, as shown in Figure 11.1b. In ASSBs, the solid-state electrolyte (SSE) decides the performance based on its ionic conductivity, electrochemical window, and compatibility with other components at the interfaces. SSEs are more advantageous in terms of safety and elevated temperature operations but have drawbacks like poor ionic conductivity and interfacial compatibility in contrast to liquid-type electrolytes.

To overcome these issues and enhance battery performance, researchers have utilized different approaches, such as the addition of different ratios of ion conductive salts, improving the structure of electrolytes, and inclusion of new additives into the components to improve the compatibility and ionic conductivity [22]. SSEs are majorly classified as solid polymer electrolytes (SPEs), solid-composite electrolytes (SCEs), and inorganic ceramic electrolytes (ICEs). SSE performance depends on conditions such as interfacial contact, dendrite restriction, thermal stability, electrochemical performance, and ionic conductivity, which are depicted in Figure 11.1c [20]. SCEs have advantages over both ICEs and SPEs.

SCEs have attracted many researchers for their advantages due to the inclusion of polymer matrix, conductive additives, and ionic-based salts/complexes. Each component of the SCEs has its own role towards the enhancement of the performance, interfacial contacts, thermal safety, and flexibility as an added advantage during the ASSBs cycling process. In SSEs, ICEs have ionic conductivity between 10^{-3}–10^{-2} S cm^{-1} under controlled conditions such as temperature and atmosphere. The presence of polymer matrix in the SSEs enhances the mechanical stability but diminishes the ionic conductivity to $\leq 10^{-4}$ S cm^{-1} at room temperature. The individual performance of the ICEs and SPEs has its own disadvantages, but the combination of ICEs and SPEs might contribute to both mechanical stability and ionic conductivity, along with boosting the safety of the ASSBs by restricting the dendrite formations [23–25].

The enhancement of the above-discussed properties of SSEs can be elevated by adding some additives along with SSEs. Studies on the inclusion of additives like metal, metal oxides, lithium salts, and 2D materials have been conducted by researchers. Among the most 2D materials, graphite and metal sulfides are used as hosts for the ions in energy storage, but they also have the additional advantage of additive materials in electrolytes. In recent studies, the 2D material MXene is attracting more attention from researchers due to its interesting and unique properties like ionic conductivity, mechanical stability, and the surface functional groups' compatibility with the functional groups and ions in the SSEs.

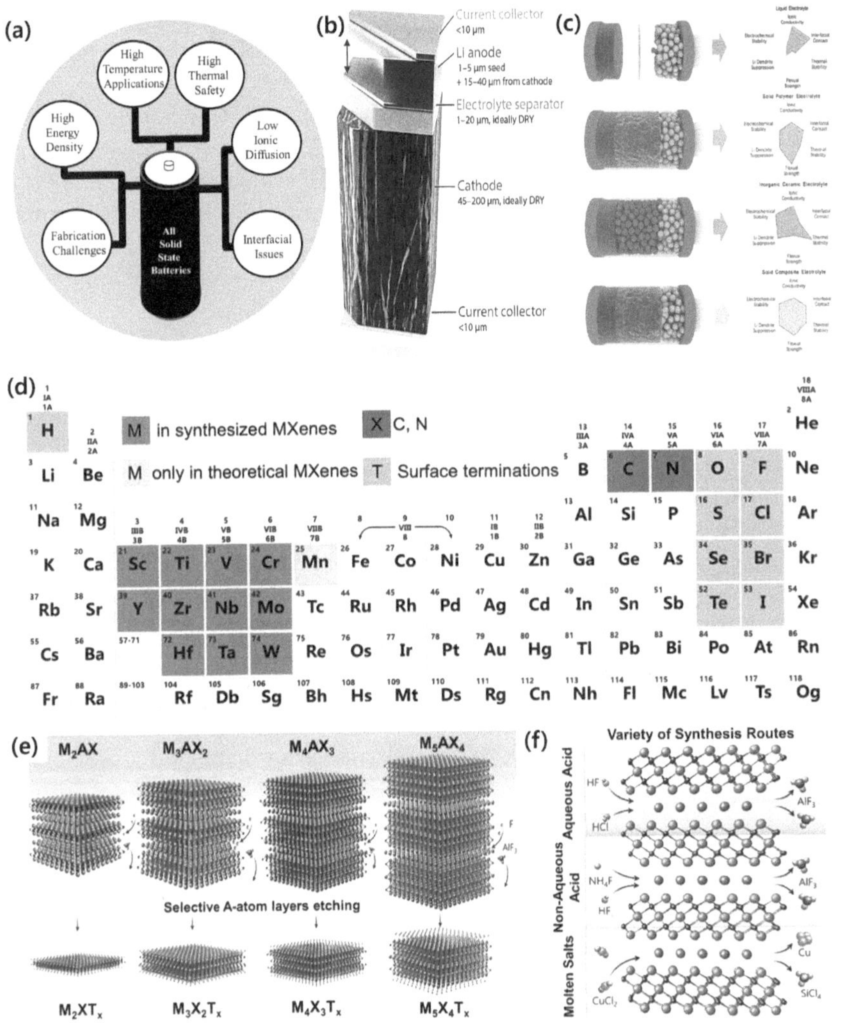

FIGURE 11.1 (a) The advantages and challenges in the all-solid-state batteries with respect to the components in the ASSBs; (b) the schematic representation of an ideal ASSB with components of exact thickness for best performance; (c) the performance estimation of SSEs depends on the different properties of ASSBs in comparison to liquid electrolytes; (d) the combination of different elements involved from the periodic table for the preparation of MXenes; (e) the schematic layered structures of various possible MXenes (M_2XT_x, $M_3X_2T_x$, $M_4X_3T_x$, and $M_5X_4T_x$) and their respective MAX phase structures ($M_{n+1}AX_n$ (n = 1–4)), where M is transition metal, X is C and/or N; T_x is the functional groups on MXenes like –H, –O, –OH, and –F, and A is p-block elements like Al and Si at the interlayers; (f) the MXene synthesis processes developed so far using aqueous acids (HF, HCl, and a mixture of HCl/HF), nonaqueous acids (mixture of NH_4F and HF), and molten salts synthesis ($CuCl_2$).

Source: (a) and (b). Reproduced with permission from ref. [19]. Copyright 2021 American Chemical Society. (c) Reproduced with permission from ref. [20]. Copyright 2020 WILEY-VCH. (d)–(f) Reproduced with permission from ref. [21]. Copyright 2022 American Chemical Society.

MXenes are one of the promising 2D materials for energy storage applications, especially as current collectors, conductive additives, coating layers on separators, and along with electrode material [26, 27]. The properties of ionic diffusion, electron conductivity, and mechanical stability of MXene have made it one of the preferential additives for electrolytes in ASSBs. In this chapter, the synthesis and characterization of MXenes have been briefly discussed prior to the SSEs containing SPEs and MXenes. SSEs made of polymers in combination with MXene have been summarized in terms of synthesis, characterization, and applications in ASSBs. A brief explanation of the factors affecting the performance of the ASSBs and improvement for interfacial compatibility was also discussed.

11.2 MXENES

MXenes are a new family of 2D transition metal carbides and/or nitrides with a general formula $M_{n+1}X_nT_x$, where M is a transition metal, X is carbon and/or nitrogen, T is the terminal functional group (–O, –OH, and/or –F), x is the number of functional groups, and n is the numerical number of X (n = 1 to 4), as shown in Figure 11.1d [21]. MXenes are synthesized from their respective precursor MAX phases, where M is a transition metal, A is a removable or etchable p-block element such as Al and Si, and X is the carbon and/or nitrogen, as shown in Figure 11.1e.

Different synthesis approaches have been reported for MXenes using aqueous, non-aqueous, and molten salt mediums, as shown in Figure 11.1f. The aqueous-based direct acid etching method is carried out by using high concentrated acids such as HF or/and HCl (Figure 11.2a). Non-aqueous medium-based synthesis approaches have been reported using NH_4F/HF. The molten salts approach also has been developed using the combination of metal salts under stochiometric conditions in order to etch different elements such as Si. After etching from the above process, a further delamination step is carried out using organic- or inorganic-based materials (like isopropyl amine, dimethyl sulfoxide, tetrabutylammonium hydroxide (TBAOH), lithium chloride, and so on) for making few/single-layered MXenes.

Apart from these approaches, the *in-situ* generation of strong acids for etching and simultaneous delamination process has been developed using the mixture of salts and acids such as LiF and HCl to obtain HF and LiF. The Li ions formed in the reaction further assists in the delamination process of MXene sheets without damaging the sheet's structure—in contrast to that of high-concentration acid-involved processes.

The LiF and HCl mixed process is named as minimally intensive layer delamination (MILD) process [29]. The detailed schematic representation of steps involved in the MXene synthesis using the MILD method is shown in Figure 11.2a and c. The different kinds of synthesized MXenes with various elemental compositions from two to four layered structures are shown in Figure 11.2b. The MXene sheet's quality and basic properties like conductivity, sheet size, and mechanical strength were good for MILD etching method synthesis in comparison to MXene sheets obtained from direct acid etching methods.

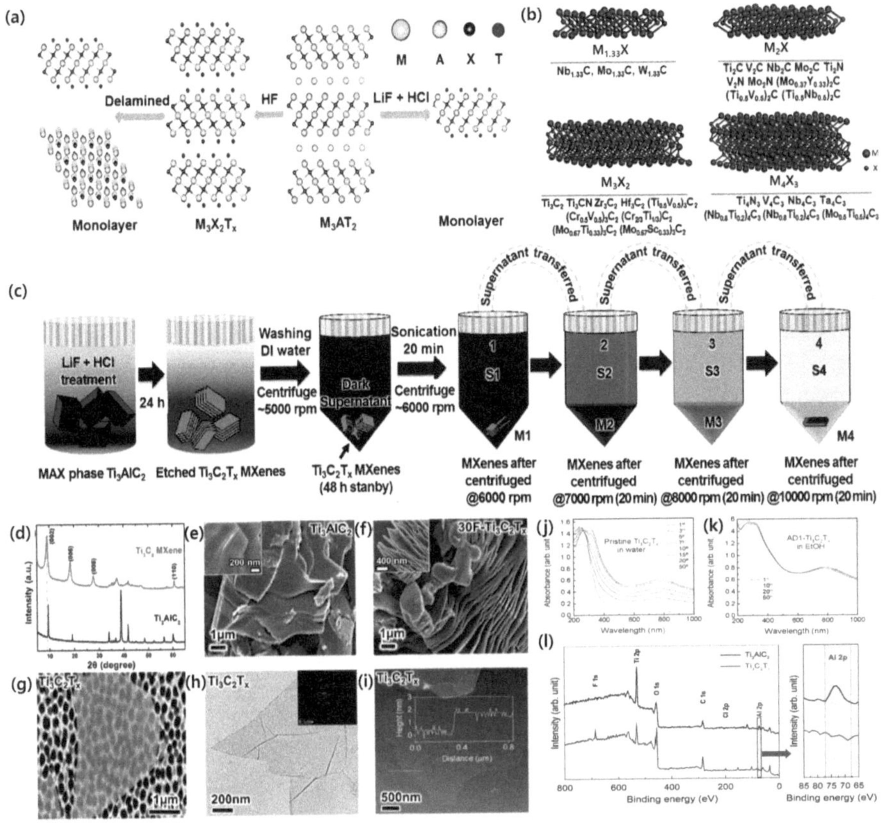

FIGURE 11.2 (a) Synthesis of MXene from MAX phase using direct acid etching method (conc. HF) and minimally intensive layer delamination (MILD) methods; (b) the various types of MXene structures with general formula; (c) the detailed process of the MILD method followed for the synthesis of $Ti_3C_2T_x$ MXene from the Ti_3AlC_2 MAX phase. Structural properties and morphology; (d) XRD of MAX phase and MXene; (e–g) SEM images of MAX phase and MXene; (h) TEM; and (i) AFM images of MXene. The optical absorption (UV-vis) spectra of (j) $Ti_3C_2T_x$ MXene in water, and (k) dopamine functionalized $Ti_3C_2T_x$ MXene in ethanol solvent as a function of storage time; (l) the XPS survey scan spectrum of Ti_3AlC_2 MAX phase (black line) and $Ti_3C_2T_x$ MXene (red line) with selective scale in the range of Al 2p binding energy.

Source: (a)–(i) Reproduced with permission from ref. [1]. Copyright 2020 Royal Society of Chemistry. (j)–(l) Reproduced with permission from ref. [28]. Copyright 2023 American Chemical Society.

Most of the above synthesis processes involve a longtime duration ranging from 2 to 72 hours with constant stirring and heating at above room temperature conditions [30]. The time taken for the synthesis of MXene has been drastically decreased using the salt-solution-based acoustic method, in which etching and exfoliation of the MAX phase. The time duration of the synthesis of MXenes has been decreased using ultrasonic-based methods, approximately 40 min (<1 h), although the MXene sheets yield and quality was lowered [31]. MXenes are hydrophilic materials due to the abundant surface functional groups, which are different from 2D graphene sheets.

Among all kinds of MXenes, $Ti_3C_2T_x$ MXene is majorly explored due to its applicability in interdisciplinary fields like photocatalysis, energy storage, and electromagnetic shielding. The surface morphology and sheet dimensions of synthesized $Ti_3C_2T_x$ MXene are characterized based on the XRD patterns, FE-SEM, TEM, and AFM images, as shown in Figure 11.2d–i. The formation of MXene sheets is identified by the peak at 8°, which is the characteristic intense peak for the (002) plane. This peak position is further shifted to a lower value (<8°) due to the increase in the interlayer distance in $Ti_3C_2T_x$ MXene sheets. The $Ti_3C_2T_x$ MXene with enhanced d-spacing value is a more favorable host for the ions used in energy storage applications such as Li, Na, Zn, and Mg.

The $Ti_3C_2T_x$ MXene sheets are commonly stored in an aqueous medium, which has dissolved oxygen and hydroxy groups that accelerate the oxidation process of Ti atoms in the $Ti_3C_2T_x$ MXene to form TiO_2. The $Ti_3C_2T_x$ MXene sheet quality and oxidation stability are determined by observing the characteristic UV-vis spectral absorption peaks at ~237, ~325, and ~785 nm, as shown in Figure 11.2j [28]. The decrease in the absorption intensity of the ~325 nm peak and the decrease in the absorption intensity of ~the 785 nm peak clearly indicates the oxidation of MXene upon prolonged storage at ambient conditions. The oxidation issues of the $Ti_3C_2T_x$ MXene can be addressed by means of surface functionalization using oxidation-resistant and hydrophilic functional groups (Figure 11.2k) such as dopamine derivatives [28] and phosphoric acid derivatives [32]. The $Ti_3C_2T_x$ MXene formation is well supported by XPS analysis, as shown in Figure 11.2l. The XPS survey scan spectrum of the Ti_3AlC_2 MAX phase shows the presence of an Al 2p peak, whereas in the case of $Ti_3C_2T_x$, has no Al 2p peak, which concludes the complete etching of the Al from the interlayers of pristine Ti_3AlC_2 MAX phase. The clear absence of the Al 2p peak is shown in the inset figure of Figure 11.2l.

11.3 APPLICABILITY OF MXENE IN ENERGY STORAGE DEVICES

MXenes have numerous applications in energy storage and conversion applications as electrodes, electrolytes, additives, and current collectors—and to improve the mechanical stability of devices [1, 33]. The role of MXene in energy storage devices as electrodes are studied majorly, especially as a host material or to improve electrical conductivity. But in recent times, MXenes are also studied as an additive in electrolytes, especially in secondary battery applications. Among all possible MXenes, $Ti_3C_2T_x$, Ti_2C, and V_2C have been studied actively in the present era [26].

11.3.1 MXene Electrolytes for All-Solid-State Batteries

The important properties to be considered for the estimation of SSEs are their ionic conductivity and ion transport number, along with the compatibility at the interfaces of the electrodes [34]. The ionic conductivity and transference numbers of the electrolytes in secondary batteries are calculated using the following equations (11.1) and (11.2).

The ionic conductivity (σ) for the electrolytes (liquid, gel, or solid) is calculated using the following equation (11.1),

$$\sigma = \frac{L}{A \times R} \tag{11.1}$$

where L and A are the thickness and surface area of the electrolyte membrane under the area of contact between the electrodes, respectively. The intersection of the semicircle from electrochemical impedance spectroscopy (EIS) analysis is fit with the axis of the real impedance part in the Nyquist plot by it as bulk resistance R. The ion transference numbers (t) of the electrolyte were measured using the symmetric cell, fabricated using the electrolyte planked between two Li metal foils. EIS was measured for that symmetric cell, before and after polarization with the DC potential. The t of the electrolyte is measured using the following formula, as shown in equation (11.2) [35, 36],

$$t = \frac{I_{ss}}{I_0} \times \frac{\Delta V - I_0 R_0}{\Delta V - I_{ss} R_{ss}} \tag{11.2}$$

where I_{ss} is the steady state current, I_0 is the initial current, R_{ss} is the resistance corresponding to the steady state current, and R_0 is the resistance corresponding to the initial resistance.

11.3.2 All Solid-State Li-ion Batteries (Li-ASSB)

In general, 2D materials have a higher specific surface area in comparison to zero and 1D materials, so they find more applications in interdisciplinary fields. Among all 2D materials such as graphene, metal sulfides, and oxides, the unique features of MXenes could easily render them into the application of SSEs. MXenes are used as the current collector and conductive additive so far in the application of LIBs. In the case of Li-ASSBs, MXene is also used as one of the additives in electrolytes to enhance ionic diffusion and good mechanical stability. The surface functional groups on MXenes make them easily compatible with the polymers and other inorganic materials in the electrolyte and as well as at the interfaces with the electrodes.

The SCEs made of polymers and inorganic materials have a hybrid nature, which is far better than their pristine forms in the application of ASSBs as electrolytes. The polymers like polyethylene oxide (PEO), polyvinylidene fluoride (PVDF), poly(vinylidene fluoride-co-hexafluoropropylene) (PVDF-HFP), poly(acrylonitrile)

(PAN), etc., are used as the SPEs [37]. These polymers have hydrophobic and hydrophilic functional groups in the polymer chains, which can easily interact with the lithium conducting salts for better diffusion of Li-ions. The hydrophilic functional groups of polymers easily interact with the MXene sheets with surface functional groups such as -OH, -F, and -O.

For instance, PEO polymer is one of the best SPEs used for energy storage and has Li-ion conductivity at high temperatures that can be enhanced by the addition of MXenes into the polymer matrix at room temperature. Figure 11.3a shows the preparation procedure for the MXene composite solid polymer electrolytes (MCPEs) from the starting materials PEO, lithium bis(trifluoromethane sulfonyl)imide (LiTFSI), and $Ti_3C_2T_x$ MXene [38]. The stoichiometric addition of $Ti_3C_2T_x$ MXene into the PEO matrix has enhanced the mechanical stability and electrochemical window. The ionic conductivity of the MCPE was 2.2×10^{-5} S m^{-1} at room temperature. The electrochemical performance of the prepared MCPE with the combination of $LiFePO_4$ cathode and Li metal foil anode has shown a specific capacity of ~140 mAh g^{-1} at 0.3C rate and 60 °C. Apart from the blending of the polymers with MXene, other fabrication methods were also followed, such as hot-press molding, dip coating, and melt-mixing.

The hot-pressing method was used to prepare the sandwich-structured solid-state polyethylene oxide (PEO)/$Ti_3C_2T_x$ MXene composite electrolyte on glass fiber cloth, as shown in Figure 11.3b. The PEO composite electrolyte containing LiTFSI and $Ti_3C_2T_x$ MXene were mixed in acetonitrile and stirred for 24 h to make a uniform solution. This solution mixture was poured onto a Teflon substrate and dried at room temperature overnight to get PEO-MXene films. The glass fiber cloth (GFC) was sandwiched between two PEO-MXene films and hot-pressed at 80 °C for 20 min to obtain the GFC@PEO-MXene SPE.

The prepared GFC@PEO-MXene SPE has good mechanical and electrochemical stabilities. The Li-ion conduction for the GFC@PEO-MXene (1 wt.%) at 60 °C was 5.01×10^{-2} S m^{-1}, which is ~200% higher than that of GFC@PEO electrolyte, which is not having MXene. The high lithium-ion transport number indicates the synergic effect of the GFC and MXene towards the PEO polymer [39]. The performance of the GFC@PEO-MXene SPE was carried out using $LiFePO_4$ as cathode and Li metal foil as anode under various temperatures. The specific capacities of the GFC@PEO and GFC@PEO-MXene SPE were 81.2 mAh g^{-1} and 110.5 mAh g^{-1} at 0.1 C rate (room temperature). The specific capacity at initial and after 200 cycles for the Li|GFC@PEO-MXene (1 wt.%)|LFP system is 166.7 mAh g^{-1} and 158.2 mAh g^{-1}, respectively, with a capacity retention of 94.9% [39]. In the case of amorphous materials like PEO-based SPEs, the diffusion of Li-ions takes place by the hopping mechanism, as shown in Figure 11.3d. The ionic conductivity of the SPEs depends on the coordination sites in the amorphous polymer chains, which allow the passage of ions in segmental motion when the electrical field is applied at above the glass transition temperature [41]. Therefore, the ionic conductivity of SPEs is majorly dependent on the crystallinity of the polymers.

In the case of crystalline materials, the ionic conductivity depends on the vacancies in the structure and permeability. The crystalline materials like ceramics and metal-organic frameworks (MOFs) such as zeolitic imidazolate framework (ZIF) are

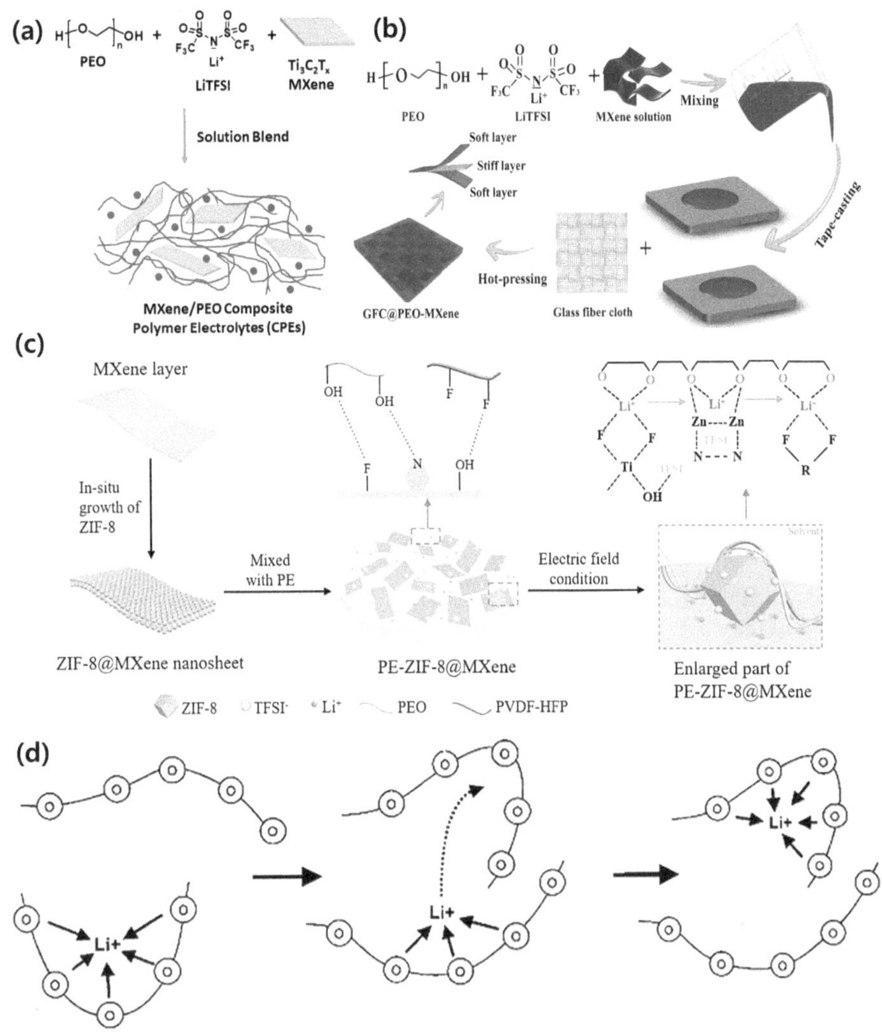

FIGURE 11.3 (a) Fabrication of MCPEs. Preparation procedure of MCPEs; (b) schematic diagram of the fabrication process of a sandwich structured GFC@PEO-MXene composite electrolyte; (c) schematic diagram of the fabrication process of ZIF-8@MXene, PE-ZIF-8@MXene, and the hydrogen-bonding interaction between ZIF-8@MXene and polymer molecules followed by the intermolecular interaction between ZIF-8@MXene, polymer molecules, and lithium salt. Schematic Illustration of Two Types of Li-ion-Conduction Mechanisms for SPEs; (d) Li-ion conduction in the amorphous phase of SPE; (e) Li-ion conduction in the crystalline phase of SPE.

Source: (a) Reproduced from reference [38] with permission from the Royal Society of Chemistry, copyright 2019. (b) Reproduced with permission from ref. [39]. Copyright 2023 Elsevier. (c) Reproduced with permission from ref. [40]. Copyright 2022 Elsevier. (d) and (e) Reproduced with permission from ref. [41]. Copyright 2018 Elsevier.

(Continued)

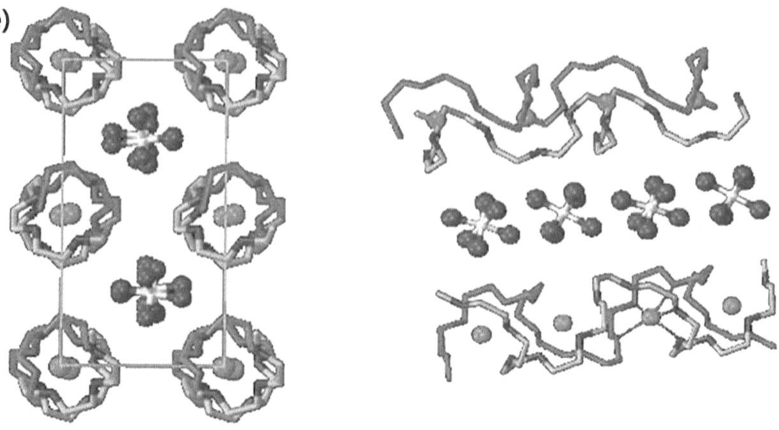

FIGURE 11.3 (Continued)

used in the SSEs, along with the ion conductive salts for enhancing the safety and mechanical stability of the batteries. For instance, Figure 11.3c shows the schematic diagram of the fabrication process of SCE (PE-ZIF-8@MXene) made from the mixture of polymers PVDF-HFP and PEO as the base polymer with the addition of ZIF-8 and MXene in the synthesis process of SCE for application in ASSBs.

The PE-ZIF-8@MXene has shown a high ionic conductivity of 4.4 mS cm^{-1} and an ionic transfer number of 0.76. The tensile strength reported was 3.77 MPa and with the additional advantage of effective restriction of the dendrite's growth during the charge-discharge process. The symmetric cells fabricated using Li on both sides of the PE-ZIF-8@MXene have shown a long cycle life of 2000 h at 0.5 mA cm^{-2} current rate application, which was 13 times longer than PE (electrolyte made of PVDF-HFP and PEO mixture). The half cells fabricated using Li|PE-ZIF-8@MXene|$LiFePO_4$ have shown a specific capacity of ~170 mAh g^{-1} and retained about 89.6% even after 500 cycles. The flame resistance, thermal stability, and mechanical stability of the PE-ZIF-8@MXene SCE are higher in comparison to the prepared control PE solid polymer electrolyte with a good porous structure. The hydrogen-bonding interaction between the ZIF-8@MXene and polymer molecules can also boost the ionic diffusion in the structures apart from the intermolecular interactions of them together with lithium salt (LiTFSI) [40].

Figure 11.3e shows the Li-ion mobility in the crystal structure of solid polymer electrolyte PEO-$LiAsF_6$. The PEO polymer chains folded together in pairs to form a cylindrical tunnel for the easy passage of Li-ions located near the ether linkages of the PEO by coordination bonds with oxygens in the structure, which allows the easy movement of Li-ions, unlike the hopping mechanism [41]. The ionic diffusion mechanism in SCEs is additionally boosted by the inclusion of lithium-based salts, which utilize the Schottky and Frenkel defects in the crystal structures. The mechanism in the crystals is further classified based on the vacancy-mediated mechanism and non-vacancy-mediated mechanism. Among both the crystal structures, the one with the least and non-defective structure needs more activation energy for the diffusion of ions in comparison to the more defective crystal structures, as the non-defective structures require higher activation energy to make a defect to increase the mobility of the ions throughout the crystal structure. This kind of defective site is easily found in MXene sheet structures due to the removal of elements during the synthesis process [21], which is more favorable for the diffusion of ions when used in the electrolytes.

MXenes are not only used upon mixing with polymers and ceramics for the preparation of electrolytes, but they are also used as surface layers on the electrolytes for enhancing the performance of ASSBs. In this regard, $LiFePO_4$ (LFP) cathode was coated with a PEO-LiTFSI mixture, and it is further coated with $Ti_3C_2T_x$ MXene solution to form MXene-PEO-LiTFSI/LFP layers, which are used against the counterpart Li anode for the fabrication of ASSBs. A similar study has been carried out by replacing LFP cathode with $LiNi_{0.8}Co_{0.1}Mn_{0.1}O_2$ (NCM811) cathode. The ASSB fabricated using Li/MXene-PEO-LiTFSI/LFP combination has shown a specific capacity of 157.1 and 142.8 mAh g^{-1} at the initial cycle and after 140 cycles at 0.5C and 60 °C [42]. The performance of this surface-coated MXene layer helps in the inhibition of TFSI anions diffusion from PEO to anode due to the electronegativity of the MXene, which results in the suppression of irreversible capacity loss and dendrites growth.

11.3.3 For Zn-Ion Batteries

Zinc ion batteries (ZIBs) are one of the widely studied aqueous-based energy storage systems due to their high safety and low price in comparison to other energy storage systems. The utility of aqueous-based electrolytes in the ZIBs makes them vulnerable to the side reactions at the Zn anode leading to the formation of Zn dendrites, corrosion, and gas formation, which results in a decrease in the self-life [43]. Aqueous-based electrolytes are currently used in ZIBs, and the addition of MXene into the electrolyte has completely restricted the formation of dendrites at the surface of the Zn anode, as shown in Figure 11.4a and b. In the case of using SSEs in Zn-ASSBs the side reactions were typically not found due to the absence of the aqueous-based components. Zn-ASSBs are also highly reliable on the electrolytes applied, similar to that of other ASSB systems for the ionic conductivity and diffusion of Zn ions. The diffusion of Zn ions is more difficult than that of other ion systems due to the strong electrostatic bonding of the divalent Zn^{2+} ions [45, 46].

The inclusion of MXenes into the electrolytes of ZIBs has enhanced ionic conductivity and mechanical stability. For instance, an SPE was synthesized using the combination of poly(vinylidene fluoride-co-hexafluoropropylene) (PVHF) matrix filled by the poly(methyl acrylate) (PMA) grafted $Ti_3C_2T_x$ MXenes (PVHF/MXene-*g*-PMA). The PVHF/MXene-*g*-PMA has shown an impressive ionic conductivity of 2.69×10^{-4} S cm^{-1} at room temperature and a high Zn^{2+} transference number of 0.16 due to the addition of MXene in the system. The PVHF/MXene-*g*-PMA SCE has shown high mechanical stability and fire resistance in comparison to the pure PVHF/MXene SCE and PVHF SPE. A highly stable and reversible Zn plating and stripping was observed for the PVHF/MXene-*g*-PMA SCEs, which attained dendrite-free cycling for ~1000 h at room temperature and 200 h at a high temperature of 55 °C [47].

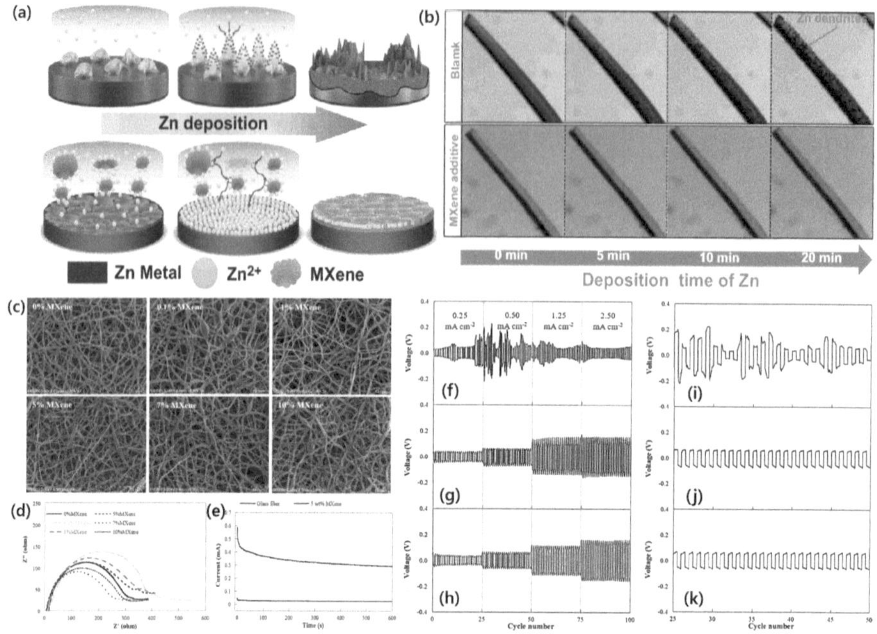

FIGURE 11.4 (a) The schematic illustration for the formation of the Zn dendrite and inhibition of dendrites on the surface of the Zn anode in the aqueous system; (b) the *in-situ* optical microscopic images depicting the Zn deposition and dendrites growth at 4 mA cm^{-2} in blank aqueous $ZnSO_4$ electrolyte and MXene-added $ZnSO_4$ electrolyte on the surface of the Zn anode; (c) the FE-SEM images showing the morphology after heating at 150 °C for 1 h; and (d) Nyquist (EIS) plots of the PAN/bio-based PU/$Ti_3C_2T_x$ MXene samples containing different weight % of $Ti_3C_2T_x$ MXene: 0 wt.%, 0.1 wt.%, 1 wt.%, 5 wt.%, 7 wt.%, and 10 wt.% MXene; (e) the representative chronoamperometry profiles for the SCE sandwiched between two Zn metal foils at room temperature with a step potential of 10 mV. The Zn symmetry cells galvanostatic charge-discharges studies of 100 cycles at 0.25, 0.5, 1.25, and 2.5 mA cm^{-2} currents for the (f) glass fiber membrane electrolyte; (g) PAN/bio-based PU; and (h) PAN/bio-based PU/MXene (5 wt.%). The voltage vs cycle profile from 25 to 50th cycle at 0.5 mA cm^{-2} current for (i) glass fiber membrane electrolyte; (j) PAN/bio-based PU; and (k) PAN/bio-based PU/MXene (5 wt.%).

Source: (a) and (b) Reproduced with permission from ref. [43]. Copyright 2021 Springer Nature. (c)–(j) Reproduced with permission from ref. [44]. copyright 2022 Springer Nature.

The PVHF/MXene-*g*-PMA SCE was used for the fabrication of full cells using manganese hexacyanoferrate (MnHCF) as cathode and Zn metal as anode. The full cell has demonstrated an HER reaction-free cycling with good cycle life performance. The full cell has shown a steady charge-discharge with a negligible amount of capacity loss at temperatures −35, −5, 25, 40, 70, and 100 °C with specific capacities of ~38, 60, 112, 121, 135, and 149 mAh g^{-1}, respectively. These results suggest that the Zn-ASSBs can work under harsh conditions such as negative temperatures to above room temperatures even after 90 days of storage in the same temperatures. These results strongly support the usage of SSEs in Zn-ASSBs due to the utilization of polymer with MXene combination, which clearly demonstrated the improved performance in terms of electrochemical stability, fire safety, and rough condition applications [47].

Similarly, PAN/bio-based polyurethane (PU)/$Ti_3C_2T_x$ MXene composite was used as the SPE and separator in Zn-ASSB for enhanced ionic diffusion and mechanical stability [44]. The fabrication of this SCE is done by preparing PAN, bio-based PU, and $Ti_3C_2T_x$ MXene all three in dimethyl formamide solvent and electrospun to obtain the solid electrolyte mat, which is further soaked into a liquid electrolyte until it is completely saturated. Later, the soaked SCE is used for the electrochemical studies. Figure 11.4c shows the FE-SEM images of PAN/bio-based PU/$Ti_3C_2T_x$ MXene composite membranes containing different weight ratios of $Ti_3C_2T_x$ MXene such as 0 wt.%, 0.1 wt.%, 1 wt.%, 5wt.%, 7 wt.%, and 10 wt.% after drying at 150 °C for 1h to remove the trapped solvents. The heating at high temperatures did not change the morphology or any damage to SCE, proving the high thermal stability of the samples. The samples after dying were soaked into the electrolyte till it was completely saturated and used for the fabrication of Zn-ASSBs.

The heating test of the fabricated SCEs has shown that thermal and fire retardant stability even at operating temperatures above 100 °C can sustain the fabricated devices without short circuits and save them from fire accidents. Figure 11.4d indicates the Nyquist (electrochemical impedance spectra) plot of PAN/bio-based PU/$Ti_3C_2T_x$ MXene SCEs, in which a sample with 5wt.% sample shows the least resistance at the interface of electrodes and electrolyte. Therefore, the addition of the MXene into the polymer mixture has enhanced the interfacial compatibility along with the ionic diffusion. The representative chronoamperometry profiles for PAN/bio-based PU/$Ti_3C_2T_x$ MXene (5 wt.%) SCEs and glass fiber membrane with liquid electrolyte for the symmetric Zn cells at room temperature with a step potential of 10 mV are shown in Figure 11.4e.

The ionic conductivity of the PAN/bio-based PU/MXene SCE was 3.35 mS cm^{-1} with a tensile strength of 1.68 MPa. The transference number for the PAN/bio-based PU/MXene SCE is 0.54, which is higher than the glass fiber membrane (0.49). Further, the Zn symmetric cells were fabricated to study cycling stability during the Zn stripping and deposition process at different current values. $(NH_4)_2V_{10}O_{25}\cdot 8H_2O$ (NVO) cathode was used as the host for the Zn ions from the Zn anode. The full cell configuration of NVO/PAN/bio-based PU/MXene/Zn was used for understanding the redox potential in cyclic voltammetry and galvanostatic charge-discharge studies [44].

Figure 11.4f–h shows the Zn symmetry cells galvanostatic charge-discharges studies up to 100 cycles at 0.25, 0.5, 1.25, and 2.5 mA cm^{-2} currents for the glass fiber membrane electrolyte, PAN/bio-based PU, and PAN/bio-based PU/MXene (5 wt.%) SCEs. Figure 11.4i–k shows the voltage vs. cycle profile from the 25 to 50th cycle at 0.5 mA cm^{-2} current for glass fiber membrane electrolyte, PAN/bio-based PU, and PAN/bio-based PU/MXene (5 wt.%) SCEs. The rate capability test of the NVO/PAN/bio-based PU/MXene (5 wt.%)/Zn has shown the highest specific capacity of ~470 mAh g^{-1} at 0.1 A g^{-1}, which is further decreased along with the increase of current rate and reached ~105 mAh g^{-1} at 5 A g^{-1}—and upon decreasing the current rate back to 0.2 A g^{-1}, showed a specific capacitance of ~380 mAh g^{-1}. To date, there are a very limited number of studies carried towards the Zn-ASSBs using MXene-based electrolytes either as an additive or as a surface layer coated on the anode side, which can restrict the passage of molecules or ions other than Zn.

11.4 LIMITATIONS IN ASSBS

ASSBs have numerous advantages over conventional batteries in terms of safety, mechanical stability, and cost reduction. Still, there are numerous issues to be addressed for ASSBs, especially at the interfaces of the electrodes and electrolytes, operating temperatures, fabrication process, and so on [48]. The components of the ASSBs have issues like chemical stability, low ionic diffusion, slow reaction kinetics, poor interfacial compatibility, high irreversible capacities, etc. [19, 49]. Figure 11.5 shows the detailed issues at the interlayers and intra-structures existing in/at the cathode and anode sides of the electrolytes.

For instance, on the cathode side, the point of contact between the electrolyte and cathode, poor contact area, compatibility of electrochemical reactions, chemical or mechanical issues, and so on, are there to address. On the other side, contact issues at the interfaces—physical factors like volume expansion, cracks in the materials, and dendrites formation and growth—are some of the challenges that need to be addressed. There are major issues to address apart from the other small issues like heat produced during the cycling, current collectors, and loss of performance due to internal resistance.

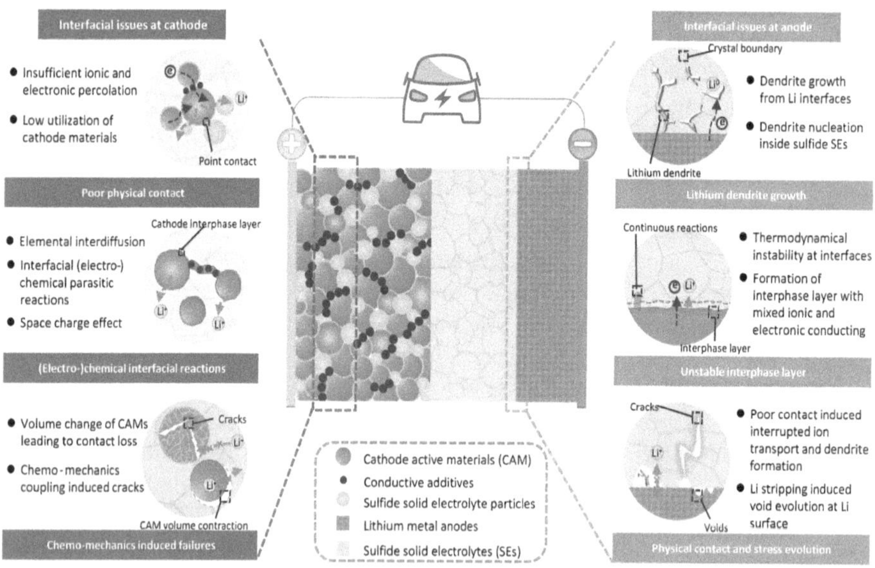

FIGURE 11.5 The most possible issues in the ASSBs at the interfaces of cathode-electrolyte and anode-electrolyte along with the internal issues like volume expansion, cracks in the materials, loss of contact within the electrodes, and with the current collectors.

Source: Reproduced with permission from ref. [48]. Copyright 2022 John Wiley and Sons.

The potential windows and operating temperatures of the ASSBs must be optimized in order to use at room temperature (RT) with wide potential windows [19, 50]. The poor chemical, electrochemical, and physical compatibility of SSEs with electrodes is more critical in determining the electrochemical performances of ASSBs, especially the power density, along with the design and structure of the devices [49, 51]. The enhancement in performance like ionic conductivity, cycling stability, high power and energy density, and compact size are some of the important factors to address in next-generation ASSBs [52]. The matching of the ideal thickness of the layers in the ASSBs during the fabrication process to reach the higher-performing devices are major issues. MXene-based materials can be an added advantage in reducing the resistance and contact issues between materials in the ASSB devices. The weight of the ASSBs is also a major issue that affects the performance of the device, which can be addressed using weight-less, low density, and with high specific surface area containing MXene as a current collector or as a conductive additive in the fabrication process. Although MXenes have advantages, they are majorly restricted for commercial use and application due to the oxidation stability and dispersion ability in different organic solvents for easy usage for fabrication. All these advantages must be implemented and limitations overcome to make commercially viable ASSBs for future applications.

11.5 FUTURE PROSPECTIVE

Current ASSB research is in its early stages and still needs a better establishment of the engineering aspects for practical applications comparable to conventional batteries. There are a few scenarios where researchers have developed and demonstrated the possibility of commercialization of ASSBs; still, it is a long-term process into the future to replace conventional batteries with ASSBs. Presently, funding agencies and companies are deploying their funds towards commercialization of ASSBs with the same or above the performance levels of conventional LIBs. Many factors influence the commercialization of ASSBs, such as interfacial compatibility, fabrication technologies with the least price, and scaling-up of the production of ASSBs as per applications. In the above all aspects, the SSE's part in the ASSBs is one of the major contributors for improving the performance which occupies more volume or space in the ASSB in comparison to the cathode and anode parts.

SSEs must overcome most of the issues discussed in this chapter to establish a pathway for commercialization. As discussed above, compatibility of the SSEs at the interfaces is the key issue—and the internal electronic and ionic conductivity is also a vital aspect in the performance enhancement of ASSBs. There are many timely issues to be addressed in SSEs, such as working conditions (temperature and pressure), interfacial issues, stability in ambient conditions, synthesis and fabrication processes, and cost of the raw materials, which are preventing rapid progress in the commercialization process of ASSBs.

MXenes are one of the cheap and viable 2D materials to overcome some of the issues mentioned above like internal ionic diffusion, electron transport, and mechanical stability in comparison to metallic conductive additives. Conventional SSEs have poor conductivity and compatibility issues with other components in ASSBs, which can be easily solved by the addition of highly conductive (ionic) and robust MXene sheets. Among all available MXenes, only $Ti_3C_2T_x$ MXene is explored more in comparison to the other MXenes containing V, Nb, Mo, etc. This area is still a vast open field for researchers to explore for an alternative to existing SSEs.

At present, researchers are only focusing on oxide and sulfur-based electrolytes, which are heavier and more expensive, along with contact issues at interfaces and internal structures. MXenes are one of the best alternatives to avoid these kinds of issues, and the addition of MXene into the electrolytes will revolutionize the production of lightweight ASSBs. Apart from the pure MXenes, surface-modified and functionalized MXene can contribute more towards the enhancement of performance in ASSBs. The surface-modified or functionalized MXenes will have suitable functional groups, which will interact well with other components of the ASSBs for easy handling and higher oxidation resistance at ambient conditions as an additional property.

Although MXene has many advantages, its complete chemistry towards ASSBs is yet to be completely explored and protocols to be established. In the near future, MXenes might replace the role of conductive additives and current collectors in battery systems especially it can reduce the weight of ASSBs, which will enhance the performance of heavy applications like electric transportation vehicles and grid storage systems. Massive applications like defense, aerospace, and space exploration need energy storage systems with the highest level of safety, high power density, and lightweight; in these aspects, MXene containing (in electrodes and electrolytes) ASSBs can be a future alternative for existing conventional battery systems with liquid type electrolytes.

11.6 CONCLUSIONS

In summary, ASSBs are one of the promising high-energy density storage devices for application in electronics and hybrid electric vehicle applications. ASSB technology is almost established and about to deliver commercial production in many electric vehicle companies; however, there are multiple issues in terms of performance, safety, cycle life, and application conditions. Therefore, advancements in the technology must be optimized to establish successful production of ASSBs. Most aspects related to ASSBs are majorly governed by the electrolytes in the systems, which are the deciding factor for performance. Apart from conventional SPEs and SCEs, conventional electrolytes with lower conductivity and without good compatibility at the interfaces are yet to be explored using MXene as an interfacial layer to fine-tune the interfacial resistance. This brief discussion on MXene-based electrolytes for ASSBs in this chapter may encourage researchers to focus more on the issues and achieve the best solution for high-energy solid-state batteries in the near future.

ACKNOWLEDGEMENTS

The Basic Science Research Program supported this research through the National Research Foundation of Korea [grant numbers 2018R1A6A1A03023788, 2022R1I1A1A01070368, 2021R1I1A1A01055790, RS-2023-00244800, and

RS-2023-00213746], and the National Research Foundation of Korea (MSIT) [grant number 2022M3J7A1062940]. This research was supported by the "Regional Innovation Strategy (RIS)" through the National Research Foundation of Korea (NRF) funded by the Ministry of Education (MOE) (2021RIS-001, grant number 1345370811).

REFERENCES

1. G. Murali, J. Rawal, J.K.R. Modigunta, Y.H. Park, J.-H. Lee, S.-Y. Lee, S.-J. Park and I. In, *Sustain. Energy Fuels*, 2021, 5, 5672–5693.
2. B. Dunn, H. Kamath and J.-M. Tarascon, *Science*, 2011, 334, 928.
3. J.F.M. Oudenhoven, L. Baggetto and P.H.L. Notten, *Adv. Energy Mater.*, 2011, 1, 10–33.
4. D.P. Dubal, O. Ayyad, V. Ruiz and P. Gómez-Romero, *Chem. Soc. Rev.*, 2015, 44, 1777–1790.
5. T. Kim, W. Song, D.-Y. Son, L.K. Ono and Y. Qi, *J. Mater. Chem. A*, 2019, 7, 2942–2964.
6. K. Liu, Y. Liu, D. Lin, A. Pei and Y. Cui, *Sci. Adv.*, 2018, 4, https://doi.org/10.1126/sciadv.aas9820
7. X. Wu, K. Song, X. Zhang, N. Hu, L. Li, W. Li, L. Zhang and H. Zhang, *Front. Energy Res.*, 2019, 7, 65.
8. S. Mahmud, M. Rahman, M. Kamruzzaman, M.O. Ali, M.S.A. Emon, H. Khatun and M.R. Ali, *Results Eng.*, 2022, 15, 100472.
9. L. Yu, J. Gu, C. Pan, J. Zhang, Z. Wei and Y. Zhao, *Compos. – A: Appl. Sci. Manuf.*, 2022, 162, 107132.
10. H. Kwak, S. Wang, J. Park, Y. Liu, K.T. Kim, Y. Choi, Y. Mo and Y.S. Jung, *ACS Energy Lett.*, 2022, 7, 1776–1805.
11. K. Takada, *Acta Mater.*, 2013, 61, 759–770.
12. R. Rajagopal and K.-S. Ryu, *J. Alloys Compd.*, 2019, 798, 235–242.
13. Q. Zhang, D. Cao, Y. Ma, A. Natan, P. Aurora and H. Zhu, *Adv. Mater.*, 2019, 31, 1901131.
14. X. Li, J. Liang, X. Yang, K.R. Adair, C. Wang, F. Zhao and X. Sun, *Energy Environ. Sci.*, 2020, 13, 1429–1461.
15. W. Xiao, J. Wang, L. Fan, J. Zhang and X. Li, *Energy Storage Mater.*, 2019, 19, 379–400.
16. Z. Zhang, R.G. Antonio and K.L. Choy, *J. Power Sources*, 2019, 435, 226736.
17. Z. Zhang, L. Zhang, Y. Liu, X. Yan, B. Xu and L.M. Wang, *J. Alloys Compd.*, 2020, 812, 152103.
18. E. Zhao, F. Ma, Y. Guo and Y. Jin, *RSC Adv.*, 2016, 6, 92579–92585.
19. P. Albertus, V. Anandan, C. Ban, N. Balsara, I. Belharouak, J. Buettner-Garrett, Z. Chen, C. Daniel, M. Doeff, N.J. Dudney, B. Dunn, S.J. Harris, S. Herle, E. Herbert, S. Kalnaus, J.A. Libera, D. Lu, S. Martin, B.D. McCloskey, M.T. McDowell, Y.S. Meng, J. Nanda, J. Sakamoto, E.C. Self, S. Tepavcevic, E. Wachsman, C. Wang, A.S. Westover, J. Xiao and T. Yersak, *ACS Energy Lett.*, 2021, 6, 1399–1404.
20. S. Li, S.-Q. Zhang, L. Shen, Q. Liu, J.-B. Ma, W. Lv, Y.-B. He and Q.-H. Yang, *Adv. Sci.*, 2020, 7, 1903088.
21. G. Murali, J.K. Reddy Modigunta, Y.H. Park, J.-H. Lee, J. Rawal, S.-Y. Lee, I. In and S.-J. Park, *ACS Nano*, 2022, 16, 13370–13429.
22. X. Zhou, H. Jiang, H. Zheng, Y. Sun, X. Liang and H. Xiang, *J. Membr. Sci.*, 2020, 603, 117820.
23. M. Keller, A. Varzi and S. Passerini, *J. Power Sources*, 2018, 392, 206–225.
24. X. Liu, X. Li, H. Li and H.B. Wu, *Chem. Eur. J.*, 2018, 24, 18293–18306.

25. M. Dirican, C. Yan, P. Zhu and X. Zhang, *Mater. Sci. Eng. R Rep.*, 2019, 136, 27–46.
26. P.A. Shinde, A.M. Patil, S. Lee, E. Jung and S. Chan Jun, *J. Mater. Chem. A*, 2022, 10, 1105–1149.
27. B. Anasori, M.R. Lukatskaya and Y. Gogotsi, *Nat. Rev. Mater.*, 2017, 2, 16098.
28. T.Y. Ko, D. Kim, S.J. Kim, H. Kim, A.S. Nissimagoudar, S.-C. Lee, X. Lin, P.T. Cummings, S. Doo, S. Park, T. Hassan, T. Oh, A. Chae, J. Lee, Y. Gogotsi, I. In and C.M. Koo, *ACS Nano*, 2023, 17, 1112–1119.
29. A. Shayesteh Zeraati, S.A. Mirkhani, P. Sun, M. Naguib, P.V. Braun and U. Sundararaj, *Nanoscale*, 2021, 13, 3572–3580.
30. M. Alhabeb, K. Maleski, B. Anasori, P. Lelyukh, L. Clark, S. Sin and Y. Gogotsi, *Chem. Mater.*, 2017, 29, 7633–7644.
31. A.E. Ghazaly, H. Ahmed, A.R. Rezk, J. Halim, P.O.Å. Persson, L.Y. Yeo and J. Rosen, *ACS Nano*, 2021, 15, 4287–4293.
32. D. Kim, T.Y. Ko, H. Kim, G.H. Lee, S. Cho and C.M. Koo, *ACS Nano*, 2019, 13, 13818–13828.
33. A. Vahid Mohammadi, J. Rosen and Y. Gogotsi, *Science*, 2021, 372, https://doi.org/10.1126/science.abf1581
34. P.R. Ilango, J.K.R. Modigunta, A.K. Parameswaran, Z. Sofer, G. Murali, I. In, Novel Design Aspects of All-Solid-State Batteries, in: N. Palaniyandy, K.P. Abhilash, B. Nalini (Eds.) *Solid State Batteries: Design, Challenges and Market Demands*, Springer International Publishing, Cham, 2022, pp. 157–191.
35. P.G. Bruce, J. Evans and C.A. Vincent, *Solid State Ion.*, 1988, 28–30, 918–922.
36. G.B. Appetecchi, G. Dautzenberg and B. Scrosati, *J. Electrochem. Soc.*, 1996, 143, 6.
37. S. Luo, X. Liu, L. Gao, N. Deng, X. Sun, Y. Li, Q. Zeng, H. Wang, B. Cheng and W. Kang, *Sustain. Energy Fuels*, 2022, 6, 5019–5044.
38. Q. Pan, Y. Zheng, S. Kota, W. Huang, S. Wang, H. Qi, S. Kim, Y. Tu, M.W. Barsoum and C.Y. Li, *Nanoscale Adv.*, 2019, 1, 395–402.
39. Y.-Q. Mao, G.-H. Dong, W.-B. Zhu, Y.-Q. Li, P. Huang and S.-Y. Fu, *Nano Mater. Sci.*, 2024, 6(1), 60–67.
40. X. Zhao, M. Zhu, C. Tang, K. Quan, Q. Tong, H. Cao, J. Jiang, H. Yang and J. Zhang, *J. Colloid Interface Sci.*, 2022, 620, 478–485.
41. S. Xia, X. Wu, Z. Zhang, Y. Cui and W. Liu, *Chem*, 2019, 5, 753–785.
42. W. Kou, Y. Zhang, W. Wu, Z. Guo, Q. Hua and J. Wang, *Green Energy Environ.*, 2024, 9, 71–80.
43. C. Sun, C. Wu, X. Gu, C. Wang and Q. Wang, *Nanomicro Lett.*, 2021, 13, 89.
44. C. Likitaporn, M. Okhawilai, P. Kasemsiri, J. Qin, P. Potiyaraj and H. Uyama, *Sci. Rep.*, 2022, 12, 19915.
45. X. Xiao, X. Xiao, Y. Zhou, X. Zhao, G. Chen, Z. Liu, Z. Wang, C. Lu, M. Hu, A. Nashalian, S. Shen, K. Xie, W. Yang, Y. Gong, W. Ding, P. Servati, C. Han, S.X. Dou, W. Li and J. Chen, *Sci. Adv.*, 2021, 7, https://doi.org/10.1126/sciadv.abl3742
46. E.J. Hansen, J. Liu, *Front. Energy Res.*, 2021, 8, 616665.
47. Z. Chen, X. Li, D. Wang, Q. Yang, L. Ma, Z. Huang, G. Liang, A. Chen, Y. Guo, B. Dong, X. Huang, C. Yang and C. Zhi, *Energy Environ. Sci.*, 2021, 14, 3492–3501.
48. Y. Liang, H. Liu, G. Wang, C. Wang, Y. Ni, C.-W. Nan and L.-Z. Fan, *InfoMat*, 2022, 4, https://doi.org/10.1002/inf2.12292
49. K. Kerman, A. Luntz, V. Viswanathan, Y.-M. Chiang and Z. Chen, *J. Electrochem. Soc.*, 2017, 164, A1731–A1744.
50. Y. Zheng, Y. Yao, J. Ou, M. Li, D. Luo, H. Dou, Z. Li, K. Amine, A. Yu and Z. Chen, *Chem. Soc. Rev.*, 2020, 49, 8790–8839.
51. Z. Ding, J. Li, J. Li and C. An, *J. Electrochem. Soc.*, 2020, 167, 070541.
52. K. Takada, N. Ohta, L. Zhang, X. Xu, B.T. Hang, T. Ohnishi, M. Osada and T. Sasaki, *Solid State Ion.*, 2012, 225, 594–597.

12 Electrolytes for Redox Flow Batteries (RFBs)

H. B. Muralidhara, B. Sharath Kumar, B. C. Gireeshkumar, Krishna Venkatesh, and Venkateswarlu Manne

12.1 INTRODUCTION

With advances in technologies and industries, there is a greater need for more efficient and environmentally friendly electrical energy storage solutions. It is highly impossible to meet the current global energy demands in today's scenario with rapidly depleting fossil fuels. Therefore, a transition towards renewable energy resources is necessary [1], and this is where energy storage systems play a big role. Energy storage solutions in the form of micro grids or smart grid level are emerging and utilizing renewable energy resources for generation of energy. However, there are several challenges, including technical difficulties for large-scale storage, as well as safety concerns and the costs incurred [2]. The development of cost-effective and environmentally friendly storage systems for the green conversion of energy will facilitate a smooth transition.

Among various energy storage technologies, Redox flow batteries (RFBs) are promising energy storage devices that utilize the reduction and oxidation reactions of redox species with variable oxidation states. They have several unique features, namely that power depends on the cell size, and capacity depends on the volume of the electrolyte—and both can be varied independently [3, 4]. The active material is stored outside the battery, and the electrodes are involved in the redox reaction. This ideally allows for charge-discharge performance without any loss in the performance of the storage system [5].

In this chapter, we will discuss the electrolytes utilized in RFBs in detail, covering their properties, preparation, and functionality. Redox-active species, such as vanadium, iron, and zinc ions, are used as electrolytes because they can undergo reversible oxidation and reduction reactions. This property allows for the storage and release of electrical energy [6].

Several emerging RFB technologies employ various electrolytes, such as aqueous, non-aqueous, organic, and inorganic type. This has opened the possibility for further research to be conducted for the benefit of humanity. A diagrammatic representation of RFB is depicted in Figure 12.1. The electrolyte tanks are kept externally on either side of the assembly. Catholyte flows through the positive half-cell, while anolyte flows through the negative half-cell where the redox reaction takes place and electrolytes are drawn back to the external tanks. The electrodes are separated by an ion exchange membrane, which allows the transfer of counter ions between them.

DOI: 10.1201/9781003441182-14

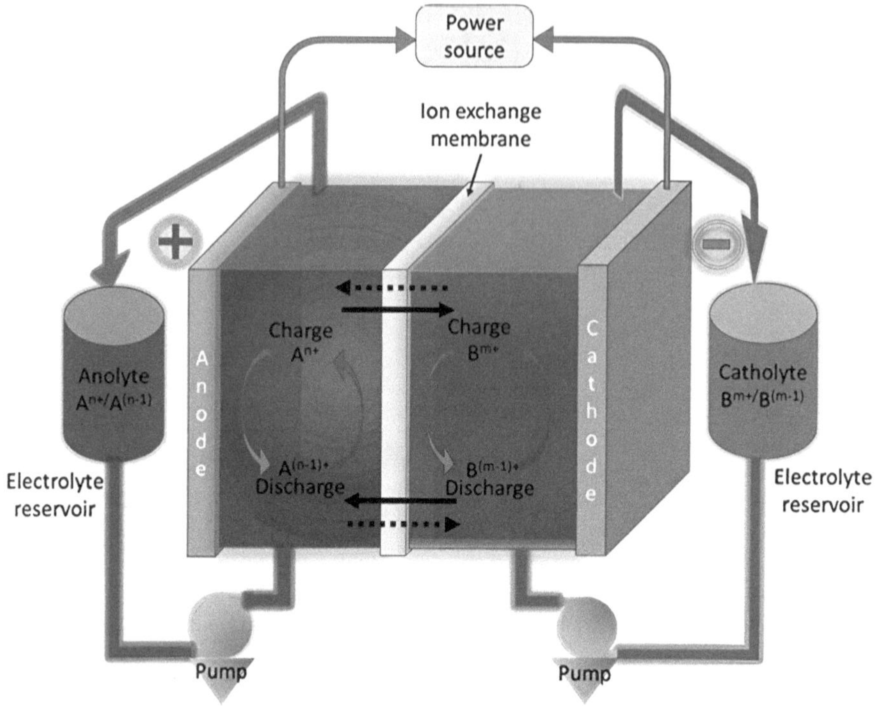

FIGURE 12.1 Schematic block diagram representing the arrangements of components in RFBs.

Care should be taken when choosing the membrane for the redox flow battery because the movement of counter ions is crucial for maintaining electrical neutrality on both sides of the flow battery and ensuring the proper functioning of the system. The membrane should fulfil two important criteria: it should be stable in the chosen electrolyte system and offer the lowest resistance to the easy movement of ions. The entire process is facilitated by a pump that ensures a consistent flow of electrolytes throughout the assembly.

12.2 THE PREPARATION OF ELECTROLYTES

The preparation of electrolytes for RFBs depends on the type of system employed. Usually, metal salts of desired concentrations are dissolved in their respective solvents, considering their solubility. For instance, vanadium electrolyte is prepared by dissolving vanadium salts such as vanadium sulphate, vanadium chloride, etc., in solvents like HCl, sulfuric acid, H_3PO_4, etc. These are also referred to as aqueous electrolytes. Similarly, if the metal salts are dissolved in organic solvents like ethylene glycol, acetyl nitrile, pentane, etc., they are termed non-aqueous electrolytes.

One of the most commonly used non-aqueous electrolytes in RFBs is lithium-polysulfide electrolyte. This electrolyte is composed of lithium polysulfides dissolved in a lithium bis(trifluoromethane sulfonyl)imide (LiTFSI) solvent and acts as a negative electrolyte. This electrolyte has a high energy density and can operate at temperatures up to 60 °C, making it a suitable option for high-temperature applications.

12.3 OPTIMAL CONDITIONS FOR THE ELECTROLYTES USED IN REDOX FLOW BATTERIES

The electrolytes used in RFBs play a crucial role in determining the Coulombic efficiency (CE) and performance of the battery. The composition of the electrolyte solution and the concentration of the ions in the solution can have a significant impact on the voltage and current of the battery. The performance characteristics of the electrolytes used in RFBs depend on several other factors, including the temperature and pH of the electrolyte.

One of the most important properties of the electrolyte solution is its conductivity, which is directly related to the mobility of the ions in the solution. A higher conductivity allows for a higher rate of ion transport and, therefore, a higher rate of charge and discharge. Another important property of the electrolyte solution is its stability. The stability of the electrolyte solution refers to its ability to maintain its properties over time. Electrolyte solutions that are stable will not degrade or change over time, which can lead to an increase in the efficiency and performance of the battery.

A higher concentration of ions in the electrolyte solution will result in a higher rate of ion transport and, therefore, a higher rate of charge and discharge. However, a higher concentration of ions in the electrolyte solution can also lead to higher internal resistance and lower efficiency. For example, in case of IRFB, the negative potential of the iron plating process produces hydrogen evolution, which changes the pH of the electrolyte. If the pH of the solution surpasses 3, the iron electrolyte forms precipitates as ferrous hydroxide.

To tackle these challenges, iron complexing with different ligands could be explored. Organic acids are excellent complexing ligands for iron, and they also lower the pH of the electrolyte. According to Hawthorne et al., adding ligand to the iron electrolyte solution increases the size of the ion, which reduces the diffusion coefficient and increases the mass transfer overpotential [7].

12.4 LIGANDS AND CONDUCTING SALTS USED IN REDOX FLOW BATTERIES

Ligands play an important role in electrochemical reactions that take place within the electrolyte, allowing for efficient energy storage and transfer. These molecules can bind to a central metal atom or ion, forming a coordination complex and thus stabilizing the metal ions in the electrolyte and preventing them from forming unwanted precipitates [8]. Conducting salts, on the other hand, are ions that can conduct electricity through the electrolyte. These ions are typically added to the electrolyte to increase its conductivity, making it easier for the electrons to flow through the system.

For example, in case an IRFB has a stable redox reaction, it is essential to maintain the pH of the iron electrolyte in the range of 1 to 3 to evade the formation of ferrous hydroxide and hydrogen evolution reaction. Various ligands were added to the iron electrolyte, and the change in the pH, density, and viscosity values were noted. The electrolytes containing ethylene diamine tetra acetic acid, malic acid, malonic acid, nitrilotriacetic acid, and diethyltriamine pentaacetic acid have resulted in the pH less than 1. The electrolyte containing glycerol resulted in pH 2.015; however, the solution was very viscous in comparison to the base electrolyte, which could result in slow movement of electrolyte through the flow channels. Thus, glycine, ascorbic acid, and citric acid were exhibiting viscosity from 2–2.6 MPa/s and stabilized the electrolyte from undergoing precipitation by maintaining the pH below 3. In summary, ligands and conducting salts are necessary components in redox flow electrolytes, as they play important roles in stabilizing the metal ions and increasing the conductivity of the electrolyte, respectively.

12.5 STUDIES ON DIFFERENT ELECTROLYTES IN REDOX FLOW BATTERIES

12.5.1 Iron-Based Redox Flow Batteries

One of the earliest researchers to study iron-based electrolytes in redox flow batteries was Dr. Guoying Chen of the University of California, Los Angeles. In a series of papers published in the early 2000s, Dr. Chen and his team investigated the use of iron ions in aqueous solutions as the active species in the negative electrode of a redox flow battery. They found that iron ions in the negative electrode had higher electrochemical stability and a lower cost compared to vanadium ions, making them a promising alternative. Hrushka and Savinell worked on iron electrolytes extensively and reported that low resistivity may be achieved by keeping the electrolyte concentration in the range of 0.8–3.9 M, and the pH change would be negligible at such concentrations [8]. Anarghya et al [9] reported that by using $FeCl_2$ electrolyte along with the glycine and ammonium chloride as supporting ligands to avoid the precipitation of iron ions during the process of charging and discharging in RFB with an electrode active surface area of 132 cm^2, they were able to conduct charge-discharge studies.

Untreated graphite felt (UGF) exhibited a limited CE of 35% (Figure 12.2), and upon modification of graphite felt (MGF) nitrogen-doped carbon derived from Bermuda grass, the electrode exhibited excellent stability during the charging and discharging process. Figure 12.2A depicts the charge-discharge response of the electrode and various current densities. A sudden voltage drop during discharging was

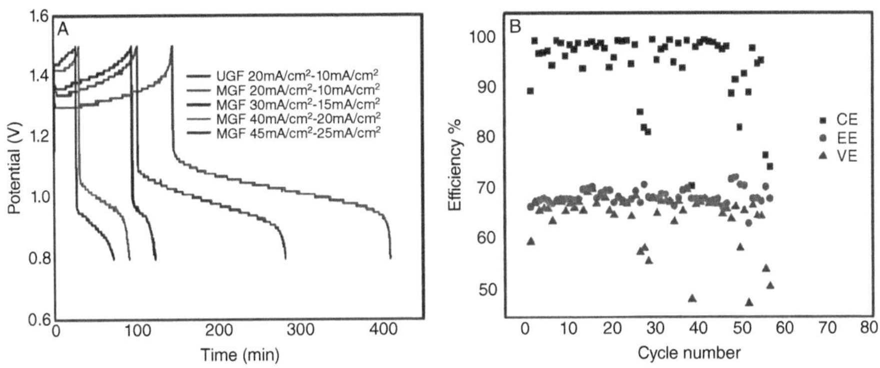

FIGURE 12.2 The images show the (a) charge-discharge curves at different current densities using unmodified graphite felt (UGF) and modified graphite felt using Bermuda grass-derived nitrogen-doped carbon as electrocatalyst (MGF); (b) efficiencies with respect to different cycle numbers in IRFB.

Source: Reproduced with permission from Ref. [9]. Copyright 2020 Elsevier.

ascribed due to the internal resistance of the battery, and it was to achieve the maximum CE of 98% at a current density of 40 mA/cm^2. The battery resulted in an average CE of 95%, a peak power density of 60 mW/cm^{-2}, and better stability during multiple cycling up to 26 cycles with modified electrodes (Figure 12.2B).

At the same current density as the modified electrode, Anantha et al. [10] were able to achieve a CE of 96.06% and a peak power density of 55.16 mW/cm^{-2} for about 25 cycles; the reason for the low cyclability of the system is due to the deposition of iron as iron hydroxide at the walls of the membrane as well as the electrode surface, which can be cleaned and reused up to 50 cycles of charge-discharge.

12.5.2 Vanadium-Based Redox Flow Battery (VRFB)

VRFB has been discovered to be the most attractive since both the anolyte and catholyte utilize the same element, preventing cross-contamination of the two half-cell electrolytes. VRFBs have emerged as the most promising and economically utilized RFB type for intermittent renewable energy storage. This is because vanadium can exist in four different oxidation states, which allows for the transfer of multiple electrons during the charging and discharging process. This makes vanadium electrolyte a good choice for systems that use renewable energy, where the amount of energy available can change a lot over time.

Different researchers have used various electrolyte solutions for VRFBs, each with its own unique properties and advantages. The most commonly used electrolyte for VRFBs is a vanadium sulphate electrolyte, which consists of vanadium ions in different oxidation states (V^{2+}, V^{3+}, V^{4+}, and V^{5+}) dissolved in sulfuric acid. This electrolyte is relatively stable, with a wide electrochemical window and a relatively low resistance to ion transport [11]. Additionally, vanadium sulphate electrolyte is stable and easy to prepare, making it a popular choice for VRFBs.

Alternatively, the electrolyte used for VRFBs is vanadium bromide electrolyte, which is composed of vanadium ions dissolved in hydrobromic acid. This electrolyte has a high vanadium ion concentration, which results in high energy density and high power density. However, the vanadium bromide electrolyte is more expensive and difficult to prepare than vanadium sulphate electrolyte [12]. Ting Long et al, were able to achieve voltaic efficiency and EE of 88.4% and 86.4% with graphene modified electrode at 20 cm^2 electrode at a higher current density of 100 mA for about 600 cycles of charge and discharging [13].

12.5.3 Iron-Chromium Batteries

The iron-chromium system (ICRFB) is a type of redox flow battery (RFB) that uses iron and chromium ions as the redox active species. The high-potential pair active ingredient, Fe^{+2}, is converted to Fe^{+3} on the positive electrode during charging, while the low-potential pair active ingredient, Cr^{+3}, is converted to Cr^{+2} on the negative electrode. During the discharge process, these reactions are reversed. NASA put a lot of effort into developing a series of 1 kW (13 kWh) Fe/Cr stacks for solar photovoltaic power generation systems between 1973 and 1982. The cost of ICRFB is expected to be $9.4/kWh, making it more cost-effective than other RFB systems.

Research has also shown that the addition of $InCl_2$ as an additive in ICRFB can significantly increase efficiency by inhibiting the evolution of hydrogen gas during the reaction; enhancing the rate of the redox couple's reaction using an optimal concentration of 0.01 M In^{3+} in the anolyte exhibits an energy efficiency of 77.0% at 200 mA cm^{-2} and a 36.3% higher capacity retention ratio than that of the pristine anolyte after 140 cycles at 160 mA cm^{-2} [14]. Further research is being conducted to identify and commercialize more cost-effective and efficient ICRFB systems by using other impurities in p-block elements as catalysts.

12.5.4 Iron-Lead Batteries

A novel RFB called the iron-lead RFB (ILRFB) has been proposed by Zeng et al. that uses cheap and abundant Pb/Pb^{2+} and Fe^{2+}/Fe^{3+} as the redox pairs for the anolyte and catholyte, respectively. The RFB uses methane sulfonic acid (MSA) as the supportive acid medium. The charge-discharge cycling process in the ILRFB involves the deposition and dissolution of lead metal in addition to the charge in ionic valence states.

The cell capacities of CE and EE at 40 mA/cm^2 are 96.2% and 86.2%, respectively, based on the charge-discharge cycle performance of ILRFB. The output power density of ILRFB is much better than other RFBs at a similar energy efficiency (EE) of roughly 80%, thanks to the outstanding kinetics of the Fe^{2+}/Fe^{3+} redox couple. It is considered a potential device for large-scale energy storage applications due to its steady cycle performance, quick electrochemical kinetics, and affordable redox materials. Further research is still needed on the lead electrodeposition improvements of the electrode structure and electrolyte composition [15].

12.5.5 The Iron-Cadmium Redox Flow Battery

The Iron-Cadmium Redox Flow Battery (ICdRFB) is a type of flow battery that uses the redox pairs Cd/Cd^{+2} and Fe^{2+}/Fe^{3+} in acid as the anolyte and catholyte, respectively. The active species cost of the ICdRFB is low, at $10 kWh^{-1}, and the capital cost of the entire system is $108 kWh^{-1}. The cost per kWh decreases with an increase in discharge duration, making the ICdRFB particularly suitable for long-term energy storage. The catholyte's Fe^{2+}/Fe^{3+} redox pair has outstanding kinetics in an acid medium, while the anolyte's Cd/Cd^{+2} redox pair has quick kinetics and a high exchange current density. The use of cadmium in ICdRFB is advantageous as it addresses the problem of hydrogen evolution and allows for efficient management of active materials and recycling. However, the blocking effect during Cd electrodeposition leads to low electrolyte usage, and it was able to achieve CE and EE of 98.7% and 80.2% at 120 mA cm^{-2} [16].

12.5.6 Iron-Vanadium Redox Flow Battery

In 2011, Wang et al. [17] presented a new redox flow battery called the Iron-Vanadium RFB (IVRFB) that uses Fe^{2+}/Fe^{3+} chloride solution as the catholyte and V^{+2}/V^{+3} chloride solution as the anolyte. This design aims to take advantage of the benefits of both VRFB and Iron-Chromium RFB (ICRFB), while avoiding their respective issues. The V^{+2}/V^{+3} redox couples have a superior electrochemical activity compared to Cr^{+2}/Cr^{+3} in ICRFB, eliminating the need for a catalyst and high working temperatures. Additionally, the operational potential of V^{+2}/V^{+3} is substantially higher than that of Cr^{+2}/Cr^{+3}, which helps to prevent hydrogen development during charging. The electrolyte for the Fe/V redox flow battery test was prepared by dissolving VCl_3 and $FeCl_2$ in concentrated HCl with 1.25 M Fe and 1.25 M V in 2.3 M HCl and was characterized using cyclic voltammetry. The redox flow cell achieved an EE of 80% and a moderate reduction in capacity that is maintained for a total of 50 cycles.

12.5.7 Tin-Iron RFB (SnRFB)

SnRFB, as suggested by Zhou et al., uses tin-iron to separate the anolyte from the catholyte (containing iron) in a redox flow battery. During charging, tin is electroplated from the anolyte to the anode, while the iron is oxidized to $FeCl_3$ on the cathode. During discharge, reversible redox processes occur at the appropriate electrode surface. Both reactions are quasi-reversible, and their kinetics can be compared to other redox reactions in RFBs. Two solutions, 0.5 M $SnCl_2$ + 3.0 M HCl and 1.0 M $FeCl_2$ + 3.0 M HCl, were used as the negative and positive electrolytes for charge-discharge testing. For current densities between 40 and 200 mA cm^{-2}, it is proved that the present battery system can attain EEs ranging from 78.5 to 93.4 %. In addition, the long-term cycling test confirms the battery's durability throughout 700 cycles. The findings indicate that the efficiency remains generally steady, and the capacity degradation rate may be as low as 0.96 per cycle [18].

12.5.8 Zinc Bromide Redox Flow Battery (ZBRFB)

A zinc bromide redox flow battery (ZBRFB) is a type of energy storage system that uses zinc bromide as the electrolyte. It stores energy in the form of chemical potential energy in the electrolyte, which can be released as electrical energy when needed. It consists of two separate tanks that store the electrolyte and a stack of electrochemical cells. During charging, electrons from an external power source reduce zinc ions at the cathode and oxidize bromide ions at the anode, storing energy. During discharging, the flow of electrons is reversed, releasing stored energy as electrical energy. ZBRFBs have a long lifespan, are relatively inexpensive to manufacture, and can store large amounts of energy over long periods, making them suitable for grid-scale energy storage applications.

Aqueous solutions of zinc bromide in the range of 1–4 M [19] in distilled water are the major electroactive species of electrolyte in ZBFB, but in reality, electrolyte is an aqueous solution containing zinc bromide and a supporting electrolyte. Due to the poor conductivity of zinc bromide solution, a supporting electrolyte is used to increase ionic conductivity and decrease internal resistance [20]. Chloride-based salts are often used as a supporting ingredient to enhance the electrolyte's conductivity and the battery's energy efficiency [21]. Additionally, the use of methane sulfonic acid can improve the kinetics and reversibility of the redox couplings, enhance the electrolyte's conductivity, and decrease the battery's internal resistance [22].

In most cases, the concentration of the supporting electrolyte in a zinc bromide battery is in the range of 20–30% by weight, but the optimal concentration may vary depending on the specific requirements of the battery. Factors such as the specific design of the battery, operating conditions, and materials used in the battery construction should all be considered when selecting the concentration of the supporting electrolyte. The research by M.C. Wu et al. [23] has shown that the use of highly ionic and conductive supporting electrolytes, such as KCl and NH_4Cl (4 M), can effectively lower the internal resistance and improve the performance of the battery. Therefore, it is important to carefully select the concentration of the supporting electrolyte in a zinc bromide battery to ensure optimal performance and a long lifespan. Several forms of RFBs have been produced throughout time, each having a unique composition of electrolyte to produce higher cell voltage and efficiency, as shown in Table 12.1.

TABLE 12.1
Types of Aqueous RFBs

Type	Electrolyte	Cell Voltage (V)	Energy Efficiency	Ref.
Fe/Cr	1 M $FeCl_2$/1 M $CrCl_3$ in 2–3 M HCl	1.18	73	[24]
All-vanadium	1.7–2 M V in 4–5 H_2SO_4	1.26	83–87	[13, 25]
Zinc-bromine	$ZnBr_2$ in excess of Br_2	1.85	69.4	[26]
Fe/V	1 M $FeCl_2$ in 2M HCl/2 M V in 4 M H_2SO_4	1.02	80	[17]
Polysulfide/ bromine (PSB)	5 M NaBr/1.3 M Na_2S_5 and 1 M NaOH	1.36	–	[27]
Mn/V	0.3 M V^{3+} in 5 M H_2SO_4/0.3 M Mn^{2+} in 5 M H_2SO_4	1.77	63	[28]
Zn/Ce	1.0 M M SA,0.152 Ce^{III}, 4 M MSA 0.152 Ce^{III}	2.2	50	[29]
Fe/Fe	$FeCl_2$ (negative) $FeCl_2$ & $FeCl_3$ (positive)	1.21	>45	[8]
Zn/I	ZnI_2 in aqueous	1.29	>80	[30]

12.6 NON-AQUEOUS ELECTROLYTES FOR REDOX FLOW BATTERIES

Non-aqueous electrolytes used in RFBs include organic electrolytes, ionic liquids, and solid electrolytes. Organic electrolytes are liquid or gel-like materials composed of small organic molecules or polymers. They are typically made of carbon, hydrogen, and oxygen atoms. Organic electrolytes are often used in lithium-ion batteries and have been investigated for use in RFBs as well. They have the advantage of being non-toxic and non-flammable, but they can be expensive, and their stability can be affected by temperature and humidity.

Ionic liquids are a type of non-aqueous electrolyte composed of positively and negatively charged ions. They are liquid at room temperature and have a low vapor pressure, which makes them attractive for use in RFBs. Ionic liquids have a wide electrochemical window and are non-flammable and non-volatile, but they can be expensive, and their viscosity can change with temperature.

All these non-aqueous electrolytes have their own advantages and challenges. Organic electrolytes are non-toxic and non-flammable, but they can be expensive, and their stability can be affected by temperature and humidity. Ionic liquids are non-flammable and non-volatile, but they can be expensive, and their viscosity can change with temperature. Solid electrolytes are non-flammable and non-volatile, but they can be difficult to fabricate, and their conductivity is often lower than that of liquid electrolytes. In recent years, researchers have been working on developing new non-aqueous electrolytes for use in RFBs. One promising area of research is the use of redox-active polymers as the electrolyte. These polymers can be synthesized to have a wide range of redox potentials and can be made to be conductive, making them a potential replacement for traditional liquid electrolytes.

Another area of research is the use of gel electrolytes. Gel electrolytes are made by mixing a liquid electrolyte with a polymer, creating a gel-like material that can be used as an electrolyte in RFBs. The Li/LiFePO4 cells assembled with the boron centered single-ion-conducting polymer (PVDF) gel electrolytes exhibit excellent rate performance and cycling performance. The initial discharge capacity is 161.3 mAh g^{-1} at 0.1 C, and a capacity of 127.7 mAh g^{-1} maintains with nearly 100% Coulombic efficiency after 100 cycles [31]. The advantage of gel electrolytes is that they can be made to be more stable and less prone to leakage than traditional liquid electrolytes.

Non-aqueous RFBs have several advantages over aqueous RFBs, including higher energy densities, wider operating temperatures, and longer lifetimes. However, non-aqueous RFBs also have some challenges, such as the need for more expensive materials and the risk of leakage and fire. The research progress on these systems leads to the development of several non-aqueous redox flow batteries, some of them are depicted in Table 12.2.

12.7 LOW-COST AND ENVIRONMENTALLY FRIENDLY ELECTROLYTES FOR REDOX FLOW BATTERIES

Redox flow batteries (RFBs) are a type of rechargeable battery that utilizes liquid electrolytes to store and release energy. They are considered a promising technology for large-scale energy storage applications, such as grid stabilization and renewable

TABLE 12.2
Types of Non-Aqueous Redox Flow Battery

Type	Electrolyte Condition	Open Cell Voltage (OCV) (V)	Ref.
$[Ru(bpy)_3]\ (BF_4)_2$	0.02M $[Ru(bpy)_3](RF_4)_2$/0.1M $TEABF_4$ in CH_3CN	2.6	[32]
$Ru(acac)_3$	0.1M $Ru(acac)_3$/1M $TEABF_4$ in CH_3CN	1.3	[33]
$V(acac)_3$	0.01M $V(acac)_3$/0.5M $TEABF_4$ in CH_3CN	2.2	[34]
$V(acac)_3$	0.05M $V(acac)_3$/0.5M $TEABF_4$ in CH_3CN	1.7	[35]
$Mn(acac)_3$	0.05M $Mn(acac)_3$/0.5M $TEABF_4$ in CH_3CN	1.1	[36]
$V(acac)_3$	0.01M $V(acac)_3$/0.05M $TEABF_4$ in CH_3CN	2.0	[37]
$V(acac)_3$	0.01M $V(acac)_3$/0.2M $TEAPF_6$ in CH_3CN	2.3	[37]
$Ru(acac)_3$	0.02M $Ru(acac)_3$/0.1M $TEABF_4$ in CH_3CN	1.75	[37]
$Ni(bpy)_3/Fe(bpy)_3$	Anolyte: 0.2M $Ni(bpy)_3$/0.5M $TEABF_4$ in propylene carbonate Catholyte: 0.4M $Fe(bpy)_3$/0.5M TEABF4 in propylene carbonate	2.25	[37]

energy integration. One of the main challenges in developing cost-effective and environmentally friendly electrolytes for RFBs is the high cost of the materials used. The most commonly used electrolytes in RFBs are vanadium and sulfur-based electrolytes, which are relatively expensive. To reduce the cost of RFBs, researchers are exploring alternative electrolytes that are less expensive yet still provide the necessary performance characteristics.

Iron-based electrolytes are one of the most promising alternative electrolytes for RFBs. Iron is a relatively cheap and abundant element, making it an attractive choice for RFBs. Additionally, iron-based electrolytes have been shown to have similar performance characteristics to vanadium- and sulfur-based electrolytes, making them a viable alternative. Another alternative electrolyte that has been explored is potassium hydroxide (KOH)-based electrolytes. KOH is a relatively cheap and readily available material, making it an attractive option for RFBs. Additionally, KOH-based electrolytes have been shown to have similar performance characteristics to vanadium- and sulfur-based electrolytes, making them a viable alternative.

In addition to cost-effective materials, environmentally friendly electrolytes for RFBs need to be non-toxic and non-corrosive. Electrolytes used in RFBs are often toxic and corrosive, which can have negative environmental impacts. Alternative electrolytes, such as potassium hydroxide and lithium-ion-based electrolytes, are less toxic and corrosive than traditional electrolytes, making them a more environmentally friendly option. Furthermore, the use of renewable and sustainable materials in RFBs is another way to make them more environmentally friendly. For example, researchers have explored the use of bi-functional electrolytes, which are made from sustainable and renewable materials such as biomass. These bi-functional electrolytes can improve the overall performance of RFBs while also reducing their environmental impact.

In conclusion, the cost and environmental impact of the electrolytes used in RFBs are significant factors that need to be addressed to make them more cost-effective and environmentally friendly. Alternative electrolytes, such as iron-based, potassium hydroxide-based, and lithium-ion-based electrolytes, are less expensive and less toxic than traditional electrolytes, making them a viable option. Additionally, the use of renewable and sustainable materials in RFBs can also improve their environmental impact. Further research and development in this area will help make RFBs more cost-effective and environmentally friendly.

12.8 FUTURE PROSPECTS

The future of electrolytes used in RFBs is likely to be shaped by multiple factors, including advances in materials science, the development of new energy storage technologies, and changes in global energy markets. One of the most important trends in the field is the ongoing search for new electrolyte materials that can improve performance and reduce the cost of RFBs. This research is driven by the need to improve the energy density, power density, and life cycle of RFBs, as well as reduce their environmental impact. However, some electrolytes will be degraded based on their stability and the environment in which they are employed.

One of the most promising developments in this area is the use of new aqueous electrolytes, such as lithium-sulfur and lithium-oxygen batteries. These new electrolytes have the potential to significantly improve the energy density and power density of RFBs, making them more competitive with other energy storage technologies, such as lithium-ion batteries. Additionally, they are non-toxic and abundant, making them more environmentally friendly than traditional aqueous electrolytes.

Another area of active research is the development of new non-aqueous electrolytes for use in RFBs. These electrolytes, which are typically composed of organic compounds, are designed to be stable at high temperatures and voltages, making them well-suited for use in high-performance applications such as electric vehicles and grid-scale energy storage. Some of the most promising non-aqueous electrolytes under development include lithium-ion electrolytes, proton exchange membrane electrolytes, and redox flow electrolytes.

In addition to advances in electrolyte materials, the future of RFBs is also likely to be shaped by the development of new energy storage technologies, such as supercapacitors, which are becoming increasingly popular for use in high-power applications. Supercapacitors are able to store and release electrical energy much faster than batteries, and they have a much longer life cycle. They may be used in RFBs as a hybrid system to improve the system's performance.

Finally, changes in global energy markets are likely to play an important role in shaping the future of RFBs. As the world continues to move towards renewable energy sources such as solar and wind power, the need for efficient and cost-effective energy storage solutions will become increasingly important. RFBs, which are well suited for use in renewable energy systems, are likely to become an increasingly important component of the global energy storage landscape in the years to come.

12.9 CONCLUSIONS

Electrolytes play a crucial role in the functioning of redox flow batteries. Various types of electrolytes have been developed and studied, including aqueous and non-aqueous options. Aqueous electrolytes, such as sulfuric acid, have been widely used in redox flow batteries due to their low cost and high conductivity. However, they can also suffer from issues such as corrosion and degradation of the electrodes over time. Non-aqueous electrolytes, such as ionic liquids, have been proposed as an alternative option due to their high stability and safety. However, they are generally more expensive and have lower conductivity compared to aqueous electrolytes.

Recent research has focused on developing new electrolyte formulations such as hybrid electrolytes to overcome the limitations of traditional electrolytes and improve the performance of redox flow batteries. Hybrid electrolytes, which combine the advantages of aqueous and non-aqueous electrolytes, have been shown to have higher stability and longer lifetimes compared to traditional electrolytes.

In addition, various types of redox flow batteries have been developed based on different electrolytes, such as vanadium redox flow batteries and zinc-bromine flow batteries. These batteries have shown to have a high energy density and high efficiency, making them suitable for large-scale energy storage applications.

Overall, the development of new and improved electrolytes is an active area of research in the field of redox flow batteries. Advancements in electrolytes are expected to play a key role in making these batteries more efficient and cost-effective, which could lead to broader adoption of this technology in various applications, including renewable energy storage and grid-scale energy storage. Therefore, the electrolytes used in redox flow batteries are an important aspect to be considered for the development of these batteries.

REFERENCES

1. H. Zhang and C. Sun, *J. Power Sources*, 2011, 493, 229445.
2. Y. Yao, J. Lei, Y. Shi, F. Ai and Y.C. Lu, *Nat. Energy*, 2021, 6, 582–588.
3. K. Zhang, J. Xiong, C. Yan and A. Tang, *Appl. Energy*, 2020, 272, 115093.
4. Y. Song, X. Li, C. Yan and A. Tang, *J. Power Sources*, 2020, 456, 228004.
5. A.H. Whitehead, T.J. Rabbow, M. Trampert and P. Pokorny, *J. Power Sources*, 2017, 351, 1–7.
6. A. Dinesh, A. Ramadas, M.S. Anantha, M.K. Umesh, K. Venkatesh, M. Kundu, H.B. Muralidhara and K.Y. Kumar, *J. Energy Storage*, 2022, 55, 105487.
7. Hawthorne, J.S. Wainright, and R.F. Savinell. *J. Electrochem. Soc.*, 2014. 161(10), 1662–1671.
8. L.W. Hruska and R.F. Savinell, *J. Electrochem. Soc.*, 1981, 128, 18–25.
9. D. Anarghya, M.S. Anantha, K. Venkatesh, M.S. Santosh, M.G. Priya and H.B. Muralidhara, *J. Electroanal. Chem.*, 2020, 878, 114577.
10. M.S. Anantha, D. Anarghya, C. Hu, N. Reddy, K. Venkatesh and H.B. Muralidhara, *J. Mater. Sci. Mater. Electron.*, 2020, 31, 15286–15295.
11. J. Noack, N. Roznyatovskaya, T. Herr and P. Fischer, *Angew. Chemie - Int. Ed.*, 2015, 54, 9776–9809.
12. F.B. Kolloquium and F. Gibbard, *Vanadium-Bromine Redox Flow Battery*, (2017).
13. T. Long, Y. Long, M. Ding, Z. Xu, J. Xu, Y. Zhang, M. Bai, Q. Sun and C. Jia, *Nano Res.*, 2021, 14, 3538–3544.
14. S. Wang, Z. Xu, X. Wu, H. Zhao, J. Zhao, J. Liu, C. Yan and X. Fan, *Electrochim. Acta*, 2021, 368, 137524.

15. Y.K. Zeng, T.S. Zhao, X.L. Zhou, L. Wei and Y.X. Ren, *J. Power Sources*, 2017, 346, 97–102.
16. Y.K. Zeng, T.S. Zhao, X.L. Zhou, L. Wei and H.R. Jiang, *J. Power Sources*, 2016, 330, 55–60.
17. W. Wang, S. Kim, B. Chen, Z. Nie, J. Zhang, G.G. Xia, L. Li and Z. Yang, *Energy Environ. Sci.*, 2011, 4, 4068–4073.
18. X. Zhou, L. Lin, Y. Lv, X. Zhang and Q. Wu, *J. Power Sourc*es, 2018, 404, 89–95.
19. G.P. Rajarathnam, *The Zn-Bromine Flow Battery*, The University of Sydney, 2016.
20. P.K. Adanuvor, R.E. White and S.E. Lorimer, *J. Electrochem. Soc.*, 1987, 134, 1450–1454.
21. G.P. Rajarathnam, A. Montoya and A.M. Vassallo, *Electrochim. Acta*, 2018, 292, 903–913.
22. M.C. Wu, T.S. Zhao, L. Wei, H.R. Jiang and R.H. Zhang, *J. Power Sources*, 2018, 384, 232–239.
23. M.C. Wu, T.S. Zhao, H.R. Jiang, Y.K. Zeng and Y.X. Ren, *J. Power Sources*, 2017, 355, 62–68.
24. M. Lopez-Atalaya, G. Codina, J.R. Perez, J.L. Vazquez and A. Aldaz, *J. Power Sources*, 1992, 39, 147–154.
25. M. Rychcik and M. Skyllas-Kazacos, *J. Power Sources*, 1988, 22, 59–67.
26. Q. Lai, H. Zhang, X. Li, L. Zhang and Y. Cheng, *J. Power Sources*, 2013, 235, 1–4.
27. C. Ponce de León, A. Frías-Ferrer, J. González-García, D.A. Szánto and F.C. Walsh, *J. Power Sources*, 2006, 160, 716–732.
28. F.Q. Xue, H.T. Zhang, C.X. Wu, T. Ning and X. Xu, *Trans. Nonferrous Met. Soc. China*, 2009, 19, s594–s599.
29. L.F. Arenas, F.C. Walsh and C.P. de León, *ECS J. Solid State Sci. Technol.*, 2015, 4, P3080–P3085.
30. G. M. Weng, Z. Li, G. Cong, Y. Zhou and Y. Lu, *Energy Environ. Sci.*, 2017, 10, 735–741.
31. Z. Xingfa, D. Linna, F. Jifang, C. Liya, Z. Jia, Z. Peisong, L. Guozhen and S. Liyi, *J. Chem. Eng.*, 2022, 428, 131100.
32. Y. Matsuda, K. Tanaka, M. Okada, Y. Takasu, M. Morita and T. Matsumura-Inoue, *J. Appl. Electrochem.*, 1988, 18, 909–914.
33. M.H. Chakrabarti, R.A.W. Dryfe and E.P.L. Roberts, *Electrochim. Acta*, 2007, 52, 2189–2195.
34. A.A. Shinkle, A.E.S. Sleightholme, L.T. Thompson and C.W. Monroe, *J. Appl. Electrochem.*, 2011, 41, 1191–1199.
35. A.A. Shinkle, T.J. Pomaville, A.E.S. Sleightholme, L.T. Thompson and C.W. Monroe, *J. Power Sources*, 2014, 248, 1299–1305.
36. A.E.S. Sleightholme, A.A. Shinkle, Q. Liu, Y. Li, C.W. Monroe and L.T. Thompson, *J. Power Sources*, 2011, 196, 5742–5745.
37. S.H. Shin, S.H. Yun and S.H. Moon, *RSC Adv.*, 2013, 3, 9095–9116.

Section III

Supercapacitors

13 Aqueous Electrolytes for Electrochemical Supercapacitors

P. Periasamy and Murthy Chavli

13.1 INTRODUCTION TO ENERGY STORAGE DEVICES

13.1.1 Pioneer of Energy Storage Devices

The energy storage device is a system that stores electrical energy for later use. Energy storage devices are essential in many applications (Figure 13.1), including renewable energy systems, electric vehicles, and portable electronics [1, 2].

There are several types of energy storage devices:

- ***Batteries***: The conversion of chemical energy into electrical energy is performed by batteries, widely used as energy storage devices in various applications such as smartphones, laptops, electric vehicles, and grid-scale energy storage systems.
- ***Capacitors***: Capacitors store electrical energy in an electric field and are often used in applications that require short bursts of power, such as camera flashes and electric vehicles.
- ***Flywheels***: Flywheels store energy in a rotating mass and are often used in grid-scale energy storage systems to provide quick bursts of power.
- ***Hydrogen storage systems***: Hydrogen storage systems store energy in the form of hydrogen, which can be used to generate electricity through a fuel cell.
- ***Thermal energy storage systems***: Thermal energy storage systems store energy in the form of heat, which can be used to generate electricity through a heat engine or a steam turbine.

The world's shift towards a cleaner and more sustainable energy system has heightened the significance of energy storage devices. They are pivotal in integrating renewable energy sources like solar and wind into the electrical grid, enhancing its efficiency and reliability. Over time, several pioneers have conducted extensive

DOI: 10.1201/9781003441182-16

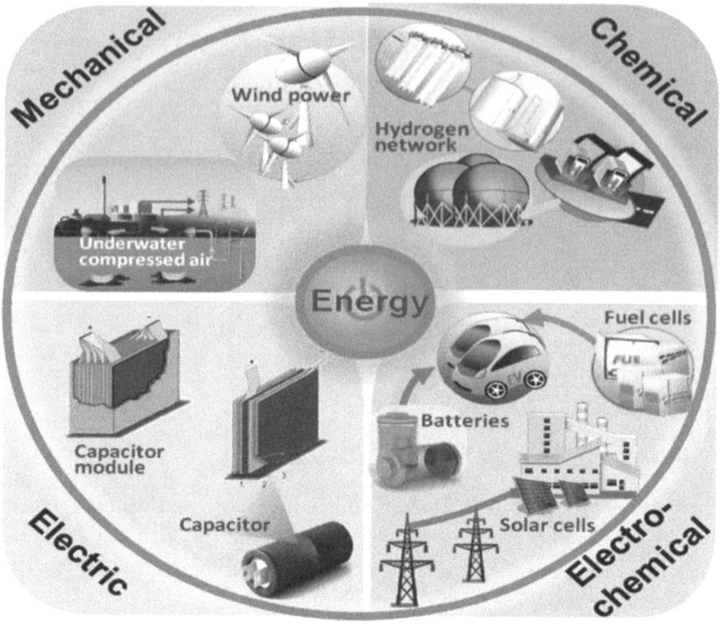

FIGURE 13.1 Energy storage devices.

Source: Reproduced with permission from ref [3]. Copyright 2018 Wiley.

research into energy storage devices [4]. However, some notable individuals who have made significant contributions in this field include:

- *John B. Goodenough*, a scientist who specialized in materials and solid-state physics, invented the lithium-ion battery in the 1980s, which has since gained widespread use in consumer electronics, electric vehicles, and energy storage systems.
- *Akira Yoshino* is a Japanese chemist who developed the first commercially viable lithium-ion battery in the 1980s while working at Asahi Kasei Corporation. His work led to the development of high-performance, lightweight, and rechargeable batteries.
- *Michael Faraday* was a scientist who significantly contributed to electromagnetism and electrochemistry. He discovered the principles of electrochemical energy storage in the early 19th century and is considered one of the pioneers of modern battery technology.
- In 1800, an Italian physicist, *Alessandro Volta*, created the first battery known as the "Voltaic pile." His invention laid the foundation for modern battery technology and helped establish the principles of electrochemistry.
- *Thomas Edison* was an American inventor best known for developing the electric light bulb. However, he also made significant contributions to the field of energy storage, including setting nickel-iron and alkaline storage batteries.

The above-mentioned individuals, among others, have played an essential role in advancing the field of energy storage devices and thereby paved the way for developing new and innovative technologies.

An electrochemical supercapacitor [1] is an energy storage device that can quickly store and release much energy. It is a type of capacitor that uses electrostatic forces to store energy, but it has a much higher capacitance than traditional capacitors due to electrochemical processes. The fundamental components of supercapacitors are two electrodes, electrolyte, and a separator. The electrodes usually comprise activated carbon or other porous materials and metal oxides like ruthenium or manganese dioxide. The electrolyte can be an aqueous or organic solvent, depending on the application. The separator is typically a thin layer of material that prevents the electrodes from coming into contact [5].

The energy in a supercapacitor is stored through two main mechanisms: electrical double-layer capacitance and pseudocapacitance [3]. Electrical double-layer capacitance (EDLC) is based on separating charges at the interface between the electrode and the electrolyte. The electrode material typically consists of a high surface area of carbon material with many pores. When a voltage is applied, ions from the electrolyte accumulate at the electrode surface, forming a double layer of charge. The second type (pseudocapacitance) occurs due to reversible redox reactions in the electrode material, leading to charge storage within the electrode. Pseudocapacitive materials such as transition metal oxides, hydroxides, and conductive polymers typically have a higher capacitance than EDLC materials since the charge is not limited to the surface of the electrode.

In addition to these two main mechanisms, supercapacitors can store energy through a combination of EDLC and pseudocapacitance, known as a hybrid supercapacitor. This hybrid mechanism leads to higher energy densities than EDLC and pseudocapacitance. Supercapacitors have several advantages over traditional batteries [5, 6]. Supercapacitors are ideal for applications that demand quick energy transfer, such as regenerative braking in hybrid vehicles, due to their ability to charge and discharge much faster than batteries. They also have a longer cycle life than batteries, which can be charged and discharged many times without significant degradation. However, they have a lower energy density than batteries, meaning that they cannot store as much energy per unit volume or weight.

In recent years, research has focused on developing new materials and designs to improve their supercapacitor performance. To increase surface area and capacitance, researchers have explored using graphene or other two-dimensional materials as electrodes [7]. Other researchers have studied hybrid devices that combine batteries' high energy density with supercapacitors' high power density. Electrochemical supercapacitors show promise as a technology for energy storage and may play a vital role in the shift toward a low-carbon energy system. These devices store electrical energy by gathering charge at the electrode-electrolyte interface. The performance of supercapacitors is comprehensively dependent on the electrode materials used, which determine specific capacitance, energy density, power density, and cycling stability [8].

Some critical factors that make electrode materials crucial in supercapacitors are:

- ***High surface area*:** Electrode materials with a high surface area provide a more extensive electrode-electrolyte interface, increasing the supercapacitor's capacitance.
- ***High electrical conductivity*:** Electrode materials with high electrical conductivity improve the charge transfer rate between the electrode and electrolyte, leading to better energy storage and faster charging/discharging.
- ***Chemical stability*:** The electrode materials should be chemically stable in the electrolyte to prevent degradation and maintain the long-term performance of the supercapacitor.
- ***Low cost and abundance*:** The price and abundance of electrode materials are essential considerations in commercializing supercapacitors.
- ***Environmental friendliness*:** The electrode materials should not contain toxic or hazardous substances that could harm the environment or human health.

Commonly used electrode materials for supercapacitors include carbon-based materials such as activated carbon, [9] graphene, [10] carbon nanotubes, and conducting polymers such as polypyrrole, polyaniline, and polythiophene. However, there is ongoing research to develop new electrode materials that can improve the performance and sustainability of supercapacitors. Electrolytes are a critical component in supercapacitors, as they provide the medium for ion transport between the electrodes, enabling energy storage and power delivery [11].

Electrolytes in supercapacitors play several vital roles, including:

- ***Conductivity*:** Electrolytes must be highly conductive to facilitate the flow of ions between the electrodes. High ionic conductivity is essential for achieving high power density, a critical requirement for many supercapacitor applications.
- ***Stability*:** Electrolytes must be stable over a wide range of operating conditions to avoid degradation of the electrodes and other components in the device. Stability is significant for high-voltage and high-temperature applications.
- ***Safety*:** Electrolytes must be non-toxic and non-flammable to ensure the device's and user's security. Safety concerns regarding flammability have been raised regarding organic electrolytes, which have driven the development of safer alternatives like aqueous and ionic liquid electrolytes.
- ***Compatibility*:** Electrolytes must be compatible with the electrode materials used in the supercapacitor to avoid unwanted reactions that can degrade performance.
- ***Cost*:** Electrolytes should be inexpensive and readily available to enable the commercialization of supercapacitors.

A wider voltage window allows for a more considerable potential difference between the electrodes, resulting in a higher energy density. The principle underlying the charge storage in a capacitor is based on the proportional relationship between the potential difference between the electrodes and the amount of charge stored. A wider voltage window can also activate more electrode surface area, increasing capacitance. This approach, however, necessitates an electrolyte that can survive at high voltage range without degrading, which might affect the performance and lifespan of the supercapacitor.

13.2 CLASSIFICATIONS OF ELECTROLYTES

Common electrolytes used in supercapacitors include aqueous, organic, and ionic liquids (Table 13.1) [12, 13]. The widespread adoption of aqueous electrolytes can be attributed to their affordability, remarkable ionic conductivity, and minimal flammability. Organic electrolytes have higher conductivity but are less stable and vastly flammable. Ionic liquid electrolytes have excellent stability and high conductivity but are more expensive than aqueous and organic electrolytes.

The selection of electrolytes for supercapacitors depends on the specific application requirements, the electrode material's properties, and the device's desired performance characteristics. Aqueous and non-aqueous electrolytes differ in chemical composition and properties, affecting their suitability in different energy storage devices, such as supercapacitors. Aqueous electrolytes typically comprise water and dissolved salts, while non-aqueous electrolytes contain organic solvents and dissolved salts.

Aqueous electrolytes generally have a lower ionic conductivity than non-aqueous electrolytes. Despite this, they possess a broader electrochemical stability range, rendering them more appropriate for high-voltage window supercapacitors. Non-aqueous electrolytes typically have a higher ionic conductivity than aqueous electrolytes but have a narrower electrochemical stability window. Selecting a suitable electrolyte can significantly influence the electrode material's compatibility with supercapacitors.

Aqueous electrolytes are generally more compatible with carbon-based electrodes, while non-aqueous electrolytes are more compatible with metal oxide-based electrodes. Solid-state electrolytes are a relatively new type of electrolyte composed of solid materials. Solid-state electrolytes have the advantage of being more stable and safer than liquid electrolytes, but their lower conductivity limits them compared to liquid electrolytes.

The electronic structure of molecules and ions could be explained through the interactions of atomic orbitals using a theoretical framework known as the molecular orbital theory. The electronic structure of the ions constituting an electrolyte solution can be comprehended using the molecular orbital theory concept within the realm of electrolytes. For example, the electronic structure of a sodium ion (Na^+) can be described using molecular orbital theory as follows:

- The atomic orbitals of the sodium atom combine to form a set of molecular orbitals in the sodium ion.
- The lowest energy molecular orbital is a bonding orbital formed by the overlap of the 3s orbital of the sodium atom with the 2p orbitals of the neighboring atoms.

TABLE 13.1
Classification of Electrolytes

Electrolytes			
Redox Active Electrolyte	**Solid State or Quasi Solid State Electrolyte**	**Liquid Electrolyte**	
		Aqueous electrolyte	Non-aqueous electrolyte
• HQ/H_2SO_4 • $VOSO_4/H_2SO_4$ • KI/H_2SO_4 • $AQS)+KBr/H_2SO_4$ • NH_4VO_3/H_2SO_4 • ARS+PPD/KOH • $K_3Fe(CN)_6/KOH$ • KBr/Na_2SO_4 • KI/IL • BMIMBr IL/ionogel	1. Polymer electrolytes • PEO • PAN • PVDF 2. Ceramic electrolytes • $LiAlO_2$ • $Li_7La_3Zr_2O_{12}$ 3. Glass electrolytes • LiPON • LiBOB 4. Gel electrolytes • Polyacrylamide • Polyethylene glycol • Polyvinyl alcohol • Ionic liquids	• H_2SO_4 • KOH • KCL • NaCl • NaOH • NH_4Cl • HNO_3 • H_3PO_4	1. Organic solvents: • $C_3H_6O_3$ • $C_5H_{10}O_3$ • $C_3H_4O_3$ • $C_4H_6O_3$ • C_2H_6OS 2. Ionic liquids • PYR13TFSI • EMImTFSI • BMImPF6 • DEMEBF4 3. Carbonate-based electrolytes • LiPF6 • LiTFSI 4. Polymeric electrolytes • PAN • PVDF 5. Ether-based electrolytes: • TEGDME • PEGDME

- The next highest energy molecular orbital is an antibonding orbital formed by the overlap of the 3s orbital of the sodium atom with the 2p orbitals of the neighboring atoms but with a phase difference that leads to destructive interference.
- The remaining molecular orbitals are non-bonding, not contributing to the bonding or antibonding interactions between the sodium ion and neighboring ions.
- Similar analyses can be performed for other ions in the electrolyte solution, such as Chloride ions (Cl^-), Sulfate ions (SO_4^{2-}), etc.

13.3 FUNCTIONING OF ELECTROLYTES IN SUPERCAPACITORS

Electrolytes play a crucial role in the functioning of supercapacitors by facilitating the charge storage and release process. The functioning of electrolytes in supercapacitors involves several key processes (Figure 13.2):

- ***Charge Separation***: Supercapacitors consist of two electrodes, typically made of porous carbon materials, immersed in an electrolyte solution. When a voltage is applied across the electrodes, charge separation occurs at the electrode-electrolyte interfaces. Positive ions (cations) are attracted to the negatively charged electrode (cathode), while negative ions (anions) are attracted to the positively charged electrode (anode).
- ***Ion Migration***: The electrolyte allows for the movement of ions between the two electrodes. Cations and anions migrate through the electrolyte to balance the charge and maintain electrical neutrality. This ion migration is facilitated by the conductivity of the electrolyte.
- ***Double-Layer Capacitance***: In supercapacitors utilizing electrostatic double-layer capacitance, the charge storage mechanism primarily involves

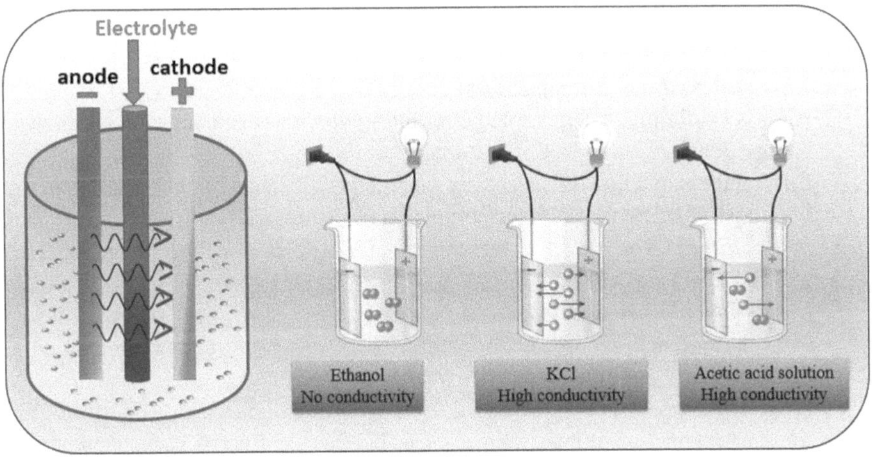

FIGURE 13.2 Functioning of electrolytes.

the formation of an electrical double layer at the electrode-electrolyte interface. This double layer consists of an inner layer of adsorbed ions on the electrode surface and an outer layer of counter-ions in the electrolyte. The separation of charges at this interface contributes to the capacitance of the supercapacitor.

- ***Pseudocapacitance*:** Some supercapacitors incorporate materials that exhibit pseudocapacitance, which involves reversible faradaic redox reactions at the electrode-electrolyte interface, leading to additional charge storage beyond the double-layer capacitance. The electrolyte plays a crucial role in facilitating these redox reactions by providing ions for the electrode surface reactions.
- ***Ion Permeability*:** The electrolyte also needs to allow for efficient ion transport between the electrode surfaces, ensuring low internal resistance and high power density of the supercapacitor. The choice of electrolyte composition and properties significantly affects ion permeability and overall device performance.

Commonly used electrolytes in supercapacitors include aqueous electrolytes (e.g., sulfuric acid solution) and non-aqueous electrolytes (e.g., organic solvents with lithium salts). The selection of an appropriate electrolyte depends on factors such as the desired voltage range, energy density, power density, and operating temperature of the supercapacitor. By enabling charge separation, ion migration, and facilitating charge storage mechanisms, electrolytes play a vital role in the efficient operation of supercapacitors, allowing them to store and deliver electrical energy rapidly. Every electrolyte type has its pros and cons, and selecting an appropriate electrolyte for a specific application necessitates meticulously evaluating the above listed characteristics.

13.4 CHALLENGES AND OPPORTUNITIES

13.4.1 Challenges

- ***Energy density*:** Supercapacitors' energy density is inferior to that of batteries, constraining their implementation in applications necessitating high energy density.
- ***Cost*:** Supercapacitors are currently more expensive than batteries due to the high cost of electrode and electrolyte materials.
- ***Scalability*:** Developing scalable manufacturing processes for supercapacitors remains a challenge.
- ***Durability*:** Supercapacitors can degrade over time due to electrode-electrolyte reactions or delamination, affecting their performance and lifespan.
- ***Standardization*:** The industry lacks standardization, making it difficult for customers to compare and choose between different supercapacitor devices.

13.4.2 Opportunities

- ***Energy storage integration*:** Supercapacitors can be combined with batteries to create hybrid energy storage systems that benefit both technologies.
- ***Electric vehicle applications*:** Supercapacitors can potentially replace or supplement batteries in electric vehicles, offering fast charging and discharging rates and improved energy efficiency.
- ***Renewable energy storage*:** Supercapacitors can store energy generated from renewable sources such as wind and solar, helping to smooth out fluctuations in power output.
- ***Wearable electronics*:** Supercapacitors have the potential to power small wearable devices due to their small size and high power density.
- ***Smart grid applications*:** Supercapacitors can provide frequency regulation and energy balancing in smart grid applications, helping to improve grid stability and efficiency.

Though supercapacitors face challenges such as lower energy density, higher cost, and durability issues, they offer several opportunities in electric vehicles, renewable energy storage, and smart grid applications. Ongoing research and development efforts aim to address these challenges and unlock the full potential of supercapacitor technology.

13.5 COMPATIBILITY OF ELECTROLYTES IN ELECTROCHEMICAL SUPERCAPACITORS TOWARD HIGH SPECIFIC CAPACITANCE

Electrolyte affects a supercapacitor's capacitance, voltage stability, and energy density. Multiple factors influence the compatibility of electrolytes in electrochemical supercapacitors, such as the electrode material type, operational voltage range, and desired device performance characteristics. Generally, the electrolyte should wet the electrode surface efficiently, provide good ionic conductivity, and exhibit good stability over the operating voltage range [14]. Figure 13.3 visually represents the reaction process between electrolytes and diverse electrodes for EDLCs, pseudocapacitors, and hybrid supercapacitors [11].

The electrochemical reactions transpiring at an electrode and electrolyte interface entail charge transfer, mass transport, and phase transformation. Wetting or superwetting behaviors at multi-phase interfaces, such as gas bubbles, oil droplets, or solid islands or films, can impede the efficacy of these reactions. The adverse effects of these behaviors can extend to the performance of electrochemical energy conversion and storage devices, industrial production, and environmental conservation [14].

Hydrogels' remarkable flexibility, ionic conductivity, and inherent safety intrigue them as electrolyte compounds for creating flexible and secure quasi-solid-state supercapacitors. A group of researchers [16] proposed bacterial cellulose (BC) aerogel infused with sodium polyacrylate/potassium hydroxide (PANa/KOH) hydrogel as a newfangled hydrogel electrolyte for flexible alkaline supercapacitors. This particular hydrogel is reinforced with potassium hydroxide. The outstanding mechanical

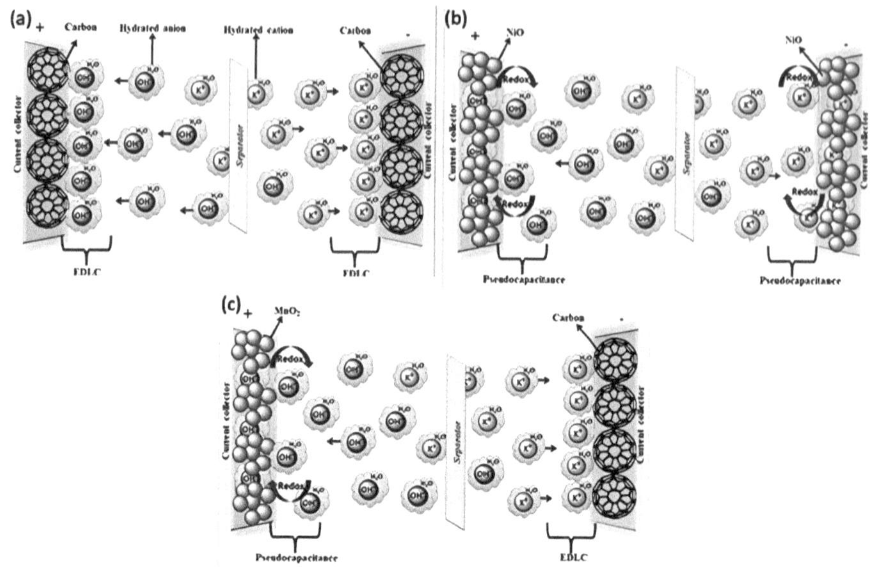

FIGURE 13.3 Graphical representation of reaction mechanism. (a) EDLCs; (b) pseudocapacitors; and (c) hybrid supercapacitors.

Source: Adopted from Ref. [11] under creative common license.

property, excellent anti-swelling performance, solid ionic conductivity, and superior anti-freezing ability were shown by the PANa/BC/KOH hydrogel electrolyte.

Using these fascinating combined properties, the PANa/BC/KOH hydrogel electrolyte and CoNi LDH@CC/AC@CC electrodes of an asymmetric supercapacitor (SC-PANa/BC/KOH) produced an excellent energy density of 25.63 Wh kg^{-1} at a power density of 1200 W kg^{-1}. They maintained superior retention of 92.05% over 10,000 cycles. The SC-PANa/BC/KOH, on the other hand, demonstrated excellent versatility and safety. It is well-suited for numerous challenging situations, including heavy-weight loads, bending, and hammering.

Deng [15] intended for the preparation of a redox-active polyvinyl alcohol (PVA)-KOH-urea-$LiClO_4$ alkaline gel electrolyte, which was later employed in the assembly of a Co oxide/Mn oxide wearable asymmetric supercapacitor (WASC). The experiment's objectives of improving redox performance and acting as a plasticizer were met by combining an alkaline KOH-PVA gel with a deep eutectic solvent, urea-$LiClO_4$. The electroplating method was used to prepare nanoporous electrodes of cobalt oxide and manganese oxide. The WASC demonstrated solid performance (186 F/g at 2 A/g) and maintained 89.5% of its capacitance after being subjected to 6000 cycles.

Possible energy storage material options include anhydrous NiC_2O_4 QDs [16]. This is due to the existence of planer oxalates anions ($C_2O_4^{2-}$), which create a quasi-zero 2d arrangement, which could make them a strange way to store energy. In an aqueous 2 M KOH electrolyte, the anhydrous NiC_2O_4 QDs electrodes demonstrated superior specific capacity, with 227.5 mAh/g at 1A/g in the 0.5 V potential window. It is thought that the substantial charge storage capacity of anhydrous

NiC_2O_4 QDs arises from the operation of the intramolecular pseudocapacitance process, akin to that of batteries. The optimum cycle stability and maximum specific energy and power were observed in 1 M Na_2SO_4 for the complete cell of NiC_2O_4 QDs in ASCs mode.

In a study conducted by Zhang & Chen [19], the charge storage mechanism and electrochemical characteristics of α-$Co(OH)_2$ in a room-temperature ionic liquid (IL) were investigated and compared with those observed in 6 M KOH. It is noteworthy that, even after undergoing 10,000 cycles at 0.15 mA cm^{-2} during a half-cell test, α-$Co(OH)_2$ in [BMPyr+][DCA] IL maintained its capacity retention at an impressive 95.9%. Scientists have successfully created a novel multifaceted polyvinyl alcohol/kappa-carrageenan alkaline gel polymer electrolyte through a green physical approach, utilizing a dual-network physical cross-linked hydrogel. Mondal et. al reported that [16] KOH was used not only as an ionic donor but also as an interagent. It is possible to construct a supercapacitor by combining an alkaline gel polymer electrolyte made of polyvinyl alcohol and kappa-carrageenan with activated carbon electrodes. The capacitor exhibits exceptional electrochemical behavior, with electrodes displaying a specific capacitance of 470 F/g at 0.5 A/g and retaining 95% or more even after undergoing 2000 charge/discharge cycles.

The microwave-assisted wet chemical method was used to synthesize the rod-shaped WO_3-V_2O_5 hybrid nanocomposites [17]. To enable straightforward comparisons, the electrochemical investigations of the fabricated composites were conducted in two distinct electrolytes: KOH and H_2SO_4. The maximum capacitance for WO_3 and WO_3-V_2O_5 composites in KOH electrolytes was found to be 44 F/g and 173 F/g. Moreover, their efficacy remains at 100% even after 5000 cycles, with capacitance stability of 126% in the same electrolyte. When tested in an H_2SO_4 electrolyte, the specific capacity of WO_3 was found to be 246 F/g, while that of the WO_3-V_2O_5 composite was significantly lower, at less than 12 F/g.

The electrochemical properties of interconnected porous carbon composites for electrode material in different electrolytes, which include acidic, alkaline, and neutral aqueous solutions, were investigated following KOH activation [18]. The electrochemical behavior of materials is influenced by several factors, including the morphology of interconnected porous carbon composites and the electrolyte's hydrated ionic radius, conductivity, and ionic mobility. These findings emphasize the importance of these factors in determining the electrochemical properties of materials. It has been revealed that ordered mesoporous carbon composites have comparatively higher capacitance and efficiency in alkaline and acidic electrolytes compared to those used in neutral electrolytes. By increasing the operating voltage to 1.7 V, OMC/G composite materials in neutral aqueous electrolytes can achieve the highest energy and power densities of 27.23 Wh kg^{-1} and 7528.5 W kg^{-1}, respectively.

Using $Mg(OAc)_2.4H_2O$ as a template, authors established a simple but successful technique to template carbonization in the current study [19], which converts folic acid into substantially nitrogenated nanoporous carbon materials. It has been observed that adding p-phenylenediamine to the KOH electrolyte as a redox additive has resulted in a significant increase in capacitance. The observed phenomenon can be mainly attributed to pseudocapacitance, generated by the redox reaction of p-phenylenediamine at the interface between the electrode and electrolyte.

Chemical co-precipitation was used to synthesize a nano-structured CoAl double hydroxide with a particle size of 60–70 nm [20]. As the positive electrode in an asymmetric hybrid supercapacitor, it was paired with an active carbon negative electrode immersed in a KOH electrolyte solution. Between 0.9 to 1.5 V, the hybrid supercapacitor achieved a specific capacitance of 77 F g^{-1} and a specific energy density of 15.5 Wh kg^{-1}. The supercapacitor retains 90% of its original capacity throughout a thousand charges and discharges, demonstrating excellent cycling performance.

The electrochemical performances of zinc oxide and nickel oxide ZnO/NiO@ MWCNT - ZNC were investigated under various electrolytes [21]. The doctor blade method was used to create improved working electrodes for ZNC NCs, and their electrochemical capabilities were evaluated using a variety of electrolytes. ZNC NC's modified active electrode had the best electrochemical performance compared to other electrolytes—as measured by its specific capacitance 1988.8 F/g in an aqueous electrolyte containing 1 M Na_2SO_4 and 0.1 M $K_4[Fe(CN)_6$ yielded better specific capacitance retention of 97.2% after 2000 cycles. Under the newly formed redox additives supplemented with Na_2SO_4 aqueous electrolyte, the developed asymmetrical type supercapacitors exhibited tremendous energy and power densities of 43.59 Wh kg^{-1} and 4000 W kg^{-1}, respectively.

The electrochemical properties of symmetric supercapacitors comprising activated carbon electrodes derived from cellulose acetate microspheres as a precursor were examined utilizing neutral aqueous sodium sulfate solutions as the electrolyte [22]. Studies of supercapacitors outside the standard room temperature settings have been very restricted, with most investigations concentrating on ionic liquids as electrolytes. This is even though electrolytes based on water may be utilized safely throughout a wide temperature range. The electrolytes were evaluated through their conductivity and viscosity at different temperatures.

Supercapacitors' stable and reliable performance using carbon electrodes in non-ambient conditions solely depends on the electrolyte's properties. The optimal electrolyte composition of such supercapacitors was found to be a neutral 1 M Na_2SO_4 solution, which yielded specific capacitances between 91 F g1 and 117 F g^{-1} at temperatures ranging from 0 to 60 °C and showed excellent cycling stability by maintaining at least 95% of the original capacitance after 2000 charging/discharging cycles. At 25 °C, energy densities of 16.6 W·h/kg and a power density of 16 kWh/kg were reported. However, at 60 °C, energy densities increase to 21.5 W·h/kg, while power densities drop to 13.7 kWh/kg. Unlike the H_2SO_4-based electrolytes used for comparison, no signs of electrochemical corrosion were observed on the battery housings during cycling.

Steam activation and chemical activation with potassium hydroxide produced four distinct types of activated carbons (ACs) [23], each with a specific surface area of about 1050 m^2 g^{-1}. The activated carbons utilized in this study were derived from the natural fir wood and pistachio shell sources. The activated carbons (ACs) derived from fir wood and treated with KOH showed the most outstanding capacitive results in H_2SO_4, primarily due to their excellent reversibility and high specific capacitance, exceeding 100 F g^{-1} in magnitude. The AgI/g-C_3N_4 nanocomposite electrode is perhaps formed by combining ultrasonic treatment and co-precipitation processes. The CV test performed on the aforementioned composite electrode demonstrates that the supercapacitor is stable and has excellent capacity retention after 2000 cycles of charge and discharge employing H_2SO_4 (85%), KOH (84%), and Na_2SO_4 (87%) electrolytes, correspondingly.

Polyethylene oxide (PEO) is initially used in conjunction with 5 wt.% potassium hydroxide (KOH) or lithium oxide to produce polymer cement electrolytes with a range of potential windows during the hydration synthesis of regular Portland cement [24]. The polymer cement electrolytes with a 2 wt.% polyethylene oxide (PEO) fundamentally display an optimal combination of flexural strength and electrochemical stability. Polymer cement electrolytes and rGO@$CuCo_2O_4$ nanowires were employed to assemble two supercapacitor devices with structural applications. As-prepared device (A) has an areal capacitance of 439.35 mF cm^{-2} and an energy density of 0.2 mWh cm^{-2}. The energy density delivered by device B is 0.74 mWh cm^{-2}, and the areal capacitance is 407.07 mF cm^{-2}. Consequently, the electrodes in devices A and B can achieve an energy efficiency of 18.1%, as well as 17.0%, accordingly. The clever combination of the electrode and PEO cement electrolyte in a structural supercapacitor device paves a balanced approach toward integrating zero-energy buildings and electrochemical energy storage.

Redox-active electrolytes and water-in-salt electrolytes improved the energy density of aqueous symmetric supercapacitors based on reduced graphene oxide (rGO-SCs) [25]. The rGO-SCs, which had an electrode mass loading of 9–10 mg cm^{-2}, demonstrated a high gravimetric specific capacitance, good cycle stability, and excellent rate in three different aqueous electrolytes (1 M H_2SO_4, 6 M KOH, and 1 M Li_2SO_4). The device's energy density was increased to 21.42 Wh kg^{-1} (at 180 W kg1) by using 0.1 M Na_2MoO_4 as the redox additive in Li_2SO_4. Reversible redox processes that occurred at the electrode surface accomplished this phenomenon. In addition, the rGO-SC obtained a high cell voltage of 2.1 V by using 11 M $NaNO_3$. This high-voltage supercapacitor maintained 98.1% of its capacitance after being subjected to 5000 cycles, resulting in an energy density of 22.87 watt-hours per kilogram (at 210 watts per kilogram).

Pharmaceutical medication wastes from the fermentation production of lincomycin hydrochloride were used to synthesize nitrogen and fluorine co-doped porous carbon nanosheets (NFPCNs) by high-temperature pyrolysis and following KOH activation [26]. The as-prepared supercapacitors electrode material displayed a remarkable energy density of 125 Wh kg^{-1} in 1-ethyl-3-methylimidazolium tetrafluoroborate electrolyte, along with an impressive gravimetric capacitance of 296 Fg^{-1} at 1 Ag^{-1}. This is in comparison to a rate capability of 65% at 20 A g^{-1}. Moreover, supercapacitors based on NFPCNS have achieved an exceptionally high energy density of 173 Wh kg^{-1} and displayed long cycle life, retaining 93% of their capacitance after 2000 cycles when employing 1-ethyl-3-methylimidazolium bis[(trifluoromethyl) sulfonyl]imide electrolyte. The retention of trifluoromethyl sulfonyl groups in the electrolyte enabled this outcome.

Co-pyrolysis and activation of a ternary biomass mixture (sugarcane bagasse, water hyacinth, and yellow oleander) produced oxygen-enriched porous carbon, denoted by the symbol PC-x [27]. The modified PC-x electrodes demonstrated a gravimetric capacitance of 251.6 F g^{-1} at a commercial electrode mass loading (10 mg cm^{-2}), high rate capability, and excellent cycle stability in a 6 M KOH electrolyte. A high-voltage aqueous symmetric supercapacitor was created in an electrolyte solution of 1 M Li2SO4, with an energy density of 22.75 Wh kg^{-1} (at 200 W kg^{-1}).

It retained 96.8% of its capacitance even after being subjected to 10,000 cycles at 5 A g^{-1} during the experiment. The energy density of the supercapacitor has been increased to 37.24 Wh kg^{-1} (at 200 W kg^{-1}) by incorporating 0.1 M Na_2MoO_4 into the Li_2SO_4 as a redox additive. Using biomass mixtures provides flexibility in selecting several types of waste biomass to use as precursors for the large-scale and environmentally friendly manufacture of carbon compounds that may be used for energy storage applications.

Cadmium selenide nanoparticles (CdSeNPs) [28] based supercapacitor electrode was fabricated by a simple and cost-effective approach based on kappa-carrageenan (-CGN). Combining activated carbon (AC) with green-prepared CdSeNPs was a simple process that resulted in outstanding electrochemical characteristics. The nanoelectrode, denoted by the notation AC@CdSe, was tested in an aqueous electrolyte containing sodium sulfate (Na_2SO_4). The AC electrode and the AC@CdSe electrode's specific capacitance (Csp) were determined to be 103 and 480 F g^{-1} when measured at 1 A g^{-1}. In addition, the symmetric supercapacitor AC@CdSe/AC@CdSe device possesses a specific capacitance of 115.5 F g^{-1}, a high specific energy of 52 Wh kg^{-1}, and a maximum specific power of 2880 W kg^{-1}. The device has maintained suitable capacitance even after 10,000 charging and discharging cycles because of its Coulombic efficiency, ranging from 82% to 100%.

A simple hard templating approach was used to synthesize defective mesoporous carbon (DMC) [29]. An aqueous solution of 1 M H_2SO_4 containing 0.01 M acid orange 45 as a redox additive dramatically enhances the performance of DMC supercapacitors. A DMC electrode synthesized through a simple hard templating method exhibits a specific capacitance of 324 F g^{-1}, and it demonstrates excellent rate capability, reaching 169 F/g at 10 Ag^{-1} with noteworthy capacitance retention of 52% in a medium containing redox-added electrolyte. The symmetric device with a specific energy of 11.5 Wh kg^{-1} was constructed when a redox-additive electrolyte was used instead of a pure H2SO4 solution at the DMC electrode. The chemical compound 4,4′-bipyridine (4,4′-by) produces an innovative electrode material that resembles honeycombs and is made of a porous carbon matrix [30].

The assisted pyrolysis of SiO_2@ZIF-8 templates and subsequent etching processes can modify the N-doping configurations. The electrochemical performance of symmetric supercapacitors using aqueous and ionic liquid gel electrolytes was affected by the 3D interconnected porous architecture of pyridinic-N and graphitic-N. Highly exposed pyridinic-N confers a strengthened pseudocapacitance contribution on the carbon electrode. The KOH/PVA electrolyte proved to be a reliable choice, as it yielded remarkable stability with almost 100% capacitance retention even after 5000 cycles at 10 Ag^{-1}. Additionally, a high specific capacitance of 436.1 Fg^{-1} was demonstrated in this configuration.

13.6 SUMMARY AND FUTURE OUTLOOK OF ELECTROLYTE FOR SUPERCAPACITORS

Hybridizing supercapacitors and batteries could provide a more efficient and reliable power source for various applications. The outlook for supercapacitors looks bright, with ongoing research and development efforts to improve their performance and reduce cost. Supercapacitors are expected to be crucial in the transition to renewable energy and transportation electrification. Here are some trends and developments expected to shape electrolyte chemistry's future:

- Solid-state electrolytes: Solid-state electrolytes are expected to become increasingly important as they offer higher energy density and safety than liquid electrolytes. Solid-state electrolytes can also enable high-energy electrode materials such as lithium metal, which is currently limited by safety concerns with liquid electrolytes.
- Ionic liquid electrolyte: Ionic liquid electrolytes are another promising class of materials for electrolytes, as they offer a wide range of tunable properties, such as high conductivity, low volatility, and thermal stability. Ionic liquids also have the potential to reduce the environmental impact of battery production and disposal.
- Redox-active organic molecules: Redox-active organic molecules, such as quinones and polyoxometalates, are being investigated as alternatives to traditional inorganic electrolytes. These molecules offer the advantages of high solubility, low toxicity, and low cost while exhibiting good stability and electrochemical properties.
- Hybrid electrolytes: Hybrid electrolytes that combine features of liquid and solid-state electrolytes are also being developed. These materials can offer the advantages of both liquid and solid-state electrolytes, such as high conductivity and flexibility while maintaining exemplary safety and stability.

The use of electrolytes is essential for the operation of supercapacitors, as they facilitate the movement of ions between the electrodes, enabling the storage and release of electrical energy. Some commonly used electrolytes include (i) Aqueous electrolytes, such as sulfuric acid and potassium hydroxide, (ii) Organic electrolytes, such as acetonitrile and propylene carbonate, and (iii) Ionic liquids, such as EMIM BF4 and BMIM PF6. Each electrolyte type has advantages and disadvantages, and the selection of the appropriate electrolyte depends on the specific application requirements. Developing advanced electrolytes with high ion conductivity, low viscosity, expansive electrochemical stability windows, and environmental compatibility is crucial for continued supercapacitor technology's continued improvement and commercialization.

REFERENCES

1. C. Liu, F. Li, M. Lai-Peng and H.M. Cheng, *Adv Mater.*, 2010, 22(8), E28–E62.
2. D.P. Dubal, N.R. Chodankar, P. Gomez-Romero and D-H. Kim, Fundamentals of Binary Metal Oxide–Based Supercapacitors. In: Deepak P. Dubal, Pedro Gomez-Romero (Eds.) *Metal Oxides in Supercapacitors*, 2017, 79–98, Elsevier, Chapter-4.
3. J. Liu, J. Wang, C. Xu, H. Jiang, C. Li, L. Zhang, J. Lin and Z. X. Shen, *Adv Sci.*, 2018, 5(1), 1700322.
4. B.E. Conway, *Electrochemical Supercapacitors: Scientific Fundamentals and Technological Applications*, 1999, 698, Springer; https://doi.org/10.1007/978-1-4757-3058-6_19

5. W. Zuo, R. Li, C. Zhou, Y. Li, J. Xia and J. Liu, *Adv. Sci.*, 2017, 4(7), 1600539.
6. B.E. Conway, *Chapter-4: The Double Layer at Capacitor Electrode Interfaces: Its Structure and Capacitance in Electrochemical Supercapacitors: Scientific Fundamentals and Technological Applications*, 1999, 698, Springer, 105–124; https://doi.org/10.1007/978-1-4757-3058-6_19
7. V. Sahu, S. Goel, A. K. Tomar, G. Singh and R.K. Sharma, *Electrochim. Acta*, 2017, 230, 255–264.
8. M.F. El-Kady, Y. Shao and R.B. Kaner, *Nat. Rev. Mater.*, 2016, 1(7), 1–14.
9. S. Ahmed, A. Ahmed and M. Rafat, *J. Saudi. Chem. Soc.*, 2018, 22(8), 993–1002.
10. K. Wang, L. Li, T. Zhang and Z. Liu, *Energy*, 2014, 70, 612–7.
11. B. Pal, S. Yang, S. Ramesh, V. Thangadurai and R. Jose, *Nanoscale Adv.*, 2019, 1(10), 3807–35.
12. S. Verma, B. Padha, P. Mahajan and S. Arya, *ECS Trans.*, 2022, 107(1), 10821–9.
13. C. Zhong, Y. Deng, W. Hu, J. Qiao, L. Zhang and J. Zhang, *Chem. Soc. Rev.*, 2015, 44 (21), 7484–539.
14. M. Li, J. Wei, L. Ren, Y. Zhao, Z. Shang, D. Zhou, W. Liu, L. Luo and X. Sun, *Cell Reports Phys. Sci.*, 2021, 2(3), 100374.
15. M.-J. Deng and Y.-S. Wu, *J. Alloys. Compd.*, 2023, 945, 169285.
16. R. Mondal, K.G. Nigam, N.K. Mishra, A. Gupta and P. Singh, *J. Energy Storage*, 2023, 60, 106549.
17. P. Periasamy, T. Krishnakumar, M. Sandhiya, M. Sathish, M. Chavali, P.F. Siril and V.P. Devarajan, *Mater. Lett.*, 2019, 236, 702–705.
18. S. Lu, Y. Song, K. Guo, X. Chen, J. Xu and L. Zhao, *J. Electroanal. Chem.*, 2018, 818, 58–67.
19. Z.J. Zhang and X.Y. Chen, *J. Electroanal. Chem.*, 2016, 764, 45–55.
20. Y.G. Wang, L. Cheng and Y.Y. Xia, *J. Power Sources*, 2006, 153(1), 191–6.
21. G. Hariharan, S. Arunpandiyan, V. Shanmugapriya, S. Bharathi, M. Babu, B. Selvakumar and A. Arivarasan, *J. Energy Storage*, 2023, 57, 106178.
22. J. Fischer, B. Pohle, E. Dmitrieva, K. Thümmler, S. Fischer and D. Mikhailova, *J. Energy Storage*, 2022, 55, 105725.
23. F.C. Wu, R.L. Tseng, C.C. Hu and C.C. Wang, *J. Power Sources*, 2005, 144(1), 302–9.
24. C. Fang, D. Zhang, *Electrochim. Acta*, 2022, 401, 139491.
25. B.J. Choudhury, K. Ingtipi and V.S. Moholkar, *J. Energy Storage*, 2022, 52, 105006.
26. C. Chen, Y. Xu, J. Shao, Y. Zhang, M. Yu, L. Sun, H. Wang, Y. Xie, G. Zhu, L. Zhang and L. Pan, *J. Colloid Interface Sci.*, 2022, 616, 413–421.
27. B.J. Choudhury, H.H. Muigai, P. Kalita and V.S. Moholkar, *Appl. Surf. Sci.*, 2022, 601, 154202.
28. L. Horáčková, K. Chmelíčková, M. Hermannová, T. Pitucha, H. Vágnerová, O. Židek, V. Velebny and J. Chmelar, *Int. J. Biol. Macromol.*, 2023, 234, 123654.
29. F. Khoramjah, M. Omidvar, M.S. Miresmaieli, S. Dalvand, A. Asghari, M. Kambarani and N. Mohammadi, *Diam. Relat. Mater.*, 2023, 132, 109590.
30. J. Gu, H. Wang, S. Li, M. Sohail Riaz, J. Ning, X. Pu and Y. Hu, *J. Colloid Interface Sci.*, 2023, 635, 254–64.

14 Non-Aqueous Electrolytes for Electrochemical Supercapacitors

Dawid Kasprzak and Jian Liu

14.1 INTRODUCTION

Electrochemical capacitors (also called supercapacitors or ultracapacitors) are energy storage devices that bridge the gap between batteries and traditional capacitors. They feature high power and moderate energy densities, excellent rate capability, extended cycle stability, and low maintenance cost. Depending on the energy storage mechanism, supercapacitors are grouped into three classes [1]:

1. Electrochemical double-layer capacitors (EDLC) store charges electrostatically *via* reversible ion electrosorption at the electrode-electrolyte interface.
2. Pseudocapacitors involve fast and reversible redox reactions occurring on the electrode surface for charge accumulation.
3. Hybrid supercapacitors combine capacitive and faradic mechanisms in a single device.

A basic supercapacitor is composed of two electrodes divided by a separator soaked with a liquid electrolyte (or a solid-state electrolyte). The applied electrodes define the primary charge storage mechanism of the supercapacitor. The electrodes could be made of carbon materials with a capacitive nature, metal oxides and conducting polymers featuring pseudocapacitive characteristics, or both. Pairing of capacitive and pseudocapacitive materials results in asymmetric supercapacitors. Capacitive/pseudocapacitive materials combined with a battery-type electrode gives the hybrid supercapacitor [2, 3]. Well-selected electrodes with a compatible electrolyte can create an efficient and high-performance energy storage device.

Besides the electrode material, an electrolyte is one of the most significant components of electrochemical capacitors. The electrolyte provides the transportation of

DOI: 10.1201/9781003441182-17

ions inside the energy storage device, allowing charge compensation on each electrode. Therefore, it contributes to the electrical double-layer (EDL) formation or fast and reversible redox reactions and is the fundamental process for charge storage in electrochemical capacitors. Basically, an electrolyte is an ionic conductor composed of mobile ions, the charge-transporting species inside the liquid or solid medium, facilitating ion transportation between the two electrodes. Liquid electrolytes, the fundamental types of electrolytes, comprise conducting salts dissolved in water or organic solvents. Alternatively, neat and liquified salts without the presence of molecular solvents can create an electrolyte. This case refers to ionic liquids or molten salts [4].

An attractive alternative to liquid electrolytes is solid-state electrolytes, which can act as both ionic conductors and separators. They provide ionic conductivity, electron insulation, dimensional stability, and leakage risk reduction. The solid matrices of these electrolytes could be made from polymers or inorganic materials (e.g., ceramic electrolytes). Polymer-based electrolytes are much more widely employed in supercapacitors and can be further classified into solid polymer electrolytes (SPEs), gel polymer electrolytes (GPEs, also known as *quasi*-solid-state electrolytes due to the presence of additional liquid phase), and polyelectrolytes. Among these, GPEs gain ground due to outstanding ionic conductivity owing to the combination of solid and liquid phases [5].

The choice of electrolyte determines the performance and safety of the supercapacitor device during operation [6]. Its nature and specific features, such as the type and size of ions; class of solvent; concentration of ions; interaction between electrolyte components (ions and solvent molecules); compatibility with the chemistry of electrode materials (active materials, binder, and current collectors); mobility of ions in the bulk electrolyte and the electrolyte/electrode interface; all have a consolidated impact on the essential parameters of supercapacitors. In addition, solid-state electrolytes contribute to the mechanical properties of portable devices, providing stability, flexibility, and wearability.

Generally, the optimization of electrolytes in supercapacitors is focused on specific requirements, including high chemical, thermal, and electrochemical stability; high ion conductivity; low viscosity and volatility; sustainability; and easy accessibility with production cost reasonability. Unfortunately, it is a very challenging task to find a universal electrolyte meeting all the prerequisites mentioned above. Each group of electrolytes is distinguished by a specific set of characteristic properties having its own advantages and disadvantages in terms of implementation for supercapacitor devices. State-of-the-art knowledge of different types of electrolytes allows for choosing an electrolyte system optimized for particular applications.

This chapter presents a brief guide to understanding the differences, benefits, and limitations of various types of electrolytes designed for electrochemical capacitors. The discussion focuses on progress and challenges in state-of-the-art systems, mainly classified as liquid and solid-state electrolytes, as shown in Figure 14.1. The outlook on future perspectives is also included in the last part of this chapter.

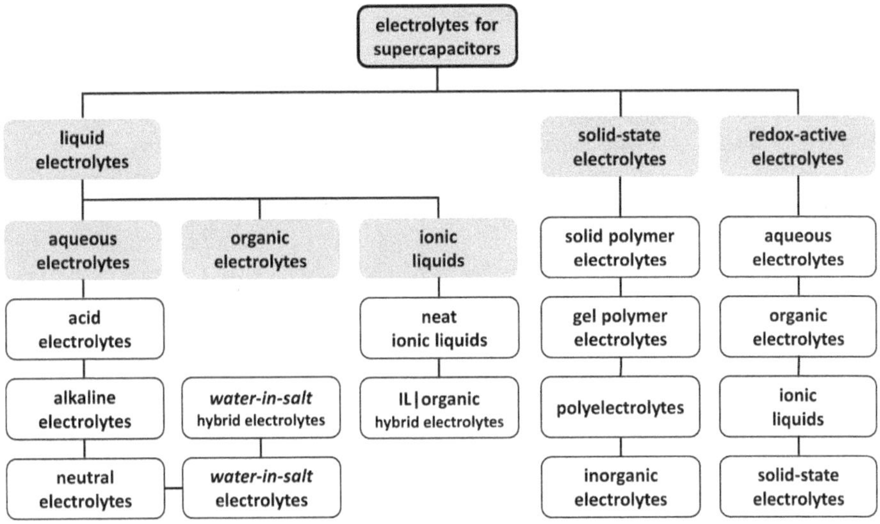

FIGURE 14.1 Classification of electrolytes for electrochemical supercapacitors.

14.2 AQUEOUS ELECTROLYTES

Aqueous electrolytes are composed of ionic compounds dissolved in water. Depending on the type of conductive species, these electrolytes can be additionally divided into acid, neutral, and alkaline aqueous systems. The most intensively studied representatives of aqueous electrolytes are H_2SO_4, Na_2SO_4, and KOH, which represent acid, neutral, and alkaline systems, respectively. In general, aqueous electrolytes are easy to handle in the laboratory environment. Compared to their organic counterparts, they do not need complicated purification procedures and apparatus providing a controlled atmosphere during supercapacitor assembling procedures.

Besides the simplification in device manufacturing, aqueous solutions exhibit additional assets, such as affordability, high conductivity, and environmental friendliness. Acidic and alkaline electrolytes can offer high EDL specific capacitance in the range of 150–400 F g^{-1} combined with excellent conductivity (e.g., 0.8 S cm^{-1} and 0.6 S cm^{-1} at 25 °C for 1M H_2SO_4 and 6M KOH, respectively). Unfortunately, the major disadvantage of these systems is their relatively narrow voltage window. The electrochemical stability of water, defined by the theoretical thermodynamic potential of 1.23 V needed for its decomposition, restricts the operating voltage of traditional aqueous electrolytes. In practice, the maximum voltage of acid and alkaline aqueous-based supercapacitors is even lower, limited to about 1.0 V for safety and stability reasons (avoiding the gas evolution of water decomposition) [7].

Neutral electrolytes could provide wider operating voltages due to reduced H^+/OH^- ions availability responsible for hydrogen/oxygen evolution reactions. For instance, a Li_2SO_4 electrolyte can offer an operating voltage of up to 2.2 V in a symmetrical EDCL system. However, this device is not long-term stable when typical current collectors (e.g., stainless steel) are used. As demonstrated, cost-efficient

EDLC systems with Li_2SO_4 aqueous electrolytes operate safely up to a voltage of only 1.5 V without significant degradation. Here, it is also worth noting that neutral aqueous electrolytes exhibit significantly lower ionic conductivities (*ca.* one order of magnitude compared to 1M H_2SO_4) and decreased capacitance values (up to 150–200 F g^{-1}) in EDLC cells compared to typical acid and alkaline systems.

Despite these issues, neutral electrolytes are still considered the most promising aqueous electrolytes due to their evaluated voltage stability and reduced corrosivity [8]. Besides Li_2SO_4-based systems, the most frequently investigated neutral aqueous electrolytes for EDLCs are [7, 8]: Na_2SO_4, $(NH_4)_2SO_4$, K_2SO_4, $MgSO_4$, $CaSO_4$, $BaSO_4$, KCl, NaCl, LiCl, CsCl, $CaCl_2$, KNO_3, $LiNO_3$, $Na(CH_3COO)$, $Li(CH_3COO)$, $Mg(CH_3COO)_2$, Na_2HPO_4, $NaHCO_3$, and $Na_2B_4O_7$.

Recently, highly concentrated aqueous solutions, named *water-in-salt* (WIS) electrolytes, have been introduced as systems that can significantly widen the working voltage of traditional aqueous electrolytes [9]. Generally, WIS electrolyte refers to an aqueous solution containing a well-soluble salt that outnumbers water in both mass and volume. Schematically, the comparison of traditional and WIS aqueous electrolytes is presented in Figure 14.2. The expanded voltage window of WIS (up to 3 V) can be explained by the following mechanisms, discussed in the representative systems containing lithium bis(trifluoromethanesulfonyl)imide (LiTFSI) salt [10].

Firstly, the anion-involved solvation sheaths of the cations (Li^+) with strongly coordinated water molecules are created in highly concentrated aqueous solutions. Conversely to traditional aqueous electrolytes, there are almost no free water molecules present in WIS electrolytes. Thus, the possibility of water electrolysis in these systems is suppressed. Furthermore, the WIS electrolytes can generate solid-electrolyte interphase (SEI) layers similar to that occurring in nonaqueous solutions. This is possible due to the unique solvation structure of WIS electrolytes. As mentioned before, anions participate in cation solvation shells of highly concentrated

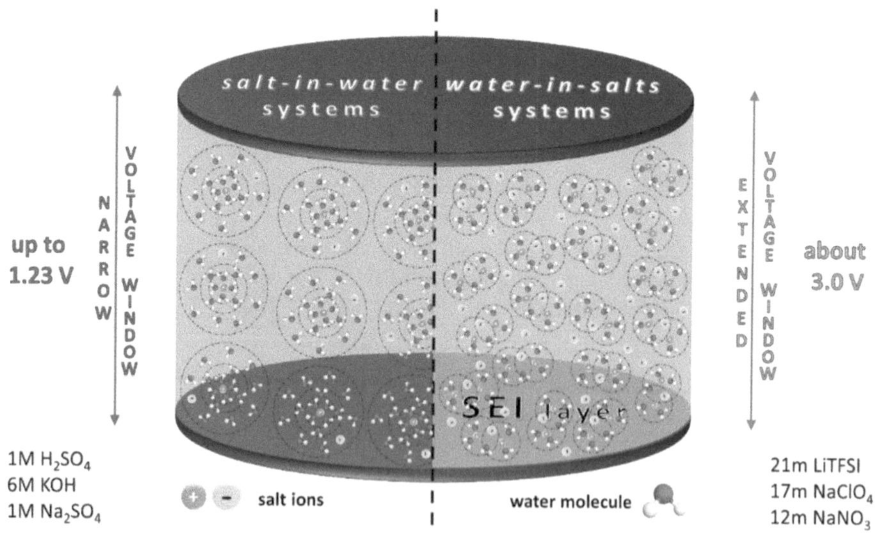

FIGURE 14.2 Comparison of traditional and WIS aqueous electrolytes.

aqueous solutions. The created ion pairs can be involved in electrode reactions, giving the products that form the SEI film.

Taking the LiTFSI-based electrolyte as an example, the reduction of $TFSI^-$ anions from the Li-TFSI ion pairs leads to the SEI layer composed of LiF products. This densely distributed SEI layer prevents the endless decomposition of both water and anions. The presented concept of WIS electrolyte improves the working voltage and, consequently, the energy density of aqueous supercapacitors. The typical WIS electrolyte system, 21 mol kg^{-1} LiTFSI aqueous solution, can usually operate above the voltage value of 2.2 V.

The low working voltage of aqueous energy storage systems limits their energy performance. Although the improved, high-voltage aqueous electrolytes, e.g., WIS systems, are already known, they are still at an early stage of development, requiring a fundamental understanding of the working mechanisms. Consequently, aqueous electrolytes are mainly used in research and development projects, giving way to organic counterparts in most commercial applications. Moreover, the operating temperature range of aqueous electrolytes is limited to their freezing and boiling points, which creates an additional restriction in the practical application of aqueous solutions in commercial supercapacitor cells.

14.3 ORGANIC ELECTROLYTES

Besides the extended interest in developing aqueous electrolytes, organic solutions remain the first choice for application in commercial electrochemical supercapacitors. Organic electrolytes owe their dominant position to high voltage stability and reasonable ionic conductivity, which impact supercapacitors' enhanced energy and power performance. Additionally, these electrolytes expand the usability of electrochemical devices to sub-zero temperatures.

A typical organic electrolyte contains conducting salt dissolved in the organic solvent. Most commercial supercapacitors employ tetraethylammonium tetrafluoroborate ($TEABF_4$) salt solutions in acetonitrile (AN) or propylene carbonate (PC). These systems can work with a high and stable operating voltage of up to 2.7–3.0 V. Although AN can dissolve more conductive salt and demonstrates higher power and better low-temperature performance than PC, it is harmful to the environment. Disadvantages of AN-based electrolytes include volatility, increased flammability, toxicity, and carcinogenic hazard potential. In this regard, PC is usually considered the preferred solvent alternative to AN, featuring less flammability and toxicity [11].

Each type of electrolyte has a set of positive and negative characteristics regarding the requirements of a supercapacitor device. Despite their state-of-the-art status in commercial applications, organic electrolytes are not devoid of shortcomings. Compared to traditional aqueous electrolytes, organic electrolytes reveal a higher price, lower specific capacitance, higher resistivity, and safety issues, including flammability, volatility, and toxicity. Moreover, to eliminate contamination possibility, these systems require purification and a restrictive atmosphere during supercapacitor assembling. For instance, moisture in organic electrolytes could significantly deteriorate the performance of electrochemical supercapacitors, including operating voltage reduction, self-discharge issues, and long-term degradation [6].

The nature of salt and solvent plays a crucial role in developing efficient electrolytes. Although a bulk solution may appear to be a simple system, the interactions between electrolyte species make it much more complex. Thus, it is noteworthy to take a closer look at the parameters of these electrolyte constituents influencing the performance of organic electrolyte-based supercapacitors. The different aspects of organic electrolyte behavior are discussed below.

Generally, organic electrolytes feature larger solvated ion sizes and lower dielectric constants than aqueous representatives. Consequently, the surface area of carbon-based electrodes is charged less efficiently in EDLC devices with organic solutions than with aqueous ones. Additionally, the contribution of the pseudocapacitance is unfavored in aprotic organic electrolytes. Both phenomena result in the reduced specific capacitance reached by electrochemical supercapacitors employing organic electrolytes.

A method for maximizing specific capacitance in organic electrolytes is based on the size adjustment of electrolyte species and electrode porosity. The matching between the sizes of electrode pores and electrolyte ions increases the electrode surface accessibility for the electrolyte and consequently improves the specific capacitance. Notably, the size factor should be analyzed for both bare and solvated ions. As suggested, the solvated ions can undergo desolvation/distortion when entering narrow pores with a size that matches bare ions' dimensions, enhancing specific capacitance. The discussed phenomenon is not universal and depends on ion-solvent interactions occurring in a particular system.

The desolvation process in electrode pores could be unfavored if the solvation shells are strongly bound to ions. Naturally, the size adjustment should be considered for both types of ions and electrodes. However, it was found that in common organic electrolytes, the capacitance depends on the size of cations rather than anions. Apparently, the small BF_4^- anion features better pore accessibility compared to the quaternary ammonium counterion [7, 8, 12].

The composition of the electrolyte also determines its electrochemical stability. This critical feature is directly linked to the operating voltage range of the supercapacitor. Organic electrolytes owe their well-established commercial position, among others, to high electrochemical stability. Since most investigation efforts on supercapacitors concern the improvement in energy performance, widening the electrolyte stability window seems the most effective strategy. According to Eq. (14.1) defining the energy density E of supercapacitors,

$$E = 1/2CV^2 \tag{14.1}$$

this parameter is a function of the capacitance C and squared voltage V of the investigated device. The electrochemical stability of an organic electrolyte is strongly dependent on the choice of its constituents, conductive salt and solvent, and the presence of any impurities. For instance, oxygen and water traces are very harmful to many organic electrochemical systems, limiting their stability by parasitic faradic reactions [4].

The ionic conductivity of organic electrolytes is usually significantly lower than that of traditional aqueous systems. Nevertheless, it can still be optimized by selecting a suitable type of salt and solvent and adjusting salt concentration. The most investigated salts are tetraalkylammonium ion-based compounds displaying good solubility and conductivity in solvents with high dielectric constants. The solvents used for organic electrolytes usually belong to the group of aprotic compounds featuring a high/intermediate dielectric constant (e.g., AN, DMF, and PC) or liquids exhibiting a low dielectric constant but strong donor character (DME). Optimizing the ionic conductivity of the organic electrolyte is crucial since it strongly influences supercapacitors' resistance and power performance [7, 8, 13].

Due to the application of nonaqueous solvents, organic electrolytes feature a wider temperature operating range compared to water-based systems, enhanced to the sub-zero region. Usually, supercapacitors with organic electrolytes can operate down to –40 °C, but their internal resistance increases with decreasing temperature. Notably, typical AN-based systems show better low-temperature characteristics compared to PC-based analogues. They feature a stable capacitance and slowly decreasing resistance in the temperature range from 60 °C to −40 °C [8, 14, 15].

Supercapacitors are energy storage devices designed for long-term applications. The electrolyte purity grade and working conditions of EDLC devices are crucial for stability and long durability. If electrodes and the electrolyte are free of electrochemically active contaminants, the degradation of supercapacitors is mainly caused by enhancing their operating voltage above the safety range. Under evaluated voltage values, the constituents of electrolytes can degrade in the form of gases or passive film products. Also, the contribution of ion intercalation processes may be enhanced at high cell voltages, causing the degradation of carbon-based electrodes. Thus, a good understanding of the electrolyte system and its interactions with electrode materials is essential before the extension of the supercapacitor operating voltage is applied [11, 16, 17].

As can be seen, the process of electrolyte optimization is very complex. Since the most common way to increase the energy density of supercapacitors is the widening of operating voltage, the greatest effort has been made to introduce high-electrochemically stable electrolytes. Some achievements regarding alternative salts (e.g., triethylmethylammonium ($TEMA^+$), 1-ethyl-1-methylpyrrolidinium ($MEPY^+$), or 1,1′-Spirobipyrrolidinium (SBP^+) cations combined with tetrafluoroborate (BF_4^-) anion and solvents (e.g., butylene carbonate (BC), γ-butyrolactone (GBL), adiponitrile (ADN), or sulfolane (TMS)) for organic electrolytes have been already made [7, 8, 18, 19]. Although several promising systems exhibit high electrochemical stability, they generally cannot compete with commercial electrolytes in terms of viscosity, conductivity, or thermal stability.

The additional problem concerns the cost-effective availability and toxicity of components comprising novel electrolytes. Consequently, the relative stagnancy of energy density in commercially available supercapacitor devices is observed. Pohlmann [20] presents in an attractive, graphical way the comparison between the state-of-the-art and fictional novel electrolytes by highlighting the achievements reached to date for each system (Figure 14.3). This broad insight could be a valuable guide for translating research results into industrial innovations.

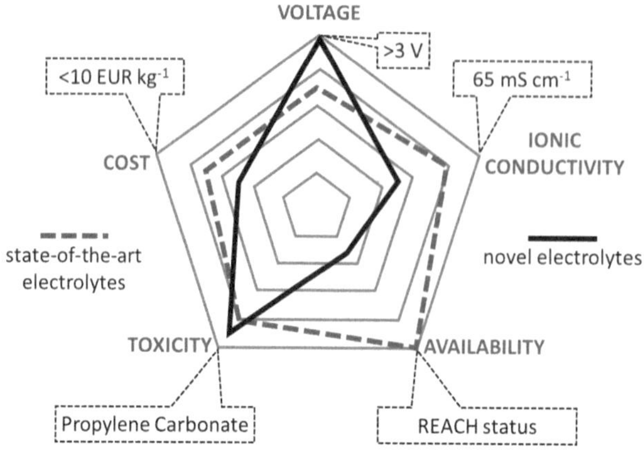

FIGURE 14.3 Comparison between the state-of-the-art and fictional novel electrolytes.

Source: Adapted from ref [20]. Copyright 2022 Springer Nature.

14.4 IONIC LIQUID-BASED ELECTROLYTES

Ionic liquids (ILs) fit into the concept of efficient and eco-friendly electrolytes for supercapacitors. This substance class is defined as compounds entirely composed of ions (organic cations and organic/inorganic anions) with melting points below 100 °C. They can act as solvent-free electrolytes possessing negligible volatility, moderate conductivity, low flammability, and high thermal, chemical, and electrochemical stability. Interestingly, they are considered *designer* compounds due to their properties' tenability, achieved by the customization of (virtually unlimited) combinations of cations and anions. The optimization of characteristics, such as conductivity, electrochemical stability window, or temperature operating range, seems the most critical strategy in seeking ionic liquid electrolytes for supercapacitors [21, 22].

The typical ILs dedicated to supercapacitors are composed of ammonium, imidazolium, pyrrolidinium, pyridinium, piperidinium, sulfonium, or phosphonium cations coupled with anions such as hexafluorophosphate (PF_6^-), tetrafluoroborate (BF_4^-), bis(fluorosulfonyl)imide (FSI^-), bis(trifluoromethanesulfonyl)imide ($TFSI^-$), or dicyanamide (DCA^-) [23–26]. Usually, they can offer a maximum working voltage of 2.5–4.0 V and a broad range of conductivity values of 0.1–18 mS cm^{-1}. The relatively low conductivity and high viscosity result from the nature of the constituents of ILs. Generally, ILs are made of large unsymmetrical ions with steric effects, giving these salts an unusual "liquified" state. However, the large and charged constituents of ILs cannot move fast in a solvent-free environment, which results in low conductivity and high viscosity of these electrolytes.

The nature of ILs also reflects on their capacitance performance in supercapacitor cells. The effective accumulation of relatively large ions in EDL of porous electrodes is troublesome, thus, the specific capacitance of IL-based capacitors is often lower than traditional systems with aqueous or organic solutions. This problem becomes severe at high current rates when the conductivity and viscosity limits of ILs are particularly apparent [19, 27]. Despite the abovementioned drawbacks, the high thermal stability of ILs is a massive practical advantage. These electrolytes can be used for high-temperature applications beyond the typical limits of aqueous and organic electrolytes. For instance, supercapacitors with the [BMIm][TFSI] electrolyte can operate at 200 °C [15].

The alternative to solvent-free ionic liquid could be the concept of hybrid electrolytes containing the organic solvent and IL mixtures [7, 8]. Typically, the solvents already used in organic electrolytes are applied to IL-based hybrid electrolytes, including AN, PC, DMF, etc. It turns out that adding organic solvents could enhance the electrochemical performance of ILs, i.e., specific capacitance, conductivity, and low-temperature performance. Although this operation usually also reduces the operating voltage range of IL-based supercapacitors, they still exhibit higher electrochemical stability than traditional organic electrolytes. For instance, pyrrolidinium-based ILs mixed with PC can work stably in supercapacitor cells up to 3.2–3.5 V. Finally, it should be noted that any new IL-based hybrid electrolyte requires the optimization of electrode surface porosity to maximize the specific capacitance. The size of solvated ions in hybrid electrolytes usually differs from that of conductive species in bare IL electrolytes [28].

The biggest issue with the practical application of ILs to supercapacitor devices relates to economic aspects. To date, there is no affordable and efficient synthesis method for obtaining high-purity ILs dedicated to electrochemical purposes. They require multi-step processes and techniques to remove impurities, such as moisture and metal or halide ions. The commercialization of IL-based processes is still in its early stages, and it requires economic feasibility before large-scale production of IL electrolytes will be possible.

As a conclusion to the above discussion, we have visualized the difference in the electrochemical performance of EDLCs assembled with various representative liquid electrolytes. Figure 14.4 presents the investigation of different supercapacitors using cyclic voltammetry (CV) (A), galvanostatic charge/discharge (GCD) (B), and electrochemical impedance spectroscopy (ESI) (C) methods. The performance of aqueous-, organic-, and IL-based systems has been shown. The differences in maximum operating voltage and power features could be observed just by visual inspection.

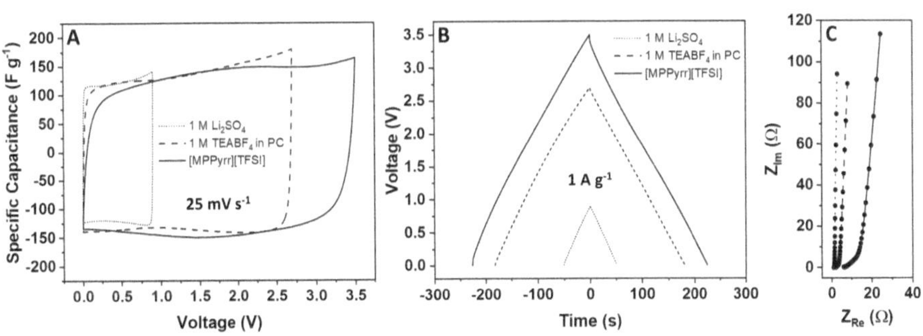

FIGURE 14.4 Electrochemical performance of EDLCs assembled with aqueous (1 M Li_2SO_4), organic (1 M $TEABF_4$ in PC), and IL ([MPPyrr][TFSI]) electrolytes.

Source: Reproduced with permission from Ref. [29]. Copyright 2023 Elsevier.

14.5 (QUASI-) SOLID-STATE ELECTROLYTES

Solid electrolyte-based supercapacitors have attracted great interest due to increasing demand for portable electronics. Solid-state electrolytes (SSEs) can offer good ionic conductivity, mechanical stability, flexibility, leakage reduction, and assembly simplification. Most SSEs are composed of polymers, and only limited research investigation relates to inorganic solid materials (e.g., ceramic).

Polymer-based SSEs can be further grouped into three types: solid polymer electrolytes (SPEs), gel polymer electrolytes (GPEs), and polyelectrolytes. SPEs consist of a polymer matrix (e.g., poly(ethyl oxide) (PEO)) and conductive salt (e.g., LiCl) without adding any solvent. The transportation of ionic species through the polymer matrix provides conductivity. In turn, GPEs are made of a polymer matrix (mainly poly(vinyl alcohol) (PVA)) swollen by the traditional liquid electrolyte, i.e., aqueous and organic solution or IL. Owing to this fact, GPEs are also called *quasi*-solid state electrolytes. The liquid electrolytes provide high ionic conductivity in GPEs, driving them to dominate the studies on solid electrolyte-based supercapacitors. Polyelectrolytes are created from immobilized and charged polymer chains with mobile and inversely charged counter ions. The mobile ions contribute to the ionic conductivity of these materials. Depending on the nature of the charges accumulated on the polymeric backbone functionalities, polyelectrolytes can be additionally divided into polyanions or polycations [30–32].

Gel polymer electrolytes are the most extensively studied SSEs because they combine the advantages of solid and liquid electrolytes. They can offer excellent conductivity, good electrode wettability, structure and shape diversity, mechanical strength, flexibility, and stretchability. Among the many materials used as polymeric hosts in GPEs, the most frequently investigated include PVA, poly(polyacrylate) (PAA), potassium polyacrylate (PAAK), poly(methyl methacrylate) (PMMA), PEO, polyacrylonitrile (PAN), poly(vinylidene fluoride) (PVDF) and poly(vinylidene fluoride-*co*-hexafluoropropylene) (PVDF-co-HFP), and poly-(ethylene glycol) blending poly(acrylonitrile) (PAN-*b*-PEG-*b*-PAN). Moreover, there is a growing interest in adapting biopolymers as matrices in GPEs to transform them into more eco-friendly components. Unlike synthetic polymers, biomaterials feature biocompatibility, biodegradability, and cost-effectiveness while being mechanically, chemically, and thermally stable. Examples of biopolymeric hosts for GPEs include cellulose, chitosan, chitin, starch, alginate, etc. [33, 34].

Traditional liquid electrolytes are used as GPE plasticizers. Aqueous-based systems are called hydrogel electrolytes. PVA-based hydrogel electrolytes [35] have so far been the most extensively studied due to advantages such as processing ability, high hydrophilicity, good film-forming properties, non-toxicity, and low cost. Typical hydrogels for supercapacitors include PVA-H_2SO_4, PVA-KOH, PVA-NaCl, PVA-Na_2SO_4, and PVA-$NaNO_3$ systems. Unfortunately, hydrogel electrolytes have similar voltage limitations as their liquid counterparts. Thus, organogel electrolytes have been introduced to increase this parameter [31, 32, 36].

Organic gel electrolytes possess more variable components than hydrogel. Besides the polymeric host and conductive salt, the type of solvent could also be adjusted. Regarding polymeric matrix, high molecular weight polymers are usually chosen, i.e., PMMA, PVDF–HFP, PAN-*b*-PEG-*b*-PAN. These are gelated with nonaqueous solutions containing conducting salt, typically $LiClO_4$, $LiPF_6$, $TBAPF_6$, or NaTFSI. The commonly studied organic solvents include PC, EC, DMC, or their mixtures, as well as ACN, DMSO, or DMF. The organogel electrolytes, similar to their liquid analogues, can work in supercapacitor cells to a maximum voltage of up to 2.5–3V.

Finally, IL-based GPEs can be obtained by incorporating an IL plasticizer into the polymeric host. IL-based GPEs offer even larger electrochemical stability than organic systems while being thermally stable, nonvolatile, and non-flammable. To date, imidazolium-based compounds with BF_4^- and $TFSI^-$ anions have been the most widely tested plasticizers for IL-GPEs. The typical polymeric hosts applied for these systems include PVDF–HFP, PVA, and PEO. Supercapacitors with IL-GPEs work up to 3.5 V exhibiting high electrochemical performance under large mechanical stress [37, 38].

The great advantage of SSEs is that they are suitable for the growing demand for wearable energy storage systems. Most SSEs can bear mechanical stress, providing flexibility, foldability, and stretchability. Besides excellent mechanical and electrochemical performance, they can also reduce the risks of leakage, flammability, and corrosion in wearable supercapacitors. Since increased interest in replacing polymeric matrices with biopolymer alternatives has been observed, SSEs are stepping stones toward *green* chemistry.

14.6 REDOX-ACTIVE ELECTROLYTES

The group of electrolytes introducing pseudocapacitance contribution to supercapacitor performance is called redox-active electrolytes. This extra capacitance usually comes from the redox mediator species found in the electrolyte [19]. The redox pair species improve charge storage capacity by electron transfer reactions on the internal and external surfaces of the supercapacitor electrodes. A typical example of a redox-active additive is the iodide/iodine pair added to traditional aqueous electrolytes. Since the capacitance increase in the iodide/iodine pair systems is only observed at the positive electrode, the concept of dual redox additives is preferred. For instance, $VOSO_4$- and KI-based electrolytes, separated by a glassy paper and a Nafion 117 membrane, enhance the capacitance of negative and positive electrodes, respectively. Excellent performance with a high capacitance of up to 500 F g^{-1} can be achieved in this EDLC system [39]. Further, heteropoly acids (e.g., silicotungstic or phosphotungstic acid) were investigated as redox couple additives for aqueous electrolytes. These compounds can provide not only the capacitance increase but also enhancement in the operating voltage (up to 1.6 V). Also, the organic redox mediators can greatly increase the specific capacitance of EDLCs. It includes humic acids, hydroquinone, methylene blue, indigo carmine, *p*-phenylenediamine, *m*-phenylenediamine, lignosulfonates, and sulfonated polyaniline [40, 41].

Additionally, metal redox couples have been considered as pseudocapacitance additives, including $Co_{(II)}/Co_{(III)}$, $Fe_{(II)}/Fe_{(III)}$, and $Cu_{(II)}/Cu_{(III)}$ [42]. Unfortunately, metal ion redox additives can accelerate the self-discharge processes of supercapacitors. It should also be noted that electrolyte redox couple additives are usually compatible with common pseudocapacitive electrode materials, i.e., RuO_2, MnO_2, $Co(OH)_2$, etc. Nevertheless, the capacitance rise in redox-mediated supercapacitors is often redeemed by poorer cycling stability compared to conventional aqueous electrolytes without redox mediators.

Redox mediator species could also be applied for nonaqueous electrolytes, including organic solutions and ILs [7, 43]. Nonaqueous redox-mediated supercapacitors can provide high energy density due to pseudocapacitance contribution and wider operating voltage compared to aqueous systems. An example of an organic redox-active electrolyte is the PC–DMC mixture solvent containing polyfluorododecaborate cluster ions. EDLC with these electrolyte features improved capacitance and protection from overcharging. Further, the mixture of 1-ethyl-3-methylimidazolium tetrafluoroborate ([EMIm][BF_4]) with its halide analogues, [EMIm][Br] or [EMIm][I], exhibits much higher specific capacitance in EDLC systems compared to the pristine [EMIm][BF_4]-based system. Also, excellent cycle stability and good power performance are observed.

Solid-state redox-active electrolytes have also been introduced in the literature [32]. Generally, GPEs with redox mediators analogous to liquid redox-active systems have attracted the most significant interest. For instance, PVA/H_2SO_4 systems with iodides, Na_2MoO_4, or KI–$VOSO_4$ mixture have been tested. Since GPE has a liquid phase trapped in the polymeric matrix, the working principles of redox mediators are similar to those in liquid redox-active electrolytes. In addition to PVA, PEO and Nafion were considered polymer hosts of these electrolytes. Moreover, the idea of organogel redox-mediated electrolytes was tested to increase the operation cell voltage of EDLCs. As an example, PEO/$LiClO_4$–AN organogels with iodide/iodine or $Fe_{(II)}$/$Fe_{(III)}$ redox pair exhibit enhanced ionic conductivity and can work in a supercapacitor cell up to 2.5–3 V and high specific capacitance of up to 209 F g^{-1} [44].

14.7 CHALLENGES AND FUTURE OUTLOOK

Although considerable efforts have been made to develop high-performance and safe electrolytes for supercapacitors, it is hard to identify one universal solution in this area. Each currently available electrolyte has its own set of benefits and drawbacks. Nevertheless, we can still try to overcome the challenges of state-of-the-art electrolytes through future research directions discussed below.

- The operating voltage is a crucial factor regarding high-energy supercapacitor devices. Therefore, an essential strategy is to explore new conductive salts, solvents, ionic liquids, or hybrid systems featured by an expanded electrochemical stability window.
- Moreover, the proposed alternative electrolytes should possess high ionic conductivity and reduced viscosity to provide low internal resistance and high-power characteristics for supercapacitor cells.
- Exploration of new electrolyte solvents (single/mixture systems) or additives could also be beneficial in terms of the increased operating temperature range, especially on the lower values.
- Another approach regards the compatibility of electrolyte and electrode materials. Optimizing the pore size of carbonaceous materials and electrolyte ion size is important in terms of both energy and power density performance. Further, reliable characterization techniques should be introduced to investigate the interaction mechanism between the constituents of the electrode material and the electrolyte.

- Also, the new redox mediators, acting as additives to electrolytes, can significantly improve the energy efficiency of supercapacitors due to pseudocapacitance contribution.
- Finally, the impact of electrolytes on the environment cannot be disregarded. The nature of electrolyte constituents, as well as their synthesis methods, should exhibit eco-friendless and cost-effectiveness. The disposal feature of whole supercapacitor devices is preferred.
- Eventually, in the era of the enormous development of portable and wearable devices, particular emphasis should be placed on exploring novel sustainable and flexible solid-state electrolytes.

14.8 SUMMARY

In this chapter, a brief overview of the development, recent technology trends, and future perspectives concerning electrolytes for electrochemical supercapacitors are presented. The discussion included different types of electrolytes grouped into liquid, solid-state, and redox-active systems. The characteristic features of each electrolyte group have been highlighted, pointing out their influence on the supercapacitor performance. Most attention has been paid to organic electrolytes due to their dominant position in commercial devices. Different aspects of the nature of organic electrolytes have been discussed, concluding insights into the issues related to the transformation of research achievements into industrial innovations.

REFERENCES

1. J. Zhao and A. F. Burke, *J. Energy Chem.*, 2021, 59, 276–291.
2. T. Brousse, D. Bélanger, and J. W. Long, *J. Electrochem. Soc.*, 2015, 162, A5185–A5189.
3. P. Simon, Y. Gogotsi, and B. Dunn, *Science*, 2014, 343, 1210–1211.
4. A. González, E. Goikolea, J. A. Barrena, and R. Mysyk, *Renew. Sustain. Energy Rev.*, 2016, 58, 1189–1206.
5. M. M. Amaral, R. Venâncio, A. C. Peterlevitz, and H. Zanin, *J. Energy Chem.*, 2022, 67, 697–717.
6. W. Raza, F. Ali, N. Raza, Y. Luo, K.H. Kim, J. Yang, S. Kumar, A. Mehmood, and E.E. Kwon, *Nano Energy*, 2018, 52, 441–473.
7. C. Zhong, Y. Deng, W. Hu, J. Qiao, L. Zhang, and J. Zhang, *Chem. Soc. Rev.*, 2015, 44, 7484–7539.
8. B. Pal, S. Yang, S. Ramesh, V. Thangadurai, and R. Jose, *Nanoscale Adv.*, 2019, 1, 3807–3835.
9. L. Suo, O. Borodin, T. Gao, M. Olguin, J. Ho, X. Fan, C. Luo, C. Wang and K. Xu, *Science*, 2015, 350, 938–943.
10. X. Tian, Q. Zhu, and B. Xu, *ChemSusChem*, 2021, 14, 2501–2515.
11. S. Biswas and A. Chowdhury, *ChemPhysChem*, 2022, 24, https://doi.org/10.1002/cphc.202200567
12. Q. Dou and H. S. Park, *Energy Environ. Mater.*, 2020, 3, 286–305.
13. Y. Wang, Y. Song, and Y. Xia, *Chem. Soc. Rev.*, 2016, 45, 5925–5950.
14. R. Kötz, M. Hahn, and R. Gallay, *J. Power Sources*, 2006, 154, 550–555.
15. R. S. Borges, Arava L. M. Reddy, M. F. Rodrigues, H. Gullapalli, K. Balakrishnan, G.G. Silva, P.M. Ajayan, *Sci. Rep.*, 2013, 3, 2572.

16. S. Ishimoto, Y. Asakawa, M. Shinya, and K. Naoi, *J. Electrochem. Soc.*, 2009, 156, A563–A571.
17. F. Béguin, V. Presser, A. Balducci, and E. Frackowiak, *Adv. Mater.*, 2014, 26, 2219–2251.
18. L. Xia, L. Yu, D. Hu, and G. Z. Chen, *Mater. Chem. Front.*, 2017, 1, 584–618.
19. C. Zhong, Y. Deng, W. Hu, D. Sun, X. Han, J. Qiao and J. Zhang, *Electrolytes for Electrochemical Supercapacitors*, 2016, 1st Edition, CRC Press.
20. S. Pohlmann, *Nat. Commun.*, 2022, 13, 1538.
21. M. Armand, F. Endres, D. R. MacFarlane, H. Ohno, and B. Scrosati, *Nat. Mater.*, 2009, 8, 621–629.
22. E. Jónsson, *Energy Storage Mater.*, 2020, 25, 827–835.
23. S. Pan, M. Yao, J. Zhang, B. Li, C. Xing, X. Song, P. Su and H. Zhang, *Front. Chem.*, 2020, 8, 261.
24. M. Galiński, A. Lewandowski, and I. Stępniak, Ionic liquids as electrolytes, *Electrochim. Acta*, 2006, 51, 5567–5580.
25. E. Kim, J. Han, S. Ryu, Y. Choi, and J. Yoo, *Materials (Basel).*, 2021, 14, 4000.
26. Y. Zheng et al., Ionic liquid electrolytes for next-generation electrochemical energy devices, *EnergyChem*, 4(3), 100075, 2022
27. K. Karuppasamy, J. Theerthagiri, D Vikraman, C-J. Yim, S. Hussain, R. Sharma, T. Maiyalagan, J. Qin and H.S. Kim, *Polymers (Basel).*, 2020, 12, 918.
28. X. Wang, M. Salari, D. Jiang, J.C. Varela, B. Anasori, D. J. Wesolowski, S. Dai, M. W. Grinstaff and Y. Gogotsi, *Nat. Rev. Mater.*, 2020, 5, 787–808.
29. D. Kasprzak and M. Galinski, *J. Power Sources*, 2023, 553, 232300.
30. Q. Abbas, M. Mirzaeian, A. G. Olabi, and D. Gibson, *Encycl. Smart Mater.*, 2021, 4, 382–392.
31. T. Ye, L. Li, and Y. Zhang, *Adv. Funct. Mater.*,2020, 30, 2000077.
32. D. P. Dubal, N. R. Chodankar, D. H. Kim, and P. Gomez-Romero, *Chem. Soc. Rev.*, 2018, 47, 2065–2129.
33. D. Kasprzak, C. C. Mayorga-Martinez, and M. Pumera, *Energy Fuels*, 2023, 37, 74–97.
34. T. Xu, K. Liu, N. Sheng, M. Zhang, W. Liu, H. Liu, L. Dai, X. Zhang, C. Si, H. Du and K. Zhang, *Energy Storage Mater.*, 2022, 48, 244–262.
35. H. N. Fard, G. B. Pour, M. N. Sarvi, and P. Esmaili, *Ionics (Kiel).*, 2019, 25, 2951–2963.
36. Y.R. Beg, G. R. Nishad, and P. Singh, *Organic Electrolytes for Flexible Supercapacitors.* Chapter 6, Flexible Supercapacitor Nanoarchitectonics, 2021, Scrivener Publishing LLC.
37. R. Jamil and D. S. Silvester, *Curr. Opin. Electrochem.*, 2022, 35, 101046.
38. K. Udaya Bhat and D. B. Panemangalore, *Ionic Liquid Electrolytes for Flexible Supercapacitors.* Chapter 19, Flexible Supercapacitor Nanoarchitectonics, 2021, Scrivener Publishing LLC.
39. E. Frackowiak, K. Fic, M. Meller, and G. Lota, *ChemSusChem*, 2012, 5, 1181–1185.
40. X. Q. X. Zhou, Y. Wang, C. Zhang, Redox electrolytes/mediators for supercapacitors, *Supercapacitor Technology*, 2019, 45–94.
41. Z. F. Wang et al., High-voltage redox mediator of an organic electrolyte for supercapacitors by lewis base electrocatalysis, *ACS Appl. Mater. Interfaces*, 2022, 14(21), 24497–24508.
42. S. Azmi and E. Frackowiak, *Electrochem. Commun.*, 2022, 138, 107289.
43. L. Sun, K. Zhuo, Y. Chen, Q. Du, S. Zhang, and J. Wang, *Adv. Funct. Mater.*, 2022, 32, 2203611.
44. J. Zhou, J. Cai, S. Cai, X. Zhou, and A. N. Mansour, *J. Power Sources*, 2011, 196, 10479–10483.

15 Ionic Liquid Electrolytes for Electrochemical Supercapacitors

Ahmed I. Abdel-Salam, Aya Mohamed Abuelftooh, Mostafa M. Omran, Mahmoud A. M. Elsaid, and Saad G. Mohamed

15.1 INTRODUCTION

Ionic Liquids (ILs), an innovative family of materials with massive potential in electrochemistry, are typically described as liquid organic salts below 100 °C made up of organic cations and organic/inorganic anions. In contrast to inorganic salts, the bulky asymmetric cation or anion of ILs can prevent intermolecular force and hinder the ionic aggregation into the low-energy crystalline state, lowering their liquidus temperature [1, 2]. ILs have several advantages over water and organic solvents, including low volatility, non-flammability, superior thermal stability, low vapor pressure, highly ionized environment, a wide working potential window of 2–6 V, and other variable characteristics like solvent miscibility and polarity. The physicochemical characteristics of ILs can be tailored and optimized via the cation and anion interaction, van der Waals interactions, and Coulombic interactions between ions (i.e., Lewis acidity/basicity of cations/anions). Therefore, ILs have become a distinctive class of innovative, adaptable materials that grab the interest of various applications in diverse areas, including synthesis, catalytic reactions, and energy storage [2–4].

Energy storage devices are one of the blooming applications that motivate the use of IL electrolytes. In this light, supercapacitors (SCs) have garnered significant interest as the energy storage technology of the next generation owing to their outstanding electrochemical characteristics like a long lifetime, safe operation, and more power density than batteries [3, 5]. SC's electrode materials are divided into three categories depending on the mechanism of charge storage as follows: 1) EDLC materials are capable of directly storing charges on the interface between electrode and electrolyte without occurring faradic reactions; 2) Pseudocapacitive materials reveal both capacitive and faradaic nature; and 3) Battery-type materials are capable of storing charges via faradaic reactions [3, 5–9].

DOI: 10.1201/9781003441182-18

ILs have been widely used in cutting-edge SC applications with long operating lives. Cyclic amines, pyrrolidinium, imidazolium, quaternary ammonium salts, and aliphatic quaternary ammonium salts are the most frequent ILs electrolytes in the SCs [10–12]. Nevertheless, at applications that require a variety of working temperatures, these salts are limited because they work at their melting state at (>−40 °C). Furthermore, to fulfill the demand for high-performance SCs, ILs with a wider working potential window and high conductivity should be synthesized.

Therefore, the following categories can be used to classify ILs: 1) Pure ILs may be employed as electrolytes for energy storage using EDLC with a broad and stable potential window—this might be due to the good ionized media, ability to remain liquid over a large temperature range, and is non-flammable; 2) designable ILs used as reactive agents or reaction media during the synthesis of the active electrode materials to control their morphology and composition, increasing the pseudocapacitive behavior of the electrode via redox reactions on the electroactive sites; 3) ILs with lower vapor pressure and characteristic molecular designability can produce functionalized carbon material via the direct carbonization process that can be employed to store the charge based on the EDLC process, producing supercapacitive behavior [3].

This chapter will cover the common types of ILs used for SC applications and their contribution to the energy storage process. Furthermore, the chapter will clarify the nature of IL interaction at the electrode surface, the advantages of the various types of ILs (in addition to the most suitable working conditions), and electrochemical system components integration. After that, each class's recent challenges and key solutions to improve their behavior during the charge/discharge process in the SC devices will be discussed. Moreover, several work samples will be provided to clarify each electrolyte type's state of the art. At the end of this chapter, critical practical recommendations and electrolyte/electrodes integration troubleshooting will be provided as effective practices in electrochemical SC fabrication.

15.2 FUNDAMENTALS OF IL ELECTROLYTES

15.2.1 TERMINOLOGY

ILs are typically defined as liquids entirely made up of ions and fluid below 100 °C. Although they are not a novel class of materials, they possess physical characteristics that have drawn more attention recently [13]. In the literature, most salts categorized as ILs are liquid at room temperature and frequently at much lower temperatures, as low as −96 °C [13, 14]. Glass transition point, not melting point, serves as the basis for IL liquidity [4]. To better understand the nature of ILs, the molten NaCl is IL, while it forms an ionic solution when dissolved in water. However, the earlier terminology, "molten salts" ("melts")—that implies elevated temperature, viscosity, as well as corrosive media—has been replaced by the term "ionic liquids [14]. The important thing to remember is that because ILs have traditionally been thought of as solvents, liquidity is crucial. However, these organic salts' ionicity rather than liquidity is the key consideration in many applications. Additionally, highly viscous ILs are unsuitable solvents, but they can be used in electrochemical systems that include a variety of electrodes or electrolytes in the solid form [4]. Therefore, the term of ILs is suitable for "ionic organic salts" since numerous salts may be employed in the same applications as ILs where ionicity is the crucial feature [4].

15.2.2 Electrochemical Potential Window

One of the essential characteristics that should be determined for the employed electrolytes in electrochemical applications is the electrochemical window (EW). The stable EW is the range of potentials where the electrolyte is not reduced at low potentials or oxidized at high potentials, i.e., the electrolyte is inert within this range. The advantage of broad EW is better than the C_s because the amount of charge stored by SCs is related to the square of the working potential, $E = 1/2\ CV^2$. The EW must thus be expanded, but this is hindered by electrolytes that become unstable at high cathodic or anodic working potentials [4, 15].

In the case of aqueous electrolytes, the theoretical EW is approximately 1.23 V between the redox potential of the water associated with oxygen and hydrogen evolution. This window drops to 1 V during the practical work due to the underpotential processes. Therefore, non-aqueous electrolytes can be used to extend the EW, which increases the SC's energy density by one order of magnitude despite the high cost and difficulties in the cell design in contrast to an aqueous electrolyte. Organic electrolytes can give a significantly wider, more stable EW than aqueous electrolytes, up to 3.5 V [16]. Acetonitrile [17] and propylene carbonate [18] are two popular organic electrolytes; however, they exhibit severe drawbacks during the practical processing of the SCs, such as flammability and extreme moisture sensitivity [19].

Therefore, ILs are promising candidates for the assembly of SC because of their superior stability, the wide working potential window of 2 to 6 V, and minimal flammability. The ILs have significant, wide, stable potential windows due to their composition, which is made up of ions that don't undergo electrochemical reactions with the active electrode material over a wide potential range [4, 10–12]. As with organic electrolytes, the ILs are moisture-sensitive, since even a tiny quantity of moisture can dramatically limit the width of the stable EW. Thus, the purity of the ILs is a critical factor that can affect the stable EW [4].

15.2.3 Ionic Conductivity

Equivalent series resistance (ESR) influences the ionic conductivity and limits the P_d and E_d of the device. The ohmic drop caused by the ESR decreases the E_d and P_d, dependent on ESR, as described in Equation: [$P_d = V^2/4 * ESR$] [20–22]. This ESR describes the bulk electrolyte, the various interfaces, and the intrinsic resistance of electrodes. Among these, the bulk electrolyte resistance is the only one that is simple to measure and control; fortunately, it also has the most significant impact. Since ILs are low in neutral species and rich in ions, one would anticipate them to have good ionic conductivity. However, the concentration of charge carriers does not alone determine this feature. This results from the IL's high viscosity, which lowers the ions' self-diffusion coefficient [21–23]. In the 1-ethyl-3-methylimidazolium (EMIM) bis(trifluoromethane) sulfonimide (TFSI) IL, for instance, the values of the cation and anion diffusion coefficients at room temperature are 0.495×10^{-10} and 0.309×10^{-10} m^2 s^{-1} [22, 24], which is lower than pure water (2.3×10^{-9} m^2 s^{-1}) by two orders of magnitude at the same temperature [22, 25].

The ILs' ionic conductivity is proportionate to the self-diffusion coefficient by neglecting any ionic correlations. However, this relation is only valid for extremely diluted salt solutions in aqueous or organic solvents. Additionally, the ILs show a lower self-diffusion coefficient at the same temperature than pure water. In contrast to traditional liquid electrolytes, this results in significantly poor ionic conduction in ILs. As a result, typical electrical conductivities in ILs at the ambient temperature are in the range from 1.0 to 10.0 mS cm^{-1} [22, 26, 27]. The high viscosity of the ILs is another factor that decreases their ionic conductivity because of their bulky ion sizes. They also often display a viscosity greater than that of regularly used organic solvents, depending on the kind of combination between cation and anion [21].

15.2.4 Thermal Stability

Thermal stability is another important parameter that is significant in evaluating electrolytes quality. An SC may be subjected to temperatures between −50 and 80 °C when used in electric vehicles under operating conditions. Consequently, in these operating conditions, temperature-related problems will mainly be attributable to the electrolyte. Compared to the ILs, the aqueous electrolyte can't reach the lowest temperatures because water freezes, even with added salt. Additionally, the organic solvents may operate at lower temperatures but frequently have too low boiling points. Compared to traditional solvents, the ILs' liquidity is stable up to significantly higher temperatures. However, they experience ionic species deterioration at high temperatures [22, 28].

The temperature at which the decomposition reactions may start can be determined using thermo-gravimetric analysis (TGA) [22, 29]. Fortunately, these decomposition temperatures are typically over 200 °C, demonstrating that using ILs in SCs at high temperatures is not troublesome. Generally, the melting points of the ILs are near room temperature, which restricts the usage of ILs in lower-temperature applications such as electric vehicles. To get over this issue and run the IL-based SCs at extremely low temperatures, Lin et al. used two ILs with different cations and the same anion [22, 30].

15.3 PERFORMANCE UNDER EXTREME CONDITIONS

Due to instability and safety concerns, conventional electrolytes are unsuitable for some applications of SCs, which need to deliver high performance at high-temperature conditions. These electrolytes can cause unwanted side chemical reactions and thermal instability while working under extreme conditions like high temperature and pressure because they are more thermally stable and non-flammable than traditional electrolytes, ILs are considered attractive options for high-temperature SCs [10, 30]. For instance, a eutectic mixture of ILs electrolytes has been used with a conventional activated carbon SC and revealed a great C_s of about 180 F g^{-1} in wide EW of about 3.5 V. Furthermore, the fabricated device can withstand a temperature between −50 and 80 °C [10]. Moreover, a number of ILs in acetonitrile have been studied by Zhang et al. in temperatures between 20 and 80 °C; the fabricated electrodes reveal straightforward temperature dependency with no non-linear behavior [31].

In addition, some oxides and other materials show high performance at higher temperatures, and these working conditions can improve their electrical and ionic conductivities. For example, the fabricated device from FeOOH and an IL reveals a high E_d of about 1.44 Wh Kg^{-1} at 200 °C [32]. Furthermore, the ILs can create supercooled liquids at temperatures under their melting points, which is a distinguishing property of practically all ILs since they can be organized in an ordered structure that facilitates diffusion and overcomes the problem of ILs' high viscosity. For example, the fabricated capacitor of activated carbon in the presence of pyrrolidinium nitrate shows high C_s of about 117 F g^{-1} at −40 °C, with electrolyte conductivity about 9 mS cm^{-1} [33]. These findings demonstrate that the ILs consider a promising choice for designing SCs that can withstand harsh working conditions. Still, careful consideration should be given when developing ILs for specific applications.

15.4 TYPES OF IL ELECTROLYTES

Using ILs in SCs is related to the physical and electrochemical properties of ILs, which depend on the chemical structure of these ILs. The melting point and viscosity strongly affect the ionic conductivity and operation temperature range; the chemical composition also strongly affects the electrochemical stability window (ESW). The aprotic ILs, which differ in their physical and electrochemical properties and are made up of separate anions and cations to provide a wide variety of ILs, attracted attention to intensively investigating ILs in SCs. This property improves the E_d without reducing the high P_d feature of SCs. The following addresses the common types of ILs based on the chemical structure of cations and their physical and electrochemical properties [1].

15.4.1 Imidazolium ILs

This type of ILs is composed of the imidazolium cation, an imidazole molecule with two protons of one and three nitrogen atoms substituted to asymmetric organic radicals and organic/inorganic anion. The two organic radicals of imidazolium cations and anion are essential factors affecting their melting temperature and viscosity, thus influencing the electrochemical properties. For instance, in two different imidazolium ILs, Chen Li et al. investigated graphene's electrochemical behavior. The 1-ethyl-3-methylimidazolium (EMIM) was the cation in both EMIM BF_4 and EMIM TMSI, which contain tetrafluoroborate (BF_4) and bis-trifluoromethylsulfonylimide (TMSI) as anions, respectively [34]. The viscosity and ionic conductivity of both ILs were 32 and 34 mPa s and 16.01 and 10.8 mS/cm for EMIM BF_4 and EMIM TMSI, respectively.

The electrochemical properties of the fabricated SC using EMIM BF_4 revealed an E_d of 106 Wh Kg^{-1} at a P_d of 10 KW Kg^{-1} and ESW of 4 V, while the fabricated device using EMIM TMSI exhibited an E_d of 135.6 Wh Kg^{-1} at the same P_d and ESW. It is quite obvious that replacing the IL anion improved the E_d of the fabricated SC. On the other hand, Yao Chen et al. investigated the behavior of reduced Graphene Oxide (rGO) active material in 1-butyl-3-methylimidazolium tetrafluoroborate (BMIM BF_4) IL. However, the only difference between BMIM BF_4 and EMIM BF_4 is substituting the ethyl radical in the imidazolium molecule with the butyl group; the ESW of BMIM BF_4 was dropped to 3 V in comparison with 4 V of EMIM BF_4.

To make a clear statement of the imidazolium ILs performance as electrolytes with respect to the conventional aqueous ionic electrolytes, Zhibin Lei et al. investigated the electrochemical properties of the mesoporous carbon intercalated into reduced graphene oxide composite rGO-CMK-5 electrodes in KOH, $LiPF_6$, and EMIM BF_4 as electrolytes (see Figure 15.1a). The IL electrolyte showed an obvious improvement in energy and power densities compared to the other electrolytes, in

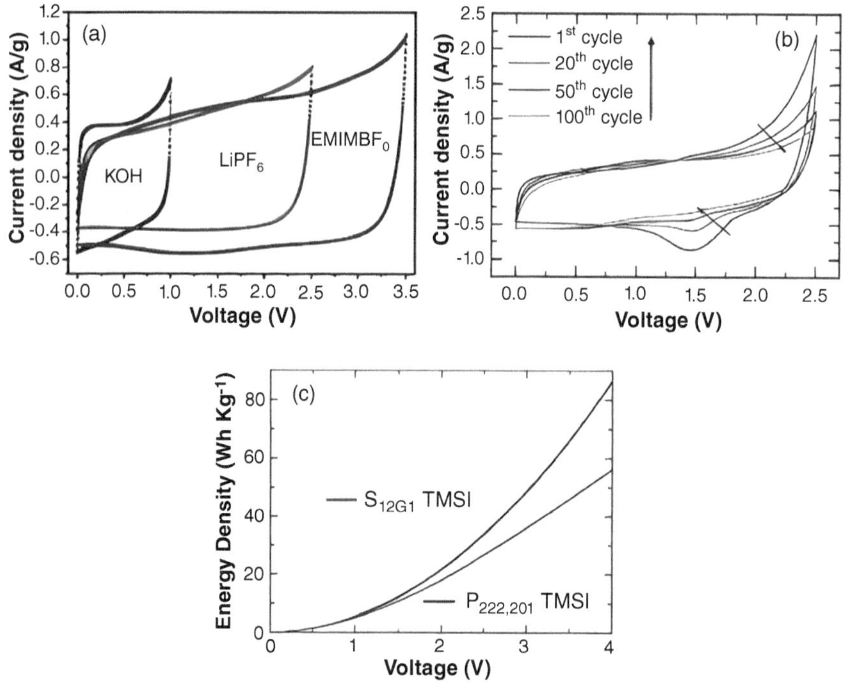

FIGURE 15.1 The performance curves of variant ILs as electrolytes for supercapacitors; (a) the cyclic voltammetry of rGO-CMK-5 composite in different electrolytes; (b) cyclic voltammetry of symmetric activated carbon supercapacitor using a mixture of Pyrr NO_3 and Pyrr TMSI ILs as electrolyte for different cycles; and (c) the relation between energy density and potential window computed for Graphene-based electrodes using Phosphonium and Sulphonium ILs sharing the same anion as electrolyte.

Source: (a) Reproduced from Ref. [35] with permission from the Royal Society of Chemistry, (b) Reproduced with permission from Ref. [37], (c) Reproduced with permission from Ref. [41].

addition to the wider potential window of 3.5 V [35]. The main advantage that makes the imidazolium family of ILs one of the most promising alternative high-performance electrolytes for SCs is its relatively high ionic conductivity compared with the other IL families, which leads to high-power SCs with enhanced E_d and wide ESW [36].

15.4.2 Pyrrolidinium ILs

The pyrrolidinium-based IL family can be stated as the opposite of the imidazolium-based IL family. Pyrrolidinium ILs offer a wider ESW up to 5–6 V but with a lower ionic conductivity, which makes this type of ILs suitable for providing high E_d with relatively limited P_d. Also, pyrrolidinium ILs suffer from low cycle stability at high potentials. For instance, L. Timperman et al. studied the electrochemical behavior of the Pyrrolidinium ILs mixture as an electrolyte for an activated carbon-based supercapacitor. This system showed a noticeable decrease in the electrochemical activity after several cycles (see Figure 15.1b) [37].

For a clearer understanding, J. H. Lee et al. studied the electrochemical stability of 1-methyl-3-propyl imidazolium bis(fluorosulfonyl)imide (MPIM FSI), N-methyl-N-propyl pyrrolidinium bis(fluorosulfonyl)imide (MPPY FSI), and N-methoxymethyl-N-methyl pyrrolidinium bis(fluorosulfonyl)imide (MMPY FSI), and it was found that MPIM FSI began exhibiting an oxidative decomposition current at (2.3 V vs. Ag/Ag^+), while MPPY FSI and MMPY FSI showed higher electrochemical stability up to (2.5 V and 2.7 V vs. Ag/Ag^+, respectively) [38]. Moreover, a stable electrochemical window for MMPY FSI was observed at 3.5 V in the cycling stability test, while stable cycling for MPIM FSI was achieved when the ESW was reduced from 3.5 to 3 V.

One strategy to improve the electrochemical properties of pyrrolidinium ILs was suggested by Pallab Bose et al. when a nanoscale hybrid electrolyte (NHE) was prepared from N-Methyl-N-propyl pyrrolidinium bis(trifluoromethanesulfone) (MPPY TFMS) with TiO_2 nanoparticle. The C_s of the designed cell increased to three times in NHE than in pure MPPY TFES. Also, higher stability for NHE was observed after 5000 cycles, which recorded a capacitance retention of 93% compared to 66% of MPPY TFES [39]. The pyrrolidinium IL family showed a promising performance as a wide ESW electrolyte; however, an in-depth study is necessary to increase ionic conductivity and stability in the wide operating potential windows.

15.4.3 Sulfonium ILs

Trialkyl sulfonium cation is the general chemical formula of the sulfonium IL family group. Sulfonium ILs are not widely investigated for SC application due to their low ionic conductivity; however, promising results were obtained from a study made by J. Anthony et al., who compared the physical characteristics and electrochemical performance of three types of sulfonium ILs (diethylmethyl sulfonium S221, triethyl sulfonium S222, and diethylpropyl sulfonium S223 cations with TMSI anion) with N-butyl-N-methyl pyrrolidinium (BMPY) cation and TMSI anion ILs [40]. Anthony and his colleagues stated that the three sulfonium ILs showed better viscosity and ionic conductivity than the studied pyrrolidinium IL.

Although the pyrrolidinium ILs showed a wider ESW of 3.8 V than the sulfonium ILs—which showed an ESW between 2.6–2.8 V—the better ionic conductivity and improved capacitance of the three sulfonium ILs provided a comparable E_d to the pyrrolidinium IL with higher P_d. This study highlights the possibility of using sulfonium ILs as adequate electrolytes for SCs in the future.

15.4.4 Phosphonium ILs

Phosphonium ILs comprise tetra alkyl phosphonium cation combined with organic/inorganic anion; this type of ILs did not get much attention until the last few years. The first fabricated SC using phosphonium ILs was reported by E. Frackowiak et al. in 2005; the assembled device from activated carbon with trihexyl(tetradecyl) phosphonium cation and TMSI anion showed a relatively high ESW of about 3.4 V and 40 Wh Kg^{-1} of Ed at room temperature. Abner Massari Sampaio et al. made a computational study to compare the phosphonium ILs to sulphonium ILs; they designed a graphene-based two-electrode system and simulated the electrochemical performance of phosphonium and sulphonium ILs sharing the same anion.

That study showed the advantage of the sulphonium over the phosphonium ILs, especially in the wide potential window (see Figure 15.1c) [41]. However, the main advantages of considering phosphonium ILs as promising candidates for SC applications are their high thermal stability, electrochemical stability, and availability in large-scale production. The better electrochemical stability of phosphonium ILs over nitrogen-based ILs is due to the slit acidity of the two carbon atoms attached to nitrogen in nitrogen-based ILs, which may lead to carbene formation. Wherefore, phosphonium ILs should be investigated more for SC application [42].

15.4.5 IL Mixtures

To overcome the challenges of low ionic conductivity of ILs and improve their physical properties, combining two different ILs can provide improved physical and electrochemical properties for better electrical energy storage performance. For example, R. Lin et al. studied the properties of N-methyl-N-propyl piperidinium bis-fluorosulfonylimide ($PIP_{13}FSI$) and N-butyl-N-methylpyrrolidinium bis-fluorosulfonylimide $PYR_{14}FSI$ mixture in the molar ratio 1:1; this mixture exhibited improved physical properties to lower the melting point to a degree lower than its constituents, which provided a wide temperature range SC between –50 and 100 °C with a potential window up to 3.5 V [30]. Furthermore, T. Tooming et al. stated that adding 5 wt.% of 1-ethyl-3-methylimidazolium iodide (EMIM I) to EMIM BF_4 increases the stored charge by 50% [43].

15.4.6 IL/Organic Electrolyte Mixtures

The presence of organic solvents with ionic liquid as a mixed electrolyte could be another strategy to obtain higher-performing SCs. The properties of ILs are temperature-dependent; in most cases, the higher operating temperatures provide better properties and electrochemical performance. On the other hand, organic solvents are less sensitive to temperature change, which may provide the mixture of IL and organic solvent electrolytes with both merits of IL and organic solvent to obtain a high-performing and less temperature-sensitive electrolyte.

For example, V. Ruiz et al. studied the effect of organic solvent added to the MPPY TMSI on the electrochemical performance [44]. Although the addition of butyronitrile (BuCN) reduced the ESW of MPPY TMSI, the C_s of the fabricated EDLC device was increased from about 45 F g^{-1} for pure MPPY TMSI to about 130 F g^{-1} for the mixture at room temperature. At the elevated temperature of 80 °C, the C_s of the device was intensively increased to about 150 F g^{-1} for the pure MPPY TMSI. Also, for the IL/organic electrolyte mixture, the C_s was slightly increased to about 135 F g^{-1}.

To conclude this part, thousands of ILs can be investigated as electrolytes for SC applications. The physical properties of viscosity and melting point are the initial factors to be presumed about the electrochemical performance of any IL type. Ionic conductivity is the second stage in IL efficiency to be used in SCs. The ionic conductivity is an initial indication of the power density ability of the IL. Once the physical properties and ionic conductivity show promising results, it is time to determine the ESW and the stable window for cycling stability. Then, the complete electrochemical profile can be entirely resolved.

15.5 ILs AS SELF-HEALING MATERIALS

Self-healing materials attracted much attention from material designing researchers due to their ability to increase the application lifetime by repairing the damage caused by intensive usage. Most Polymerized Ionic Liquids (PILs) are considered a perfect choice as a self-healing material because of their exceptional ability to perform a reversible reaction such as coordination bonding, Deals-Alder reaction, or electrochemical reaction that regenerates the initial properties of the material. For example, hydrophilic hydrogels are well-known materials to be used as self-healing electrolytes that can absorb moisture to prevent electrochemical cell drying and, subsequently, cell deactivation. This property of super hydrophilicity of specific types of hydrogels provides a longer time of wetness to the electrochemical cell, which keeps the internal resistance and ionic transportation in their steady state for a long time even if extreme conditions are applied [45].

As a self-healing material, PILs are usually mixed with polymeric electrolytes and forming hydrogen bonding together—like ureido-pyrimidinone, to improve their thermal and mechanical properties and self-heal the mechanical damages caused by extreme mechanical conditions—for use in flexible supercapacitors. Polymers like acylhydrazone-based materials are widely used as self-healing materials for mechanical applications, but they are not the best choice to be used as electrolytes due to their limited conductivity. On the other hand, using the hydrogen bond polymers/PILs mixture as self-healing electrolytes provides both better conductivity and higher stability [46]. Hui Peng et al. mixed PVA/agar with $EMIMBF_4$ and Li_2SO_4 and used this mixture as a self-healing electrolyte for a flexible supercapacitor. Hui Peng and his colleagues mentioned that adding EMIMBF4 and Li_2SO_4 to the dual-network PVA/Agar led to improved conductivity and wider operating temperature [47].

15.6 IONIC LIQUIDS ELECTROLYTE/ACTIVE ELECTRODE MATERIAL INTERACTION

SC electrode material can store charges either by the formation of an EDL, as in the case of EDLC electrode materials or by redox reaction at the electrode surface, in the case of pseudocapacitor or at both the bulk and the surface, as in the case of the battery-type electrode. Notably, in the case of pseudocapacitor and battery-type electrode, the energy is also stored as a result of EDL formation; this can be summarized at Figure 15.2a. Therefore, it is expected that a pseudocapacitor and battery-type electrode have higher charge capacity than an EDLC electrode, but less cycling stability during charge and discharge [5, 9, 48]. The definite charge storing mechanism is determined as a result of the electrode-electrolytes interaction, which can be affected by both the nature of electrode material "such as its elemental composition, morphology, particle size, porosity, other factors," and electrolyte nature such as "its conductivity, ion mobility, viscosity, electrochemical and thermal stability, and other parameters" [7, 49].

ILs have drawn interest as electrolytes due to their unique properties like wide EW (up to 4.5 V) if compared to the aqueous electrolytes, in addition to the high thermal stability, little volatility, high conductivity, and less flammability if compared with the organic electrolytes. Practically, it has many drawbacks such as high viscosity and low conductivity if compared with the aqueous electrolytes. This property limits its capacitance and represents a critical challenge that limits its application in SC devices [50–52].

The most popular charge-storing mechanism in case of ILs electrolyte would be the EDLC due to the large ion volume and restricted ion mobility and other parameters related to the composition of these liquids. Also, in some cases, redox mechanisms take place at the electrode surface via a redox reaction between the electrode surface and ions of IL electrolyte [3, 52].

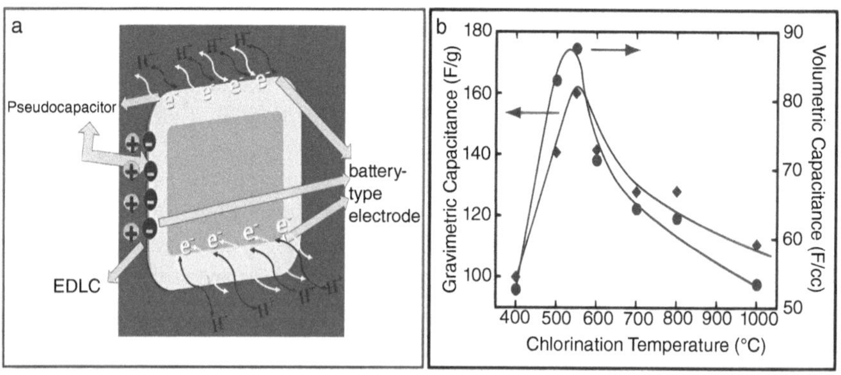

FIGURE 15.2 (a) The electrode-electrolyte interaction at different energy-storage mechanisms of SCs electrode materials; (b) specific gravimetric (F g^{-1}) and volumetric (F cm^{-3}) capacitances change depending on the chlorination temperature for CDC electrodes tested in neat [EMI][TFSI] electrolytes at 60 °C.

Source: (b) is Reproduced with permission from Ref. [53].

15.7 EDLC BASED ON IL ELECTROLYTES

Most IL electrolytes, as most organic electrolytes, interact with the electrode surface via the EDLC mechanism due to their nature. The interaction between the electrode and the electrolyte determines the electrochemical performance, which can be affected by different factors. One of the most important factors is the relationship between the electrode's pore size and the electrolyte's ion size, which controls, in some way, the efficiency of the formed electric double layer [21].

The relation between the electrode's pore size and the electrolyte's ion size was demonstrated by Largeot et al. [53]. They use (1-ethyl-3-methylimidazolium bis(trifluoromethylsulfonyl)imide) [EMI][TFSI] as an electrolyte [EMI] and [TFSI] with approximately equal ion size (~0.7 nm). The capacitance was measured when the electrodes' pore size became smaller than, equal to, and larger than the ion size of [EMI][TFSI]. This work operated by controlling chlorination temperature since the pore size increased to 0.65–1.1 nm depending on the chlorination temperature range, as shown in Figure 15.2b.

At a chlorination temperature of 600 °C, the highest specific gravimetric or volumetric capacitances were attained. The mean pore size at 600 °C was 0.72 nm, quite near the [EMI][TFSI] ion size. In this instance, the ions adsorbed on both pore walls implies that there isn't enough space for more than one ion per pore (the distance between one pore and another). They revealed that the most effective technique to promote ion adsorption was reducing the empty spare space if the pore size was comparable to the ion size. The C_s fell below 600 °C when the CDCs' pore size was lower than the [EMI][TFSI] ions' size.

The pore size was too tiny to allow ions to pass through, especially at 400 °C, which caused a significant drop in the specific gravimetric or volumetric capacitance. Additionally, the specific capacitance dropped beyond 600 °C because the pore size was larger than the ion size. The separation between both the pore walls and the centers of the ions grew as the pore size rose, even though there was still room for one ion per pore, which resulted in decreased volumetric capacitance [21, 53].

Based on their analysis of conditions when the pore size was larger than the ion size, Soavi et al. stated the functionality of the carbon electrode and chemistry of electrolyte in SCs. They proposed that the carbon porosity and IL characteristics control the capacitive response, whereas an electric double layer was readily formed once the pore size was wider than the ion size and the number of accessible ions grew as the pore size rose. Therefore, IL electrolyte characteristics such as conductivity, viscosity, and solvent polarity had little impact on electrode response qualities. Thus, instead of the bulk properties of ILs, the pore-to-ion size affects the EDLC's performance [21, 54].

15.8 PSEUDOCAPACITOR BASED ON IL ELECTROLYTES

Recently, a perspective has emerged to deal with the issues of IL electrolytes by examining new electrolytes and mixing them with additives that support the redox process to improve the C_s due to the contribution of redox current coming from the surface reaction in addition to the current coming from EDL formation; also it would dimension the charge-transfer resistances. Notably, the challenge would be to do this without sacrificing the other positive properties of IL electrolytes and preserving the excellent cyclability of EDLC electrode materials [3]. For example, Navalpotro et al. used para-Benzoquinone (p-BQ) solution in the N-butyl-N-methyl pyrrolidinium bis(-trifluoromethanesulfonic)imide IL as an electrolyte for Pica and Vulcan carbon, which have various characteristics. Therefore, it is an excellent chance to study the effect of this addition at those electrodes. Figure 15.3a and b demonstrate the cyclic voltammetry (CV) curves for Pica and Vulcan carbon electrodes in the absence and presence of (p-BQ), respectively.

When p-BQ was present, the CV curves showed two redox peaks with larger integral areas, which indicated a higher current than when p-BQ was absent.

The galvanostatic charge/discharge (GCD) curves in Figure 15.3c and d indicate that the pica carbon electrode has characteristic pseudocapacitor behavior in the presence of (p-BQ), while the Vulcan carbon electrode showed battery-type behavior which might be due to the more diffusion to some extent, see Figure 15.3d [55].

Also, Sathyamoorthi et al. used a combination of triethylammonium bis (trifluoromethane)sulfonimide and hydroquinone with a charcoal-based electrode; the effect of this addition was estimated via CV and GCD in Figure 15.4a and b, as the redox couple, the enhanced C_s of 72.0 F g^{-1} and E_d 31.22 Wh kg^{-1} [11]. Additionally, Fleischmann et al. added soluble alkali ions to 1-methyl-1-propylpyrrolidinium

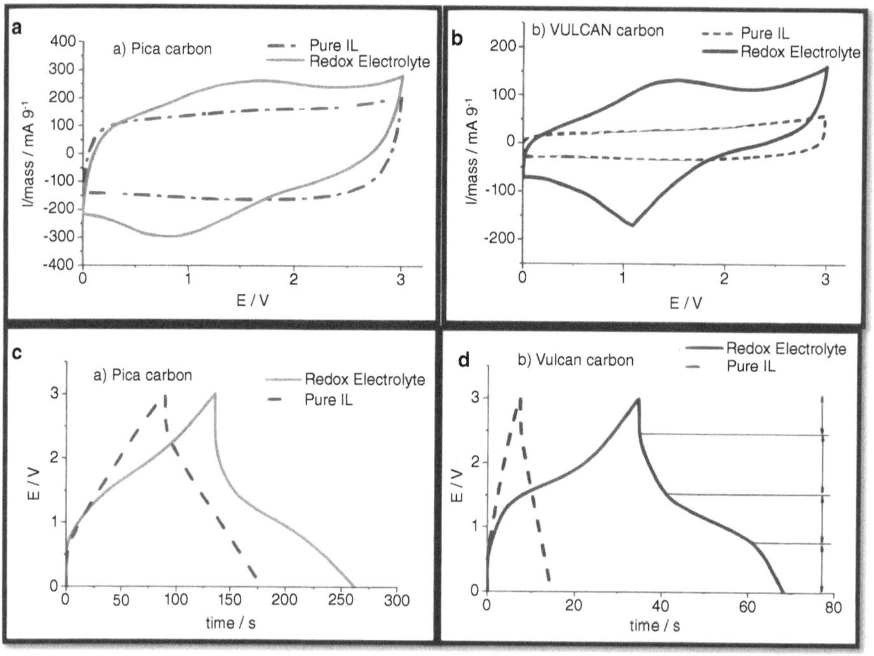

FIGURE 15.3 (a) CV at a scan rate of 5 mV^{-1} for Pica activated carbon; (b) CV at a scan rate of 5 mV^{-1} for Vulcan carbon; (c) voltage profiles plot up to 3 V at current density 10 mA cm^{-2} of Pica carbon; and (d) voltage profiles plot up to 3 V at current density 10 mA cm^{-2} of Vulcan carbon. At both electrolytes: – (0.4 M p-BQ in PYR14TFSI) and pure PYR14TFSI. at 60 °C.

Source: Reproduced with permission from Ref. [55].

bis(trifluoromethyl sulfonyl)imide; this produced an SC device with an EW of 4 V and output E_d of 100 W h kg^{-1} with P_d more than 2 kWh kg^{-1} [56]. Recently, more reports came out concerning maximizing the benefit of the ILs electrolytes via using additives and ILs mixtures.

Some reports analyzed the surface and non-surface currents and reported that there were some contributions of diffusion current as a result of a limited bulk redox reaction, similar to some extent to the aspect of the battery-type electrode. Medagedara et al. studied the contribution of diffusion and surface reaction to the total output current (see Figure 15.4c). They synthesized IL electrolyte based on triethylammonium cation and thiocyanate anion and electronically conducted activated carbon derived from cleaned coconut shells as electrodes [57].

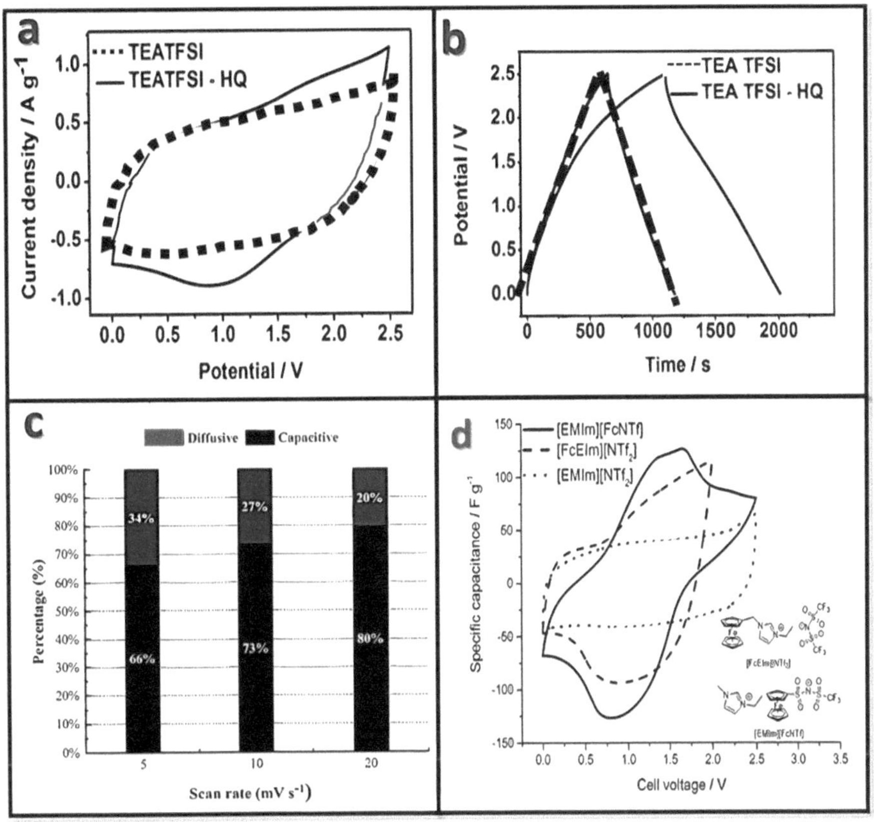

FIGURE 15.4 (a) CVs of the SC at 10 mV s^{-1}; and (b) GCD at 0.57 mA cm^{-2} in neat TEATFSI and TEATFSI/HQ (0.3 M) of activated charcoal-based SC; (c) capacitive and diffusive contributions to the total current at different scan rates at triethylammonium cation and thiocyanate anion as electrolyte and activated carbon derived from cleaned coconut shells as electrode; and (d) Cyclic voltammograms recorded with two-electrode cells with 80 wt.% of the ionic liquid in acetonitrile. Each carbon electrode weighed 3.5 mg and contained 80 wt.% of activated carbon. The curves were obtained at a scan rate of 10 mV s^{-1} at a temperature of 25 °C

Source: (a) and (b) Reproduced with permission from Ref. [11]. (c) Reproduced with permission from Ref. [57]. (d) Reproduced with permission from Ref. [60].

Other studies reported the addition of some redox additives to modify the electrical behavior of ionic liquids. Lee et al. studied the addition of vanadium sulfate and tin sulfate to the traditional ionic liquid and reported the enhancement of the capacitive behavior [58, 59]. Furthermore, by including a faradaic component in the charge storage mechanism of activated carbon electrodes and the ionic liquid electrolyte system, Xie et al. presented another example of the prospective design of ionic liquids that supplied higher specific energy. Ferrocene was directly linked to the ionic liquid's molecular structure, enabling concentrations of up to 2.4 M ferrocenyl units in acetonitrile, increasing the device's specific energy by 83% (vs. unmodified ionic liquid). The Fc-anion [FcNTf] allowed for the total inhibition of self-discharge at the positive electrode, whereas redox ionic liquid electrolyte capacitors with Fc connected to the cation offered a considerable self-discharge rate, also the specific energy and stability's rate dependence might be made better [60]. Figure 15.4d reveals the cyclic voltammograms of the 2-electrode system at the three electrolytes. As shown, the highest current was obtained in case of the Fc-anion [FcNTf].

15.9 CHALLENGES AND FUTURE MOTIVATION FOR IL ELECTROLYTES

ILs, which differ from traditional aqueous and organic electrolytes, have drawn a lot of interest in energy storage applications, particularly LIBs and SCs. Although ILs can increase the EW of LIBs and SCs, improving specific capacitance is not possible because of the difficulties associated with ion separation. This challenge must be addressed in further research. An SC may be subjected to a wide range of temperatures between −50 and 80 °C when used in electric vehicles and for some military purposes [61–63]. Consequently, under these operating conditions, temperature-related problems will mostly be attributable to the electrolyte.

So, the performance of the SCs for these applications should be improved in these harsh conditions. Unfortunately, most ILs have high viscosity at a lower temperature, which deteriorates the ionic conductivity, as well as the charge transportability of the device [64, 65]. Consequently, a lot of research must be performed to improve the ILs' properties via increasing the EW and safety, and decreasing their viscosity for energy storage. These drawbacks of the ILs can be overcome by mixing several IL types or using ILs as electrolyte additions; this could decrease the ILs' viscosity at a lower temperature and improve the operational temperature range [65].

When the ILs mixed with organic electrolytes, the operational temperature ranges, as well as the ionic conductivity, were improved. However, the working EW and the E_d of the SCs might be reduced. The second solution of mixing different types of ILs is considered a good candidate to improve the range of operational temperatures, as well as the EW and the E_d of the SCs. For example, a eutectic IL mixture of 1-propyl-3-methylpyrrolidinium bis(trifluoromethylsulfonyl)imide with (EMIM)(TFSI) has been developed by R. Newell et al. This IL mixture reveals a wide EW of about 3.5 V, temperatures between −70 and 80 °C, which is suitable for SCs applications [52, 66]. Recent research studies are focused on new alternative electrolytes composed of ionogel electrolytes based on poly (ionic liquids) (PILs) and conventional polymers due to their safe high-voltage applications and self-healing properties [52, 67, 68]. The common polymers that have been studied in the ionogel electrolytes are poly (vinylidene fluoride) (PVDF), poly(ethyl oxide) (PEO), and poly(vinylidene fluoride-co-hexafluoropropylene) (PVDF-HFP) [52, 69, 70].

15.10 SUMMARY

ILs are characterized by significant mobile and flexible ions, which offer wonderful prospects in the area of electrochemistry. The uses of ILs in SCs and LIB are way beyond those of simple solvents or electrolytes, even though they were first thought of as alternatives. Therefore, ILs have been regarded as promising electrolytes for SCs. Due to the remarkable physicochemical features, like their low vapor pressure, outstanding thermal and chemical stability, and molecular designability, ILs are predicted to grow significantly to develop high-energy performance SCs. In this chapter, we have summarized the principles and fundamentals of ILs for using in SCs in terms of their terminology, wide potential window, high thermal stability, and high performance under extreme conditions. In addition, the chapter deals with the common types of ILs, such as Imidazolium, Pyrrolidinium, Sulfonium, and Phosphonium ILs, as well as the ILs' performance when mixed with organic electrolyte as an electrolyte mixture to allow for the charge storage process within broad ESW. Moreover, it discusses the uses of ILs in state-of-the-art SC systems such as EDLC and PCs.

There is still a significant number of possibilities for ILs in SC applications, and future research directions are suggested to improve the ILs' performance in harsh conditions and obtain SC devices with high-efficiency and durable energy storage. Moreover, the chapter deals with some research studies that have been performed to overcome the drawbacks of ILs, such as either mixing several IL types or using ILs as electrolyte additions, which could decrease the ILs' viscosity at a lower temperature and improve the operational temperature range. Furthermore, the chapter references recent research studies that have focused on new alternative electrolytes composed of ionogel electrolytes based on poly (ionic liquids) (PILs) and conventional polymers due to their safe high-voltage applications and self-healing properties, as well as the common polymers that have been studied in ionogel electrolytes.

REFERENCES

1. A. Brandt, S. Pohlmann, A. Varzi, A. Balducci and S. Passerini, *MRS Bull.*, 2013, 38, 554–559.
2. H. Liu and H. Yu, *J. Mater. Sci. Technol.*, 2019, 35, 674–686.
3. L. Miao, Z. Song, D. Zhu, L. Li, L. Gan and M. Liu, *Energy Fuels*, 2021, 35, 8443–8455.
4. A. Eftekhari, *Energy Storage Mater.*, 2017, 9, 47–69.
5. A. I. Abdel-Salam, S. Y. Attia, F. I. El-Hosiny, M. A. Sadek, S. G. Mohamed and M. M. Rashad, *Mater. Chem. Phys.*, 2022, 277, 125554.
6. A. Noori, M. F. El-Kady, M. S. Rahmanifar, R. B. Kaner and M. F. Mousavi, *Chem. Soc. Rev.*, 2019, 48, 1272–1341.

7. S. Y. Attia, S. G. Mohamed, Y. F. Barakat, H. H. Hassan and W. A. Zoubi, *Rev. Inorg. Chem.*, 2022, 42, 53–88.
8. C. R. Mariappan, V. Gajraj, S. Gade, A. Kumar, S. Dsoke, S. Indris, H. Ehrenberg, G. V. Prakash and R. Jose, *J. Electroanal. Chem.*, 2019, 845, 72–83.
9. A. I. Abdel-Salam, S. Y. Attia, S. G. Mohamed, F. I. El-Hosiny, M. A. Sadek and M. M. Rashad, *Int. J. Hydrogen Energy*, 2023, 48, 5463–5477.
10. W.-Y. Tsai, R. Lin, S. Murali, L. L. Zhang, J. K. McDonough, R. S. Ruoff, P.-L. Taberna, Y. Gogotsi and P. Simon, *Nano Energy*, 2013, 2, 403–411.
11. S. Sathyamoorthi, V. Suryanarayanan and D. Velayutham, *J. Power Sources*, 2015, 274, 1135–1139.
12. M. Hantel, A. Płatek, T. Kaspar, R. Nesper, A. Wokaun and R. Kötz, *Electrochim. Acta*, 2013, 110, 234–239.
13. M. E. Van Valkenburg, R. L. Vaughn, M. Williams and J. S. Wilkes, *Thermochim. Acta*, 2005, 425, 181–188.
14. R. D. Rogers and K. R. Seddon, *Science*, 2003, 302, 792–793.
15. M. Hayyan, F. S. Mjalli, M. A. Hashim, I. M. AlNashef and T. X. Mei, *J. Ind. Eng. Chem.*, 2013, 19, 106–112.
16. K.-W. Nam, C.-W. Lee, X.-Q. Yang, B. W. Cho, W.-S. Yoon and K.-B. Kim, *J. Power Sources*, 2009, 188, 323–331.
17. Q. Dou, S. Lei, D.-W. Wang, Q. Zhang, D. Xiao, H. Guo, A. Wang, H. Yang, Y. Li and S. Shi, *Energy Environ. Sci.*, 2018, 11, 3212–3219.
18. A. Brandt, C. Ramirez-Castro, M. Anouti and A. Balducci, *J. Mater. Chem. A*, 2013, 1, 12669–12678.
19. Y. Ito and T. Nohira, *Electrochim. Acta*, 2000, 45, 2611–2622.
20. A. González, E. Goikolea, J. A. Barrena and R. Mysyk, *Renew. Sustain. Energy Rev.*, 2016, 58, 1189–1206.
21. E. Kim, J. Han, S. Ryu, Y. Choi and J. Yoo, *Materials*, 2021, 14, 4000.
22. M. Salanne, *Ion. Liq. II*, 2017, 29–53.
23. B. E. Conway, *Electrochemical Supercapacitors: Scientific Fundamentals and Technological Applications*, Springer Science & Business Media, 2013.
24. H. Tokuda, K. Hayamizu, K. Ishii, M. A. B. H. Susan and M. Watanabe, *J. Phys. Chem. B*, 2004, 108, 16593–16600.
25. K. Krynicki, C. D. Green and D. W. Sawyer, *Faraday Discuss. Chem. Soc.*, 1978, 66, 199–208.
26. M. Salanne, C. Simon, H. Groult, F. Lantelme, T. Goto and A. Barhoun, *J. Fluorine Chem.*, 2009, 130, 61–66.
27. M. Levesque, V. Sarou-Kanian, M. Salanne, M. Gobet, H. Groult, C. Bessada, P. A. Madden and A.-L. Rollet, *J. Chem. Phys.*, 2013, 138, 184503.
28. M. J. Earle, J. M. Esperança, M. A. Gilea, J. N. Canongia Lopes, L. P. Rebelo, J. W. Magee, K. R. Seddon and J. A. Widegren, *Nature*, 2006, 439, 831–834.
29. Z. Xue, L. Qin, J. Jiang, T. Mu and G. Gao, *Phys. Chem. Chem. Phys*, 2018, 20, 8382–8402.
30. R. Lin, P.-L. Taberna, S. Fantini, V. Presser, C. R. Pérez, F. Malbosc, N. L. Rupesinghe, K. B. Teo, Y. Gogotsi and P. Simon, *J. Phys. Chem. Lett.*, 2011, 2, 2396–2401.
31. L. Zhang, K. Tsay, C. Bock and J. Zhang, *J. Power Sources*, 2016, 324, 615–624.
32. B. Shen, R. Guo, J. Lang, L. Liu, L. Liu and X. Yan, *J. Mater. Chem. A*, 2016, 4, 8316–8327.
33. M. Anouti and L. Timperman, *Phys. Chem. Chem. Phys*, 2013, 15, 6539–6548.
34. C. Li, X. Zhang, K. Wang, X. Sun, G. Liu, J. Li, H. Tian, J. Li and Y. Ma, *Adv. Mater.*, 2017, 29, 1604690.

35. Z. Lei, Z. Liu, H. Wang, X. Sun, L. Lu and X. Zhao, *J. Mater. Chem. A*, 2013, 1, 2313–2321.
36. Y. Chen, X. Zhang, D. Zhang and Y. Ma, *Mater. Lett.*, 2012, 68, 475–477.
37. L. Timperman, A. Vigeant and M. Anouti, *Electrochim. Acta*, 2015, 155, 164–173.
38. J. H. Lee, J.-B. Ryu, A. S. Lee, W. Na, H.-S. Yoon, W.-J. Kim and C. M. Koo, *Electrochim. Acta*, 2016, 222, 1847–1852.
39. P. Bose and S. Bhattacharya, *J. Electrochem. Soc.*, 2017, 164, H788.
40. A. J. R. Rennie, V. L. Martins, R. M. Torresi and P. J. Hall, *J. Phys. Chem. C*, 2015, 119, 23865–23874.
41. A. M. Sampaio, G. F. L. Pereira, M. Salanne and L. J. A. Siqueira, *Electrochim. Acta*, 2020, 364, 137181.
42. K. J. Fraser and D. R. MacFarlane, *Aust. J. Chem.*, 2009, 62, 309–321.
43. T. Tooming, T. Thomberg, L. Siinor, K. Tõnurist, A. Jänes and E. Lust, *J. Electrochem. Soc.*, 2014, 161, A222.
44. V. Ruiz, T. Huynh, S. R. Sivakkumar and A. G. Pandolfo, *RSC Adv.*, 2012, 2, 5591–5598.
45. W. Zhang, H. Jiang, Z. Chang, W. Wu, G. Wu, R. Wu and J. Li, *J. Mater. Sci.*, 2020, 55, 13543–13558.
46. C. Li, R. Bhandary, A. Marinow, D. Ivanov, M. Du, R. Androsch and W. H. Binder, *Polymers*, 2022, 14, 4090.
47. H. Peng, X. Gao, K. Sun, X. Xie, G. Ma, X. Zhou and Z. Lei, *Chem. Eng. J.* 2021, 422, 130353.
48. J. Xie, P. Yang, Y. Wang, T. Qi, Y. Lei and C. M. Li, *J. Power Sources*, 2018, 401, 213–223.
49. Y. Shao, M. F. El-Kady, J. Sun, Y. Li, Q. Zhang, M. Zhu, H. Wang, B. Dunn and R. B. Kaner, *Chem. Rev.*, 2018, 118, 9233–9280.
50. S. Shahzad, A. Shah, E. Kowsari, F. J. Iftikhar, A. Nawab, B. Piro, M. S. Akhter, U. A. Rana and Y. Zou, *Global Challenges*, 2019, 3, 1800023.
51. A. Balducci, *Ion. Liq.*, 2017, II, 1–27.
52. A. Ray and B. Saruhan, *Materials*, 2021, 14, 2942.
53. C. Largeot, C. Portet, J. Chmiola, P.-L. Taberna, Y. Gogotsi and P. Simon, *J. Am. Chem. Soc.*, 2008, 130, 2730–2731.
54. M. Lazzari, M. Mastragostino, A. G. Pandolfo, V. Ruiz and F. Soavi, *J. Electrochem. Soc.*, 2011, 158, A22.
55. P. Navalpotro, J. Palma, M. Anderson and R. Marcilla, *J. Power Sources*, 2016, 306, 711–717.
56. S. Fleischmann, M. Widmaier, A. Schreiber, H. Shim, F. M. Stiemke, T. J. S. Schubert and V. Presser, *Energy Storage Mater.*, 2019, 16, 391–399.
57. A. D. T. Medagedara, N. M. Waduge, T. M. W. J. Bandara, I. G. K. J. Wimalasena, M. Dissanayake, K. Tennakone, R. M. G. Rajapakse, C. P. Rupasinghe and G. R. A. Kumara, *J. Energy Storage*, 2022, 55, 105628.
58. J. Lee, B. Krüner, A. Tolosa, S. Sathyamoorthi, D. Kim, S. Choudhury, K.-H. Seo and V. Presser, *Energy Environ. Sci.*, 2016, 9, 3392–3398.
59. L. Zhang, S. Yang, J. Chang, D. Zhao, J. Wang, C. Yang and B. Cao, *Front. Chem.*, 2020, 8. 413.
60. H. J. Xie, B. Gélinas and D. Rochefort, *Electrochem. Commun.*, 2016, 66, 42–45.
61. J. J. Xu, H. Ye and J. Huang, *Electrochem. Commun.*, 2005, 7, 1309–1317.
62. A. Ray and B. Saruhan, *Materials*, 2021, 14, 2942.
63. I. A. Khan, Y.-L. Wang and F. U. Shah, *J. Energy Chem.*, 2022, 69, 174–184.
64. S. Zhang, J. Li, N. Jiang, X. Li, S. Pasupath, Y. Fang, Q. Liu and D. Dang, *Chem. Asian J.*, 2019, 14, 2810–2814.

65. H. Niu, L. Wang, P. Guan, N. Zhang, C. Yan, M. Ding, X. Guo, T. Huang and X. Hu, *J. Energy Storage*, 2021, 40, 102659.
66. R. Newell, J. Faure-Vincent, B. Iliev, T. Schubert and D. Aradilla, *Electrochim. Acta*, 2018, 267, 15–19.
67. K. Karuppasamy, J. Theerthagiri, D. Vikraman, C.-J. Yim, S. Hussain, R. Sharma, T. Maiyalagan, J. Qin and H.-S. Kim, *Polymers*, 2020, 12, 918.
68. A. J. D'Angelo and M. J. Panzer, *Chem. Mater.*, 2019, 31, 2913–2922.
69. M. Que, Y. Tong, G. Wei, K. Yuan, J. Wei, Y. Jiang, H. Zhu and Y. Chen, *J. Mater. Chem. A*, 2016, 4, 14132–14140.
70. S. Pan, M. Yao, J. Zhang, B. Li, C. Xing, X. Song, P. Su and H. Zhang, *Front. Chem.*, 2020, 8, 261.

16 Electrolytes for Hybrid Supercapacitors

Malaya K. Sahoo

16.1 INTRODUCTION

An electrolyte is an important factor that influences the performance of a supercapacitor as well as it plays a part in the electric double layer formation (in EDLCs), and sometimes in the charge storage via reversible redox process (in pseudocapacitors). Desirable electrolyte properties consist of a broad voltage range, excellent electrochemical stability, high ionic concentration with a small solvated ionic radius, low resistivity and viscosity, as well as low volatility and toxicity. Additionally, they should be available in high purity and cost-effective [1].

Supercapacitors generally possess two types of electrolytes: conventional liquid electrolytes and solid-state electrolytes—whereas electrolytes in liquid form can be classified into two major groups: aqueous and non-aqueous electrolytes. Each electrolyte has its own advantages and disadvantages. For instance, aqueous electrolytes are far cheaper, easy to handle, and provide high values of both conductivity and capacitance, but the only problem with them is their working voltage, i.e., restricted to 1 V, as water decomposition takes place at 1.23 V. Organic electrolytes like tetraethyl ammonium tetra fluoroborate ($TEABF_4$) in the acetonitrile (ACN) solvent displays potential window range >2.5 V, and ionic liquids (ILs) such as 1-ethyl-3 methyl imidazolium tetra fluoro borate give wider potential window (>4 V). They also suffer from much lesser ionic conductivity. On the other hand, solid-state electrolytes such as PVA/H_2SO_4 and PVA/H_3PO_4 are better, compared to their liquid counterparts. Nowadays, it's becoming more fashionable to design conventional supercapacitors that are small, thin, lightweight, and flexible to meet the needs of the increasing modern population. With solid-state electrolytes, potential leakage problems may be avoided, as in the case of liquid electrolytes. However, solid-state electrolyte lacks high conductivity.

Redox-active electrolytes are the recent eye-catchers that provide extra capacitance to the supercapacitor. Here, pseudocapacitance originates from the electrode materials as well as from the electrolyte solution, i.e., from the reduction/oxidation of the redox mediator in the electrolyte. The following reactions show the pseudocapacitive capacitance contributed by the redox active electrolyte potassium iodide (KI) [2]:

$$3I^{-1} \leftrightarrow I_3^{-1} + 2e^{-1}$$

DOI: 10.1201/9781003441182-19

$$2I^{-1} \leftrightarrow I_2 + 2e^{-1}$$

$$2I_3^{-1} \leftrightarrow 3I_2 + 2e^{-1}$$

$$I_2 + 6H_2O \leftrightarrow 2IO_3^{-1} + 12H^+ + 10e^{-1}$$

Many kinds of redox-active electrolytes, like KI, hetero poly acids, hydroquinone (HQ), p-phenylenediamine (PPD), and $K_3Fe(CN)_6$, have been used in supercapacitor applications, which improved their specific capacitance as well as energy/power densities [2–6]. Table 16.1 demonstrates their cell voltage, cost, ease of handling, and hazardous nature. Compared with the traditional electrolytes used in supercapacitors, the performance of supercapacitors can be enhanced by adding redox additives into the electrolytes. In doing this, the properties of ionic conductivity, potential window, and redox behavior are improved, thereby improving the capacitive performance and energy density. This is because the pseudocapacitance is not only contributed by electrode but can also be contributed from the redox additives' electrolyte. Here, we will discuss the latest research on different electrolytes for redox active or hybrid ion capacitors.

16.2 PROGRESS ON REDOX ADDITIVES IN AQUEOUS ELECTROLYTE

The addition of redox additives to an aqueous solution is a good way to improve the capacitive behavior of different electrode materials. Some typical redox-active electrolytes used so far in literature includes KI, Na_2MoO_4, $Ce_2(SO_4)_3$, Fe^{3+}/Fe^{2+}, viologen substances, $AgNO_3$, 1,4-dihydroxyanthraquinone, hydroquinone (HQ), and

TABLE 16.1
Comparison of Different Electrolytes

Electrolytes	Cell Voltage (V)	Cost	Safety
Aqueous electrolytes	1 to 1.6 V	Low price	Low toxicity, environmental friendliness
Organic electrolytes	1 to 2.8 V	Expensive	Flammability, volatility, and toxicity
Ionic liquid electrolytes	2 to 6 V	Expensive	High viscosity, low volatility
Solid or quasi-solid electrolytes	2 to 6.5 V	Expensive	Simplification in packaging and fabrication
Polymer electrolytes	2 to 6.5 V	Expensive	Flammable, degrade over time
Aqueous and organic	1 to 3 V	Depends on salt	Non-flammability
Aqueous and gel polymer	1 to 2.5 V	Depends on salt	Direct contact with human skin
ILs and organic	2 to 4 V	Expensive	No corrosiveness
ILs and gel polymer	2 to 4 V	Expensive	Non-flammability
Redox-active electrolytes	2 to 5 V	Expensive	Flammable, toxic degradation products

so on [7–13]. In 2005, Tanahashi et al. reported the enhancement of electrochemical capacitive properties of activated carbon fiber cloth electrodes by using a redox additive of aqueous electrolyte, combining 5.0 M H_2SO_4 and 1.0×10^{-1} M $AgNO_3$ solution [12]. The activated carbon fiber cloth electrode displays a high specific capacitance of 248 F g^{-1} (an increment of about 300%) by increasing the concentration of $AgNO_3$ additive in H_2SO_4 solution. They proposed that the enhancement of capacitance is owing to the pseudocapacitance associated with the redox reaction of the Ag^+/Ag redox couple.

Later, Li et al. reported that the capacity and stability of porous carbon electrodes can be improved significantly by introducing Cu^{2+} and Fe^{2+} into the H_2SO_4 electrolyte [13]. Their cyclic voltammetry results showed that the redox reaction of Cu^{2+}/Cu and Fe^{3+}/Fe^{2+} ion pairs in electrolytes produce the faradaic pseudocapacitance and delivered an enhanced specific capacity of 223 mAh g^{-1} at a current density of 0.1 A g^{-1}. The pseudocapacitive mechanism can be expressed as:

$$\text{Positive electrode}: \mathrm{Fe}^{2+} + \mathrm{Cu} \leftrightarrow \mathrm{Fe}^{3+} + \mathrm{Cu}^{2+} + 3\mathrm{e}^-$$

$$\text{Negative electrode}: x\mathrm{Fe}^{3+} + y\mathrm{Cu}^{2+} + zC + (x+2y)\mathrm{e}^- \leftrightarrow \left(\mathrm{Fe}_x\mathrm{Cu}_y\right)_{\mathrm{ads}}^{2x+} C_z$$

The outstanding performance of the porous carbon electrode is attributed to the improved reversibility of the Cu^{2+}/Cu redox reaction developed from the formation of complexes between $(Fe_xCu_y)^{2x+}$ and the edge site of the carbon atom of the porous carbon electrode. Xu et al. used multiple redox additives (KI and Na_2MoO_4) in an H_2SO_4 electrolyte solution to enhance the energy density and capacitive performance of carbon-based electrodes [8]. Their research found that the molar ratio of redox additives is crucial to enhance capacitive performance. The optimal molar ratio of Na_2MoO_4:KI = 1:1 exhibits a higher capacitance, which is 17.4 times higher than other systems due to the simultaneous redox reaction between Na_2MoO_4 and KI at the overlapping redox voltage windows. During the electrochemical reaction, a complex substance $(Mo_xI_yO_{4x})^{n-}C_z$ is formed, which can promote the redox reaction of Mo^{n+} and I^{n-} at the electrode-electrolyte interface, and the complex formation is represented as:

$$zC + x\mathrm{MoO}_4^{2-} + y\mathrm{I}^- \leftrightarrow \left(\mathrm{Mo}_x\mathrm{I}_y\mathrm{O}_{4x}\right)^{n-} C_z$$

Sathyamoorthi et al. investigated the charge storage performance of activated charcoal in 1.0 M H_2SO_4 electrolyte with the addition of viologen (1,1′-diethyl-4,4′-bipyridinium bromide)-based redox active electrolyte [10]. It was found that viologen exhibits dual redox behavior (cathodic/anodic), which is responsible to improve the performance of both positive and negative electrodes simultaneously through the redox behavior of bromide and 1,1′-diethyl-4,4′-bipyridinium ions. Their results demonstrate that the π–π interaction between the viologen and carbon material results continuously increased the specific capacitance during charge/discharge cycling, and after 1000 cycles, there was an observed increase of 30%.

Wang and colleagues examined the charge storage capability of composite containing vanadium oxide nanoribbons and 3D-graphene (VO*x*-3DG) in $VOSO_4$ + KCl electrolytes. Their investigation revealed that through electrolyte optimization, the VO*x*-3DG electrode's specific capacitance increased by 7.2 times after 500 charge/discharge compared to its initial value. It was found that the abnormal enhancement of charge storage performance is ascribed to the double role of $VOSO_4$ during the charge/discharge cycles; namely, the capacitance increases by its redox reactions and vanadium oxide electrodeposition occurring on the surface of the electrode [14]. Many efforts have also been made to use redox additives in alkaline [15–17] and neutral electrolytes [18, 19] to improve the capacitance and stability of the electrodes.

16.3 PROGRESS ON REDOX ADDITIVES IN ORGANIC ELECTROLYTES

Organic electrolytes have a wide electrochemical stability window (around 3 V), which is beneficial to improve the energy density of supercapacitors. Kim and coworkers employed a redox active additive of HQ in an organic solvent of poly(methyl methacrylate)-propylene carbonate-lithium perchlorate (PMMA-PC-$LiClO_4$) electrolyte to enhance the specific capacitance as well as the energy density of multiwalled carbon nanotube-based flexible micro-supercapacitor [20]. Their results revealed that the addition of HQ generates additional pseudocapacitance at the electrode-electrolyte interface.

Further, the fabricated device displays a wide stable potential window up to 1.2 V and higher specific capacitance, which is 35 times higher than that of the device without HQ. The introduction of each HQ molecule reversible generates and stores two electrons and two protons as a pseudocapacitive material, which is responsible for the improvement of capacitance and energy density via the generation of additional pseudocapacitance as follows (see Figure 16.1).

Zhang et al. used ferrocene and 4-oxo-2,2,6,6-tetramethylpiperidinooxy as redox additives in an organic electrolyte to extend the operating voltage up to 1.5 V, enhance the capacitance up to 363 F g^{-1} and energy density of 27.4 Wh kg^{-1} of poly(3,4-ethylenedioxythiophene) (PEDOT) electrode [21]. Both ferrocene and 4-oxo-2,2,6,6-tetramethylpiperidinooxy are soluble in various organic solvents and display excellent reversibility in redox reactions. During the electrochemical reaction, the redox mediators take/give electrons from and to the corresponding electrodes and added extra pseudocapacitance. Poochai et al. reported nitrogen-doped reduced graphene

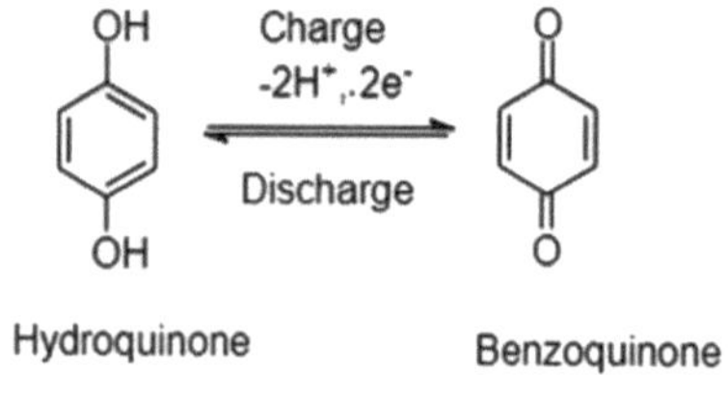

FIGURE 16.1 Electrochemical redox reaction of hydroquinone.

oxide (N-rGO) base coin-cell and pouch-cell supercapacitors by employing para-phenylenediamine (PPD) as redox additive in 1 M tetraethyl ammonium tetrafluoroborate ($TEABF_4$)/acetonitrile solution [22].

Here, the role of PPD is to improve the specific capacitance and specific energy by amplifying the redox reaction rate. During the electrochemical reaction, the interaction between N/O functional groups on N-rGO and the amino groups on the PPD molecule offers the faradaic electron transfer and enhances the charge storage performance. Their electrochemical analysis indicates that the symmetric coin cell delivers a high specific capacitance of 563 F g^{-1} with an energy and power density of 149.4 Wh·kg^{-1} and 1382 W·kg^{-1}, respectively. Moreover, the fabricated pouch cell delivers a specific capacitance, energy, and power density of 340 F g^{-1}, 77.2 Wh·kg^{-1}, and 1278 W·kg^{-1}, respectively.

16.4 PROGRESS ON REDOX ADDITIVES IN IONIC LIQUID ELECTROLYTES

Ionic liquids (ILs) have been projected as novel electrolytes for supercapacitor applications owing to their wide electrochemical potential window, high electrochemical stability, good oxidation resistance, and so on [23]. In recent years, Ma et al. used ferrocene methanol redox additive in 1-butyl-1-methylpyrrolidinium dicyanamide ionic liquid to enhance both specific energy and power, as well as the stability of N-doped reduced graphene oxide aerogel electrochemical capacitor [24]. Their research demonstrates that the ferrocene methanol not only offers the faradaic contribution but that also it promotes itself as an extra active ion to compensate for the charge via the electrical double layer.

To improve the energy density of an activated charcoal-based supercapacitor, Sathyamoorthi et al. added HQ to a protic IL triethylammonium bis(trifluoromethane) sulfimide ([Tea][TFSI]) electrolyte [25]. The redox additive HQ has better compatibility with ILs, and thus contributes extra pseudocapacitance by losing electrons to form quinone-oxygen bonds. In 2016, Navalpotro added 0.4 M para-Benzoquinone (p-BQ) as a redox additive in N-butyl-N-methylpyrrolidinium bis(-trifluoromethane-sulfonyl) imide ($PYR_{14}TFSI$) ionic liquid to study the charge storage performance of Pica carbon and Vulcan carbon-based hybrid supercapacitors [26]. The addition of p-BQ in $PYR_{14}TFSI$ ionic liquid gives two quasi-reversible redox peaks by two one-electron reduction steps (Figure 16.2a) as bellows.

$$p-\mathrm{BQ}+\mathrm{e}^{-} \leftrightarrow p-\mathrm{BQ}^{-}$$

$$p-\mathrm{BQ}^{-}+\mathrm{e}^{-} \leftrightarrow p-\mathrm{BQ}^{2-}$$

The comparative CV curves of Pica activated carbon in $PYR_{14}TFSI$ ionic liquid and p-BQ modified $PYR_{14}TFSI$ ionic liquid in Figure 16.2b indicate that the CV behavior is slightly distorted with redox electrolyte. It is evident that faradaic contribution coupled with the redox reactions experienced by the addition of p-BQ in $PYR_{14}TFSI$.

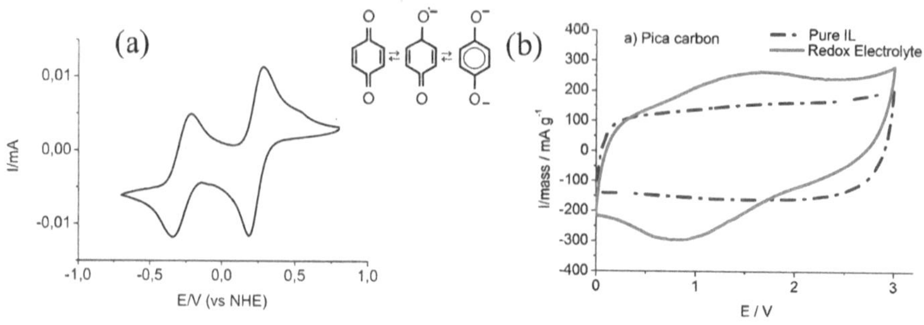

FIGURE 16.2 Cyclic voltammetry of (a) 20 mM solution of p-BQ in $PYR_{14}TFSI$ in three electrode cells; and (b) pica carbon in $PYR_{14}TFSI$ and p-BQ modified $PYR_{14}TFSI$ ionic liquid.

16.5 PROGRESS ON REDOX ADDITIVES IN GEL ELECTROLYTES

The gel is a kind of material that lies between solid and liquid. Gel electrolytes possess higher ionic conductivity and flexibility than solid electrolytes, which makes them a suitable choice over solid electrolytes. Many redox additives including HQ, Alizarin red S (ARS), Na_2MoO_4, and $K_3Fe(CN)_6$ – $K_4Fe(CN)_6$, have been widely used in gel electrolytes to fabricate flexible electrodes for supercapacitor applications [27–30]. Potphode et al. reported a solid-state high energy density symmetric supercapacitor in redox mediated gel electrolyte. The electrolyte was prepared by adding HQ as a redox additive to a mixture of Polyvinyl alcohol (PVA) and H_2SO_4 [27].

It was found that the redox gel electrolyte offers high ionic conductivity and leakage safety over liquid electrolytes and improved the charge storage performance without impeding the redox activity of the HQ. Due to the redox reaction of HQ-based gel electrolyte, the specific capacitance of the symmetric device was increased by 2.5 times that of the device in 1 M H_2SO_4 electrolyte. Sun et al. introduced alizarin red S (ARS) into a PVA–H_2SO_4 gel polymer system to improve the energy density and charge-discharge stability of activated carbon [28]. With the addition of ARS to polymer electrolyte, the conductivity was raised to 33.3 mS cm^{-1}.

Their studies noted that the charge storage performance of activated carbon in PVA–H_2SO_4–ARS electrolyte showed improved capacitance of 441 F g^{-1} with a high energy density of 39.4 Wh kg^{-1}, as compared to ARS free electrolyte (160 F g^{-1}). This high performance was attributed to the coexistence of the double layer capacitance of porous carbon materials and pseudocapacitance contribution from the faradaic processes of the redox mediator. Senthilkumar et al. used Na_2MoO_4 as the redox additive in PVA–H_2SO_4 gel electrolyte to improve the capacitive performance of activated carbon [29]. They suggested that molybdate species (MoO_4^{2-}) tends to form a polymeric ion (H_2MoO_4) in an acidic medium (oxidation state of Mo is +VI), which can be involved in redox reactions as follows:

$$H_2Mo(VI)O_4 + e^- + H^+ \leftrightarrow HMo(V)O_3 + H_2O$$

$$H_2Mo(VI)O_4 + 2e^- + 2H^+ \leftrightarrow H_2Mo(IV)O_3 + H_2O$$

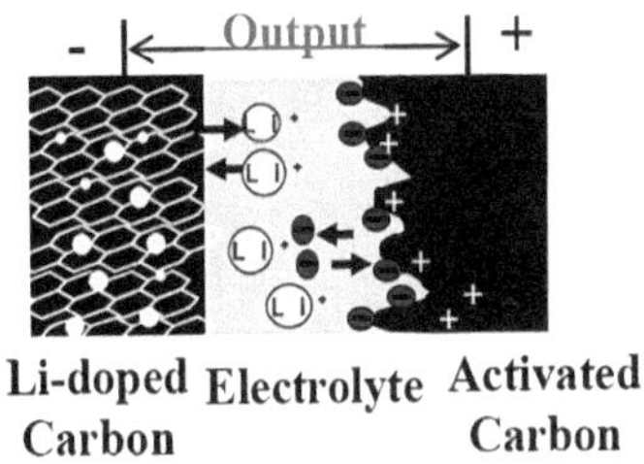

FIGURE 16.3 Working mechanism of Li-ion supercapacitors.

Due to the above redox reaction, Na_2MoO_4–PVA–H_2SO_4 gel electrolyte showed 52% and 12% higher areal capacitance compared to pristine PVA–H_2SO_4 and H_2SO_4 electrolyte, respectively. In addition, Na_2MoO_4–PVA–H_2SO_4 gel electrolyte delivered high specific capacitance and energy density of 648 F g^{-1} at 1.56 A g^{-1} and 14.4 W h kg^{-1} at 625 W kg^{-1}, while that of PVA–H_2SO_4 and PVA–Na_2MoO_4 gel electrolytes delivered only 412 F g^{-1}, 9.16 Wh kg^{-1} and 364 F g^{-1}, 8 Wh kg^{-1}, respectively.

The above results prompted the research community to improve both the energy density and power density, bridging the gap between supercapacitors and batteries by assembling a hybrid configuration of battery and supercapacitor called hybrid capacitors (HCs). The HCs consist of a positive electrode, negative electrode, electrolyte, and separator. These HCs are named based on the metal ion used as charge carriers, such as Li-ion HCs (Li ion as charge carrier), Na-ion HCs (Na ion as charge carrier), and so on [31–52]. The working mechanism of Li-HCs is shown in Figure 16.3. In this section we would like to discuss the development of different metal ions including lithium, sodium, and potassium as charge carriers on the electrochemical capacitive performances.

16.6 LITHIUM-ION HYBRID CAPACITOR

The charge storage mechanism of Li-HC depends on the reversible intercalation/de-intercalation of Li ions in the anode and the adsorption/desorption of ions on the surface of the cathode. Typically, its power density depends on how quickly the charge adsorbs/desorbs on the surface of the cathode, while the energy density depends on the lithiation ability. The working potential window of Li-HC is expanded by working the two electrodes in different potential ranges, thus improving the energy density and power density.

In 1987, Yata et al. first reported the lithium-ion intercalation into polyacenic semiconductor in 1 M $LiClO_4$/sulfolane + γ-butyrolactone electrolyte [32]. Then later, other researchers made great efforts on modifying the electrolytes or electrodes to expand the working voltage. Li et al. suggested a double (alkaline-neutral) electrolyte separated with a K^+ conductive membrane to construct a Li-HC by coupling $LiMn_2O_4$ nanorods as the cathode and nitrogen-rich biomass carbon (NBC) as the anode. The electrolyte is composed of 2 M KOH in the anode segment and 1 M Li_2SO_4 in the cathode segment, both containing 0.3 M K_2SO_4 aqueous solution separated by a K^+ ion conductive membrane. Here, the role of the K^+ ion conductive membrane is to avoid the inter-mixing between Li_2SO_4 and KOH solutions, and K_2SO_4 was added to maintain charge balance during the charging/discharging process [33].

Electrochemical analysis revealed that the assembled NBC//$LiMn_2O_4$ Li-HC with the alkaline-neutral electrolyte showed much better rate capability with a working voltage of 2.3 V and two times higher energy and power density than conventional Li-HCs. In 2019, Maurya et al. prepared poly(vinylidene fluoride-co-hexafluoropropylene) (PVDFHFP)/$Li_{7.1}La_3Ba_{0.05}Zr_{1.95}O_{12}$ nanohybrid membrane electrolytes (esHPMEs) with ionic conductivity of 3.304×10^{-3} S cm^{-1} and stable potential window of 4.6 V for high-energy Li-ion capacitors [36]. The Li-HC delivers the highest energy and power density of 43.77 Wh kg^{-1} and 7.973 kW kg^{-1}, respectively, with 83% capacity retention even after 1000 cycles. The synergistic effect between PVDFHFP and $Li_{7.1}La_3Ba_{0.05}Zr_{1.95}O_{12}$ plays an important role to improve the energy storage performance as well as the electrochemical potential window.

Chiou et al. added PVdF-HFP gel into $LiPF_6$ organic electrolyte to improve the device reliability and cycle stability at the expense of power. They found that when the assembled Li-HC worked in a low voltage window (2.0/2.5 V), the energy and power in cycling were very stable, while broadening the working voltage up to 3.5/4.0 V, a high irreversible capacity or evil cycle performance caused by the electrolyte decomposition at the negative electrode. Addition of PVdF-HFP gel greatly reduces the electrolyte decomposition and improves cyclic stability as well as Coulombic efficiency [34].

16.7 SODIUM-ION HYBRID CAPACITOR

In recent years, sodium hybrid capacitors (Na-HCs) have attracted research interest owing to their high energy and power density. In particular, the high abundance of Na reservoir and low cost of Na resources makes them a suitable alternative to Li-HCs. The working principle of Na-HCs involves the migration of Na ions between the cathode and anode during charging/discharging. However, there are many obstacles, including sluggish redox reaction kinetics and large volume expansion, that impede achieving its landmark electrochemical properties [37–41]. In this regard, different approaches such as the use of different anode/cathode materials, tuning their nanostructure, and tuning the electrolyte compositions has aimed to improve the performance of Na-HCs.

Since organic electrolyte possesses wider working voltage (~4 V) and reasonable conductivity, research on organic electrolytes has attracted extensive attention. The organic electrolytes composed of $NaClO_4$ or $NaPF_6$ in organic solvents (propylene carbonate, ethylene carbonate, ethyl methyl carbonate, dimethyl carbonate, and diethyl carbonate) have been widely used in Na-HCs [42–44]. To improve the passivation and suppress the side reaction, some researchers added 5% fluoroethylene carbonate as an additive into the above organic systems [45].

During the electrochemical process, Na^+ cations were stored through the insertion process and ClO_4^-/PF_6^- anions were adsorbed on the cathode surface, simultaneously. Recently, Chowdhury et al. enhanced the charge storage performance of Na-HC by adding potassium ferricyanide and potassium iodide as redox additives into conventionally used NaOH electrolyte [46]. Ganesan et al. used a hybrid mixed electrolyte in Na-HC to reduce capacity loss and stability loss issues. Due to the larger ionic size, the kinetics of Na intercalation/deintercalation is very slow in the anode. Therefore, the authors have used a hybrid mixed electrolyte ($NaClO_4$ + $LiPF_6$) containing both Na and Li ions. The fundamental behind such electrolytes is that larger ions occupy larger spaces, whereas smaller ions can be seated in smaller gaps and voids.

The electrochemical results demonstrated that the fabricated device displays high energy density and stable cyclic performance (up to 40000 cycles) when conventional sodium-ion electrolyte was replaced with hybrid mixed electrolytes [47]. Fleischmann et al. fabricated an asymmetric device by using lithium titanate as an anode and activated carbon as cathode electrodes in Na-ion containing ionic liquid electrolyte. The fabricated Na-HC operated between 1.0 V and 4.0 V cell voltage range with a high energy efficiency of 75%. It was worth noting that the specific energy increases significantly from about 35 Wh kg^{-1} to 44 Wh kg^{-1} during the first 50 cycles, and the maximum calculated specific energy was 90 Wh kg^{-1} [48].

16.8 POTASSIUM ION HYBRID CAPACITOR

Potassium ion hybrid capacitors (K-HCs) have also emerged as promising energy storage devices like Na-HCs because of their low cost and wide abundance. Further, K ion displays a lower reduction potential (−2.936 V vs. K^+/K) as compared to Na ion (−2.714 V vs. Na^+/Na), which could help to widen the operating voltage and boost the energy density. Additionally, potassium does not form an alloy with Al, which allows the use of the Al current collector for both the anode and the cathode [41].

To enhance the energy density of K-HCs, most research was focused on organic electrolytes in which potassium salts are dissolved. The most common potassium salts used for electrolytes are potassium hexafluorophosphate (KPF_6), Potassium bis(-fluorosulfonyl)imide (KFSI), potassium bis(tri-fluoromethanesulfonyl)imide (KTFSI), and Potassium perchlorate ($KClO_4$). Recently, Wang et al. optimized potassium salts in the electrolyte to achieve highly stable K-ion storage [49]. They found that KFSI possesses stronger solvation than KPF_6, which decreases the amount of free solvent in the electrolyte and reduces the side reactions between K metal and solvent molecules. Hence, the symmetric K cells with the KFSI-based electrolyte enable superior stability for K-HCs than that with the KPF_6-based electrolyte. Benefiting from the KFSI-based electrolyte, the K-HCs deliver a high capacity of 420.7 mAh g^{-1} after 100 cycles with a capacity retention of 95.3%.

Babu et al. proposed a new type of electrolyte containing a methacrylate-based polymer matrix and a diglyme-based liquid electrolyte (DOKBn-GPE) for K-HCs [50]. This polymer-gel electrolyte shows high ionic conductivity (2.84 mS cm^{-1}) and good mechanical and electrochemical stability (> 5.0 V). The fabricated K-HC demonstrates a high energy density of 40 Wh kg^{-1} and high cyclic stability (capacity retention of 78% after 2000 cycles). It has been observed that some researchers added Fluoroethylene carbonate (FEC) into the electrolyte to improve the cyclic stability and Coulombic efficiency, and suppress dendrite formation [51, 52]. We believe that these results will guide the research communities for further design of high energy density or hybrid ion capacitors.

16.9 CONCLUSIONS AND FUTURE OUTLOOK

Electrolyte is a key component in capacitors, which significantly affects the electrochemical performance (including cycling stability, rate performance, energy density, etc.). In this chapter, we have summarized the recent progress of various electrolytes or different additives in electrolytes for high-energy supercapacitors and hybrid ion capacitors. Each electrolyte has its own advantages or disadvantages. The limited working voltage of aqueous-based electrolytes hinders the application of high-energy supercapacitors. Therefore, the addition of some redox additives can extend the working voltage as well as contribute additional pseudocapacitance to the device. Further, metal ion-based HCs possess high energy and power density, which combine the merits of SCs and batteries.

To further improve energy storage performance, research should focus on the interaction mechanism between electrolyte and electrode and find out some novel electrolytes with superior electrochemical stability with a wide potential window. We hope that this chapter may provide some ideas to develop electrolytes for high energy density capacitors.

REFERENCES

1. L. Zhang, S. Yang, J. Chang, D. Zhao, J. Wang, C. Yang and B. Ca, *Front. Chem.*, 2020, 8, 1–7.
2. C. Zhong, Y. Deng, W. Hu, J. Qiao, L. Zhang and J. Zhang, *Chem. Soc. Rev.*, 2015, 44, 7484–7539.
3. H. Xie, Y. Zhu, Y. Wu, Z. Wu, E. Liu, *Mater. Res. Bull.*, 2014, 50, 303–306.
4. L. Su, L. Gong, H. Lu and Q. Xu, *J. Power Sources*, 2014, 248, 212–217.
5. S.T. Senthilkumar, R. K. Selvan, M. Ulaganathan and J. S. Melo, *Electrochim. Acta*, 2014, 115, 518–524.
6. S. Maiti, A. Pramanik and S. Mahanty, *ACS Appl. Mater. Interfaces*, 2014, 6, 10754–10762.
7. Y. Zhang, L. Zu, H. Lian, Z. Hu, Y. Jiang, Y. Liu, X. Wang and X. Cui, *J. Alloys Compd.*, 2017, 694, 136–144.
8. D. Xu, W. Hu, X. N. Sun, P. Cui and X. Y. Chen, *J. Power Sources*, 2017, 341, 448–456.
9. P. Díaz, Z. González, R. Santamaría, M. Granda, R. Menéndez and C. Blanco, *Electrochim. Acta*, 2015, 168, 277–284.
10. S. Sathyamoorthi, M. Kanagaraj, M. Kathiresan, V. Suryanarayanan and D. Velayutham, *J. Mater. Chem. A*, 2016, 4, 4562–4569.
11. D. Xu, X. N. Sun, W. Hu and X. Y. Chen, *J. Power Sources*, 2017, 357, 107–116.
12. I. Tanahashi, *Electrochem. Solid-State Lett.*, 2005, 8, A627–A629.
13. Q. Li, K. Li, C. Sun and Y. Li, *J. Electroanal. Chem.*, 2007, 611, 43–50.
14. A. Y. Wang, M. Chaudhary and T. W. Lin, *Chem. Eng. J.*, 2019, 355, 830–839.
15. G. K. Veerasubramani, K. Krishnamoorthy and S. J. Kim, *J. Power Sources*, 2016, 306, 378–386.
16. C. Lamiel, Y. R. Lee, M. H. Cho, D. Tuma and J. J. Shim, *J. Colloid Interface Sci.*, 2017, 507, 300–309.
17. Z. J. Zhang, Y. Q. Zhu, X. Y. Chen and Y. Cao, *Electrochim. Acta*, 2015, 176, 941–948.
18. A. Singh and A. Chandra, *Sci. Rep.*, 2016, 6, 25793.
19. K. Fic, M. Meller and E. Frackowiak, *J. Electrochem. Soc.*, 2015, 162, A5140–A5147.
20. D. Kim, G. Lee, D. Kim, J. Yun, S. S. Lee and J. S. Ha, *Nanoscale*, 2016, 8, 15611–15620.

21. H. Zhang, J. Li, C. Gu, M. Yao, B. Yang, P. Lu and Y. Ma, *J. Power Sources*, 2016, 332, 413–419.
22. C. Poochai, C. Sriprachuabwong, N. Srisamrarn, Y. Chuminjak, T. Lomas, A. Wisitsoraat and A. Tuantranont, *Appl. Surf. Sci.*, 2019, 489, 989–1001.
23. M. K. Sahoo and G. Ranga Rao, *Nanoscale Adv.*, 2021, 3, 5417–5429.
24. N. Ma, S. Kosasang, N. Phattharasupakun and M. Sawangphruk, *J. Electrochem. Soc.*, 2019, 166, A695–A703.
25. S. Sathyamoorthi, V. Suryanarayanan and D. Velayutham, *J. Power Sources*, 2015, 274, 1135–1139.
26. P. Navalpotro, J. Palma, M. Anderson and R. Marcilla, *J. Power Sources*, 2016, 306, 711–717.
27. D. D. Potphode, L. Sinha and P. M. Shirage, *Appl. Surf. Sci.*, 2019, 469, 162–172.
28. K. Sun, F. Ran, G. Zhao, Y. Zhu, Y. Zheng, M. Ma, X. Zheng, G. Ma and Z. Lei, *RSC Adv.*, 2016, 6, 55225–55232.
29. S. T. Senthilkumar, R. K. Selvan, J. S. Melo and C. Sanjeeviraja, *ACS Appl. Mater. Interfaces*, 2013, 5, 10541–10550.
30. A. Kundu and T. S. Fisher, *ACS Appl. Energy Mater.*, 2018, 1, 5800–5809.
31. P. Naskar, D. Kundu, A. Maiti, P. Chakraborty, B. Biswas and A. Banerjee, *ChemElectroChem*, 2021, 8, 1393–1429.
32. S. Yata, Y. Hato, K. Sakurai and T. Osaki, *Synth. Met.*, 1987, 18, 645–648.
33. C. Li, W. Wu, S. Zhang, L. He, Y. Zhu, J. Wang, L. Fu, Y. Chen, Y. Wu and W. Huang, *J. Mater. Chem. A*, 2019, 7, 4110–4118.
34. Y. D. Chiou, D. S. Tsai, H. H. Lam, C. Chang, K. Y. Lee and Y. S. Huang, *Nanoscale*, 2013, 5, 8122–8129.
35. Y. Firouz, N. Omar, J. -M. Timmermans, P. V. Bossche and J. V. Mierlo, *Energy*, 2015, 83, 597–613.
36. D. K. Maurya, V. Murugadoss and S. Angaiah, *J. Phys. Chem. C*, 2019, 123, 30145–30154.
37. X. Liu, Y. Sun, Y. Tong, X. Wang, J. Zheng, Y. Wu, H. Li, L. Niu and Y. Hou, *Nano Energy*, 2021, 86, 106070.
38. Y. Yuan, C. Wang, K. Lei, H. Li, F. Li and J. Chen, *ACS Cent. Sci.*, 2018, 4, 1261–1265.
39. W. Feng, R. R. Maca and V. Etacheri, *ACS Appl. Mater. Interfaces*, 2020, 12, 4443–4453.
40. Z. Chen, T. Yuan, X. Pu, H. Yang, X. Ai, Y. Xia and Y. Cao, *ACS Appl. Mater. Interfaces*, 2018, 10, 11689–11698.
41. Y. Wu, Y. Sun, Y. Tong, X. Liu, J. Zheng, D. Han, H. Li and L. Niu, *Energy Storage Mater.*, 2021, 41, 108–132.
42. N. R. Kim, Y. S. Yun, M. Y. Song, S. J. Hong, M. Kang, C. Leal, Y. W. Park and H. J. Jin, *ACS Appl. Mater. Interfaces*, 2016, 8, 3175–3181.
43. G. Dong, H. Wang, W. Liu, J. Shi, S. Sun, D. Li, H. Zhang, Y. Yang and Y. Cui, *ACS Appl. Energy Mater.*, 2018, 1, 5636–5645.
44. S. K. Park, S. H. Kwon, S. G. Lee, M. S. Choi, D. H. Suh, P. Nakhanivej, H. Lee and H. S. Park, *ACS Energy Lett.*, 2018, 3, 724–732.
45. X. Zhao, W. Cai, Y. Yang, X. Song, Z. Neale, H. E. Wang, J. Sui and G. Cao, *Nano Energy*, 2018, 47, 224–234.
46. A. Chowdhury, S. Biswas, T. Singh and A. Chandra, *Electrochem. Sci. Adv.*, 2022, 2, 1–14.
47. B. K. Ganesan, R. Thangavel, M. Moorthy, S. J. Lee, W. S. Yoon and Y. S. Lee, *J. Power Sources*, 2021, 500, 229918.
48. S. Fleischmann, M. Widmaier, A. Schreiber, H. Shim, F. M. Stiemke, T. J. S. Schubert and V. Presser, *Energy Storage Mater.*, 2019, 16, 391–399.
49. H. Wang, D. Yu, X. Wang, Z. Niu, M. Chen, L. Cheng, W. Zhou and L. Guo, *Angew. Chem. Int. Ed.*, 2019, 58, 16451–16455.

50. B. Babu, C. Neumann, S. Muench, M. Enke, L. Medenbach, C. Leibing, A. L. Balducci, A. Turchanin, U. S. Schubert and A. Balducci, *Energy Storage Mater.*, 2023, 56, 342–350.
51. M. Tang, Y. Wu, Y. Chen, C. Jiang, S. Zhu, S. Zhuo and C. Wang, *J. Mater. Chem. A*, 2019, 7, 486–492.
52. G. He and L. F. Nazar, *ACS Energy Lett.*, 2017, 2, 1122–1127.

17 Advanced Electrolytes for Electrochemical Supercapacitors

R. Jagadeesh, M. Manoj Cristopher, and S. Arumugam

17.1 INTRODUCTION

An electrolyte is a medium containing ions that electrically conduct through the movement of those ions but do not conduct electrons. It is called an electrolyte. Most salts, acids, and bases are soluble when dissolved in polar solvents like water. The material divides into cations and anions during dissolution, which scatter evenly throughout the solvent. There are other solid-state electrolytes. The dissolved material is an electrolyte in medicine and occasionally in chemistry [1, 2]. In this chapter, we focus our discussion on the aqueous electrolyte-based supercapacitor, which is expected to be the storage technology of the next generation, which is to achieve a broad voltage potential window with a huge energy density [3]. The development of cleaner and more effective energy storage systems has been prompted by the lack of fossil fuels, environmental crises, and rising global energy consumption [4]. Supercapacitors are a novel power/energy storage source type because of their quick charge and discharge time, high power and energy densities, and long cyclic stability behavior. The electrolyte and electrode material are essential elements that define the charge-storage process in supercapacitor applications [5].

The voltage windows of the IL and OE, non-aqueous as solvents, are significantly more expansive, averaging 3.0 V, to increase the current density [6]. However, the low conductivity and high viscosity of IL and the flammable, volatile, poisonous nature of OE have limited their use for SCs to a greater extent. Aqueous electrolytes provide many advantages over IL and OE, including excellent ionic conductivity, low cost, and non-toxic (see Figure 17.1).

Due to its corrosive nature and the direction of many metallic nature elements of a supercapacitor device, such as the connections, containers, and current collector, H_2SO_4 as an electrolyte in aqueous electrolytes (conventional "salt-in-water") has significant technical limitations that prevent its practical deployment. Compared to an alkaline electrolyte, the neutral electrolyte (Na_2SO_4 solution) has a more considerable stability potential for biomass carbon. In terms of aqueous electrolytes, the

DOI: 10.1201/9781003441182-20

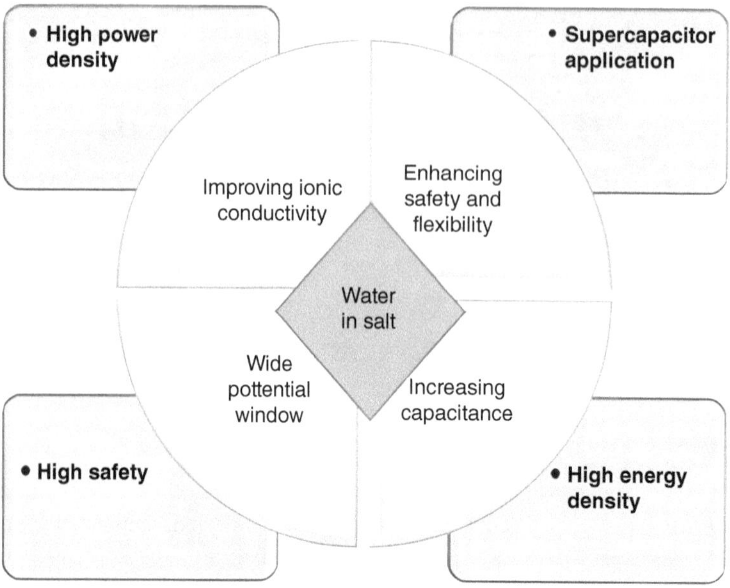

FIGURE 17.1 Advantages of WiSE electrolytes.

neutral electrolyte is the chief option. However, due to the spontaneous small operating voltage window, the neutral-based electrolyte produces inadequate energy densities of 1.23 V with thermodynamic conditions [7].

Research has demonstrated that a novel aqueous electrolyte known as "water-in-salt electrolyte" (WIS) can overcome this limited operating voltage window. The water in the salt electrolytes can produce an extraordinarily high-level voltage of >3.0 V because almost every H_2O molecule tightly aligns to metal ion cations, reducing these molecules' chemical activity [8, 9]. However, the neutral-based electrolyte has insufficient energy densities because of the naturally small operating voltage window. Recent studies have shown that a particular aqueous electrolyte called a "water-in-salt electrolyte" (WiSE) can be used to circumvent this confined working voltage window [10]. Since all these H_2O molecules tightly coordinate with metal ions, the WiSE electrolyte can generate an unusually high cell voltage (>3.0), which lowers the chemical activity of H_2O molecules.

17.2 PARAMETERS FOR DESIGNING ELECTROLYTES FOR SCs

17.2.1 Electrolyte Conductivity

Electrolytes are essential for high energy and power density, long cycle lifetimes, flexibility, and protecting supercapacitors. Ionic conductivity and mobility are the two most critical factors to consider when evaluating the behavior of electrolytes. Aqueous electrolytes have a higher ionic conductivity than organic and solid-based aqueous

TABLE 17.1
Types of Electrolytes

Liquid Electrolyte	Redox-Active Electrolyte	Quasi-Solid-State Electrolyte
1. Water in salt electrolytes	1. Aqueous	1. Dry solid-state
2. Aqueous electrolytes	2. Organic	2. Polymer
Acid	3. Ionic liquid	3. Gel polymer
Alkaline	4. Gel polymer	4. Inorganic
Neutral		
3. Non-aqueous electrolytes		
Organic		
Ionic liquid		
Mixtures		

electrolytes because they have a lower dynamic viscosity [11]. The conductivity depends upon the types of electrolytes listed in (Table 17.1). The charge-carrier concentration (ni), the elementary charge carrier (e), and the size of the valence state of mobility ion charges all have an essential effect on the conductivity (s) of a species (zi).

17.2.2 Electrochemical Stability

The kinetics, rather than thermodynamics, of the first reaction against reductions and oxidations often serve as a good indicator of how inert electrolytes should be on cathode and anode materials. The safety and life cycle of SCs are closely related to electrochemical stability. The lower and upper bounds of the electrolytes' oxidation and reduction reactions represent the electrochemical stability window [12]. Electrochemical stability is influenced by the electrolyte's composition and its compatibility with the electrode.

The most often used measurement techniques are LSV and CV, despite significant variances in test results when using a variety of test methodologies. Temperature-dependent electrochemical stability of electrolytes must be thoroughly investigated [13]. These investigations have demonstrated that solvent systems, salt type, and working electrodes affect the electrochemical potential stability window, specifically the unique decomposition potentials of various vital electrolytes.

17.2.3 Thermal Stability

The electrode and electrolyte relationship is the charge and discharge process that causes electrolytes to degrade. The primary reasons are the increasing operational and heat emissions, and the rise in temperature also harms the SCs. The supercapacitor electrolyte's heat stability is the most crucial variable to investigate. The thermal stability of the supercapacitor device is influenced by the interaction of the electrode and electrolyte [14, 15].

17.2.4 Solvent Effect

Liquid conductivity and gel-based electrolytes are primarily influenced by the solvent components. The dielectric constant facilitates salt dissociation, while viscosity affects the salt's ionic mobility. The viscosity dielectric and dielectric constant are the primary features of the solvents that affect the electrolyte conductivity [16]. High-viscosity-based solvent has a high dielectric constant, which enhances the interactions between the electrolyte and the electrode. As a result, combining a solvent with a low viscosity and a high dielectric constant is the most popular technique for creating a better solvent system for a supercapacitor device [17]. Numerous research teams have conducted complete analyses of the conductivity of electrolytes and various solvents for SCDs. This investigation shows that binary solvent electrolytes have a significantly higher conductivity than single-component solvent electrolytes.

17.2.5 Salt Effect

The ion conductivity of various salts in the same solvent might vary because of the interaction between anion and cation and the volume differences of different anions. The ions' conductivity varies with salt content in the same solution [18]. Ionic conductivity can also be increased by the ideal salt concentration since the quantity in low salt concentrations is dominated by free ions. The conductivity is exceptionally high when the equilibrium between viscosity and free ions is achieved. The anions and cations in the solvent will strongly contribute to neutral-based ions if the salt concentration in the solvent is very high, thereby decreasing the conductivity of the electrolyte. However, conductivity in an ionic liquid of high viscosity systems often decreases monotonically as solution concentration is higher [19, 20].

17.3 WATER-IN-SALT (WIS) ELECTROLYTE

WIS has recently emerged as a prospective substitute, significantly increasing stability, safety, and cost-effectiveness compared to organic electrolytes in supercapacitor applications. These improved electrochemical stability windows (ESWs) in the ultra-concentrated aqueous electrolytes may allow lithium-based WiSEs to perform up to 4.0 V [21]. A new family of super-concentrated electrolytes called WIS has a prominent electrochemical characteristic used for a supercapacitor application. Although they have been the subject of several investigations in supercapacitors and Li-ion batteries, little is known about their interfacial properties. Here, we use a constant applied potential window voltage to model the contact between highly concentrated water in salt and working electrodes [22].

We have demonstrated how the capacitance between the negative and positive electrodes differs significantly. By selecting salts with suitable composition, salt and water can be mixed in quantities greater than 1 in terms of mass and volume. These electrolytes exhibit super-concentrated aqueous solutions because of the lack of free water, which limits water molecules' activity. As a result, the finite ESW of water (1.23 V) in conventional aqueous electrolytes can be expanded, resulting in a potential window that increases up to 4.0 V [23].

Several experimental works, including a few capacitance measurements, have been reported for the advantages of WIS to EDLCs. The exploration of the electrodes in the WIS system using water-based electrolytes mainly focuses on their directions (as shown in Figure 17.2): energy density, cycling stability, sustainability, and environmental friendliness. The interfacial structure of these electrolytes' close-by electrodes has also been the subject of a few simulated studies, although their capacitive behavior has not yet been investigated. If water and salt act as ionic solutions or liquids, that will be an essential problem to be answered.

Electrode absorption experiments have demonstrated that hydrophobic electrolyte additions create a shielding interphase. A wider voltage window is produced due to the protective interphase's ability to repel H_2O molecules and increase the resistance toward the decomposition of electrolytes. To increase the aqueous electrolyte's OER limit, it has been demonstrated that 1% vinylene carbonate, disodium propane-1,3-disulfonate in a mixture of Na_2SO_4 and $ZnSO_4$ electrolyte at the cathode absorbs and produces a water-repellent barrier.

Suo et al. were the first to demonstrate how changing from an SIW system to a WIS system with an appropriate salt that contains an anion can produce strong kinetic protection through stable interphase formation at the anode in an aqueous electrolyte. The WIS electrolyte is a "solvent-in-solute" electrolyte where the volume and mass of the dissolved salt are more significant than that of the water molecules. Due to the solvation structure, the charge carrier of the metal ions is primarily associated with the solvent molecules in a dilute electrolyte [24]. The mixture of free ions solvent molecules and, therefore, the solvent-separated ion pairs of the metal salts are often concentrated because of the physical makeup of such an electrolyte.

On the other hand, when an electrolyte solution is concentrated, a rise in salt ions has an impression on how well the metal ion coordinates with the solvent molecule. The ionic aggregates of the anion and cation of the electrolyte salt, which have a low number of free molecules of solvent, become more prevalent as the salt concentration rises (the "solvent-in-solute" condition). The solid electrolyte interface is often developed by the anion of the ions pairing degrading at the anode surface. Electrically insulating and ionically conducting describe the SEI layer [25, 26]. The event of a

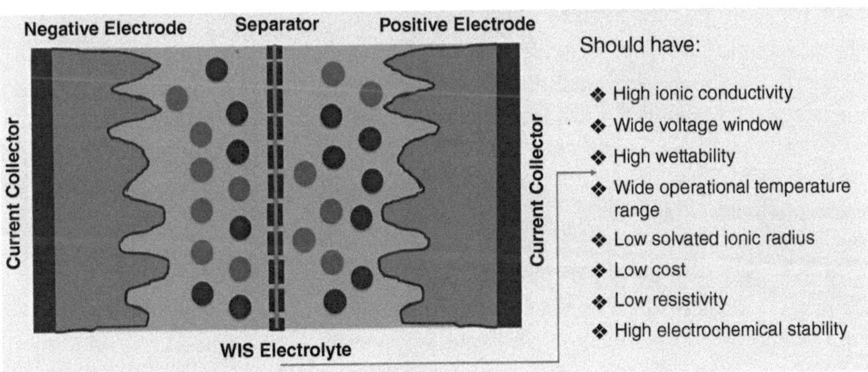

FIGURE 17.2 Electrolyte properties of the supercapacitor device.

stable anode surface lowers the onset potential of hydrogen evaluation reaction within the water-in-salt electrolyte. The electrolysis additionally suffers from the smaller amount of free water molecules, which involves a big over potential and eventually leads to a protracted electrochemical stability window of >3 V. The idea behind water in salt has been broadened to incorporate an extended electrochemical stability window in different aqueous electrolytes, like hydrate-melt electrolytes, water-in-basalt electrolytes (WiBSE), water-in-ionomers, and hybrid aqueous-non-aqueous electrolytes (HANE), pushing the electrochemical stability window up to 4.1 V.

Concentrated aqueous electrolytes have recently gained research attention. As an illustration, Borodin et al. focused on super-concentrated electrolytes and described their numerous characteristics, including ions transport characteristics, interfacial structure, and chemistry. The recent perspective on the electrochemical stability of WIS was developed by Ruben-Simon Kühnel et al. [27]. The impact of the cut-off current density on the apparent electrochemical stability window of WiSEs was emphasized. It was proposed that employing a higher cut-off current density could reveal the true ESW for the WiSEs. [28] delved into the physical, chemical, and electrical properties of WiSEs, underscoring their importance in electrochemical energy storage systems such as supercapacitors and batteries.

17.4 RESEARCH PROGRESSES OF WATER-IN-SALT ELECTROLYTES

The ion transport process in WIS electrolytes and its molecular/mesoscale structural underpinnings have attracted current research. Consistent with one emerging picture, WIS-based electrolytes are highly heterogeneous, with water and ion molecules self-assembling into nanoscale domains. In these electrolytes, there are many ion-water interfaces, and as a result, ion transport is controlled by the structure and dynamics of those interfaces. Recent research techniques and contributions of several WIS electrolytes for flexible-based WIS electrolytes supercapacitors and non-lithium-based rechargeable batteries (Na^+, K^+, Zn^{2+}, Mg^{2+}, and Al^{3+}).

17.4.1 Metal Salt-Based Water-in-Salt Electrolytes

17.4.1.1 Single Salt System

The main factors in single-salt electrolytes were the salt's high solubility and chemical thermal stability. So, since the first Lithium bis(trifluoromethanesulfonyl)imide, single salt materials were introduced, a few other WiSEs with similar properties have been developed [29]. Specifically, cheap $NaClO_4$ and $NaNO_3$ could effectively solve the high TFSI-based WiSE cost, and they are regarded as a promising alternative for WiSE commercialization.

17.4.1.2 Double Salt System

Double salt contains two or more different cations or anions. In addition to inhibiting Li dendrite formation and stabilizing the interphase of Li metal, the addition of LiDFOB as the double salt to the electrolyte also inhibits $LiPF_6$ from dissolving, passivates the cathode surface, and lowers the interfacial resistance [30]. For instance, the dual-salt electrolyte $LiPF_6$ + LiDFOB was created to enhance the electrochemical behavior of Li metal batteries when used with the high-capacity NCM811 cathode, which has a more extensive voltage range of 3.0 V to 4.5 V.

17.4.1.3 Water-in-Salt Electrolytes-Derived Electrolytes

Water-in-salt electrolytes-derived Electrolytes have high viscosity and low conductivity due to the strong coordination of cations and anions, which leads to poor rate performances of aqueous EES devices. Additionally, the temperature range of the WiSEs will be constrained by salt's propensity to precipitate at low temperatures. As a result, it is tough to fix these problems while maintaining the safety aspects and a broad ESW. Researchers have made several attempts to accomplish this goal in recent years.

17.4.1.4 Aqueous/Non-Aqueous Electrolytes

When it comes to electrolytes, aqueous-based electrolytes (such as KOH, H_2SO_4, Na_2SO_4 aqueous solution, and so on) show an advantage over organic-based electrolytes, which are said to be less conductive, high-cost, frequently flammable, and more toxic. They also have higher ionic conductivity and are less harmful and corrosive. However, compared to organic electrolytes, aqueous electrolytes possess a much smaller voltage window of 1.2 V [31]. Low energy density is the cause of low potential windows, which restricts market use. As seen in the section on symmetric supercapacitors, adding redox-active substances to aqueous electrolytes can significantly increase capacitance through additional redox reactions between the electrolyte and electrode.

17.4.2 Macromolecular Crowding Electrolyte

The development of liquid-based electrolytes not harmful to the environment with a large voltage window is necessary to achieve cost-effective, water-based electrolytes and secure high-energy-density supercapacitors (SCs). Molecular crowding electrolytes are prepared to simulate the crowded environment in natural cells. Reducing the molecular weight of the crowding agent can significantly improve ion transport in the molecular crowding electrolyte, polyethene glycol (PEG). PEG200, for instance, can maintain a considerable voltage window and considerably improve ionic conductivity [32]. These benefits allow commercial activated carbon-based SCs to operate at 2.5 V while exhibiting exceptional energy density, rate performance, and stability for more than 15,000 cycles [33]. PEG200's suitability for wide-voltage aqueous electrolytes is demonstrated by developing three molecular crowding electrolytes utilizing salts such as sodium perchlorate, lithium perchlorate, and sodium trifluoromethanesulfonate.

17.5 WORKING MECHANISM OF WIS

It is essential to understand how the water in salt electrolytes functions. To explain the expanded ESWs of the WIS electrolytes and create high-performance executing electrolytes for supercapacitors, it is crucial to comprehend how they function. In energy storage devices, the electrolytes undergo a few chemical reactions, including the salt effect on ion solvation, the transfer of solvated ions, and the formation of SEI on the electrode surface. The water in salt electrolytes exhibits a drastic difference in ionic solvation structure to traditional diluted aqueous-based electrolytes [34]. One example is comparing the ionic solvation structures of electrolytes on LiC2F6NO4S2 at various doses (see Figure 17.3a).

The salt anions (TFSI) and solvent molecules (H_2O) compete to coordinate in the Li^+ electrolytes when $LiC_2F_6NO_4S_2$ is combined with water [35]. As a result, TFSI anions cannot touch Li^+ ion, which is why they are known as solvent-separated ion pairs. In the diluted $LiC_2F_6NO_4S_2$ aqueous electrolytes, Li^+ is completely solvated by four molecules of H_2O in the main solvation shell (5 mol kg^{-1}) (Figure 17.3b). There are also significant amounts of free H_2O molecules in this scenario, which are simple to break down [36]. The number of H_2O molecules drops as the concentration of $LiC_2F_6NO_4S_2$ (510 mol kg^{-1}) in the electrolyte rises; the anions of TFSI begin to interact with Li^+ ion and take part in the solvation shell of Li^+, producing contact ion pair (CIP). Nearly all H_2O molecules in the WIS electrolyte of 21 mol kg^{-1} $LiC_2F_6NO_4S_2$ are moved in the Li^+ ion solvation shell; there are very few free H_2O molecules.

Additionally, the ionic aggregates are produced by large TFSI anions directly coordinating with Li^+ (AGGs). Some large-size ion clusters with four TFSI surrounding each Li^+ and TFSI anions often shared by more Li^+ ions start forming the molecules. The solvation ion structure surrounding the Li^+ ions changes from SSIPs to the concurrence of CIPS, AGGs, and SSIPs, increasing $LiC_2F_6NO_4S_2$ concentration in the aqueous electrolyte [37, 38]. AGGs will take up a more significant percentage of space in an electrolyte with a higher concentration. When AGGs and CIPs predominate as the WIS electrolytes ionic species, the TFSI anions are suitable for participating in the significant solvation shell of Li^+ ions. TFSI anion that coordinates with more Li^+ ions also has a long-range reduction potential because of the electron transfer stability.

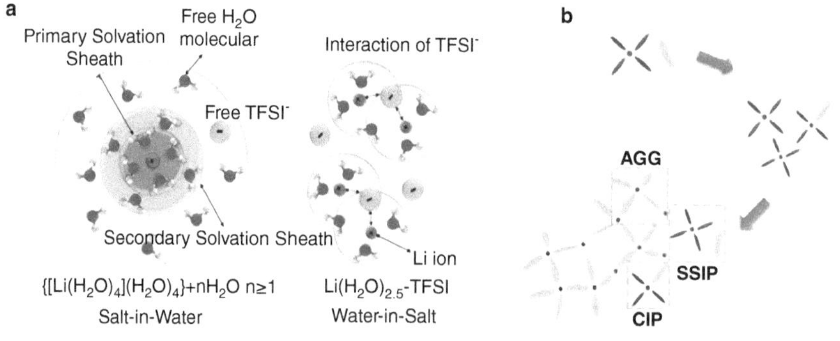

FIGURE 17.3 Illustration of the Li+ solvation structures in salt-in-water and WIS electrolytes. (b) The Li+ solvation structure from dilute to WIS electrolytes.

(a) Reproduced with permission from ref [36]. Copyright 2015 AAAS. (b) Reproduced with permission from ref [34]. Copyright 2021 Wiley-VCH.

In WIS electrolytes, the cathodic process allows the TFSI anions to degrade preferentially over the bound H_2O. At the electrode-electrolyte contact, the decomposition products may deposit and create an SEI layer, preventing further interaction between the water molecules and TFSI anions. Generally, acknowledging the evolution of H_2 has significantly simpler kinetics than the evolution of O, which poses the biggest obstacle to using aqueous electrolytes [39]. WIS electrolytes have higher ESWs due to the stable solvation ionic structure of AGGs, CIPs, and cathodic stability.

17.6 ENHANCEMENT OF THE WIS ELECTROLYTES IN SUPERCAPACITORS APPLICATION

An aqueous solution where the volume and mass are more significant than the salt of the solvent is known as a WIS electrolyte. The difference in their diluted counterparts, the water in salt electrolytes, can implement a substantially wider ESW. Because of this distinctive electrochemical property, the WIS electrolytes have been employed in supercapacitors and secondary batteries for high power voltage. These devices maintain the benefits of high-power density, securing the aqueous electrolyte systems while showing substantially higher cell voltage than conventional aqueous devices. The high energy cell voltage, long cyclic performance, and therefore high power and energy densities of the water in salt electrolytes support the creation of high-effectiveness aqueous supercapacitors application.

17.6.1 Safety and Flexibility

Energy storage systems should always prioritize safety. WIS liquid electrolytes, which are non-flammable and eco-friendly, have a wide ESW, giving supercapacitors a high level of security and an energy density comparable to that of organic systems. However, regarding practical use, the risk of liquid leakage remains a severe problem. One effective solution to the leakage issue is to disperse water in salt-aqueous electrolytes in an inorganic matrix or polymer to create a gel-based electrolyte.

Additionally, flexible supercapacitors can be built using gel-type electrolytes [40]. Using inorganic particles as a lattice to contain the water in salt electrolytes and create gel-based electrolytes is also possible. They use silicon spheres and a 5 mol/L of LiTFSI H_2O solution to produce a SiO_2-LiTFSI gel electrolyte referred to in Qin et al. They are assembled an asymmetrical all-solid-state micro-SCs with an m-MnO_2 nanosheet structure representing both a positive electrode and a porous structure of vanadium nitride nanosheets representing the negative electrode. The energy density was 21.6 mWh cm^3, and the device's cell voltage was 2.0 V. The device showed outstanding structural stability and mechanical flexibility even when bent 180 degrees, exhibiting an essentially constant CV curve and retaining 98% of its initial capacitance [41]. The gel-type WIS electrolyte is the potential method for developing adaptable and safe all-solid-state supercapacitors. However, most of the studies concentrate only on one: high-performance-oriented supercapacitors require the fabrication of improved electrolytes and electrodes. Future research should be more concerned with the synergistic effect between the electrode and the electrolyte.

17.6.2 Operating Temperature Range

Even though the cell voltage and supercapacitor energy density can be effectively increased when WIS electrolytes are used, they typically have a limited operating temperature range. At low temperatures, the salts tend to precipitate a concentration of WIS electrolytes, consequently reducing the temperature range over which the supercapacitors containing WIS electrolytes can operate. We manufactured and operated symmetric electric double-layer supercapacitors with different concentrated LiTFSI water in salt electrolytes at temperatures ranging from −20 to 100 °C to improve the impact of the concentration of salt in the electrolytes on the electrochemical behavior of supercapacitors at various temperature ranges [42]. In the SCs, carbon with an SSA of 3384 $m^2\ g^{-1}$ was used as the working electrode material. The supercapacitor with capacitances for 5 mol kg^{-1} LiTFSI electrolyte was 25 and 51 F g^{-1} at 5 and 0.5 A g^{-1}, respectively.

According to electrochemical test findings, the supercapacitor with 20 mol/kg LiTFSI electrolyte exhibited only 34 F/g at 0.5 A g^{-1} and failed to function at 4 A g^{-1} [43]. A capacitance gap shrunk as the temperature increased to 25 °C. When the temperature reached 60 °C, the supercapacitor's capacitance performed better than the 5 and 10 mol kg^{-1} and 63 F g^{-1} at 0.5 A g^{-1} LiTFSI electrolytes [44]. As a result, it can be said that water in salt electrolytes with minimal salt content concentrations is appropriate for lower working temperatures. Still, WIS electrolytes with high salt concentrations are preferable for high temperatures. Optimizing electrolyte concentration is critical to improving electrochemical performance over a wide temperature range.

17.6.3 Specific Capacitance

Electrolytes are typically considered inactive electrochemically, meaning there are no oxidation or reduction reactions during charge and discharge. It is generally known that quick redox processes on or near the surface of pseudocapacitive electrodes can produce high capacitance. This has motivated many researchers to focus on the redox-active electrolyte to provide more capacitance for energy storage devices. It is interesting to observe that the unexpected WIS electrolyte can serve as the redox species solvents. In most cases, the water in salt electrolytes has enlarged the ESW by reducing the number of molecules in water. The capacitance of the supercapacitors can be significantly increased with only a minor loss of electrochemical stability window by incorporating the properly diluted WIS electrolyte with water-soluble redox-active species, producing energy density performance; for example, 1 mol/kg of LiCl was added to a 21 mol/kg of LiTFSI electrolyte, compared to the theoretical capacity of 766 mAh/g to perform the extra redox processes of $Cl_2/2Cl^-$.

Due to the decreasing volume, the WIS using dissociated ions exhibits more significant ion-ion interaction and a lower dielectric constant (er). The ionic liquid based on BMPy exhibits the highest viscosity and the lowest conductivity and diffusivity, highlighting the impact of "soft" and "hard" cations on mass transport parameters [45]. The size of ions must be considered because it impacts the conductivity of electrolytes. The highest conductivity and diffusivity can be achieved by pairing a small anion with a "soft" cation. $EMImBF_4$ exhibited the best charge and discharge

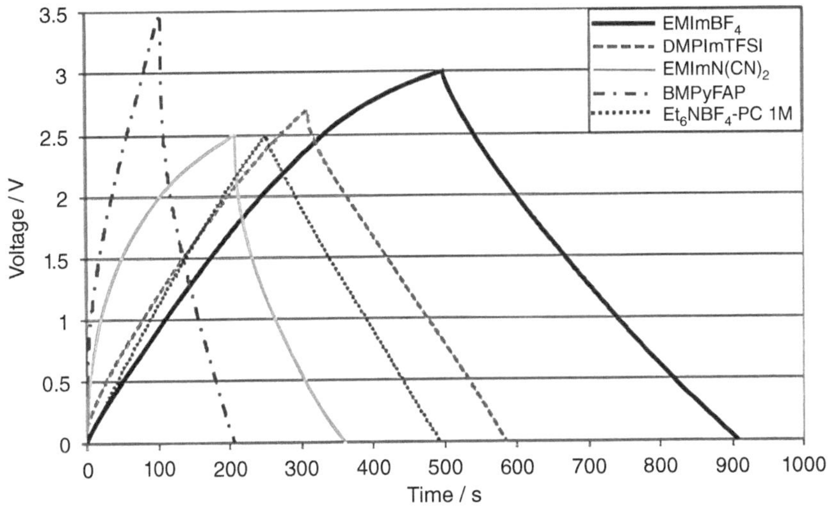

FIGURE 17.4 Charge and discharge curves concerning different WIS electrolytes.

Reproduced with permission from Ref. [45]. Copyrights 2023 Royal Society of Chemistry.

performance, as shown in Figure 17.4. The ions' small size, reduced viscosity, and relatively high conductivity achieve the best performance.

An oriented hydrogel film-based graphene served as the cathodic capacitance in a Li-ion capacitor. This capacitor was constructed using the electrolyte and a hydrogel film's TiO2/rGO anode-type battery. The redox-active LiCl in the electrolyte enabled a gain of 169 mAh g^{-1} in cathode capacity. This represents a significant increase from the cathode's 54 mAh/g capacity in LiCl-free water using the salt electrolyte [45]. Furthermore, when the electrolyte was diluted from 1 mol/kg LiCl/21 mol/kg LiTFSI to 1 mol/kg LiCl/10 mol/kg LiTFSI, the electrochemical stability window slightly decreased from 2.7 volts to 2.6 volts. However, the Li+ ion capacitor exhibited an improved drop voltage from 2.1 to 2.2 volts due to the significantly reduced ion/electron activity [46]. The Li-ion capacitor achieved a power density of 880 W/kg and an energy density of 74.6 Wh/kg by utilizing a salt electrolyte consisting of 1 mol/kg of LiCl and 10 mol/kg of LiTFSI in water, as reported in [47]. This finding suggests that the energy density gap between non-aqueous and aqueous supercapacitors can be reduced by incorporating redox additives into the WIS when 0.5 mol/kg of KI is added to 8 mol/kg of NaTFSI water in the salt electrolyte, the symmetric supercapacitors made of activated carbon exhibit an energy density of 37.8 Wh/kg at a cell voltage of 1.8 V [48].

17.6.4 Ionic Conductivity

LiTFSI saltwater electrolytes have lower ionic conductivity than dilute aqueous electrolytes, resulting in poor rate performance due to increased viscosity caused by the high concentration of 21 mol/kg. An easy and efficient way to promote ion dynamics is by reducing viscosity and enhancing the ionic conductivity of the electrolyte. This can be achieved by adding a concentrated WIS electrolyte, which can then be diluted using solvents. This method can enhance the capacity and rate capability of supercapacitors. This suggests that the recommended 21 mol/kg volume for the LiTFSI-based WIS electrolytes in supercapacitors may need to be revised. The optimal concentration for various devices should be chosen based on the specific circumstances.

For a supercapacitor that utilizes monolithic electrodes made of activated carbon and is symmetrically designed, the optimal concentration of LiTFSI in an aqueous dilute electrolyte is 18.7 mol/kg [49]. The symmetric SCs intended for industrial use have a maximum working voltage of 2.4 V and a high power and energy density. The ionic conductivity of the 18.7 mol/kg LiTFSI-WIS electrolytes has been improved compared to the 21 mS cm^{-1} reported in reference [50]. The effect of LiTFSI in the electrolyte solution on the performance of an EDLC supercapacitor was studied. The results showed that a LiTFSI electrolyte with a concentration of 7 mol/kg provided the best overall performance in terms of voltage (3 V), energy density (25 Wh/kg), power density (195.5 W/kg), and cycling stability.

17.7 WIS-BASED ELECTROLYTES IN SUPERCAPACITORS

Supercapacitors and aqueous metal-ion batteries have been investigated using water-based electrolytes with inexpensive perchlorate salts, sulfates, and nitrates. Due to its high solubility in water and ability to create a low-viscosity, highly concentrated aqueous electrolyte (with a viscosity of 5.0 mm^2 s^{-1} at 17 mol/kg) that enhances conductivity (up to 64.2 mS cm^{-1} at 17 mol/kg), $NaClO_4$ is a commonly used salt in batteries and supercapacitors. These properties are superior to a 21 mol/kg LiTFSI electrolyte in water. However, the cost of using salt in $NaClO_4$ is significantly higher than that of LiTFSI [51]. Therefore, using $NaClO_4$-electrolytes based on the critical factor of WIS would be highly beneficial for supercapacitors.

The advantages of electrolytes based on LiTFSI with ion solvation structures of 2 mol/kg and 17 mol/kg $NaClO_4$ were compared for WIS. Among the salts that can replace LiTFSI, such as K_2CO_3, $NaClO_4$, and Li_2SO_4, $NaNO_3$ is the most cost-effective option. The $NaNO_3$-WIS electrolyte has a high ionic conductivity of 198 mS/cm at a concentration of 12 mol/kg, surpassing other electrolytes such as 17 mol/kg $NaClO_4$ in water (64.2 mS/cm) and 21 mol/kg LiTFSI (8.2 mS/cm). Using 12 mol/kg $NaNO_3$ water as a salt electrolyte has numerous advantages. These include ionic solid conductivity, low cost, significant ESW, and environmental responsibility [52]. As a result, this solution has considerable potential for use in various industrial applications.

An asymmetric supercapacitor was developed to test the idea's viability using YP-50 Faraday electrodes and a cell voltage of 2.1 V [53]. The 12 mol kg^{-1} $NaNO_3$ WIS electrolyte procedure resulted in the highest capacitance and energy density, making it the most competitive option. This was in comparison to those that use electrolytes containing 17 mol kg^{-1} of $NaClO_4$ and 21 mol kg^{-1} of LiTFSI [54]. However, further research is still necessary to fully comprehend its potential application in asymmetric supercapacitors.

Jiang et al. explored an aqueous anode consisting of perylene tetracarboxylic diimide and a cathode comprising K1.85 Fe 0.33 Mn 0.67 [Fe $(CN)_6$] 0.980.77 H_2O, with a 20 m KCF_3SO_3 (KOTf) WIS electrolyte. The whole battery demonstrated exceptional performance regarding reversible capacities and rate behaviors. Zhang et al. investigated the potential for a novel low-cost WIS electrolyte of 30m ZnCl2 to deliver a sizeable electrochemical window of 2.3V. This electrolyte has fewer side reactions and low oxidative Cl^-. However, the $ZnSO_4$ electrolyte for AZIBs always suffers due to its lower Zn stripping/plating efficiency and inherently low solubility. The Zn $(CF3SO_3)_2$ electrolyte, on the other hand, exhibits lower polarization and higher CE values, making it suitable for extensive use in aqueous ZIBs (Xie et al.). Chen et al. reported that electrolytes containing 30 mM KFSI demonstrate an electrochemical stability window ranging from 1.55 to 2.42 V (vs. Hg/Hg_2Cl_2). This characteristic enables the stable operation of b-perylene-tetracarboxylic dianhydride (b-PTCDA) and even b-PTCDA-based complete batteries in high-concentration electrolytes without experiencing hydrogen evolution or material dissolution.

17.8 CONCLUSIONS

Electrolytes most significantly influence the performance of electrochemical supercapacitors. Aqueous-based electrolytes have multiple benefits for supercapacitor applications, including fast ion movement, safety, affordability, and friendly environments; however, they have a narrow ESW. The aqueous electrolytes from the highly concentrated water in salt, where more salt is abundant than the solvent in both volume and mass, increase the electrochemical stability window of aqueous-based electrolytes to roughly 3 Voltage, indicating the possibility of manufacturing high cell voltage and high energy density aqueous-based supercapacitors.

Water in salt electrolytes functions distinctly from traditional dilute aqueous electrolytes. While nearly all the H_2O molecules in the WIS electrolytes are contained within the salt cations' solvation sheath, there is a significant quantity of free H_2O ions in the dilute electrolytes. The TFSI anions also tightly coordinate with Li^+ cations, especially for the LiTFSI-based WIS electrolytes, causing preferential breakdown during the cathodic process and developing the SEI layer. The SEI and the comparatively stable bound H_2O both help to broaden the ESW of the WIS electrolytes. More investigation into the mechanisms at work and the causes of the wide electrochemical stability window is still anticipated for directing the water's design in salt electrolytes.

Water development in salt electrolyte supercapacitors is still in its early stages; however, they combine the advantages of a wide electrochemical stability window, highly secured, and fast ion transport. Scrutinizing the mechanism of cation conduction and the current condition of water, theory, and more experiments are crucial. Understanding the operating principles of diverse WIS systems, enhancing performance thoroughly, and developing low-cost electrolytes are achievable goals. Advancements in these directions give supercapacitors exceptional power and energy densities, the promise to be environmentally safe, and low cost, illustrating the encouraging future of water in salt electrolytes in SC applications.

17.9 FUTURE DIRECTIONS

Future research should concentrate on the electrochemical stability of the electrolyte and how it affects cycle life, energy, and power efficiency. To determine the practical usefulness of novel cell components, a thorough evaluation of each element should be conducted regarding all pertinent performance parameters. For instance, batteries with highly concentrated electrolytes have a somewhat lower gravimetric energy density than those with more typical electrolytes that are more diluted. For example, 1 M $LiPF_6$ in ethylene carbonate: dimethyl carbonate 1:1 (by weight) has a density of 1.30 g cm^3 at 24 °C, whereas 27.8 m Li (TFSI) 0.7 (LiBETI) 0.3 has a density of 1.78 g cm^3 at 25 °C. We mainly focus on WIS electrolytes and their recent advancements in various non-lithium aqueous rechargeable metal-ion batteries and thin film-based flexible supercapacitors.

ABBREVIATIONS

WIS Water in Salt
ESW Electrochemical Stability Window
SCs Supercapacitors
LSV Linear Sweep Voltage
CV Cyclic Voltage
EDLC Electric Double Layer Capacitor
OER Oxygen Evolution Reaction
SIW Solvent in Water
SEI Solid Electrolyte Interface
IL Ionic Liquid
OE Organic Electrolyte

REFERENCES

1. B. Pal, S. Yang, S. Ramesh, V. Thangadurai and R. Jose, *Nanoscale Adv.*, 2019, 1, 3807–3835.
2. Q. Li, J. Chen, L. Fan, X. Kong and Y. Lu, *GEE*, 2016, 18–42.
3. Y. Shen, B. Liu, X. Liu, J. Liu, J. Ding, C. Zhong and W. Hu, *Energy Storage*, 2022, 34, 461–474.
4. C. Rajesh Kumar and M. A. Majid, *Energy Sustainability Soc.*, 2020, 2, 1–36.
5. S. H. Nagarajarao, A. Nandagudi, R. Viswanatha, B.M. Basavaraja, M.S. Santhosh, B.M. Praveen and A. Pandith, *ChemEngineering*, 2022, 6, 5–34.
6. S. Pan, M. Yao, J. Zhang, B. Li, C. Xing, X. Song, Peipei Su and H. Zhang, *Front. Chem.*, 2020, 8, 45–74.
7. M. Pang, S. Jiang, J. Zhao, S. Zhang, R. Wang, N. Li, R. Liu, Q. Pan, W. Qua and B. Xing, *RSC Adv.*, 2020 10, 35545–35556.
8. H. Gao, K. Tang, J. Xiao, X. Guo, W. Chen, H. Liu and G. Wang, *J. Energy Chem.*, 2022, 69, 84–99.
9. P. Kulkarni, D. Ghosh and R. G. Balakrishna, *Sustain. Energy Fuels*, 2021, 5, 1619–1654.
10. T. Liang, R. Hou, Q. Dou, H. Zhang and X. Yan, *Adv. Funct. Mater.*, 2020, 2006749.
11. K. Xu, *Chem. Rev.*, 2004, 104, 4303–4417.
12. A. Hayashi, K. Noi, A. Sakuda and M. Tatsumisago, *Nat. Commun.*, 2012, 3, 856.

13. H. Che, S. Chen, Y. Xie, H. Wang, K. Amine, X.Z. Liao and Z. Ma, *Energy Environ. Sci.*, 217, 10, 1075–1101.
14. L. Chancelier A. O. Diallo, C. C. Santini, G. Marlair T. Gutel, S. Mailleyb and C. Len, *Phys. Chem. Chem. Phys.*, 2014, 16, 1967–1976.
15. J. Jiang, J.R. Dahn, *Electrochim. Acta*, 2004, 49, 4599–4604.
16. V. Vladimir Shcherbakov, M. Yuliya Artemkina, Irina A. Akimova and Irina M. Artemkina, *Materials*, 2021, 14, 5617.
17. H. Che, S. Chen, Y. Xie, H. Wang, K. Amine, X. Z. Liao and Z. Ma, *Energy Environ. Sci.*, 2017, 10, 1075–1101.
18. F. Wu, N. Zhu, Y. Bai, L. Liu, H. Zhou and C. Wu, *ACS Appl. Mater. Interfaces*, 2015, 7, 4, 2626–2631.
19. C. Ding, T. Nohira and R. Hagiwara, *J. Power Sources*, 2017, 231, 412–416.
20. Y. Zhang and T. Peter Cummings, *ACS Appl. Mater. Interfaces*, 2019, 11 (45), 42680–42689
21. M. Chen, G. Feng and R. Qiao, *J. Colloid Interface Sci.*, 2020, 47, 99–110.
22. Z. Li, G. Jeanmairet, T. Mendez-Morales, B. Rotenberg and M. Salanne, *J. Phys. Chem. C*, 2018, 122, 23917–23924.
23. P. Inoue, E. Fileti and T. Malaspina, *J. Phys. Chem. B*, 2020, 124, 5685–5695.
24. C. Long, L. Miao, D. Zhu, H. Duan, Yaokang Li, L. Li, M. Liu and L. Gan, *ACS Appl. Energy Mater.*, 2021, 4, 5727–5737.
25. J. Park, J. Lee and W. Kim, *ACS Energy Lett.*, 2021, 6, 769–777.
26. Daniel P. Leonard, Z. Wei, G. Chen, F. Du and X. Ji, *ACS Energy Lett.*, 2018, 3, 373–374.
27. R. Simon Kühnel, D. Reber and C. Battaglia, *J. Electrochem. Soc.*, 2020, 167, 070544.
28. P. Ruschhaupt, S. Pohlmann, A. Varzi and S. Passerini, *Batteries & Supercaps*, 2017, 8, 698–707.
29. Y. P. Yang, A. C. Huang, Y. Tang, Y.C. Liu, Z. H. Wu and H. L. Zhou, *Polymers*, 2021, 13, 707.
30. R. Miao, J. Yang, X. Feng, H. Jia, J. Wang and Y. Nuli, *J. Power Sources*, 2014, 271, 291–297.
31. C. Zhong, Y. Deng, W. Hu, J. Qiao, L. Zhang and J. Zhang, *Chem. Soc. Rev.*, 2015, 44(21), 7431–7920.
32. L. Katherine, V. Aken, M. Beidaghi and Y. Gogotsi, *Angew. Chem. Int. Ed.*, 2015, 7, 54.
33. F. Sun, H. B. Wu, X. Liu, F. Liu, R. Han, Z. Qu, X. Pi, L. Wang, J. Gao, Y. Lu, *Mater. Today, Energy*, 2018, 9, 428–439.
34. X. Tian, Q. Zhu and B. Xu, *Chem Sus Chem*, 2021, 15, 825.
35. A. Eftekhari, *Adv. Energy Mater.*, 2018, 12, 1801156.
36. L. Suo, O. Borodin, T. Gao, M. Olguin, J. Ho, X. Fan, C. Luo, C. Wang, K. Xu, *Science*, 2015, 350, 6263.
37. D. Sun, M. Okubo and A. Yamada, *Chem. Sci.*, 2021, 12, 4450–4454.
38. C. Yang, J. Chen, T. Qing, X. Fan, W. Sun and A. V. Cresce, *Joule*, 2017, 1, 122–132.
39. J.C. Liu, Z. H. Huang and T. Y. Ma, *Small Struct.*, 2020, 1, 2688–4062.
40. L. Daia, O. Arcelusc, L. Suna, H. Wanga, J. Carrascoc, H. Zhanga, W. Zhang and J. Tang, *J. Mater. Chem. A*, 2019,7, 24800–24806.
41. Z. Song, H. Duan, D. Zhu, Yaokang Lv, W. Xiong, T. Cao, L Li, M Liu and L Gan, *J. Mater. Chem. A*, 2019, 7, 15801–15811.
42. L.W. Le Fevre, A. Ejigu, R. Todd, J. A. Forsytha and R.A. W. Dryfe, *Chem. Commun.*, 2021,57, 5294–5297
43. J. Zhu, Y. Sun, J. Gao, Z. Qin, Y. Zhou, R. Tian and Y. Gao, *Chem. Commun.*, 2022, 58, 5861–5864
44. Q. Dou, S. Lei, D.W. Wang, Q. Zhang, D. Xiao and H. Guo, *Energy Environ. Sci.*, 2012, 00, 1–3.

45. F.B. Sillars, S. I. Fletcher, M. Mirzaeian and P. J. Hall, *Phys. Chem. Chem. Phys.*, 2012, 14, 6094–6100.
46. B. Akinwolemiwa, C. Peng and G.Z. Chen, *J. Electrochem. Soc.*, 2015, 162 (5) A5054–A5059.
47. W Li, B Song and A Manthiram, *Chem. Soc. Rev.*, 2017, 46, 3006–3059.
48. D. Reber, R. S. Kühnel and C. Battaglia, *Sustain. Energy Fuels*, 2017, 1, 2155–2161
49. H. Avireddy, B. Byles, D. Pinto, J. M. Delgado Galindo and J. J. Biendicho, *Nano Energy*, 2019, 64, 103961.
50. G. Hasegawa, K. Kanamori, T. Kiyomura, H. Kurata, T. Abe and K. Nakanishi, *Chem. Mater.*, 2016, 28, 11, 3944–3950
51. M. H. Lee, S. J. Kim, D. Chang, J. Kim, S. Moon and K. Oh, *Mater. Today*, 2019, 29, 26–36.
52. Y. Guo, W. Maxwell, W. Terban, I. Moudrakovski, A. Münchinger, R. Dinnebier, J. Popovic and J. Maier, *J. Mater. Chem. A*, 2022, 11, 3427–3436
53. M. Zhang, W. Wang, X. Liang, C. Li, W. Deng, H. Chen and R. Li, *CCL*, 2021, 32, 2217–2221.
54. M. Zhanga, S. Makinoa, D. Mochizukia and W. Sugimoto, *J. Power Sources*, 2018, 396, 498–505.

Index

Pages in *italics* refer to figures and pages in **bold** refer to tables

M

N

O

P

FSC
www.fsc.org
MIX
Papier | Fördert gute Waldnutzung
FSC® C083411

Kolibri 360 GmbH
Ferdinand-Jühlke-Straße 7
99095 Erfurt, Deutschland
produktsicherheit@kolibri360.de